Energy, Environment, and Sustainability

This books series publishes cutting edge monographs and professional books focused on all aspects of energy and environmental sustainability, especially as it relates to energy concerns. The Series is published in partnership with the International Society for Energy, Environment, and Sustainability. The books in these series are editor or authored by top researchers and professional across the globe. The series aims at publishing state-of-the-art research and development in areas including, but not limited to:

- Renewable Energy
- Alternative Fuels
- Engines and Locomotives
- Combustion and Propulsion
- Fossil Fuels
- Carbon Capture
- Control and Automation for Energy
- Environmental Pollution
- Waste Management
- Transportation Sustainability

More information about this series at http://www.springer.com/series/15901

Saptarshi Basu · Avinash Kumar Agarwal
Achintya Mukhopadhyay
Chetankumar Patel
Editors

Droplets and Sprays

Applications for Combustion and Propulsion

Springer

Editors
Saptarshi Basu
Department of Mechanical Engineering
Indian Institute of Science Bangalore
Bangalore, Karnataka
India

Achintya Mukhopadhyay
Department of Mechanical Engineering
Jadavpur University
Kolkata, West Bengal
India

Avinash Kumar Agarwal
Department of Mechanical Engineering
Indian Institute of Technology Kanpur
Kanpur, Uttar Pradesh
India

Chetankumar Patel
Department of Mechanical Engineering
Indian Institute of Technology Kanpur
Kanpur, Uttar Pradesh
India

ISSN 2522-8366 ISSN 2522-8374 (electronic)
Energy, Environment, and Sustainability
ISBN 978-981-10-7448-6 ISBN 978-981-10-7449-3 (eBook)
https://doi.org/10.1007/978-981-10-7449-3

Library of Congress Control Number: 2017959311

Printed on acid-free paper

This Springer imprint is published by Springer Nature
The registered company is Springer Nature Singapore Pte Ltd.
The registered company address is: 152 Beach Road, #21-01/04 Gateway East, Singapore 189721, Singapore

Preface

Energy demand has been rising remarkably due to increasing population and urbanization. Global economy and society are significantly dependent on the energy availability because it touches every facet of human life and its activities. Transportation and power generation are major examples of energy. Without the transportation by millions of personalized and mass transport vehicles and availability of 24×7 power, human civilization would not have reached contemporary living standards.

The first international conference on 'Sustainable Energy and Environmental Challenges' (SEEC-2017) was organized under the auspices of 'International Society for Energy and Environmental Sustainability' (ISEES) by the 'Center of Innovative and Applied Bioprocessing' (CIAB), Mohali, from 26 to 28 February 2017. ISEES was founded at IIT Kanpur in January 2014, with the aim of spreading knowledge in the fields of energy, environment, sustainability and combustion. The society's goal is to contribute to the development of clean, affordable and secure energy resources and a sustainable environment for the society and to spread knowledge in the above-mentioned areas and awareness about the environmental challenges, which the world is facing today. ISEES is involved in various activities such as conducting workshops, seminars and conferences in the domains of its interest. The society also recognizes the outstanding works done by the young scientists and engineers for their contributions in these fields by conferring them awards under various categories.

This conference provided a platform for discussions between eminent scientists and engineers from various countries including India, the USA, South Korea, Norway, Malaysia and Australia. In this conference, eminent speakers from all over the world presented their views related to different aspects of energy, combustion, emissions and alternative energy resource for sustainable development and cleaner environment. The conference started with four mini-symposiums on very topical themes, which included (i) New Fuels and Advanced Engine Combustion, (ii) Sustainable Energy, (iii) Experimental and Numerical Combustion and (iv) Environmental Remediation and Rail Road Transport. The conference had 14 technical sessions on topics related to energy and environmental sustainability and panel discussions on 'Challenges, Opportunities and Directions of Technical Education and Research in the Area of Energy, Environment and Sustainability' to wrap up the three-day technical extravaganza. The conference included 2 plenary talks, 12 keynote talks, 42 invited talks from prominent scientists, 49 contributed talks and 120 posters. A total of 234 participants and speakers attended this three-day conference, which hosted Dr. V. K. Saraswat, Member, NITI Aayog, India, as a chief guest for the award ceremony of ISEES. This conference laid out the road map for technology development, opportunities and challenges in this technology domain. The technical sessions in the conference included Advances in IC Engines and Fuels; Conversion of Biomass to Biofuels; Combustion Processes; Renewable Energy: Prospects and Technologies; Waste to Wealth—Chemicals and Fuels; Energy Conversion Systems; Numerical Simulation of Combustion Processes; Alternate Fuels for IC Engines; Sprays and Heterogeneous Combustion of Coal/ Biomass; Biomass Conversion to Fuels and Chemicals—Thermochemical Processes; Utilization of Biofuels; and Environmental Protection and Health. All these topics are very relevant for the country and the world in the present context. The society is grateful to Prof. Ashok Pandey for organizing and hosting this conference, which led to the germination of this series of monographs, which included 16 books related to different aspects of energy, environment and sustainability. This is the first time that such a voluminous and high-quality outcome has been achieved by any society in India from one conference.

The editors would like to express their sincere gratitude to the authors for submitting their work in a timely manner and revising it appropriately at short notice. We would like to express our special thanks to Prof. Swarnendu Sen, Prof. Koushik Ghosh, Prof. Pallab Sinha Mahapatra and Prof. Ranjan Ganguly, who reviewed various chapters of this monograph and provided their valuable suggestions to improve the manuscripts. We acknowledge the support received from various funding agencies and organizations for successfully conducting of the first ISEES conference SEEC-2017, where these monographs germinated. These include Department of Science and Technology, Government of India (special thanks to Dr. Sanjay Bajpai); TSI, India (special thanks to Dr. Deepak Sharma); Tesscorn, India (special thanks to Sh. Satyanarayana); AVL, India; Horiba, India; Springer (special thanks to Swati Mehershi); CIAB (special thanks to Dr. Sangwan).

Combustion of liquid fuel is a major source of energy production in today's world. Liquid fuel provides several advantages like high specific energy on a volumetric basis and relative ease of storage. Fuel is supplied to these devices in the form of a spray of tiny droplets. The ever-increasing demand for energy and depleting reserves of fossil fuels has also led to exploration of newer grades of liquid fuels and fuel blends. The combustion and emission characteristics of the liquid-fuelled devices depend to a large extent on the behaviour of fuel droplets individually and collectively in a spray. Critical combustion phenomena like ignition, extinction, flame stability and emission strongly depend on the mean and range of droplet size in a spray. Consequently, behaviour of individual droplets and sprays continues to form a major focus of combustion and propulsion research.

The present monograph contains 13 different chapters discussing various aspects of current research on droplets and sprays relevant to combustion and propulsion applications. They include fundamental studies related to heating, evaporation and combustion of individual droplets and basic mechanisms of spray formation and also extend to latest analytical, numerical and experimental techniques for investigating behaviour of spray in devices like combustion engines and gas turbines. In addition, They also include several emerging areas like interaction between sprays and flames and dynamic characteristics of spray combustion systems on the fundamental side and development of novel fuel injectors for specific devices on the application side.

The monograph contains leading international researchers in the field of spray combustion and propulsion and covers a diverse range of topics spanning fundamentals and applications, modelling and experiments. The chapters have been developed in a pedagogical manner starting with the basics of the relevant topics and background material and extending up to the latest developing developments. Thus, we hope the monograph would be useful to practising engineers, researchers and graduate students working in the area of spray combustion and propulsion.

Bengaluru, India — Saptarshi Basu
Kanpur, India — Avinash Kumar Agarwal
Kolkata, India — Achintya Mukhopadhyay
Kanpur, India — Chetankumar Patel

Contents

Editors and Contributors

About the Editors

Prof. Saptarshi Basu is currently an Associate Professor in the Department of Mechanical Engineering at Indian Institute of Science, Bangalore. He received his M.S. and Ph.D. in Mechanical Engineering from University of Connecticut in 2004 and 2007, respectively. He was a Tenure-Track Faculty Member in the Department of Mechanical, Materials and Aerospace Engineering at the University of Central Florida, USA, from 2007 to 2010. He joined Indian Institute of Science, Bangalore, as Faculty Member in 2010. His current research interests include combustion instability, flame–vortex interaction, droplet-level transport, multiphase combustion, spray atomization and break-up, water transport characteristics in fuel cells and general areas of heat and mass transfer. He has published over 105 articles in different frontline journals. He is the recipient of DST Swarnajayanti Fellowship from Government of India in Engineering Sciences. He has also received the K. N. Seetharamu Medal from Indian Society of Heat and Mass Transfer for his contributions in multiphase transport. He is a member of ASME, ISHMT and Combustion Institute. He is a Fellow of the Indian National Academy of Engineering.

Prof. Avinash Kumar Agarwal joined IIT Kanpur in 2001 and is currently a Poonam and Prabhu Goyal Endowed Chair Professor. He was at ERC, University of Wisconsin, Madison, USA, as a Postdoctoral Fellow (1999–2001). His areas of interest are IC engines, combustion, alternative fuels, hydrogen, conventional fuels, lubricating oil tribology, optical diagnostics, laser ignition, HCCI, emission and particulate control and large bore engines. He has published more than 160 peer-reviewed international journals and conference papers. He is Associate Editor of ASME Journal of Energy Resources Technology and International Journal of Vehicle Systems Modelling and Testing. He has edited 'Handbook of Combustion' (5 volumes; 3168 pages), published by Wiley VCH,

Germany. He is a Fellow of SAE (2012), a Fellow of ASME (2013) and a Fellow of INAE (2015). He is a recipient of several prestigious awards such as NASI-Reliance Industries Platinum Jubilee Award-2012; INAE Silver Jubilee Young Engineer Award-2012; Dr. C. V. Raman Young Teachers Award-2011; SAE International's Ralph R. Teetor Educational Award-2008; INSA Young Scientist Award-2007; UICT Young Scientist Award-2007; INAE Young Engineer Award-2005. He is the recipient of prestigious Shanti Swarup Bhatnagar Award-2016 in Engineering Sciences. He is the first combustion/IC engine researcher to get this honour.

Prof. Achintya Mukhopadhyay is a Professor of Mechanical Engineering at Jadavpur University, Kolkata (Calcutta), India. He has served as Professor of Mechanical Engineering at Indian Institute of Technology Madras. He has also held visiting positions at Technical University of Munich where he was Alexander von Humboldt Fellow and University of Illinois, Chicago. He obtained his bachelor's, master's and doctoral degrees from Jadavpur University; Indian Institute of Science, Bangalore; and Jadavpur University, all in Mechanical Engineering. His teaching interests include thermodynamics, heat transfer, combustion, multiphase flows and design and analysis of thermal systems. His major research interests are chemically reacting flows, multiphase flow and heat transfer and microscale flow and heat transfer. His current research activities include droplet and spray combustion, structure and dynamics of partially premixed flames, nonlinear dynamics and chaos in combustion systems, instability of liquid sheets and atomization, spray impingement heat transfer and electrokinetic and thermomagnetic microscale flows. He has over 250 research publications including 94 international journal publications and has advised a number of master's and doctoral theses. He has also served as reviewer for a number of international journals. He is a Fellow of the West Bengal Academy of Science and Technology, Life Member of Indian Society of Heat and Mass Transfer, Indian section of Combustion Institute, Indian Society for Energy, Environment and Sustainability and Member of the Society of Automotive Engineers (India).

Dr. Chetankumar Patel is currently a Project Scientist in Engine Research Laboratory, IIT Kanpur, after completing his Ph.D. from the same place in 2016. He obtained his bachelor's degree in Mechanical Engineering from L. D. College of Engineering, Ahmedabad, in 2002. Thereafter, he completed his master's degree in Mechanical Engineering with specialization in I.C. engines and automobiles from the same institute in 2007. He taught Mechanical Engineering for 4 years in few engineering institutions in India. His primary areas of research include microscopic and macroscopic spray investigations, in-cylinder spray and combustion visualization, in-cylinder combustion investigations, emissions, noise and vibration investigations and biofuels. He has published several peer-reviewed papers in high-impact journals and conferences.

Contributors

Avinash Kumar Agarwal Department of Mechanical Engineering, Indian Institute of Technology Kanpur, Kanpur, Uttar Pradesh, India

S. Balachandar Department of Mechanical Engineering, University of Florida, Gainesville, FL, USA

Saptarshi Basu Department of Mechanical Engineering, Indian Institute of Science Bangalore, Bengaluru, India

Michele Battistoni University of Perugia, Perugia, Italy

Nilanjan Chakraborty School of Mechanical and Systems Engineering, University of Newcastle, Newcastle Upon Tyne, UK

Souvick Chatterjee Department of Mechanical Engineering, Jadavpur University, Kolkata, West Bengal, India

Khanh D. Cung Mechanical Engineering-Engineering Mechanics, Michigan Technological University, Houghton, MI, USA

Amlan Garai Department of Mechanical Engineering, Jadavpur University, Kolkata, India

Shinjan Ghosh Department of Mechanical and Aerospace Engineering, University of Central Florida, Orlando, FL, USA

Srinibas Karmakar Department of Aerospace Engineering, Indian Institute of Technology, Kharagpur, West Bengal, India

Ranganathan Kumar Mechanical Engineering Department, Indian Institute of Science, Bengaluru, Karnataka, India

Seong-Young Lee Mechanical Engineering-Engineering Mechanics, Michigan Technological University, Houghton, MI, USA

Sean P. Malkeson MMI Engineering Ltd., Warrington, UK

M. Manish Department of Mechanical Engineering, Indian Institute of Technology Madras, Chennai, India

Ahmed Abdul Moiz Mechanical Engineering-Engineering Mechanics, Michigan Technological University, Houghton, MI, USA

Achintya Mukhopadhyay Department of Mechanical Engineering, Jadavpur University, Kolkata, West Bengal, India

Chetankumar Patel Project Scientist, Indian Institute of Technology Kanpur, Kanpur, Uttar Pradesh, India

Binita Pathak Mechanical Engineering Department, Indian Institute of Science, Bengaluru, Karnataka, India

Samadhan A. Pawar Department of Aerospace Engineering, IIT Madras, Chennai, India

Kuppuraj Rajamanickam Department of Mechanical Engineering, Indian Institute of Science, Bangalore, India

D. Chaitanya Kumar Rao Department of Aerospace Engineering, Indian Institute of Technology, Kharagpur, West Bengal, India

Swapneel Roy Department of Mechanical Engineering, Indian Institute of Science, Bangalore, India

Kaushik Saha Argonne National Laboratory, Lemont, IL, USA

Srikrishna Sahu Department of Mechanical Engineering, Indian Institute of Technology Madras, Chennai, India

Sergei S. Sazhin Sir Harry Ricardo Laboratories, School of Computing, Engineering and Mathematics, Advanced Engineering Centre, University of Brighton, Brighton, UK

Swarnendu Sen Department of Mechanical Engineering, Jadavpur University, Kolkata, West Bengal, India

S. K. Som Department of Aerospace Engineering, Indian Institute of Technology, Kharagpur, West Bengal, India

Sibendu Som Argonne National Laboratory, Lemont, IL, USA

Shankar Subramaniam Department of Mechanical Engineering, CoMFRE: Multiphase Flow Research and Education, Iowa State University, Ames, Iowa, IA, USA

R. I. Sujith Department of Aerospace Engineering, IIT Madras, Chennai, India

Daniel H. Wacks School of Engineering and Computing Sciences, Durham University, Durham, UK

Part I
Multiphase Flow Fundamentals

Chapter 1
Introduction to Droplets and Sprays: Applications for Combustion and Propulsion

Saptarshi Basu, Avinash Kumar Agarwal, Achintya Mukhopadhyay and Chetankumar Patel

Abstract Spray characteristics play an important role in determining efficiency of gas turbine, rocket combustors and internal combustion engines. Droplet atomization and break-up forms the fundamental building block in many spray-based applications. Detailed analysis of the internal and the near-nozzle flow of fuel injectors is a necessity for a comprehensive understanding of any internal combustion engine performance. Understanding turbulence is one of the most difficult topics in science and engineering. In liquid fuelled combustion, the interaction of spray droplets with surrounding turbulent air flow is crucial since it influences the evaporation rate of the fuel droplets and the process of air and fuel vapour mixture preparation. For detail understanding on the droplet–turbulence interaction mechanisms as well as to validate numerical simulations of sprays, simultaneous measurement of both dispersed and carrier phases of the spray is essential. Turbulent spray combustion involves many areas of physics and chemistry, which accompany a variety of mathematical challenges. This monograph deals with all above aspects to enhance the understanding of spray characteristics involved with different applications.

S. Basu (✉)
Department of Mechanical Engineering, Indian Institute of Science Bangalore, Bengaluru 560012, India
e-mail: sbasu@mecheng.iisc.ernet.in

A. K. Agarwal
Department of Mechanical Engineering, Indian Institute of Technology Kanpur, Kanpur, Uttar Pradesh, India
e-mail: akag@iitk.ac.in

A. Mukhopadhyay
Department of Mechanical Engineering, Jadavpur University, Kolkata, West Bengal, India
e-mail: achintya.mukho@gmail.com

C. Patel
Department of Mechanical Engineering, Indian Institute of Technology Kanpur, Kanpur, Uttar Pradesh, India
e-mail: chetanpatel.iitk@gmail.com

S. Basu et al. (eds.), *Droplets and Sprays*, Energy, Environment, and Sustainability,
https://doi.org/10.1007/978-981-10-7449-3_1

Keywords Spray · Turbulent characteristics · Multiphase flow
Droplet atomization

Spray characteristics are the most important aspects related to power sector such as gas turbines, internal combustion (IC) engines. Gas turbines have wide applications in transportation and power generating sectors, most of which are currently driven by fossil fuels. Utilization of these fossil fuels and associated issues are highly dependent on fuel spray atomization characteristics. This book is divided into five subsections in which each subsection discusses the different aspects related to fuel spray characteristics. First subsection discusses combined deterministic and statistical approaches to modelling dispersed multiphase flows. The goal of modelling dispersed multiphase flows is to predict spatial distributions of the volume fraction, velocities and temperature of the dispersed and continuous phases in geometries.

Second section of this book discusses modelling of droplet heating and evaporation. It is very important to understand the behaviour of droplet evaporation and combustion in an actual condition. However, it is not always possible to conduct the experiment. In this context, modelling and simulation techniques gain popularity for the atomization process in non-evaporative and evaporative condition. Several approaches for the modelling of droplet heating and evaporation have also been discussed. Discrete and Multi-dimensional Quasi-discrete model and its applications for biodiesel, diesel and gasoline fuel droplets are discussed. This chapter also discusses the changes in droplet shapes and its surfaces due to heating and evaporation. In addition to that kinetic model for droplet heating and evaporation described along with new approaches for the estimation of evaporation coefficient to account quantum-chemical effects were discussed. Next chapter describes the combustion of pure component fuel or multicomponent fuel droplets. This chapter comprises of three sections. First part of the chapter briefly summarizes the fundamental understanding of the underlying mechanism of evaporation and combustion of pure component droplet. Second part of the chapter discusses several numerical and experimental study on combustion of multicomponent droplet like miscible and immiscible fuel blends. Third part of the chapter discusses the challenges in evaporation and combustion if nanofuel droplets.

Spray and combustion studies in internal combustion engine and gas turbine combustors gain momentum in last few decades. However, it suffers with higher density and viscosity compared to conventional diesel fuel which significantly affect fuel injection system, process and fuel atomization. Next chapter discusses the primary atomization mechanisms in spray nozzle to the propulsive devices. They also discussed difficulties associated with the experimental characterization of near-nozzle break-up. Spray characteristics play an important role in determining efficiency of gas turbine or rocket combustors. Current availability of powerful computing tools has made computational fluid dynamics (CFD) along with other analytical and numerical techniques a viable tool for design of combustors which reduced the requirement of expensive experimental studies in the design process.

A system which incorporates a liquid jet that breaks up in the presence of a strong swirling field and sandwiched between two swirling air flow streams has been discussed in the next chapter. This work includes computational fluid dynamics, analytical technique (linear stability) and statistical tool (entropy formulation) to model the spray characteristics in the form of break-up length and droplet distribution from nozzle geometry and inlet kinematic conditions.

Detailed analysis of the internal and the near-nozzle flow of fuel injectors is a necessity for a comprehensive understanding of any internal combustion engine performance. For gasoline direct injection engines, under part-load conditions, the in-cylinder pressure can be subatmospheric when the high-temperature fuel is injected, resulting in flash boiling. This chapter discusses three-dimensional computational fluid dynamics (CFD) simulations for the flash boiling phenomenon. A spray nozzle which can produce extremely small droplet size finds relevance in drying and liquid fuelled engines. Many of the nozzles which adopt standalone techniques (pressure, air-assist) fail to produce finer size droplets. Efforts have been made to hybridize two or more techniques to achieve the finer atomization. For instance, standard air-assist atomizer can be combined with effervescent/ultrasonic means to achieve further reduction in droplet size. This chapter presents the comprehensive aspects of such type of hybrid atomizers. Features such as mode of operation, benefits, shortcomings, areas of application are discussed in greater details.

Gas turbines have wide applications in transportation and power generating sectors which are currently driven by fossil fuels. The problem of air pollution can be associated with the use of conventional fuels, and their prolonged use has caused the fuel reserves to get depleted gradually. Addition of ethanol in conventional fossil fuel leads to better spraying characteristics and decreases air pollution as well. Next chapter describes the spray characteristics of pure kerosene, pure ethanol and ethanol-blended kerosene (10 and 20% ethanol-blended kerosene by volume) by using a hybrid atomizer. In liquid fuelled combustion, the interaction of spray droplets with surrounding turbulent air flow is crucial since it influences the evaporation rate of the fuel droplets and the process of air and fuel vapour mixture preparation. For detail understanding on the droplet–turbulence interaction mechanisms as well as to validate numerical simulations of sprays, simultaneous measurement of both dispersed and carrier phases of the spray is essential. This chapter focuses on the advances in the optical diagnostics (especially the planar techniques) in the last few decades to meet these requirements.

Another chapter begin with a general discussion on turbulence. This chapter covers concepts involving higher fidelity in description of turbulent combustion. This chapter also discusses premixed combustion and new advanced combustion modes such as partially premixed combustion (PPC) and multiple injections as well as numerical simulation. Most industrial flows are turbulent, and the modelling of turbulent flow remains a challenging task. The advancement of high-performance computing has made computational fluid dynamics (CFD) simulations a viable alternative to expensive experimentation. This complexity is augmented in turbulent droplet combustion simulations by the complex interaction of heat and mass transfer associated with evaporation, fluid dynamics, combustion and heat release. Next chapter focuses on the effects of droplet diameter a_d and droplet equivalence ratio ϕ_d on the statistical behaviours of the variances of the fuel mass fraction fluctuations, the mixture fraction fluctuations and co-variance and their transport.

Droplet atomization and break-up forms the fundamental building block in many spray-based applications. Next chapter explains the different modes of atomization and break-up induced by heat and acoustics. In addition to that this chapter also discusses different phases of instabilities induced in bicomponent system of different volatilities at lower and higher concentrations. Thermoacoustic instability is a plaguing problem in confined combustion systems, where self-sustained periodic oscillations of ruinous amplitudes cause serious damage and performance loss to propulsive and power generating systems occur. This chapter shows the recent developments in understanding the transition route to thermoacoustic instability in gaseous combustion systems and describes a detailed methodology to detect this route in a two-phase flow combustion system.

This research monograph presents both fundamental science and applied innovations on several key and emerging technologies involving fossil and alternate fuel utilization in power and transport sectors. Specific topics covered in the manuscript include:

- Deterministic and Stochastic Aspects of Dispersed Multiphase Flow Modelling
- Droplet Heating and Evaporation
- Multicomponent Droplet Combustion
- Basic Principles of Primary Atomization
- Comprehensive Modelling of Primary Atomization Processes
- Modelling of Flash Boiling in Fuel Injector Nozzles
- Development of Novel Injectors for High-Speed Combustors
- Spray Characteristics of Ethanol, Kerosene and Ethanol-Kerosene Blends in Gas Turbine Hybrid Atomizers
- Optical Techniques for Spray Diagnostics
- Turbulent Spray Combustion
- Spray–Flame Interaction
- Dynamics of Droplet Break-up
- Intermittency in Spray Combustion Systems.

These topics are organized in five different sections: (i) Multiphase Flow Fundamentals, (ii) Droplet and Spray Combustion, (iii) Atomization Principles and Injection Strategies, (iv) Turbulent Spray Combustion and (v) Droplet and Spray Dynamics.

Chapter 2
Towards Combined Deterministic and Statistical Approaches to Modeling Dispersed Multiphase Flows

Shankar Subramaniam and S. Balachandar

Abstract The goal of modeling dispersed multiphase flows is to predict spatial distributions of the volume fraction, velocities and other properties of interest of the dispersed and continuous phases in geometries of practical interest. In this chapter we summarize key advances that have enabled progress towards this goal by starting with the development of interphase momentum transfer models for flow past single particles with a view to extending them to multiparticle systems. Particle-resolved direct numerical simulation of microscale (particle scale) phenomena has emerged as a powerful tool to propose statistical models for mesoscale and macroscale simulations of dispersed multiphase flow. Key advances in modeling of interphase momentum as well as particle and fluid velocity fluctuations are briefly reviewed. Building on this foundation we examine deterministic models for multiparticle effects that are based on the generalized Faxén theorem which provides a rigorous theoretical foundation for their development. In particular, the recently proposed pairwise interaction extended point-particle (PIEP) model is presented as a means of systematically including fluid-mediated particle-particle interactions in a deterministic framework. The performance of the PIEP model is assessed in different validation tests. Although the deterministic modeling approach provides physical insight and mechanistic interpretation of model terms, its limitation is it does not fully account for all the fluctuations that occur at the microscale. Therefore it is important to complement the deterministic model with a stochastic component to properly account for the statistical variability that is naturally present in all dispersed multiphase flows. We review key advances in statistical modeling of dispersed multiphase flows in

S. Subramaniam (✉)
Department of Mechanical Engineering, CoMFRE: Multiphase Flow Research and Education, Iowa State University, Ames, Iowa, IA 50010, USA
e-mail: shankar@iastate.edu

S. Balachandar
Department of Mechanical Engineering, University of Florida, Gainesville, FL 32611, USA
e-mail: bala1s@ufl.edu

S. Basu et al. (eds.), *Droplets and Sprays*, Energy, Environment, and Sustainability,
https://doi.org/10.1007/978-981-10-7449-3_2

terms of both the two-fluid theory as well as the Euler-Lagrange approach, with an emphasis on the latter, where we consider both Reynolds-averaged Navier–Stokes (RANS) and large eddy simulation (LES) descriptions of the carrier phase. Finally, we propose a new paradigm for modeling dispersed multiphase flows that effectively combines the best elements of the deterministic and statistical modeling approaches as the most promising direction for development of predictive models.

2.1 Introduction

Computer simulations hold the promise of predicting and controlling multiphase flows such as sprays and fluidized beds to meet the needs of engineering design. They are particularly useful in problems where experimental measurements are difficult and also for rapid exploration of a design space. However, in order for simulations to be reliable, they must be accurate. One way to guarantee accuracy and reliability of prediction is to perform the simulations from first principles. For example, in the case of fluidized bed, this would require resolving the motion of every particle within the bed and solving the complete details of the flow around all the particles over the entire region of the fluidized bed occupied by the fluid. With present-day computational power, this level of resolution demanded by the first-principle simulation is possible only in small systems consisting of $O(10^4)$–$O(10^5)$ particles with the Reynolds number of the flow being modest. Industrial-scale fluidized beds that involve billions of particles and high Reynolds number flows are well out of reach for this approach. Even a single fully resolved simulation of this magnitude is impossible in the foreseeable future, let alone using this approach as a design tool. The goal of multiphase flow modeling is to develop computationally cheaper reduced-order simulation approaches that are of sufficient accuracy that they can still be used as reliable engineering tools. The key to success is to inject as much physics as possible into the reduced-order approaches so that they preserve much of the accuracy of the fully resolved approach. Here, we present a strategy where by encapsulating the information contained in accurate and computationally expensive fully resolved simulations in terms of closure models and using them in faster reduced-order simulations, sufficient accuracy for engineering design is achieved. In this chapter, we focus on the modeling challenges in dispersed multiphase flows, i.e., flows in which dispersed elements such as particles, droplets, or bubbles are distributed with a continuous gas or liquid medium, which we will refer to as 'particles' and 'fluid.' Our discussion will primarily focus on solid particles.

In fully resolved direct numerical simulation (FR-DNS) of particle-laden flows, the Navier–Stokes equations are solved by fully resolving the particle by imposing boundary conditions at each particle's surface (Bagchi and Balachandar 2003; Uhlmann 2005; Elghobashi and Truesdell 1992; Zhang and Prosperetti 2005; Quan and Schmidt 2007; Gorokhovski and Herrmann 2008; Herrmann 2008; Esmaeeli and Tryggvason 1998; Tenneti et al. 2011; Xu and Subramaniam 2010). Because FR-DNS produces three-dimensional time-dependent velocity and pressure fields, all Eulerian and Lagrangian flow statistics can be extracted, thus making it a powerful tool for model development and validation. Therefore, FR-DNS of flow past particles is a first-principle approach to developing accurate closure models at all levels of statistical closure. Such simulations provide accurate representation of flow around each particle but are limited to idealized geometries and small domains due to computational cost. The objective of this chapter is to show how the information obtained from FR-DNS can be used to develop models for particle behavior in less computationally expensive simulation approaches that can be used for simulating industrial devices involving complex geometry and larger domain sizes.

There are many different approaches to modeling dispersed multiphase flows (Tenneti and Subramaniam 2014; Subramaniam 2013), but in this chapter we will focus on the approach in which the particles are represented in a Lagrangian frame of reference, while they interact with an Eulerian field representation of the ambient fluid.[1] There are several variants in this Euler–Lagrange (EL) approach, which arise depending on the statistical representation of the particles and the fluid. At one extreme, when a single realization of the particle field is coupled to a single realization of the fluid phase, and the evolution of the multiphase system is obtained by solution of the Navier–Stokes equations by fully resolving the particle by imposing boundary conditions at each particles surface, we have the model-free FR-DNS approach already referred to. This results in a complete description of the multiphase flow which we first explain in detail.

2.1.1 *Fully Resolved (FR) Approach*

A complete description of gas–solid flow specifies information about the state of every particle and fluid point at every time instant, and it completely determines the future time evolution of the gas–solid flow. The state of gas–solid flow $S(t)$ at any instant t can be represented in terms of the state of the particles $S_p(t)$ and that of the fluid $S_f(t)$. As a specific example, we consider a simple gas–solid flow with smooth, monodisperse spheres but complete descriptions of more complex gas–solid flows are also possible. The set of positions and velocities $\left\{\mathbf{X}^{(i)}, \mathbf{V}^{(i)}, i = 1, \ldots, N(t)\right\}$ of $N(t)$ particles characterizes the state of the particles $S_p(t)$. The state of the fluid is characterized by the knowledge of the fluid velocity field $\mathbf{u}(\mathbf{x}, t)$ and the pressure field $p(\mathbf{x}, t)$. This complete description of the gas–solid flow at the scale of the particles is denoted a *microscale description*.
The evolution of $S(t)$ is given by

[1]The Eulerian-Eulerian approach where the dispersed phase is also represented by Eulerian fields of volume fraction and mean velocity is not relevant to the deterministic modeling approach which is discussed in this work.

$$\frac{dS}{dt} = \mathscr{L}\left[S(t)\right], \tag{2.1}$$

where $\mathscr{L}$ represents a nonlinear operator operating on the state $S(t)$ of the gas–solid flow. For incompressible flows, this operator becomes the mass and momentum conservation equations for the fluid phase:

$$\frac{\partial u_i}{\partial x_i} = 0, \tag{2.2}$$

$$\rho_f \frac{\partial u_i}{\partial t} + \rho_f \frac{\partial u_i u_j}{\partial x_j} = -g_i + \frac{\partial \tau_{ji}}{\partial x_j}, \tag{2.3}$$

and the position and velocity evolution equations for the solid phase:

$$\frac{d\mathbf{X}^{(i)}(t)}{dt} = \mathbf{V}^{(i)}(t), \tag{2.4}$$

$$m^{(i)} \frac{d\mathbf{V}^{(i)}(t)}{dt} = \mathbf{B} + \mathbf{F}_h^{(i)}(t) + \sum_{\substack{j=1 \\ j \neq i}}^{N(t)} \mathbf{F}_{\text{int}}^{(i,j)}(t), \tag{2.5}$$

where particle rotation has been omitted for simplicity. In Eq. 2.3, ρ_f is the thermodynamic density of the fluid phase, $\mathbf{g}$ represents body forces (e.g., hydrostatic pressure-gradient, gravitational force) acting throughout the volume of an infinitesimal fluid element, and $\boldsymbol{\tau}$ represents the surface stresses (both pressure and viscous stresses) acting on the surface of an infinitesimal fluid element. Equations 2.2 and 2.3 are subject to no-slip and no-penetration velocity boundary conditions on the surface of the particles. In the velocity evolution equation for the particles (Eq. 2.5), $m^{(i)}$ is the mass of the ith particle, $\mathbf{B}$ is any external body force, $\mathbf{F}_h^{(i)}$ is the hydrodynamic force (due to the pressure and viscous stresses at the particle surface), and $\mathbf{F}_{\text{int}}^{(i,j)}$ is any interaction force (e.g., contact force due to collisions, cohesion) between particles i and j. This coupled set of equations together with the boundary conditions governs the state of the gas–solid flow at the microscale. At the microscale, FR-DNS provides a numerical solution to the evolution of such a complete description of gas–solid flow.

2.1.2 Point-Particle (PP) Approach

In one class of simpler reduced-order methods, the particles are treated as points and thereby the details of the flow around each particle at the scale of the particles are erased. In this point-particle (PP) approach, the mass, momentum, and energy exchange between the particle and the surrounding flow must be modeled. We will briefly describe the PP approach before we start to address how FR-DNS can be used to develop reliable coupling models for the PP approach.

In all the EL methods, the carrier-phase information on the scale of a particle diameter d is implicitly averaged. In other words, flow variation on scales smaller than a particle diameter is removed. Furthermore, the EL approach is on sound theoretical footing when the particle size and the disturbance flow that they generate are much smaller than the grid resolution (i.e., when $\Delta x \gg d$). In this limit, the particle is much smaller than the smallest flow scales that are resolved in the carrier phase and this allows particle to be represented as a point in its interaction with the surrounding fluid. We will refer to this class of multiphase flow simulations as the point-particle (PP) approach. The majority of EL simulations to date have been using the PP approach. However, in recent years, EL methods are being extended to finite-size particles, whose diameter is comparable to the mesh spacing (Capecelatro and Desjardins 2013). In the case of finite-size EL methods, there are still outstanding questions as to how to couple the dispersed and carrier phases. Here, we will restrict attention to the PP approach.

We can further classify the PP approach based on the magnitude of grid resolution compared to the length scales of the flow. It is useful to think of an *undisturbed macroscale flow* that would exist in the limit when the particles are all absent. Such a flow can be turbulent with a wide range of scales from the Kolmogorov scale η to the integral length scale (in general, let η be the smallest length scale associated with the equilibrium macroscale flow). In comparison, the disturbance flow generated around the particles also has a range of scales, but they are of the same order as the particles. This perturbation flow due to the wake around the particles is often referred to as *pseudo-turbulence*.

Now, two different regimes can be considered which have been discussed in (Balachandar 2009; Balachandar and Eaton 2010). In the first regime, the Kolmogorov scale of the equilibrium macroscale flow is much larger than the particle size (i.e., $\eta \gg d$) and therefore larger than the scales of pseudo-turbulence. In this regime, (Squires and Eaton 1991b; Elghobashi and Truesdell 1993; Boivin et al. 1998; Sundaram and Collins 1999; Mashayek and Taulbee 2002), the grid resolution can be chosen to be $\Delta x \sim \eta \gg d$, so that all the scales of macroscale turbulence are fully resolved. Only the scales of pseudo-turbulence are averaged, and their effect must be modeled. This approach can be termed PP-DNS, since macroscale turbulence is fully resolved (Squires and Eaton 1990, 1991a, b; Wang et al. 2009; Mashayek 1998; Mashayek and Jaberi 1999; Sundaram and Collins 1997; Miller and Bellan 1999). In many applications, the Reynolds number of macroscale turbulence is quite large and it is not computationally feasible to resolve the entire range of scales down to the Kolmogorov scale, in which case we resort to partial resolution with $\Delta x \gg \eta \gg d$. This requires large eddy closure for macroscale turbulence, and the approach can be termed PP-LES (Almeida and Jaberi 2008; Apte et al. 2009, 2003; Okong'o and Bellan 2000, 2004).

We now consider the second regime where the particle diameter is comparable or larger than the Kolmogorov scale of macroscale turbulence (i.e., the regime where $d \gtrsim \eta$). Here, the restriction of PP approach that $\Delta x \gg d$ allows only partial resolution of macroscale turbulence. Here, both the unresolved macroscale turbulence and pseudo-turbulence are of comparable scale (see (Balachandar 2009; Balachandar and Eaton 2010; Ling et al. 2013, 2016)), and both must be taken into account through LES closure. Thus, in this regime of larger-than-η particles, the only viable point particle approach is PP-LES. It should be pointed out that the limit of Reynolds-averaged simulation of the carrier phase with point particles gives rise to the PP-RANS approach.

The above discussion centered on different levels of averaging of the carrier phase. An averaging of the dispersed phase is also often employed with the EL method. If all the physical particles are faithfully represented and their time evolution is tracked, then there is no averaging of the dispersed phase and any error in the corresponding PP-DNS approach will be due to inaccuracies in the interphase mass, momentum, and energy coupling models. In case of PP-LES, there will be additional error associated with LES closure. In many applications involving very large number of particles, it is impossible to track each and every particle in the simulation. Only a statistical sample of particles are tracked (these are termed *computational particles*) with each representing a cloud of physical particles. In this approach, each computational particle is an average over the cloud of physical particles that they represent.

2.1.3 *Euler–Lagrange Interphase Coupling Models*

Euler–Lagrange (EL) models are obtained by systematically removing information from the carrier phase and from the dispersed phase, which can be considered as reduced-order projection of the fully resolved information. As a result in the EL approach, the mass, momentum, and energy coupling between the phases must then be modeled.

2.1.3.1 EL Approach with PP-DNS

In PP-DNS, the interaction between particles and fluid, which was treated exactly through boundary conditions in FR-DNS, is now replaced by a modeled hydrodynamic force on the particles. This force on the particle is also transferred back to the fluid. In PP-DNS, the principal challenges are as follows: (1) developing a model for the hydrodynamic force exerted on the particle in terms of the motion of the particle and the motion of the surrounding fluid, (2) applying the interphase transfer of momentum from the particles back to fluid in the carrier-phase momentum conservation equation, and (3) representing the close-range and collisional interactions between the particles.

Among these, here we address only the modeling of the hydrodynamic force $\mathbf{F}_h$ in terms of available flow quantities in the context of the PP approach. In the most general representation, the force $\mathbf{F}_h^{(i)}$ on the *i*th particle is a function of (1) parameters that characterize the particle motion (i.e., particle position $\mathbf{X}^{(i)}$, particle velocity $\mathbf{V}^{(i)}$, particle diameter, particle density), (2) fields that characterize the flow around the particle (i.e., flow velocity $\mathbf{u}$, and flow stress tensor $\boldsymbol{\tau}$, typically evaluated at the particle location from the macroscale flow computed with the PP approach), and (3) parameters associated with all the other particles (i.e., their position, velocity, and size). In functional form, this dependence can be expressed as

$$\mathbf{F}_h^{(i)} = f\left[\left\{\mathbf{X}^{(i)}, \mathbf{V}^{(i)}, i = 1, \ldots, N(t)\right\}, \mathbf{u}(\mathbf{x}, t), p(\mathbf{x}, t)\right], \tag{2.6}$$

where $\left\{\mathbf{X}^{(i)}, \mathbf{V}^{(i)}, i = 1, \ldots, N(t)\right\}$ represent the position and velocity of the particles in the PP model.

Although specifying the interphase momentum transfer from particles to fluid and the representation of collisional interactions are also important in determining model accuracy, these aspects are discussed elsewhere (Garg et al. 2007, 2009; Schmidt and Rutland 2000). In the case of PP-DNS, the general form given in Eq. 2.6 is usually simplified to

$$\mathbf{F}_h^{(i)} = f\left[\mathbf{V}^{(i)}, \mathbf{u}(\mathbf{X}^{(i)}, t)\right]. \tag{2.7}$$

In other words, the modeled force on a particle depends only on that particle's velocity and the modeled fluid velocity field interpolated to that particle's location. Note that the position and velocity of neighboring particles does not affect the modeled force, nor does the modeled fluid pressure. This is a *deterministic* model for the modeled hydrodynamic force acting on a particle in a particular (modeled) realization of the carrier flow field. Recent developments in multiphase flow modeling propose improvements to this basic model by incorporating information from neighboring fluid nodes (Horwitz and Mani 2016) and neighbor particles (Akiki et al. 2017). Note that this modeled force appears in a dynamical equation for the particle velocity (the modeled counterpart of Eq. 2.5) which is then used to update the particle position via its trajectory evolution $d\mathbf{X}/dt = \mathbf{V}$.

2.1.3.2 EL Approach with PP-LES

In many problems, the resolution requirement of PP-DNS becomes prohibitive and it is expedient to not represent the entire range of scales in carrier fluid motion. If the carrier fluid-phase equations are spatially filtered, then we obtain the PP-LES approach that couples a Lagrangian description of the dispersed phase with large eddy simulation (LES) of the carrier phase. These are typically hybrid methods (save that of Carrara and Desjardins (Carrara and Desjardin 2006) where both phase equations are filtered) in that the filtering operation is not usually applied to the dispersed phase. Instead, several variants emerge depending on the treatment of the dispersed phase (Almeida and Jaberi 2008; Apte et al. 2009, 2003; Okong'o and Bellan 2000, 2004). Here, the modeling problem expressed by Eq. 2.6 is modified to

$$\mathbf{F}_h^{(i)} = f\left[\left\{\mathbf{X}^{(i)}, \mathbf{V}^{(i)}, i = 1, \ldots, N(t)\right\}, \tilde{\mathbf{u}}(\mathbf{x}, t), \tilde{p}(\mathbf{x}, t)\right] \approx f\left[\mathbf{V}^{(i)}, \tilde{\mathbf{u}}(\mathbf{X}^{(i)}, t)\right], \quad (2.8)$$

where $\tilde{\mathbf{u}}$ is the carrier velocity field resulting from the LES solution.

2.1.3.3 EL Approach Using PP-RANS

If the dispersed phase is treated as Lagrangian point particles coupled to an Eulerian carrier flow description based on Reynolds-averaged Navier–Stokes (RANS) equations, then we obtain a PP-RANS model. These are commonly found in both research (Amsden 1989) and commercial codes, although these typically involve a statistical representation of the dispersed phase.

2.1.3.4 Physical Particles and Parcels or Computational Particles

So far we have discussed the problem of how the hydrodynamic force experienced by a physical particle should be modeled using a computational point particle, as the information content of the carrier fluid is systematically reduced by filtering or averaging. In such formulations, a *deterministic* force model is appropriate. Statistical representations of the particle field emerged as a means to efficiently represent a large number of particles in the physical system with a fewer number of computational particles which do not bear a one-to-one correspondence with their physical counterparts.

If the computational point particles in PP-DNS are not intended as a one-to-one representation of physical particles, then the correspondence between the two systems exists only in a statistical sense (Subramaniam 2013), i.e., at the level of the number density function (NDF) and its moments. In the context of modeling the hydrodynamic force on a particle, this means that the correspondence between the computational system and its physical counterpart should be established at the level of mean interphase momentum transfer and second moments as described in detail elsewhere (Pai and Subramaniam 2009; Subramaniam 2013).

This is also related to the third issue in Sect. 2.1.3.1 regarding the treatment of collisions. The choice of how to treat collisions depends on whether the point particles in the PP-DNS are conceived as a one-to-one representation of physical particles, or whether they are intended to represent physical particles in a statistical sense. If one-to-one correspondence is intended between the computational point particles in PP-DNS, then a *deterministic* collision model is appropriate. For the latter representation, the collision model can be statistical, and this can be computationally efficient.

2.2 Single-Particle Drag

The essence of Euler–Euler (EE) and Euler–Lagrange (EL) formulations of dispersed multiphase flows is in the modeling of mass, momentum, and energy coupling between the dispersed and the continuous phases. Since the details of the flow at the microscale around each individual particle will not be resolved in the EE and EL formulations, the net momentum transfer between the particle and the surrounding flow is modeled in terms of the hydrodynamic force on the particle. Similarly, mass and energy exchange between the phases are modeled as mass and heat transfer coefficients. The discussion to follow in this section will primarily focus on force modeling. Although less well studied, similar ideas apply for mass and heat transfer modeling as well.

The goal is then to express the force on a particle in terms of its motion and the surrounding continuous phase flow it is subjected to. Note that in the EE and EL formulations the continuous phase flow approaching each particle has been averaged or filtered at the microscale (i.e., averaged on the scale of the particle). The flow at the microscale and consequently the force on the particle can be substantially influenced by the presence of other particles in the neighborhood. This flow-mediated particle–particle interaction introduces significant complication to force modeling. In this section, we avoid this complication by considering an isolated particle without the presence of nearby neighbors. However, we pose no further restriction to the flow approaching the particle. The effect of neighbors on force modeling will be considered in Sect. 2.3.

There is substantial body of research on interphase force modeling, and in this paper, we will briefly describe this development in the following three steps: (i) force modeling in the Stokes flow (i.e., zero Reynolds number regime), (ii) finite Reynolds number correction, and (iii) turbulence effects. The interphase coupling models in the Stokes and low Reynolds number limits have been developed through theoretical analysis. However, FR-DNS of a isolated particles has been quite valuable in extending the coupling models to finite Reynolds numbers and to turbulence effects. This aspect of FR-DNS contributing to point-particle models at the level of an individual particle will be the focus of this section.

The celebrated Stokes drag $\mathbf{F} = 6\pi \mu_f R\, \mathbf{u}$ on a spherical particle applies in the limit of a steady uniform flow $\mathbf{u}$ past a stationary particle of radius R at zero Reynolds number (here, μ_f is the dynamic viscosity of the fluid). The first significant extension of the Stokes drag law relaxes the steady flow restriction. The resulting Basset–Boussinesq–Oseen (BBO) equation for the hydrodynamic force on a particle (Basset 1888; Boussinesq 1885; Oseen 1927) is an exact expression for force on a spherical particle in arbitrary motion in a spatially uniform time-dependent ambient flow, in the limit of zero Reynolds number. The BBO equation for force can be written as

$$\mathbf{F}(t) = V_p\, \nabla \cdot \boldsymbol{\tau} + 6\pi \mu_f R(\mathbf{u} - \mathbf{v}) + C_M m_f \left[\frac{d\mathbf{u}}{dt} - \frac{d\mathbf{v}}{dt} \right] + \int_{-\infty}^{t} K_{BH} \left[\frac{d\mathbf{u}}{dt} - \frac{d\mathbf{v}}{dt} \right]_{@\xi} d\xi\,, \tag{2.9}$$

where V_p is the volume of the particle, m_f is the mass of fluid displaced by the particle, and ρ_f and ν_f are the density and kinematic viscosity of the fluid. Also, $\mathbf{v}$ is the velocity of the particle, and $\mathbf{u}$ is the velocity of the fluid, which is considered to be time-dependent, but spatially uniform.

In the above, $\boldsymbol{\tau}$ is the total stress tensor of the undisturbed flow (i.e., stress tensor of the flow in the absence of the particle), and from the Navier–Stokes equation of the undisturbed flow, we have $\nabla \cdot \boldsymbol{\tau} = \rho_f (d\mathbf{u}/dt - \mathbf{g})$, where $\mathbf{g}$ is the acceleration due to gravity. Thus, the first term on the right is the force on the particle due to the undisturbed ambient flow, and it is also known as the pressure-gradient force or the Archimedes force. The second term on the right is the Stokes drag for a moving particle. The third term on the right is the added-mass force, and it is proportional to the relative acceleration between the particle and the surrounding undisturbed ambient flow. The added-mass coefficient for a sphere is $C_M = 1/2$. The last term on the right is the Basset history force, and it depends on the past history of relative acceleration weighted by the history kernel $K_{BH} = 6R^2 \rho_f \sqrt{\pi \nu_f}/\sqrt{t - \xi}$. The kernel decays as one over square root of time, and the relative acceleration within the square parenthesis is evaluated at time ξ.

Another significant extension of the Stokes drag formula is by Faxén (1923). While still limited to the zero Reynolds number limit, Faxén obtained the formula $\mathbf{F} = 6\pi \mu_f R\, \overline{\mathbf{u}}^s$, which remarkably is the exact drag on a stationary particle subjected to steady undisturbed ambient flow, which can now be spatially varying in any arbitrary manner. As can be seen, the Stokes drag still applies but with the fluid velocity now replaced by $\overline{\mathbf{u}}^s$, where the overbar with superscript 's' denotes an average over the surface of the particle.

There have been several attempts at combining the Faxén's form with the BBO equation to obtain a force expression that is valid for a particle in motion in an ambient flow that is both spatially and temporally varying, but still in the Stokes limit. These efforts culminated in the works of Maxey and Riley (1983) and Gatignol (1983) who obtained the following Maxey–Riley–Gatignol (MRG) equation for hydrodynamic force:

$$\mathbf{F}(t) = V_p\, \overline{\nabla \cdot \boldsymbol{\tau}}^v + 6\pi \mu_f R(\overline{\mathbf{u}}^s - \mathbf{v}) + C_M m_f \left[\overline{\frac{d\mathbf{u}}{dt}}^v - \frac{d\mathbf{v}}{dt} \right] + \int_{-\infty}^{t} K_{BH} \left[\overline{\frac{d\mathbf{u}}{dt}}^s - \frac{d\mathbf{v}}{dt} \right]_{@\xi} d\xi\,, \tag{2.10}$$

Although they appear quite similar, there is a fundamental difference between the BBO and MRG equations. In the former, since the undisturbed ambient flow is spatially uniform, the flow-related quantities such as $\mathbf{u}$ and $\boldsymbol{\tau}$ can be evaluated anywhere within the particle (for convenience often evaluated at the particle center). In contrast, in a spatially varying flow, application of (2.9) will involve the ambiguity of where the fluid-related quantities in the equation are to be evaluated. The Faxén form given in (2.10) removes this ambiguity, since all fluid-related quantities are specified either as surface average $\overline{()}^{s}$ or as volume average $\overline{()}^{v}$.

2.2.1 Reynolds Number Effects

The advantage of the force expression given in (2.10) is that it is divided into physically motivated individual force contributions. This allows meaningful extension of each contribution to finite Reynolds number as appropriate. However, it must be cautioned that while Eq. (2.10) is rigorous, its finite Reynolds number extension is necessarily empirical.

The undisturbed flow force given by the first term on the right in (2.10) remains valid at all Reynolds number. So no correction to this term is required. There are many finite Reynolds number corrections to the quasi-steady drag represented by the second term on the right in (2.10). One popular model that is widely used is (Schiller and Naumann 1933; Clift et al. 1978)

$$\mathbf{F}_{qs} = 6\pi\mu_f R(\overline{\mathbf{u}}^{s} - \mathbf{v})\left(1 + 0.15\,\mathrm{Re}^{0.687}\right) , \tag{2.11}$$

where the Reynolds number is based on particle diameter and relative velocity as $\mathrm{Re} = 2R|\overline{\mathbf{u}}^{s} - \mathbf{v}|/\nu_f$. This model is good for Reynolds numbers less than about a thousand. But there are other models that extend to higher Re and the reader is referred to Crowe et al. (1999). Although the Reynolds number corrections have been experimentally obtained, these have been verified with the use of FR-DNSs.

The added-mass force arises from the inviscid perturbation flow due to the no-penetration boundary condition imposed on the surface of the particle. It thus does not depend on the Reynolds number, and as a result, $C_M = 1/2$ even at finite Reynolds number. In contrast, the history force arises from the viscous perturbation flow due to the no-slip boundary condition imposed on the surface of the particle. At finite Reynolds number, the Basset history kernel (K_{BH}) is accurate only for small values of $(t - \xi)$ and must be modified for large values of $(t - \xi)$. In particular, there exists a time beyond which the kernel must decay faster than $1/\sqrt{t - \xi}$ (Mei and Adrian 1992). No matter how small we take Re to be, there exists a Stokes length beyond which inertial effects become important. Mei and Adrian (1992) performed a matched asymptotic analysis where they solved the unsteady Stokes equation in the inner region of $r < L$ and Oseen equations in the outer region of $r > L$ and matched the two solutions. The resulting viscous-unsteady force can still be represented by the general convolution integral given in (2.10), but the revised viscous-unsteady kernel is given by

$$\text{Mei}-\text{Adrian}: \quad K_{vu}(t) = \frac{1}{\tau_{vu}} \frac{6R^2 \rho_f \sqrt{\pi \nu_f}}{\left[(t/\tau_{vu})^{1/4} + t/\tau_{vu}\right]^2}, \tag{2.12}$$

where the viscous-unsteady time scale is defined as

$$\tau_{vu} = \left(\frac{4}{\pi}\right)^{1/3} \frac{d_p^2}{\nu_f} \left(\frac{0.75 + 0.105\,\mathrm{Re}}{\mathrm{Re}}\right)^2 . \tag{2.13}$$

For $t \ll \tau_{vu}$, the above kernel approaches $1/\sqrt{t}$ behavior of the Basset kernel. But for $t \gg \tau_{vu}$, the kernel decays faster as $1/t^2$. Thus, for finite Re, no matter how small, there exists a $t = \tau_{vu}$ beyond which the kernel will decay faster than the Basset kernel. But the finite Reynolds number history kernel given by Mei and Adrian is not universal. Time-dependent vorticity diffusion due to relative acceleration is far more complex and depends on the nature of relative acceleration (Lovalenti and Brady 1993a, b). Other forms of history kernels that are appropriate for varying relative motion have been proposed by Kim et al. (1998).

Since the pioneering work of Saffman (1965) in 1965, it is now recognized that the particle when subjected to ambient shear (or vorticity) experiences a lift force in addition to the force contributions presented in (2.10). The lift force depends both on Reynolds number Re based on relative velocity and on the Reynolds number based on ambient vorticity, defined as $\mathrm{Re}_\omega = 4|\boldsymbol{\omega}|R^2/\nu_f$, where $\boldsymbol{\omega}$ is the vorticity of the undisturbed flow at the particle in a direction perpendicular to the direction of relative velocity. Note that any vorticity component that is parallel to the direction of relative velocity will not contribute to force. The following theoretical expression of the lift force by Saffman applies in the limit $\mathrm{Re} \ll \sqrt{\mathrm{Re}_\omega} \ll 1$:

$$\mathbf{F}_L = C_L \frac{\pi}{2} R^2 \rho_f |\mathbf{u} - \mathbf{v}|^2 \mathbf{e}_\perp \quad \text{where} \quad C_L = \frac{12.92}{\pi} \frac{\sqrt{\mathrm{Re}_\omega}}{\mathrm{Re}}, \tag{2.14}$$

where $\mathbf{e}_\perp$ is unit vector in the direction $\boldsymbol{\omega} \times (\mathbf{u} - \mathbf{v})$. As with the history force, vorticity-induced lift force on the particle is also dependent on the ambient flow condition in a complex way. For example, the above Saffman lift force expression is applicable when the ambient vorticity $\boldsymbol{\omega}$ is in the form of a linear shear flow. Instead, if the particle is subjected to a rotating flow of uniform vorticity $\boldsymbol{\omega}$, then the corresponding theoretical expression for the lift coefficient is given by Herron (1975).

$$C_L = 5.091 \frac{\sqrt{\mathrm{Re}_\omega}}{\mathrm{Re}} \quad \text{for} \quad \mathrm{Re}, \mathrm{Re}_\omega \ll 1 \,. \tag{2.15}$$

These vorticity-induced lift forces must be combined with the Magnus lift force that arises due to particle rotation in the presence of a cross-flow. Since the angular velocity $\boldsymbol{\Omega}$ of the particle plays an analogous role as the ambient vorticity, the Magnus lift force will depend on Re and the rotational Reynolds number $\mathrm{Re}_\Omega = |\boldsymbol{\Omega}|d/\nu_f$. In the limit $\mathrm{Re}_\Omega \ll \mathrm{Re} \ll 1$, the theoretical investigation of Rubinow and Keller (1961) yields the same expression (2.14) for the lift force with the lift coefficient given by $C_L = \mathrm{Re}_\Omega / \mathrm{Re}$. As shown by Saffman (1965), in the low Reynolds number limit the lift contributions from ambient vorticity and particle rotation linearly superpose. There is now a large body of literature that extends the above vorticity and rotation-induced lift expressions to finite Reynolds numbers. However at finite Reynolds number, their interaction is far more complex and simple linear superposition will not be accurate. FR-DNS has been very useful in extending the theoretical results to finite Reynolds numbers.

2.2.2 *Turbulence Effects*

Past experimental measurements on the effect of carrier-phase turbulence on force on a particle have been generally inconclusive, with some predicting drag increase, while others observing drag reduction (Uhlherr and Sinclair 1970; Zarin and Nicholls 1971; Brucato et al. 1998; Clamen and Gauvin 1969; Wu and Faeth 1994). For additional discussion on these experimental measurements of turbulence effect, see (Crowe et al. 1999).

The definition of the undisturbed flow becomes easy and straightforward in the case of particle diameter much smaller than the smallest scales of carrier-phase turbulence (i.e., $d \ll \eta$). For such particles, the ambient flow appears as nearly uniform and the undisturbed ambient flow that would have existed in the absence of the particle can be well approximated. As observed in the direct numerical simulations of a small particle subjected to isotropic turbulence (Bagchi and Balachandar 2003), or a small particle at the center of a turbulent pipe flow (Merle et al. 2005), provided the time history of the carrier-phase velocity and acceleration at the particle location are known, the finite Reynolds number version of the BBO equation can be used to accurately compute the time evolution of force. In fact, for such small particles, the dominant contribution to force comes from the quasi-steady term and the unsteady contributions are typically negligible.

Particle–turbulence interaction becomes complex and far more interesting, when particle size becomes comparable or larger than the carrier flow scales. Since there is no scale separation, the ambient can no more be simply considered as a time-varying nearly uniform flow. Ambient flow scales much larger than the particle can still be treated as time-varying nearly uniform large eddy field. If we denote $\mathbf{u}_>$ to be the ambient velocity contribution from only these large scales, then Eq. (2.10) with $\mathbf{u}_>$ replacing $\mathbf{u}$ can be used to evaluate force due to these larger-than-particle scales of ambient flow. On the other hand, it can be anticipated that turbulent eddies of size much smaller than the particle may not influence the mean particle motion. It can be assumed that eddies of size an order of magnitude smaller than the particle may be considered sufficiently small for heavy particles, while they need to be much smaller for the case of bubbles (Balachandar and Eaton 2010). Eddies of size comparable to the particle interact with the particle in a complex way.

In the spectral description of particle–turbulence interaction, a dispersed particle interacts with a range of eddies which in turn corresponds to a range of wave numbers in the fluid-phase TKE spectrum. One may define a Stokes number St_κ as the ratio of the particle response timescale τ_p to the timescale τ_κ corresponding to the eddies of wave number κ. Some eddies (say, type A) in this range have a timescale such that $St_\kappa > 1$, while the others (say, type B) in this range have a timescale such that $St_\kappa < 1$. The particle responds immediately to eddies that have a timescale such that $St_\kappa < 1$, while it responds slowly to eddies that have a timescale such that $St_\kappa > 1$. In the former case, the timescale for interphase TKE transfer is influenced more by the timescale corresponding to the eddies with wave number κ, while in the latter case the timescale for interphase TKE transfer is influenced more by the particle response timescale τ_p. Thus, the effective timescale for particle–turbulence interaction is obtained by integrating the effects of the two wave number ranges identified above, over the energy spectrum of fluid-phase turbulence in the two-phase flow. The multiscale interaction timescale $\langle \tau_{\text{int}} \rangle$ presented elsewhere (Pai and Subramaniam 2007, 2012) is a single-point analogue of the above spectral model.

The problem of a finite-sized particle of diameter larger than the Kolmogorov scale subjected to isotropic turbulence has been considered by Bagchi and Balachandar (2003; 2004), Merle et al. (2005), and Burton and Eaton (2005). These early efforts have been extended to multiple particles and moving particles. Together, these studies provide us insight into the effect of carrier-phase turbulence on the momentum transfer between the particle and the surrounding turbulent flow, which can be summarized as:

(i) As far as time-averaged mean drag force on the particle, there is no systematic effect of ambient turbulence. Standard drag law is sufficient to capture the mean drag force.
(ii) Only in case of particles of size smaller than the Kolmogorov scale, the standard drag relation accurately predicts the instantaneous time evolution of the force on the particle.

(iii) For larger particle ($d \gg \eta$), slow variations in particle force arising from scales larger than the particle can be well described by the standard drag law. However, rapid variations in the instantaneous force arising from scales comparable or smaller than the particle cannot be accurately captured by the standard drag or other forms of force expressions given in 2.10.

(iv) Self-induced vortex shedding becomes important above a certain critical particle Reynolds number. The critical Reynolds number for the onset of vortex shedding in a quiescent ambient flow is about 280, and the critical Reynolds number decreases with increasing level of turbulence in the oncoming flow.

(v) Self-induced vortex shedding introduces force fluctuations, which are most pronounced in the lift component and are not correlated to the oncoming turbulent ambient flow.

Based on the above outlined understanding of turbulence effect on the drag and lift forces on a particle, the following picture emerges. For particles of size smaller than the carrier flow scales, force expressions such as (2.10) are adequate to predict instantaneous force on the particle. Recent work by Horwitz et al. (2016) has led to comparison of PP-DNS with FR-DNS, and the finding is that their correction for the undisturbed flow velocity is needed to get good match of turbulence statistics such as kinetic energy and the probability density function (PDF) of particle acceleration. It must be pointed out that the Reynolds number of such small particles is expected to be small and self-induced vortex shedding will not be present (Balachandar 2009). For particles of size that match the ambient flow scales, the best strategy will be to consider the force on the particle to comprise of a deterministic and a stochastic component. The deterministic component will account for the effect of ambient flow scales larger than the particle, and force expressions of the form (2.10) can be used for this purpose. Contribution to force from smaller scale eddies can be best accounted in terms of a stochastic component. In the context of RANS simulations of carrier phase, Langevin-based models have been used to account for the effect of unresolved turbulence on particle motion. This approach can be adapted to account for the small scales, whose effect is not included in the deterministic component. Such modeling of turbulence effect is fundamentally different from the deterministic approach of Uhlerr and Sinclair (1970), Clamen and Gauvin (1969) and others, which included relative turbulence intensity as an additional parameter in the drag correlation. If the stochastic component is appropriately modeled, the proposed modeling approach can account for the dispersion of particles that arise from the effect of the smaller scales of ambient motion. However, it should be noted that with the proposed approach we admit our inability to precisely predict the trajectory of large particles in turbulent flows and resort to statistical description.

2.3 Beyond Single-Particle DNS

Although single-particle DNS is very useful in understanding flow physics and various regimes of multiphase flow phenomena, practical applications such as sprays involve multiparticle interactions. Therefore, DNS of assemblies of particles or freely evolving suspensions of particles is performed to quantify these multiparticle interactions. These simulations have been performed under different conditions with varying objectives. While there is a growing body of literature on FR-DNS of freely evolving particle suspensions in turbulent flow, these are usually performed in the absence of mean slip velocity between the particles and fluid and are therefore of somewhat limited value in developing EL force models which must be applicable over a range of mean slip velocities. Concurrently, Meharabadi et al. (2015) have shown that the presence of a nonzero mean slip between particles and gas generates pseudo-turbulent fluctuations which would typically mask the level of preexisting turbulence at the microscale for typical parameter ranges encountered in industrial applications. Therefore, from the standpoint of developing EL force models, the FR-DNS setup with nonzero mean slip velocity between gas and freely moving particles is more useful, and such studies are the focus of this section.

As noted earlier, Lagrangian models for the hydrodynamic force experienced by a particle typically take the form

$$\mathbf{F}_h^{(i)} = f\left[\mathbf{V}^{(i)}, \mathbf{u}(\mathbf{X}^{(i)}, t)\right] = m^{(i)} \Omega_p \left(\mathbf{u}(\mathbf{X}^{(i)}, t) - \mathbf{V}^{(i)}\right), \tag{2.16}$$

that subsumes the vast majority of models (Sundaram and Collins 1999; Amsden 1989; Ormancey and Martinon 1984; Brown and Hutchinson 1979; Gosman and Ioannides 1983) in the literature. Here, **u** represents the modeled velocity of the carrier flow (which can be either from PP-DNS, LES, or RANS), and Ω_p is a characteristic frequency associated with the particle momentum response time. This model is applicable to solid particles (although it neglects unsteady acceleration effects) and is often also used for droplets. Note that in the case of droplets the effect of vaporization (Dwyer and Dandy 1990; Dwyer 1989; Warnica et al. 1995a, b) and droplet deformation (Quan et al. 2009; Helenbrook and Edwards 2002) could also be important, but these are not represented in this simple model.

Multiparticle effects essentially manifest as the change in surface pressure and deviatoric stress fields resulting from the presence of neighbor particles. Multiparticle DNS can be used to quantify these effects, elucidate flow phenomena (Glowinski et al. 2001; Fortes et al. 1987; Mehrabadi et al. 2015), and propose improved models (Tenneti and Subramaniam 2014).

2.3.1 *FR-DNS of Random Arrays*

These FR-DNSs are usually performed by establishing a steady flow past a statistically homogeneous fixed bed or freely evolving suspension of particles by specifying a constant mean pressure gradient in the fluid phase, which at steady state results in a constant mean slip velocity between the particles and fluid (Garg et al. 2011; Tenneti et al. 2011). In FR-DNS methodologies that solve the incompressible NS equations, the drag force on a particle can be obtained by integrating the pressure and viscous stresses exerted by the fluid on the particle surface. The average drag force on particles in a homogeneous suspension for the μth realization is computed as

$$\left\{\mathbf{F}_{\mathrm{d}}\right\}_{\mathcal{V}}^{\mu} = \frac{1}{N_p}\left\{ - \langle \mathbf{g} \rangle_{\mathcal{V}} V_s - \oint_{\partial \mathcal{V}_{\mathrm{s}}} p \mathbf{dA} + \mu_{\mathrm{f}} \oint_{\partial \mathcal{V}_{\mathrm{s}}} \nabla \mathbf{u} \cdot \mathbf{dA} \right\}. \tag{2.17}$$

In Eq. 2.17, the first term on the right-hand side is the force due to the mean pressure gradient, the second term is the pressure contribution, and the third term is the viscous stress contribution. Drag correlations can be proposed for the total fluid-particle force (Hill et al. 2001a; Tenneti et al. 2011) or only for the force from surface stresses (van der Hoef et al. 2005; Beetstra et al. 2007). Because the mean interphase momentum transfer represents an average over all particle configurations corresponding to the same volume fraction and pair correlation function, the drag from a single realization (see Eq. 2.17) should be averaged over multiple independent realizations:

$$\left\{\mathbf{F}_{\mathrm{d}}\right\}_{\mathcal{V},\mathcal{M}} = \frac{\sum_{\mu=1}^{\mathcal{M}} \left\{\mathbf{F}_{\mathrm{d}}\right\}_{\mathcal{V}}^{\mu}}{\mathcal{M}}. \tag{2.18}$$

This ensemble-averaged drag force is reported as a normalized average drag force given by

$$F = \frac{\left|\left\{\mathbf{F}_{\mathrm{d}}\right\}_{\mathcal{V},\mathcal{M}}\right|}{F_{\mathrm{Stokes}}}, \tag{2.19}$$

where $F_{\mathrm{Stokes}} = 3\pi \mu_{\mathrm{f}} D \left(1 - \phi\right) |\langle \mathbf{W} \rangle|$ is the Stokes drag acting on an isolated sphere moving with a slip velocity of $\left(1 - \phi\right) |\langle \mathbf{W} \rangle|$. The number of multiple independent simulations $\mathcal{M}$ and the number of particles N_p should be chosen to minimize statistical variability.

2.3.2 *Computational Drag Laws*

The normalized average force acting on a suspension corresponding to a solid volume fraction ϕ and mean slip Reynolds number Re_m is reported as a drag law

$F(\phi, \mathrm{Re}_m)$. The advantage of PR-DNS methodologies that solve the NS equations over methods such as the LBM is that the relative contribution of pressure and viscous forces to the total drag force can be quantified in a straightforward manner. Moreover, the scaling of these forces with ϕ and Re_m provides physical insight into the formulation of the drag law. Tenneti et al. (2011) showed that the average pressure and viscous forces in a monodisperse gas–solid suspension are collinear, and hence, the magnitude of the total force can be expressed as the sum of their magnitudes. Simple scaling arguments dictate that the pressure force varies as the square of the slip velocity, while the viscous force varies linearly with the slip velocity. When the magnitudes of pressure and viscous forces are normalized by F_{Stokes}, the pressure force varies linearly with Re_m, while the viscous force is independent of Re_m. This simple scaling argument led Hill et al. (2001a) to propose the following form of the drag law for monodisperse spheres at moderate Reynolds numbers:

$$F\left(\phi, \mathrm{Re}_m\right) = F_2(\phi) + F_3(\phi)\,\mathrm{Re}_m \tag{2.20}$$

where $F_2(\phi)$ represents the viscous contribution, and $F_3(\phi)$ represents the volume fraction dependence of the pressure contribution. However, Tenneti et al. (2011) showed that the viscous contribution is a sublinear power of Re_m and that the pressure contribution is approximately linear for $\mathrm{Re}_m > 40$. Based on these observations and an analysis of the average pressure and viscous force variation along the particle surface, Tenneti et al. (2011) proposed the following functional form for the normalized average force for monodisperse particles:

$$F\left(\phi, \mathrm{Re}_m\right) = \frac{F_{\mathrm{isol}}\left(\mathrm{Re}_m\right)}{(1-\phi)^3} + F_\phi(\phi) + F_{\phi,\mathrm{Re}_m}\left(\phi, \mathrm{Re}_m\right). \tag{2.21}$$

Here, F_{isol} is the drag force acting on an isolated sphere moving in an unbounded medium, F_ϕ is purely a function of the solid volume fraction, and F_{ϕ,Re_m} is a separable function that is linear in Re_m. This drag correlation is obtained from highly resolved PR-DNSs that have been shown to be numerically convergent and whose solutions agree well with methods that employ boundary-fitted grids. Similar correlations have been proposed using FR-DNS of fixed particle assemblies (Hill et al. 2001a, b; Wylie et al. 2003; Yin and Sundaresan 2009a, b; van der Hoef et al. 2005; Beetstra et al. 2007; Deen et al. 2012; Sarkar et al. 2009).

Although Tenneti et al. (2011) decomposed the drag into pressure and viscous contributions, to date it has been difficult to recast the FR-DNS data from multiparticle simulations along the lines of the force decomposition for single particles given in Sect. 2.2. However, it is quite common in the bubbly flow literature to encounter EL models for the hydrodynamic force experienced by a bubble in a suspension that is obtained by modifying the coefficients corresponding to the different terms in the single-particle force model (cf. Eq. 2.10) by simply introducing a dependence on the average volume fraction to incorporate multiparticle effects.

In most of these studies, the multiparticle effect on drag or other force contributions only manifests through the average volume fraction. Chiu et al. (1997) have developed models to incorporate neighbor droplet or particle effects, but their accuracy is difficult to establish in the absence of validation data. It is worth noting that recently Mehrabadi et al. (2016) showed that changing the pair correlation function (which changes the second-order structure of the particle field) can dramatically influence the average drag experienced by a random particle array even if the average volume fraction is held constant. This seems to indicate that modifying the EL closure model for particle force by simply introducing a dependence on the average volume fraction may not be sufficient to capture multiparticle effects.

2.3.3 *Developing LE Particle Force Models from FR-DNS*

An important question is how to use these FR-DNS correlations for the average drag force in EL closure models such as Eq. 2.16. Note that the EL model requires a model for the instantaneous hydrodynamic force acting on a particle, while the drag correlations provide a closure for the average drag in terms of the average slip velocity between particles and fluid. It is customary to simply replace the average particle velocity in the drag correlation with the instantaneous velocity of the particle when applying it to EL models. Also, most of the models to date propose a correction to the particle momentum response frequency (cf. Eq. 2.16) by introducing a dependence on the average dispersed-phase volume fraction which corresponds to the chosen drag law for average force on a particle, i.e.,

$$\Omega_p = \frac{\mathbf{F}_h^{(i)}}{m^{(i)}\left(\mathbf{u}(\mathbf{X}^{(i)}, t) - \mathbf{V}^{(i)}\right)}. \tag{2.22}$$

Recent findings by Tenneti et al. (2010), however, indicate that such treatments are problematic. Figure 2.1 shows a scatter plot of the streamwise component of fluctuations in particle acceleration $\mathbf{A}''$ versus the streamwise component of fluctuations in particle velocity $\mathbf{v}''$. It is clear that the joint statistics of the particle acceleration and particle velocity observed in PR-DNS are not reproduced by such EL models. Kriebitzsch et al. (2013) arrived at a similar conclusion by making a direct comparison between PR-DNS data and discrete element method simulations. In Fig. 2.1, the particles with $\mathbf{A}'' \cdot \mathbf{v}'' > 0$ contribute to the source of particle granular temperature,

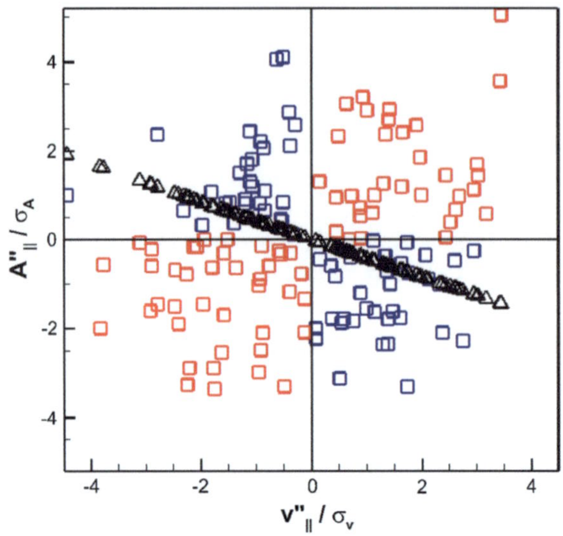

Fig. 2.1 Scatter plot of the streamwise component of particle acceleration fluctuations $\mathbf{A}''$ (normalized by the standard deviation in the particle acceleration distribution σ_A) versus the streamwise component of particle velocity fluctuations $\mathbf{v}''$ (normalized by the standard deviation in the particle velocity distribution σ_v) in a homogeneous gas–solid suspension. The fluctuations in the particle acceleration and velocity are defined with respect to their corresponding mean values. The square symbols denote the fluctuations in the particle acceleration obtained from particle-resolved direct numerical simulation (PR-DNS) of a freely evolving gas–solid suspension corresponding to a solid volume fraction of 0.2, mean flow Reynolds number of 20.0, and solid-to-fluid density ratio of 1,000. The triangles denote the fluctuations in the particle acceleration predicted using a model for the fluid-particle force of the form $\mathbf{F}_h^{(i)} = m^{(i)}\Omega_p \left(\langle \mathbf{u} \rangle (\mathbf{X}^{(i)}, t) - \mathbf{V}^{(i)}\right)$, where the angle brackets denote an ensemble average

whereas the particles with $\mathbf{A}'' \cdot \mathbf{v}'' < 0$ contribute to the dissipation of granular temperature. The granular temperature is the variance in the particle velocity distribution and is given by

$$T = \frac{1}{3}\left\langle \mathbf{v}'' \cdot \mathbf{v}'' \right\rangle, \tag{2.23}$$

where $\mathbf{v}''$ are the fluctuations in the particle velocity with respect to the mean particle velocity $\langle \mathbf{v} \rangle$. The particle velocity variance affects dispersion of particles in flow.

The scatter observed in the particle acceleration suggests a stochastic contribution to the fluid-particle force that arises because of the effect of the neighbor particles and fluid velocity fluctuations, which should model the PDF of hydrodynamic force obtained from FR-DNS (see Fig. 2.2).

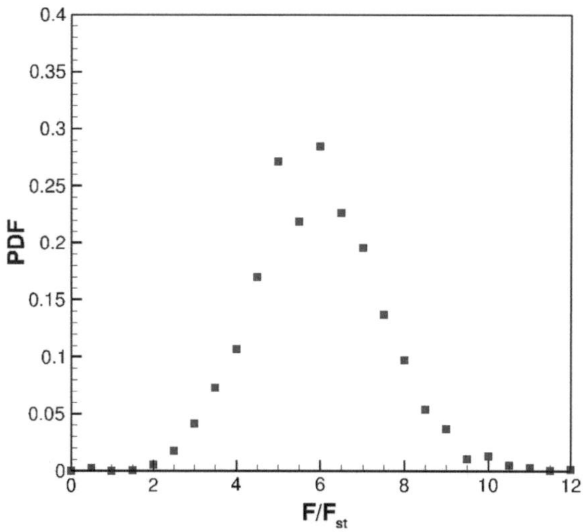

Fig. 2.2 Probability density function of the streamwise component of hydrodynamic force experienced by a particle $\mathbf{F}_h$ in a freely evolving gas–solid suspension. The square symbols denote the fluctuations in the particle acceleration obtained from particle-resolved direct numerical simulation (PR-DNS) of a freely evolving gas–solid suspension corresponding to a solid volume fraction of 0.2, mean flow Reynolds number of 20.0, and solid-to-fluid density ratio of 1,000

2.4 Deterministic Coupling Models

One of the advantages of the EL method is that the precise location of the Lagrangian particles is known. In the case where each computational particle corresponds to a real particle, there is no averaging over the dispersed phase. In this case, we have the advantage of each particle knowing the precise location of all its neighbors. We recall Eq. (2.6) where the most general expression for the force depends not only on the particle parameters and the local carrier phase, but also on the relative location and motion of all the neighbors. Nevertheless, virtually every one of the existing point-particle models of force have ignored the influence of neighboring particles in the calculation of force on each particle. The only way in which the neighbors affect the force is through the average local volume fraction dependence, such as that given in equation (2.21). In what follows, we will first establish that it is not adequate to account for the influence of neighbors through an average local volume fraction—it critically matters where the neighbors are located: upstream, downstream, or on the side. We will then briefly introduce the pairwise interaction extended point-particle (PIEP) model that attempts to deterministically account for the influence of the neighbors. For additional information, the reader is referred to Akiki et al. (2017).

2.4.1 *Neighbors Matter*

In order to evaluate the manner in which neighboring particles influence the force on any chosen particle, we resort to the detailed information on drag forces obtained from the FR-DNSs of nominally uniform flow over an array of particles discussed in Sect. 2.3. Our first investigation will be to explore the dependence on local volume fraction. For example, in FR-DNSs of 44% mean particle volume fraction, if we define the local particle volume fraction around each particle to be based on only neighbors within a sphere of few diameters, such a local particle volume fraction is observed to vary between 38 and 52%. The mean drag averaged over all the particles was very well captured by the drag formula given in (2.6) with ϕ being the mean volume fraction of 44%. However, the drag on the individual particles substantially varied from the mean substantially with a rms value of about 17–20% of the mean. In fact, there were particles whose drag was only 50% of the mean, while there were others whose drag was 50% higher than the mean.

We will first explore the possibility if this variation can be explained in terms of local volume fraction. A scatter plot of the local volume fraction around each sphere as the abscissa and the drag on that sphere as the ordinate is plotted. Using linear regression, a line is fitted through the scatter plot. The correlation coefficient R2 is then computed to serve as the indication of the extent of correlation between the local volume fraction parameter and the actual drag. In all the simulation cases considered, R2 reached a maximum of only 0.089, with its value being mostly in the range of 10^{-3}. Figure 2.3 shows a typical scatter plot of the local volume fraction versus the drag obtained from the DNS of 44% mean volume fraction. It is clear that there is

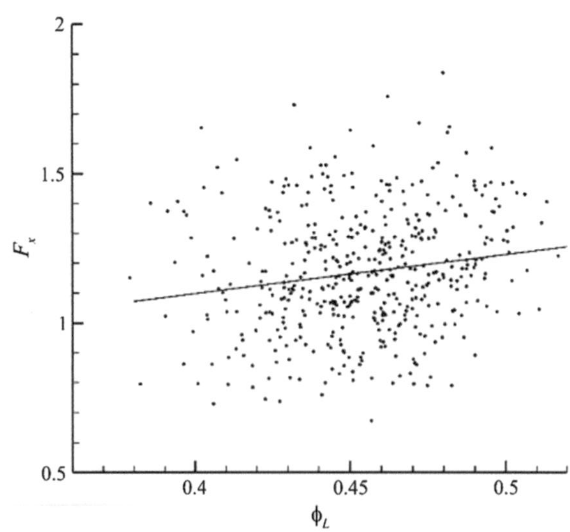

Fig. 2.3 Scatter plot of drag vs the local volume fraction for case vf3re060 and the linear fit of the data. $R^2 = 0.029$ and $R_L = 2$

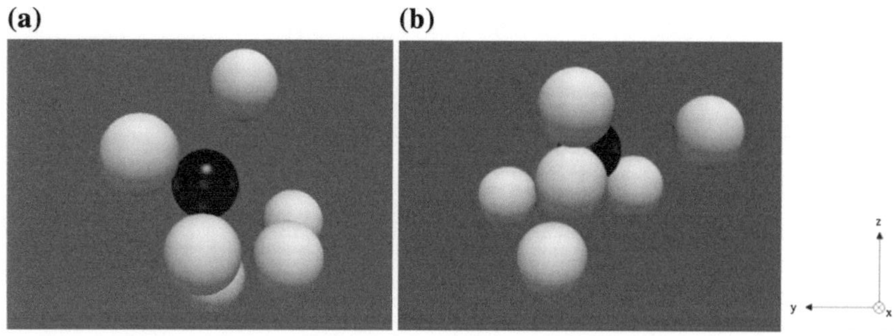

Fig. 2.4 Upstream view of the scatter of the six closest neighbors for two different spheres with the same 6th-nearest-neighbor parameter

no significant correlation—particles of higher drag do not correspond to higher local volume fraction, and particles of lower drag do not correspond to lower local volume fraction.

From the above discussion, it is clear that simple measures of local volume fraction and the nearest-neighbor distance are unable to predict the deviation of the actual force on the particle about the mean value. It should be noted that the present simulations, in agreement with past findings, obtained a strong correlation between mean drag and mean volume fraction for the random array as a whole. Based on this, one might expect positive correlation between local volume fraction and increase in drag above the mean value. But such correlation is not observed in any of the cases considered. For the same local volume fraction, we observe both increase and decrease in drag force depending on other parameters that go beyond the above simple measures of local volume fraction.

Figure 2.4 shows two sample particles (black color) with their six closest neighbors. They are chosen from the random distribution of particles within the periodic box. These two particles have nearly the same local volume fraction. The central particle of Fig. 2.4a has a drag coefficient of 6.03, while the one in Fig. 2.4b has a drag coefficient of 1.48, while the mean drag of the whole cluster in this case of 11% volume fraction is 4.69 (case vf1re60 of Akiki et al. (2017)). The particles are visualized in Fig. 2.4 from an observer located directly upstream. Even though the distances of the surrounding particles are comparable, the particle in Fig. 2.4b is shielded from the incoming flow, thus having a drag on the lower end of the drag distribution of the cluster. In contrast, the surrounding particles in Fig. 2.4a seem to funnel the flow toward the central sphere which results in a higher than average drag.

From our understanding of flow over an isolated particle, it can be seen that a particle lying upstream of another particle will likely experience a higher pressure field in its wake. Similarly, a particle lying downstream of another particle will likely experience a lower pressure at the front stagnation point. This suggests that one can expect the role of a neighboring particles located downstream to be quite different from one located upstream. The perturbation vorticity field around a particle on the other hand has a smaller radius of influence than the pressure field. Its main effects are concentrated in regions that are different than the pressure field. The highest vorticities are located at an angle upstream where pressure has its mildest influence. A wider influence of the vorticity field can be seen at an angle downstream with little effect on locations directly downstream. When it comes to clusters of particles, the pressure and vorticity effects of all neighboring spheres affect the departure of the drag and lift forces from the mean. This shows the complexity of the problem and that it is nearly impossible to predict the drag on a particle exactly without accounting for the complex effect of all the neighboring particles with a FR-DNS simulation.

2.4.2 PIEP Model

The PIEP model is built upon two basic ideas. First, the undisturbed flow at each particle location, defined as the flow that would exist at a particle location in the absence of that particle but with all other particles present, is separated into two parts: a macroscale flow that accounts for the collective action of all the particles and a microscale flow that accounts for the presence of all the neighbors taken one at a time (this is the pairwise interaction approximation). In a sense, the latter accounts for the effect of pseudo-turbulence generated by the neighboring particles. Second, the above-defined undisturbed flow is used to calculate the net aerodynamic force and torque on the particle using the Faxén form of the quasi-steady, added-mass, Basset history, and vorticity-induced (lift) forces relation (and similarly for the torque), since the undisturbed flow obtained from the first step is non-uniform.

The important aspect of the PIEP model is it attempts to systematically account for the microscale flow induced by the neighbors by making use of their precise location, information which is readily available in EL simulations. By accounting for the precise location of neighbors, the PIEP model goes beyond the mean neighborhood information of local particle volume fraction and distinguishes the influence of upstream, downstream, and laterally located neighbors. Such particle–particle interaction information is critical in order to capture phenomenon such as collision and close-range interactions of particles.

It is important to note that PIEP model retains the computational efficiency of the standard Euler–Lagrange PP approach. The microscale perturbation field induced by a neighbor can be easily computed for a range of Reynolds number and ambient conditions. This perturbation field can then be used to precompute the pairwise interactions and stored as PIEP maps. These maps are then used repeatedly to calculate the microscale contribution to hydrodynamic force and torque.

The implementation of the complete PIEP model requires the use of force and torque maps. The force maps are again separated into streamwise and transverse components. Here, streamwise direction is along the macroscale ambient flow seen by the moving particle, and transverse direction is perpendicular in the plane formed with the neighbor (see Akiki et al. (2017) for details of the model and notations).

Each map is a function of streamwise and transverse distance between the reference particle and the neighbor, whose effect on the reference particle is being computed.

2.4.3 PIEP Results

We present results from a test where eighty spheres are randomly released to settle in a large periodic computational box. This problem was simulated using fully resolved DNS using the immersed boundary method (Esmaeeli and Tryggvason 1998; Tenneti et al. 2011) using a grid resolution of 20 points per particle diameter. The problem is also simulated using EL methodology with PIEP model where we only solved for the translational and rotational motion of the 80 spheres. As a result, the PIEP model simulation is orders of magnitude faster than the DNS. In the PIEP model, we limit the number of neighbors used to 30. Figure 2.5 shows a comparison of the spatial dispersion of the 80 spheres at time $t = 46$. In comparison, an EL simulation using the standard drag would predict all the spheres settling at the same rate without any interaction or dispersion. We can observe very good comparison between the FR-DNS and PIEP simulation. We note that in this case of only a few particles there was no need to simulate the macroscale flow (which is taken to be stagnant), and thus, in the PIEP simulation only 80×6 ODEs for the translational and rotational

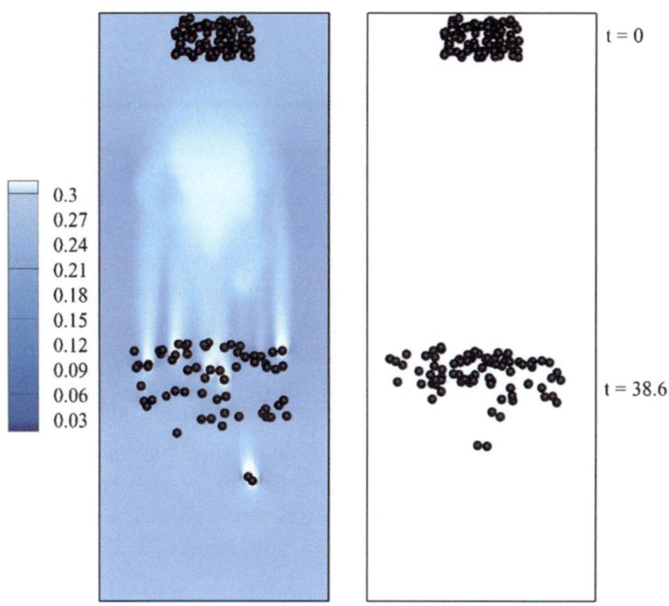

Fig. 2.5 Side view of 80 sedimenting spheres starting from a random initial distribution at $t = 38.6$. Both the DNS results as well the PIEP model prediction are shown

motion of the 80 particles needed to be simulated. As a result, while the FR-DNS took about 10 hrs of simulation on a 16 core cluster, the PIEP model took only a few seconds on MATLAB. This example demonstrates the power of PIEP model. While it is not perfect and does not replicate the FR-DNS results, it seems to capture the essential aspects of neighbor–neighbor interaction, which was completely ignored in the traditional PP models.

2.5 Stochastic Models

Stochastic models in the EL context can be used to introduce random terms in the position or velocity of the particles. Each of these random terms has a different physical significance. Adding a random term converts each of these ordinary differential equations (ODEs) into a stochastic differential equation (SDE), which implies an evolution equation for the joint probability density function (JPDF) of particle position and velocity (following the correspondence between the Langevin equation and the Fokker–Planck equation). The evolution of the JPDF in turn implies evolution equations for all the moments, and therefore, the model coefficients in the SDEs must be chosen carefully in order to match the evolution of moments obtained from FR-DNS. In the following, we discuss the physical motivation for stochastic modeling and briefly describe some recent models.

2.5.1 Particle Dispersion and Turbulence Modulation in Dilute Turbulent Flow

Dispersion of particles and the modulation of turbulence in the ambient gas by the dispersing particles are two coupled phenomena that are closely linked to the evolution of global multiphase flow characteristics, such as the spreading rate of the multiphase flow. PP-DNS of turbulent gas flows laden with sub-Kolmogorov size particles, in the absence of gravity, report that dispersion statistics and turbulent kinetic energy (TKE) evolve on different timescales.

Particles with high Stokes number lose energy faster than particles with low Stokes number in freely decaying turbulence (Sundaram and Collins 1999). On the other hand, particles with high Stokes number lose correlation with their initial velocities slower than particles with low Stokes number in stationary turbulence (Mashayek and Jaberi 1999; Squires and Eaton 1991b). The disparate behavior of the velocity autocorrelation and TKE time scales affects the dispersion characteristics of a spray.

2.5.1.1 Position Evolution Model

The simplest position evolution model is

$$\frac{d\mathbf{X}}{dt} = \mathbf{V}$$

but in some works it is modified to include a random term (O'Rourke 1989). The justification for adding this random term is to represent the effect of particle dispersion due to random motion of the turbulent eddies. It is well known that adding a random term (Wiener process increment, to be precise) makes the position evolution equation a stochastic differential equation (SDE). The corresponding change to the modeled number density function (NDF) evolution equation is the addition of a term representing diffusion in physical space (as in the Fokker–Planck equation corresponding to the SDE (Gardiner 1985; Pope 2000; Dreeben and Pope 1998)). Note that in homogeneous flows this term does not affect the particle velocity variance and its effect on turbulence modulation. Therefore, a stochastic position evolution model alone is insufficient to capture the complex coupling between dispersion and turbulence modulation observed in PP-DNS. In inhomogeneous flows, the enhanced dispersion will affect turbulence modulation indirectly.

Furthermore, it is well established from analyses of the system of SDEs that arise in both turbulent single-phase flows and Brownian dynamics, that diffusion arises from the effect of velocity autocorrelation. In the limit of rapid momentum relaxation, the system of equations can be simplified to a Langevin equation for position (Ermak and McCammon 1978; Markutsya 2008). The assumption of fast momentum relaxation is generally not applicable to particles because that would imply that the particle velocity distribution relaxes to an equilibrium Maxwellian distribution, which is obviously not true for the strongly non-equilibrium situation in multiphase flows. These observations motivate the development of stochastic velocity models which are capable of capturing the physical effects of particle dispersion as well as particle velocity variance evolution and turbulence modulation.

2.5.1.2 Coupled Stochastic Model for Particle and Fluid Velocity

The coupled stochastic model (CSM) for homogeneous turbulent two-phase flows consists of two coupled stochastic differential equations (SDEs) for the modeled fluctuating Lagrangian gas-phase velocity **u** and fluctuating Lagrangian dispersed-phase velocity **v**. This model possesses a unique feature that the implied turbulent kinetic energy (TKE) and velocity autocorrelation in each phase evolve on different timescales. Consequently, this model has the capability of simultaneously predicting the disparate Stokes number trends in the evolution of dispersion statistics, such as velocity autocorrelations, and TKE in each phase. Predictions of dispersion statistics and TKE from the new model show good agreement with published PP-DNS of non-evaporating and evaporating droplet-laden turbulent flow.

The proposed system of SDEs in CSM is

$$du'^{*}_{f,i} = -\left[\frac{1}{2\tau_1} + \left(\frac{1}{2} + \frac{3}{4}C_0\right)\frac{\varepsilon_f}{k_f}\right] u'^{*}_{f,i} dt + \left[C_0 \varepsilon_f + \frac{2}{3}\frac{k_f}{\tau_1} + \frac{2}{3}\left(\frac{k^e_f - k_f}{\tau_2}\right)\right]^{1/2} dW^u_i \tag{2.24}$$

$$dv'^{*}_i = -\frac{1}{2\tau_3} v'^{*}_i dt + \left[\frac{2}{3}\frac{k_d}{\tau_3} + \frac{2}{3}\left(\frac{k^e_d - k_d}{\tau_4}\right)\right]^{1/2} dW^v_i, \tag{2.25}$$

where τ_1, τ_2, τ_3, and τ_4 are timescales that appear in the drift and diffusion coefficients[2] of each SDE, while dW^u_i and dW^v_i are independent Wiener processes (Kloeden and Platen 1992). The subscript i denotes the Cartesian components. The TKE in the dispersed phase is denoted k_d, and the TKE in the gas phase is denoted k_f, with a superscript 'e' to denote their 'equilibrium' values (the concept of 'equilibrium' is explained in (Xu and Subramaniam 2006; Pai and Subramaniam 2006)).[3] Also, ε_f is the gas–phase dissipation enhanced by the presence of the dispersed phase. The constant $C_0 = 2.1$, which is identical to that used in the simplified Langevin model (SLM) (Pope 2000). Mean velocity and hence mean slip in either phase are assumed to be zero for simplicity, although this is not an inherent limitation of CSM. The fluid-phase SDE can be viewed as an extension of the SLM (Pope 2000; Haworth and Pope 1986) to two-phase flows, but with an important difference being the introduction of drift and diffusion timescales that are different from each other. Also, additional terms involving k^e_f and k^e_d (in parentheses) that represent interphase interactions have been added. The coupling between the two phases is only through moments of the velocities in each phase such as TKE (k_f and k_d) and the dissipation ε_f, and not explicitly through the instantaneous values of $u'^{*}_{f,i}$ and v'^{*}_i.

Note that for widely used EL models, the interphase TKE transfer evolves on the particle response timescale τ_p, which was found to be inadequate to capture the multiscale nature of particle–turbulence interaction (Pai and Subramaniam 2006). The specification of the drift timescales τ_1 and τ_3 in Eqs. 2.24 and 2.25 is summarized here. Detailed justification for these choices is given in (Pai and Subramaniam 2012, 2007). The specification for the drift timescale τ_3 is

$$\frac{1}{\tau_3} = 2\left[\frac{1}{2\tau_1} + \left(\frac{1}{2} + \frac{3}{4}C_0\right)\frac{1}{\tau}\right]\frac{1}{1 + St_\eta C_3}, \tag{2.26}$$

where C_3 is a model constant ($C_3 = 0.1$), and $\tau = k_f/\varepsilon_f$ is the fluid-phase eddy turnover timescale. This specification for τ_3 obeys the correct limiting behavior in the limit of zero Stokes number ($St_\eta \to 0$), where the particles or droplets respond

[2] The terms 'drift' and 'diffusion' are used in the sense of stochastic differential equation theory.

[3] The subscript f stands for the gas phase or fluid phase, and the subscript d stands for the dispersed phase.

immediately to the surrounding fluid and the fluid-phase velocity autocovariance and the dispersed-phase velocity autocovariance must match. The drift timescale τ_1 is prescribed to be

$$\frac{1}{\tau_1} = \frac{C_1 \phi}{\tau},$$

where C_1 is a model constant ($C_1 = 0.5$). In the limit of zero mass loading, the timescale τ_1, which essentially represents the modification to the fluid velocity autocorrelation timescale due to the presence of dispersed phase, should tend to infinity. In this limit, the drift timescale in Eq. 2.24 approaches the specification for the single-phase simplified Langevin model (Pope 2000).

The timescales τ_2 and τ_4 govern the evolution of TKE in each phase, which are given by

$$\frac{dk_f}{dt} = -\frac{k_f - k_f^e}{\tau_2} - \varepsilon_f \tag{2.27}$$

$$\frac{dk_d}{dt} = -\frac{k_d - k_d^e}{\tau_4}. \tag{2.28}$$

In accordance with the 'equilibration of energy' concept, and to introduce the capability to capture the multiscale nature of a turbulent two-phase mixture into CSM, the timescales τ_2 and τ_4 are chosen to be equal to $\tau_\pi = \langle \tau_{\text{int}} \rangle / C_\pi$, where $\langle \tau_{\text{int}} \rangle$ is a multiscale interaction timescale for interphase TKE transfer first proposed by (Pai and Subramaniam 2006). It was shown in (Pai and Subramaniam 2006) that the new timescale accurately captures the dependence of the interphase TKE transfer on St_η. This timescale has been successfully employed in the context of EE two-phase turbulence modeling by (Xu and Subramaniam 2006). The constant C_π is chosen to be 2.5.

2.5.2 Neighbor Effects in Non-dilute Gas–Solid Flow with Mean Slip

As noted earlier in Sect. 2.3, particle force fluctuations are also observed in FR-DNS of gas–solid flow at non-dilute volume fractions (>0.1) in a non-turbulent flow with finite mean slip velocity, and these are attributed to neighbor particle effects.

Tenneti et al. (2016) proposed the following stochastic model for the increment in the particle velocity:

$$dv_i = -\beta \langle W_i \rangle \, dt - \gamma v_i'' dt + B d\mathscr{W}_i. \tag{2.29}$$

In this equation, which is an isotropic form of a general Langevin model, dv_i is the increment in the particle velocity, v_i'' is the fluctuation in the particle velocity, and

$d\mathscr{W}_i$ is a Wiener process increment. The first term on the right-hand side of Eq. 2.29 accounts for the effect of the mean slip velocity between the solid and the gas–phase, the second term accounts for the fluctuation in the particle velocity, and the last term models the effect of the hydrodynamic interaction with neighboring particles and fluid velocity fluctuations. In a slightly different context, Goswami et al. (2010) found that the effect of fluid velocity fluctuations on high Stokes number particles in a turbulent channel flow also can be modeled using a white noise term. The Langevin model in Eq. 2.29 has been verified (Tenneti et al. 2016; Garzo et al. 2012) by considering the decay of the particle velocity autocorrelation from PR-DNS.

The coefficients of the Langevin model are functions of the solid volume fraction (ϕ), mean slip Reynolds number (Re_m), and solid-to-fluid density ratio (ρ_p/ρ_f). They have been extracted from the source and dissipation of the granular temperature, which are computed from PR-DNS using the method developed by Tenneti et al. (2016). The Langevin model for the velocity increment (see Eq. 2.29) implies a Fokker–Planck equation for the NDF and thus closes the kinetic equation. Based on this correspondence, Garzó et al. (2012) used this Langevin model to develop an Enskog kinetic theory for gas–solid flow in the Stokes flow regime. It has subsequently been extended to higher Reynolds number and yields excellent predictions of particle velocity variance (Tenneti et al. 2016).

2.6 Outlook and Challenges

We will first summarize the key points of this paper. While fully resolved direct numerical simulation (FR-DNS) of dispersed multiphase flow provides the complete and accurate information, it is limited to small systems and will remain computationally impossible for most problems of practical interest. The focus of this paper has been on approximate reduced-order approaches, which are computationally cheaper and can be used as practical tools for solving many multiphase flow problems of practical interest. In particular, we consider the Euler–Lagrange approach with point particle models that couple the dispersed and continuous phases. This gives rise to three classes of point-particle simulations: PP-DNS, PP-LES, and PP-RANS, whose difference is in the level of resolution employed for the carrier fluid.

We outline approaches by which results from FR-DNSs of smaller systems at the microscale can be used to improve the accuracy of models that need to be employed. Usually, the results of the FR-DNSs are used to obtain closure models for the average force, which are then used in the PP framework to calculate the instantaneous force on a particle. It must be cautioned that naive implementation of such EL-PP closure models can mispredict the particle velocity variance, which is crucial in determining particle dispersion statistics. FR-DNS also shows evidence of force fluctuations over and above the average value in two different scenarios.

One is that force fluctuations can arise in dilute turbulent flows. In PP-DNS, these force fluctuations can be fairly accurately reproduced even by a *deterministic* model with the corrected expression for the undisturbed fluid velocity (Horwitz and Mani

2016). However, if we pursue the PP-LES approach, where the scales of carrier-phase turbulence are only partially resolved, then we have to model the effect of subfilter scale motions on the particle force and this is still an open question. A coupled system of Langevin equations with a dual timescale model was introduced by Pai et al. (2007, 2012) to accurately capture the decay of particle velocity variance (as well as fluid velocity variance—TKE) and reproduce dispersion statistics. Variants may be useful in the PP-LES problem, but this direction is yet to be explored.

The second scenario involves problems where particles are at higher volume fraction in a non-turbulent flow with nonzero mean slip, and the force fluctuations observed in FR-DNS are attributed to neighbor particle effects. One of the advantages of the EL model is that each particle knows precisely the location of all the neighboring particles, and thus, in principle it should be possible to account for the force fluctuation about the average value that arises due to the influence of neighbors. Here, the pairwise interaction extended point-particle (PIEP) model gives a deterministic framework to incorporate these effects, and it has shown promise in early investigation.

The effect of neighbors when viewed in a statistical sense manifests in the pair correlation function. An open question is how to use this information in EL models. While Chiu et al. (1997) proposed models that incorporate the pair correlation function, those ideas have not been extended to EL models. Tenneti et al. (2016) have proposed a stochastic model to account for neighbor particle effects through a random term. But this model does not give physical insight into the decomposition of the total force into individual contributions as given in a generalized Faxén form of the force expression (Annamalai and Balachandar (2017)). Establishing a connection between the different terms of such a model and the stochastic model could be useful.

There are several avenues to continue the progress made during the past two decades. The deterministic closure models rely upon our understanding of the multiphase flow physics and our ability to translate this understanding to models. But such an approach has inherent limitations. For example, the PIEP model is built upon the assumption of pairwise interaction and therefore ignores higher-order interactions between the particles. Thus, PIEP model can be expected to be less than perfect. On the other hand, data-driven statistical approaches are proving to be very useful in many fields as rapid progress is being made in areas such as machine learning. There is immense potential for combining the deterministic and statistical approaches.

Furthermore, the focus of this short chapter has been mostly on particle-laden multiphase flows. Although much of past research has been on rigid particle, there has been significant research in droplet-laden and bubbly flows. Though there are similarities, there are also fundamental differences. Most importantly, droplets and bubbles deform, break up, and agglomerate. The flow in and around a droplet or bubble can be significantly different that neighbor–neighbor interaction can vary from that of particle–particle interaction. Such differences must be accomodated in both PIEP-like deterministic and other stochastic models.

References

Akiki G, Jackson TL, Balachandar S (2017) Pairwise interaction extended point-particle model for a random array of monodisperse spheres. J Fluid Mech 813:882–928

Almeida T, Jaberi F (2008) Large-eddy simulation of a dispersed particle-laden turbulent round jet. Int J Heat Mass Transfer 51(3–4):683–695

Amsden AA, O'Rourke PJ, Butler TD (1989) KIVA–II: a computer program for chemically reactive flows with sprays. Technical Report LA–11560–MS, Los Alamos National Laboratory

Annamalai S, Balachandar S (2017) Faxén form of time-domain force on a sphere in unsteady spatially varying viscous compressible flows, 816:381–4111

Apte SV, Mahesh K, Gorokhovski M, Moin P (2009) Stochastic modeling of atomizing spray in a complex swirl injector using large eddy simulation. Proc Combust Inst 32(2):2257–2266

Apte SV, Mahesh K, Moin P, Oefelein JC (2003) Large-eddy simulation of swirling particle-laden flows in a coaxial-jet combustor. Int J Multiphase Flow 29(8):1311–1331

Bagchi P, Balachandar S (2003) Effect of turbulence on the drag and lift of a particle. Phys Fluids 15(11):3496

Bagchi P, Balachandar S (2004) Response of the wake of an isolated particle to an isotropic turbulent flow. J Fluid Mech 518:95–123

Balachandar S (2009) A scaling analysis for point-particle approaches to turbulent multiphase flows. Int J Multiphase Flow, 35(9)

Balachandar S, Eaton JK (2010) Turbulent dispersed multiphase flow. Ann Rev Fluid Mech 42:111–133

Basset A (1888) A Treatise on hydrodynamics. Deighton, Bell and Company, London

Beetstra R, van der Hoef MA, Kuipers JAM (2007) Drag force of intermediate Reynolds number flows past mono- and bidisperse arrays of spheres. AIChE J 53:489–501

Boivin M, Simonin O, Squires KD (1998) Direct numerical simulation of turbulence modulation by particles in isotropic turbulence. J Fluid Mech 375:235–263

Boussinesq J (1885) Sur la r esistance quoppose un liquide ind efini au repos au mouvement vari e dune sph'ere solide. CR Acad Sci Paris 100:935–937

Brown DJ, Hutchinson P (1979) The Interaction of solid or liquid particles and turbulent fluid flow fields—a numerical simulation. J Fluids Eng Trans ASME, 101:265–269

Brucato A, Grisafi F, Montante G (1998) Particle drag coefficients in turbulent fluids. Chem Eng Sci 53:3295

Burton TM, Eaton JK (2005) Fully resolved simulations of particle-turbulence interaction. J Fluid Mech 545:67–111

Capecelatro J, Desjardins O (2013) An Euler-Lagrange strategy for simulating particle-laden flows. J Comput Phys 238:1–31

Carrara MD, Desjardin PE (2006) A filtered mass density function approach for modeling separated two-phase flows for LES I: mathematical formulation. Int J Multiphase Flow 32(3):365–384

Chiu HH, Su SP (1997) Theory of droplets (II): states, structures, and laws of interacting droplets. Atom Sprays 7(1):1–32

Clamen A, Gauvin WH (1969) Effect of turbulence on the drag coefficients of spheres in a supercritical flow regime. AIChE J 15:184

Clift R, Grace JR, Weber ME (1978) Bubbles. In: Drops and particles. Academic Press

Crowe CT, Sommerfeld M, Tsuji Y (1999) Multiphase flows with droplets and particles. CRC Press, Boca Raton, FL

Deen NG, Kriebitzsch SHL, van der Hoef MA, Kuipers JAM (2012) Direct numerical simulation of flow and heat transfer in dense fluid-particle systems. Chem Eng Sci 81:329–344

Dreeben TD, Pope SB (1998) Probability density function Monte Carlo simulation of near-wall turbulent flows. J Fluid Mech 357:141–166

Dwyer HA, Dandy DS (1990) Some influences of particle shape on drag and heat transfer. Phys Fluids A, 2(12):2110–2118

Dwyer HA (1989) Calculations of droplet dynamics in high temperature environments. Progress Energy Combust Sci 15:131–158

Elghobashi SE, Truesdell GC (1992) Direct simulation of particle dispersion in a decaying isotropic turbulence. J Fluid Mech 242:655–700

Elghobashi SE, Truesdell GC (1993) On the two-way interaction between homogeneous turbulence and dispersed solid particles. I: turbulence modification. Phys Fluids A 5:1790–1801

Ermak DL, McCammon JA (1978) Brownian dynamics with hydrodynamic interactions. J Chem Phys 69(4):1352–60

Esmaeeli A, Tryggvason G (1998) Direct numerical simulation of bubbly flows. Part 1. Low Reynolds number arrays. J Fluid Mech 377:313–345

Faxén H (1923) Die Bewegung einer starren Kugel lngs der Achse eines mit zher Flssigkeit gefllten Rohres

Fortes AF, Joseph DD, Lundgren TS (1987) Nonlinear mechanics of fluidization of beds of spherical particles. J Fluid Mech 177

Gardiner CW (1985) Handbook of stochastic methods, 2nd edn. Springer, Berlin

Garg R, Narayanan C, Lakehal D, Subramaniam S (2007) Accurate numerical estimation of interphase momentum transfer in lagrangian-eulerian simulations of dispersed two-phase flows. Int J Multiphase Flow 33(12):1337–1364

Garg R, Narayanan C, Lakehal D, Subramaniam S (2009) A numerically convergent Lagrangian-Eulerian simulation method for dispersed two-phase flows. Int J Multiphase Flow 35(4):376–388

Garg R, Tenneti S, Mohd-Yusof J, Subramaniam S (2011) Direct numerical simulation of gas-solids flow based on the immersed boundary method. In: Pannala S, Syamlal M, O'Brien TJ (eds) Computational gas-solids flows and reacting systems: theory, methods and practice, pp 245–276. IGI Global

Garzo V, Tenneti S, Subramaniam S, Hrenya CM (2012) Enskog kinetic theory for monodisperse gas-solid flows. J Fluid Mech 712:129–168

Gatignol R (1983) The Faxén formulae for a rigid particle in an unsteady non-uniform Stokes flow. J Mech Theor Appl 1:143–160

Glowinski R, Pan T, Hesla T, Joseph D, Periaux J (2001) A fictitious domain approach to the direct numerical simulation of incompressible viscous flow past moving rigid bodies: application to particulate flow. J Comput Phys 169:363–426

Gosman AD, Ioannides E (1983) Aspects of computer simulation of liquid fueled combustors. J Engine Res 6(7):482–490

Goswami PS, Kumaran V (2010) Particle dynamics in a turbulent particle-gas suspension at high Stokes number. Part 2. The fluctuating-force model. J Fluid Mech 646:91–125

Haworth DC, Pope SB (1986) A generalized Langevin model for turbulent flows. Phys Fluids 29:387–405

Helenbrook BT, Edwards CF (2002) Quasi-steady deformation and drag of uncontaminated liquid drops. Int J Multiphase Flow 28(10):1631–1657

Herrmann M (2008) A balanced force refined level set grid method for two-phase flows on unstructured flow solver grids. J Comput Phys 227(4):2674–2706

Herron IH, Davis SH, Bretherton FP (1975) On the sedimentation of a sphere in a centrifuge. J Fluid Mech 68:209–234

Hill RJ, Koch DL, Ladd AJC (2001a) Moderate-Reynolds-number flows in ordered and random arrays of spheres. J Fluid Mech 448:243–278

Hill RJ, Koch DL, Ladd AJC (2001b) The first effects of fluid inertia on flows in ordered and random arrays of spheres. J Fluid Mech 448:213–241

Horwitz JAK, Mani A (2016) Accurate calculation of stokes drag for pointparticle tracking in two-way coupled flows. J Comput Phys 318:85–109

Kim I, Elghobashi S, Sirignano WA (1998) On the equation for spherical-particle motion: effect of reynolds and acceleration numbers. J Fluid Mech 367:221–253

Kloeden P, Platen E (1992) Numerical solution of stochastic differential equations. Springer, New York

Kriebitzsch SHL, van der Hoef MA, Kuipers JAM (2013) Fully resolved simulation of a gas-fluidized bed: a critical test of DEM models. Chem Eng Sci 91:1–4

Ling Y, Parmar M, Balachandar S (2013) A scaling analysis of added-mass and history forces and their coupling in dispersed multiphase flows. Int J Multiphase Flow 57:102–114

Ling Y, Balachandar S, Parmar M (2016) Inter-phase heat transfer and energy coupling in turbulent dispersed multiphase flows. Phys Fluids, 28(3)

Lovalenti PM, Brady JF (1993a) The force on a sphere in a uniform flow with small-amplitude oscillations at finite Reynolds-number. J Fluid Mech 256:607–614

Lovalenti PM, Brady JF (1993b) The hydrodynamic force on a rigid particle undergoing arbitrary time-dependent motion at small Reynolds-number. J Fluid Mech 256:561–605

Markus U (2005) An immersed boundary method with direct forcing for the simulation of particulate flows. J Comput Phys 209(2):448–476

Markutsya S, Subramaniam S, Dennis Vigil R, Fox RO (2008) On Brownian dynamics simulation of nanoparticle aggregation. Ind Eng Chem Res 47(10):3338–3345

Mashayek F (1998) Direct numerical simulations of evaporating droplet dispersion in forced low Mach number turbulence. Int J Heat Mass Transfer 41(17):2601–2617

Mashayek F, Jaberi FA (1999) Particle dispersion in forced isotropic low-Mach-number turbulence. Int J Heat Mass Transfer 42(15):2823–2836

Mashayek F, Taulbee DB (2002) Turbulent gas-solid flows, part I: direct numerical simulations and Reynolds stress closures. Numer Heat Transfer Part B Fund 41(1):1–29

Maxey MR, Riley JJ (1983) Equation of motion for a small rigid sphere in a nonuniform flow. Phys Fluids 26:883–889

Mehrabadi M, Murphy E, Subramaniam S (2016) Development of a gassolid drag law for clustered particles using particle-resolved direct numerical simulation. Chem Eng Sci 152:199–212

Mehrabadi M, Tenneti S, Garg R, Subramaniam S (2015) Pseudo-turbulent gas-phase velocity fluctuations in homogeneous gas-solid flow: fixed particle assemblies and freely evolving suspensions. J Fluid Mech 770:210–246

Mei R, Adrian RJ (1992) Flow past a sphere with an oscillation in the free-stream velocity and unsteady drag and finite Reynolds number. J Fluid Mech 237:323–441

Merle A, Legendre D, Magnaudet J (2005) Forces on a high-Reynolds-number spherical bubble in a turbulent flow. J Fluid Mech 532:53–62

Mikhael G, Marcus H (2008) Modeling primary atomization. Annu Rev Fluid Mech 40(1):343–366

Miller RS, Bellan J (1999) Direct numerical simulation of a confined three-dimensional gas mixing layer with one evaporating hydrocarbon-droplet laden stream. J Fluid Mech 384:293–338

Okong'o NA, Bellan J (2000) A priori subgrid analysis of temporal mixing layers with evaporating droplets. Phys Fluids 12(6):1573–1591

Okong'o NA, Bellan J (2004) Consistent large-eddy simulation of a temporal mixing layer laden with evaporating drops. Part 1. Direct numerical simulation, formulation and a priori analysis. J Fluid Mech 499:1–47

Ormancey A, Martinon J (1984) Prediction of particle dispersion in turbulent flows. PhysicoChem Hydrodyn 3/4(5):229–244

O'Rourke PJ (1989) Statistical properties and numerical implementation of a model for droplet dispersion in a turbulent gas. J Comput Phys 83(2):345–360

Oseen CW (1927) Hydrodynamik. Akademische Verlagsgesellschaft, Leipzig

Pai MG, Subramaniam S (2012) Two-way coupled stochastic model for dispersion of inertial particles in turbulence. J Fluid Mech 700:29–62

Pai MG, Subramaniam S (2006) Modeling interphase turbulent kinetic energy transfer in Lagrangian-Eulerian spray computations. Atom Sprays 16(7):807–826

Pai MG, Subramaniam S (2007) Modeling droplet dispersion and interphase turbulent kinetic energy transfer using a new dual-timescale Langevin model. Int J Multiphase Flow 33(3):252–281

Pai MG, Subramaniam S (2009) A comprehensive probability density function formalism for multiphase flows. J Fluid Mech 628:181–228

Pope SB (2000) Turbulent flows. Cambridge University Press, Port Chester, NY

Quan S, Schmidt D (2007) A moving mesh interface tracking method for 3D incompressible two-phase flows. J Comput Phys 221(2):761–780

Quan S, Schmidt DP, Hua J, Lou J (2009) A numerical study of the relaxation and breakup of an elongated drop in a viscous liquid. J Fluid Mech 640:235

Rubinow SI, Keller JB (1961) The transverse force on a spinning sphere moving in a viscous fluid. J Fluid Mech 11:447–459

Saffman PGT (1965) The lift on a small sphere in a slow shear flow. J Fluid Mech 22:385–400

Sarkar S, van der Hoef MA, Kuipers JAM (2009) Fluid-particle interaction from lattice Boltzmann simulations for flow through polydisperse random arrays of spheres. Chem Eng Sci 64(11):2683–2691

Schiller L, Naumann AZ (1933) A drag coefficient correlation. Z Ver Deutsch Ing, 318–320

Schmidt DP, Rutland CJ (2000) A new droplet collision algorithm. J Comput Phys 164(1):62–80

Squires KD, Eaton JK (1990) Particle response and turbulence modification in isotropic turbulence. Phys Fluids A 2:1191–1203

Squires KD, Eaton JK (1991a) Lagrangian and Eulerian statistics obtained from direct numerical simulations of homogeneous turbulence. Phys Fluids 3(1):130–143

Squires KD, Eaton JK (1991b) Measurements of particle dispersion obtained from direct numerical simulations of isotropic turbulence. J Fluid Mech 226:1–35

Subramaniam S (2013) Lagrangian-Eulerian methods for multiphase flows. Progress Energy Combust Sci 39:215–245

Sundaram S, Collins LR (1997) Collision statistics in an isotropic particle-laden turbulent suspension. Part I. Direct numerical simulations. J Fluid Mech 379:75–109

Sundaram S, Collins LR (1999) A numerical study of the modulation of isotropic turbulence by suspended particles. J Fluid Mech 379:105–143

Tenneti S, Garg R, Hrenya CM, Fox RO, Subramaniam S (2010) Direct numerical simulation of gas-solid suspensions at moderate Reynolds number: Quantifying the coupling between hydrodynamic forces and particle velocity fluctuations. Powder Technol 203(1):57–69

Tenneti S, Garg R, Subramaniam S (2011) Drag law for monodisperse gas-solid systems using particle-resolved direct numerical simulation of flow past fixed assemblies of spheres. Intl J Multiphase Flow 37(9):1072–1092

Tenneti S, Mehrabadi M, Subramaniam S (2016) Stochastic Lagrangian model for hydrodynamic acceleration of inertial particles in gas-solid suspensions. J Fluid Mech 788:695–729

Tenneti S, Subramaniam S (2014) Particle-resolved direct numerical simulation for gas-solid flow model development. Ann Rev Fluid Mech 46:199–230

Uhlherr PHT, Sinclair CG (1970) The effect of freestream turbulence on the drag coefficients of spheres. Proc CHEMECA 1:1

van der Hoef MA, Beetstra R, Kuipers JAM (2005) Lattice-Boltzmann simulations of low-Reynolds-number flow past mono- and bidisperse arrays of sphere: results for the permeability and drag force. J Fluid Mech 528:233–254

Wang L-P, Rosa B, Gao H, He G, Jin G (2009) Turbulent collision of inertial particles: point-particle based, hybrid simulations and beyond. Int J Multiphase Flow 35(9):854–867

Warnica WD, Renksizbulut M, Strong AB (1995) Drag coefficients of spherical liquid droplets part 2: turbulent gaseous fields. Exp Fluids 18:265–276

Warnica WD, Renksizbulut M, Strong AB (1995) Drag coefficients of spherical liquid droplets part 1: quiescent gaseous fields. Exp Fluids 18:258–264

Wu J-S, Faeth GM (1994) Effect of ambient turbulence intensity on sphere wakes at intermediate reynolds numbers. AIAA J 33:171–173

Wylie JJ, Koch DL, Ladd AJC (2003) Rheology of suspensions with high particle inertia and moderate fluid inertia. J Fluid Mech 480:95–118

Xu Y, Subramaniam S (2006) A multiscale model for dilute turbulent gas-particle flows based on the equilibration of energy concept. Phys Fluids 18:033301(1–17)

Yin X, Sundaresan S (2009a) Fluid-particle drag in low-Reynolds-number polydisperse gas-solid suspensions. AIChE J 55(6):1352–1368

Yin X, Sundaresan S (2009b) Drag law for bidisperse gas-solid suspensions containing equally sized spheres. Ind Eng Chem Res 48(1):227–241

Ying Xu, Subramaniam S (2010) effect of particle clusters on carrier flow turbulence: a direct numerical simulation study. Flow Turbul Combust 85(3–4):735–761

Zarin NA, Nicholls JA (1971) Sphere drag in solid rocketsnoncontinuum and turbulence effects. Combust Sci Technol 3:273

Zhang Z, Prosperetti A (2005) A second-order method for three-dimensional particle simulation. J Comput Phys 210(1):292–324

Part II
Droplet Evaporation and Combustion

Chapter 3
Modelling of Droplet Heating and Evaporation

Sergei S. Sazhin

3.1 Introduction

The modelling of droplet heating and evaporation has been extensively studied since the beginning of the last century, and the results have been summarised in numerous reviews and monographs including Sazhin (2006, 2014, 2017). These studies have been mainly motivated by engineering, environmental and pharmaceutical applications of the results of this modelling. These processes are an integral part of the processes leading to autoignition of the automotive fuel vapour/air mixture in Diesel engines (Heywood 1998). The following analysis will concentrate primarily on the modelling of automotive fuel droplets, although most of the results may have a much wider range of application. There will be some overlap with the results presented in Sazhin (2006, 2014, 2017), and the analysis of the most recent publications, not included in Sazhin (2006, 2014, 2017), will be reported. As in the above-mentioned publications, some topics related to droplet heating and evaporation will not be covered. These include heating and evaporation of droplets during their interaction with walls, and the Soret effect (Rahman and Saghir 2014). The analysis of purely experimental papers and papers focused on multidimensional simulations of these processes will be limited. This chapter is intended to be complementary to reviews (Sirignano 2014; Aggarwal 2014), which look mainly at the ignition and combustion of individual droplets and arrays of droplets.

The approaches to modelling of non-evaporating droplets are discussed in Sect. 3.2. The hydrodynamic models for heating and evaporation of mono-component droplets are presented in Sect. 3.3. The hydrodynamic heating and evaporation models for multi-component droplets are summarised in Sect. 3.4. Section 3.5 considers the kinetic and molecular dynamics models.

S. S. Sazhin (✉)
Sir Harry Ricardo Laboratories, School of Computing, Engineering and Mathematics, Advanced Engineering Centre, University of Brighton, Brighton BN2 4GJ, UK
e-mail: S.Sazhin@brighton.ac.uk

S. Basu et al. (eds.), *Droplets and Sprays*, Energy, Environment, and Sustainability,
https://doi.org/10.1007/978-981-10-7449-3_3

3.2 Heating of Non-evaporating Droplets

The most widely used model for convective droplet heating is the one based on the assumption that liquid thermal conductivity is infinitely large. In this model, the evolution of droplet temperature with time is inferred from the energy balance equation. This approach is almost universally used in research and commercial Computational Fluid Dynamics (CFD) codes and many original investigations of the problem of droplet heating, including the most recent ones (Yadav et al. 2017; Wang et al. 2017; Wittig et al. 2017; Bojko and DesJardin 2017).

The simplest way to take into account the effects of temperature gradient within droplets was based on the solution to the one-dimensional (1D) heat transfer equation, assuming that the heating process is spherically symmetric. This equation was solved either numerically (Olguin and Gutheil 2014) or analytically (Sazhin 2014). It was shown that the analytical solution to this equation in the liquid phase with Robin boundary conditions (see Solution (3.10)) is particularly useful for practical applications in Computational Fluid Dynamics codes. This approach was implemented into the ANSYS Fluent CFD code using user-defined functions (UDF) (Rybdylova et al. 2016).

The above-mentioned solution is strictly valid only for stationary spherical droplets, but it was generalised to the case of moving droplets based on the so-called effective thermal conductivity (ETC) model (Sazhin 2014). In this model, the effects of droplet motion on the heat transfer processes inside the droplet were taken into account by replacing the liquid thermal conductivity (k_l) with the effective thermal conductivity $k_{\text{eff}} = \chi_T k_l$, where the values of χ_T varied from 1 for stationary droplets to 2.72 for fast-moving droplets. This model could predict the average surface temperature of the droplets, which is particularly useful for many engineering applications. In the limiting case where liquid thermal conductivity is infinitely large, the effects of temperature gradient inside droplets and the effect of recirculation can be ignored. This model is known as the infinite thermal conductivity (ITC) model.

The solution to the 1D heat transfer equation for spherical droplets shows that the dependence of temperature on the distance from the droplet centre is close to parabolic (except at the very beginning of the heating process). This allows one to assume that this dependence is parabolic and is characterised by two temperatures: at the centre and at the surface of the droplet. This model is known as the parabolic model (Sazhin 2014). The values of these temperatures were obtained from analysis of the energy balance equation at the surface of the droplet. The modification of this model made it applicable both at the beginning of the heating process and at times when the temperature profiles inside droplets are close to parabolic (Sazhin 2014).

The limitations of the parabolic model and the complexity of the model based on the rigorous analytical/numerical solutions to the heat transfer equation inside droplets stimulated efforts to develop new models. These were more accurate than the parabolic model and more simple than the models based on the rigorous solutions to the heat transfer equation. One such model, known as the power law approximation, was suggested in Brereton (2013) and further investigated in Snegirev (2013).

This model is based on the assumption that the temperature profile inside the droplet can be approximated as:

$$T(R) = c_{p0} + c_{pp}\left(\frac{R}{R_d}\right)^p, \tag{3.1}$$

where R is the distance from the droplet centre, p is the parameter adjusted to replicate temperature profiles, c_{p0} and c_{pp} are inferred from the values of the heat flux at the surface of the droplets and their average temperature. For $p = 2$, Expression (3.1) describes the parabolic temperature profile.

To describe the transient process, it was assumed that p is time dependent. At the initial stage of heating, the values of p were in the range 10–100, and then these values decreased with time, approaching 2, with $p \geq 2$ at all times.

The model developed in Xiao et al. (2016) is also based on (3.1) but with $c_{p0} = T_c$ and $c_{pp} = T_s - T_c$, where T_c and T_s are temperatures at the centre and surface of the droplet, respectively. An empirical formula for p was obtained based on the distribution of temperature predicted by the rigorous 1D solution. One of the limitations of this approximation is that it can predict only monotonic temperature profiles. This limitation was overcome by the so-called polynomial approximation, originally suggested in Subramanian et al. (2005), and further investigated in Snegirev (2013). This approximation is based on the following presentation of the temperature profile inside the droplet:

$$T(R) = c_{p0} + c_{p2}\left(\frac{R}{R_d}\right)^2 + c_{pp}\left(\frac{R}{R_d}\right)^p, \tag{3.2}$$

where $p > 2$.

The most important limitation of the power law and polynomial approximations stems from the assumption that the temperature profile is instantly established in the whole droplet volume, when one would expect that initially only a thin layer close to the droplet surface is affected by the external heat, and that this heat gradually penetrates inside the droplet. These processes are taken into account in the heat balance integral method. The method is based on the introduction of the thermal layer of time-dependent thickness $\delta(t)$ (Snegirev et al. 2012; Snegirev 2013). Inside this layer, temperature is approximated by the parabolic profile, while the temperature outside this layer is assumed equal to the initial temperature:

$$T(t)\begin{cases} c_{h0} + c_{h2}\left(\frac{R-(R_d-\delta)}{R_d}\right)^2, & R_d - \delta < R \leq R_d \\ T_0 & 0 \leq R \leq R_d - \delta \end{cases} \tag{3.3}$$

The droplet average temperature is found from the heat balance equation for the whole droplet; thickness δ is estimated by iterations of the following equation:

$$\delta^{(i)} = R_d \sqrt{\frac{2k_{\mathrm{eff}}\left(T_0 - \overline{T}\right)}{q_s R_d}\left[1 - \frac{1}{2}\frac{\delta^{(i-1)}}{R_d} + \frac{1}{10}\left(\frac{\delta^{(i-1)}}{R_d}\right)^2\right]^{-1}}, \quad (3.4)$$

where $i = 1, 2, 3, \ldots$ is the iteration number, q_s is the heat flux at the droplet surface, k_{eff} is the droplet effective thermal conductivity. In the case when $\overline{T} > T_0$, we expect that $q_s < 0$. In the limiting case when $\delta^{(i)} = \delta^{(i-1)} = R_d$, Eq. (3.4) reduces to:

$$T_0 = \overline{T} + \frac{3q_s R_d}{10k_{\mathrm{eff}}}. \quad (3.5)$$

This expression coincides with the one predicted by the parabolic model for $R = R_d$. It was shown that the thermal layer expands to $\delta = R_d$ when the Fourier number $\mathrm{Fo} = k_{\mathrm{eff}}t/(c_l \rho_l R_d^2)$ (c_l and ρ_l are specific liquid heat capacity and density respectively) reaches 0.1. This method, as well as the power law and polynomial approximations, was verified in Snegirev (2013) based on the analytical solution to the heat transfer equation inside a droplet using the Neumann boundary condition. The limits of applicability of this solution have not been investigated. A solution to this equation based upon the Robin boundary condition (Sazhin 2014) would have been a more rigorous approach.

The approaches to droplet heating discussed so far are based on the assumption that the heat conduction equation is linear and the heat conduction process follows the Fourier law (see Moore and Jones 2015; Mierzwiczak et al. 2015; Hristov 2015; Feng et al. 2016 for possible approaches to the solution to the nonlinear problem and Qi and Guo 2014; Khayat et al. 2015; Zhang and Shang 2015; Borukhov and Zayats 2015; Zhukovsky 2016a, b; Li and Cao 2016 for the analysis of non-Fourier models of heat transfer).

The heat supplied to the droplets from the gas phase is characterised by convection heat transfer coefficient h. In the case of stationary droplets, $h = k_g/R_d$, where k_g is gas thermal conductivity and R_d is the droplet radius. In many practically important cases, the convective heating of droplets is described by the Nusselt number $\mathrm{Nu} = 2hR_d/k_g$. Several correlations were suggested for the estimation of Nu for the moving droplets, including (Sazhin 2014):

$$\mathrm{Nu} = 2 + \beta_c \mathrm{Re}^{1/2}\mathrm{Pr}^{1/3}, \quad (3.6)$$

where Re and Pr are Reynolds and Prandtl numbers based on gas properties and the relative velocity of droplets, $\beta_c = 0.6$ (Ranz and Marshall correlation) and $\beta_c = 0.552$ (Frossling correlation) (see Aissa et al. 2015 for a discussion of other similar correlations). An alternative correlation for Nu was suggested by Clift et al. (1978):

$$\mathrm{Nu} = 1 + (1 + \mathrm{RePr})^{1/3} \max\left[1, \mathrm{Re}^{0.077}\right] \quad (3.7)$$

for Re $\leq$ 400. Correlation (3.7) was recommended in Abramzon and Sirignano (1989). Correlations for Nu, inferred from experimental studies, are discussed in Will et al. (2017).

The most widely used model for radiative heating of droplets is based on the assumption that they are opaque grey spheres with emissivity ϵ. This approach allows us to consider the effect of radiative heating of droplets as a surface phenomenon: radiative heat fluxes are added to the convective heat fluxes at the droplet surface. This approach is used in all CFD codes we are aware of, and even in a number of original recent studies (e.g. Borodulin et al. 2017). The main assumption of this model contradicts a simple observation that one can see the bottom of a glass filled with Diesel or gasoline fuel. One might anticipate that these fuels are at least partially transparent in the infrared part of the spectrum. Thus, one would expect that droplet radiative heating takes place not at their surface but via the absorption of thermal radiation penetrating inside the droplets.

The rigorous approach to the calculation of absorption of external thermal radiation inside fuel droplets should be based on the solution to the Maxwell equations, with boundary conditions at the droplet's surface (Sazhin 2014). This solution was obtained in the well-known Mie theory. Direct application of the formulae predicted by this theory is limited by the complexity of relevant calculations. In most practical applications, however, we are interested not in the details of the distribution of thermal radiation absorption inside droplets but in the integral absorption of this radiation in the whole volume of droplets. This integral absorption is characterised by the efficiency factor of absorption Q_a (the ratio of radiative power absorbed in a droplet to the radiative power illuminating the droplet). The results of Mie calculations of Q_a at a certain wave length λ for a typical Diesel fuel were approximated as:

$$Q_a = \frac{4n}{(n+1)^2}\left[1 - \exp(-2\tau_0)\right], \tag{3.8}$$

where $n \equiv n_\lambda \approx 1.46$ is the index of refraction, $\tau_0 = a_\lambda R_d$ is the optical thickness of droplets, and a_λ is the absorption coefficient.

Using the experimentally measured values of the index of absorption $\kappa_\lambda = a_\lambda \lambda/(4\pi)$, it was found that a reasonably good approximation of the average values of Q_a ($\overline{Q}_a$) in the ranges $5\,\mu\text{m} \leq R_d \leq 50\,\mu\text{m}$ and $1000\,\text{K} \leq \theta_R \leq 3000\,\text{K}$ can be described by the following expression:

$$\overline{Q}_a = aR_d^b, \tag{3.9}$$

where a and b are polynomials (or quadratic functions) of the radiative temperature θ_R (external temperature in the case of optically thin media), R_d is in μm.

Approximation (3.9) appears to be particularly useful for implementation into CFD and research numerical codes (e.g. Yin 2015). It was shown that the predictions of typical droplet heating and evaporation based on (3.9) are very close to those based on a more complex model taking into account the difference in thermal radiation absorption in different areas inside droplets (Abramzon and Sazhin 2006). Classical

Mie theory can be applied to spherical droplets only. In the case of illumination of droplets of more complex shapes, more advanced mathematical tools, including the generalised Lorenz–Mie theories (Gouesbet and Grehan 2011), would be needed.

3.3 Hydrodynamic Models (Mono-component Droplet Heating and Evaporation)

In a number of papers, including the most recent ones (Bouchenna et al. 2017; Ashna and Rahimian 2017; Chen et al. 2017), the problem of heating and evaporation of droplets was solved based on direct numerical solution of transport equations in the vicinity of individual droplets. This approach, however, cannot be applied in CFD codes and will not be considered in this chapter.

In a series of our earlier papers, summarised in Sazhin (2014), the heating of mono-component evaporating spherical droplets was described by the following analytical solution to the heat transfer equation inside them:

$$T(R,t) = \frac{R_d}{R} \sum_{n=1}^{\infty} \left\{ q_n \exp\left[-\kappa_R \lambda_n^2 t\right] - \frac{\sin \lambda_n}{\| v_n \|^2 \lambda_n^2} \mu_0(0) \exp\left[-\kappa_R \lambda_n^2 t\right] - \right.$$

$$\left. - \frac{\sin \lambda_n}{\| v_n \|^2 \lambda_n^2} \int_0^t \frac{\mathrm{d}\mu_0(\tau)}{\mathrm{d}\tau} \exp\left[-\kappa_R \lambda_n^2 (t-\tau)\right] \mathrm{d}\tau \right\} \sin\left[\lambda_n \left(\frac{R}{R_d}\right)\right] + T_{\mathrm{eff}}(t), \quad (3.10)$$

where λ_n are solutions to the equation:

$$\lambda \cos \lambda + h_0 \sin \lambda = 0, \quad (3.11)$$

$$\| v_n \|^2 = \frac{1}{2}\left(1 - \frac{\sin 2\lambda_n}{2\lambda_n}\right) = \frac{1}{2}\left(1 + \frac{h_0}{h_0^2 + \lambda_n^2}\right),$$

$$q_n = \frac{1}{R_d \| v_n \|^2} \int_0^{R_d} \tilde{T}_0(R) \sin\left[\lambda_n \left(\frac{R}{R_d}\right)\right] \mathrm{d}R, \quad \kappa_R = \frac{k_l}{c_l \rho_l R_d^2}, \quad \mu_0(t) = \frac{h T_g(t) R_d}{k_l},$$

$h_0 = (hR_d/k_l) - 1$, $\tilde{T}_0(R) = RT_{d0}(R)/R_d$; the solution to Eq. (3.11) gives a set of positive eigenvalues λ_n numbered in ascending order ($n = 1, 2, \ldots$);

$$T_{\mathrm{eff}} = T_g + \frac{\rho_l L \dot{R}_d}{h}, \quad (3.12)$$

L is the specific heat of evaporation, and the value of $\dot{R}_d$, the derivative of the droplet radius with respect to time, describes the rate of droplet evaporation; remembering that in numerical codes, Solution (3.10) was applied at each time step, the value of $\dot{R}_d$ was taken from the previous time step (it was taken as zero at the first time step).

The effect of evaporation on the Nusselt number for stationary spherical droplets can be described by the following equation (the Stefan–Fuchs model):

$$\mathrm{Nu} = \mathrm{Nu}_0 \frac{\ln(1 + B_T)}{B_T}, \tag{3.13}$$

where $\mathrm{Nu}_0 = 2h_0 R_d k_g = 2$, $h_0 = k_g/R_d$ is the convective heat transfer coefficient for a non-evaporating sphere, B_T is the Spalding heat transfer number

$$B_T = \frac{c_{pv}(T_g - T_s)}{L(T_s) - (|\dot{q}_d|/\dot{m}_d)}, \tag{3.14}$$

c_{pv} is the specific heat capacity of fuel vapour at constant pressure, T_s is the droplet surface temperature, $|\dot{q}_d|$ is heat spent on raising droplet internal energy. The droplet evaporation rate $\dot{m}_d \le 0$ for stationary droplets can be estimated as:

$$\dot{m}_d = -4\pi R_d D_v \rho_{\mathrm{total}} \ln\left(1 + B_M\right), \tag{3.15}$$

where D_v is the binary diffusion coefficient of fuel vapour in air, $\rho_{\mathrm{total}} = \rho_v + \rho_g$ is the density of the mixture of vapour and ambient air, B_M is the Spalding mass transfer number defined as

$$B_M = \frac{\rho_{vs} - \rho_{v\infty}}{\rho_{gs}}, \tag{3.16}$$

where subscript $_s$ refers to the surface of the droplet, subscript $_\infty$ refers to ambient conditions. When deriving (3.15), it was assumed that ρ_{total} does not depend on the distance from the surface of the droplet. This is expected to be a serious limitation of the model for strongly evaporating droplets with high surface temperatures. A model in which this assumption is relaxed was developed in Tonini and Cossali (2012).

Note that ρ_{vs} is controlled by the droplet surface temperature. This leads to a strong link between (3.15) and the corresponding equation for droplet heating.

Expression (3.15) can be presented in a more compact form:

$$\dot{m}_d = -2\pi R_d D_v \rho_{\mathrm{total}} \mathrm{Sh} B_M, \tag{3.17}$$

where

$$\mathrm{Sh} \equiv \frac{2h_m R_d}{D_v} = \mathrm{Sh}_0 \frac{\ln\left(1 + B_M\right)}{B_M} = \frac{2h_{m0} R_d}{D_v} \frac{\ln\left(1 + B_M\right)}{B_M} = 2\frac{\ln\left(1 + B_M\right)}{B_M} \tag{3.18}$$

is the Sherwood number, h_m is the convective mass transfer coefficient, subscript $_0$ indicates non-evaporating droplets.

An alternative expression for $\dot{m}_d$ for stationary droplets was derived as (Sazhin 2014):

$$\dot{m}_d = -\frac{4\pi k_m R_d}{c_{pv}} \ln(1 + B_T), \tag{3.19}$$

where k_m is thermal conductivity of the mixture of ambient gas and fuel vapour (in the case of weak evaporation, $k_m \approx k_g$), and B_T is defined by (3.14).

Expressions (3.13), (3.15), (3.18) and (3.19) could be generalised to the case of moving evaporating droplets using the so-called 'film theory' (Abramzon and Sirignano 1989). The key concepts of this theory are film thicknesses δ_T and δ_M. Ignoring the Stefan flow, they can be estimated as (Abramzon and Sirignano 1989):

$$\delta_{T0} = \frac{2R_d}{\mathrm{Nu}_0 - 2}, \quad \delta_{M0} = \frac{2R_d}{\mathrm{Sh}_0 - 2}. \tag{3.20}$$

For stationary droplets, $\mathrm{Nu}_0 = \mathrm{Sh}_0 = 2$. Hence, $\delta_{T0} = \delta_{M0} = \infty$.

The effect of droplet motion on Nu_0 for non-evaporating droplets is described by Eqs. (3.6) and (3.7) (subscripts $_0$ in these equations need to be added to indicate non-evaporating droplets). Similar expressions were obtained for Sh_0 (Abramzon and Sirignano 1989):

$$\mathrm{Sh}_0 = 2 + \beta_c \mathrm{Re}^{1/2} \mathrm{Sc}^{1/3}, \tag{3.21}$$

$$\mathrm{Sh}_0 = 1 + (1 + \mathrm{ReSc})^{1/3} \max\left[1, \mathrm{Re}^{0.077}\right] \tag{3.22}$$

for $\mathrm{Re} \leq 400$, where $\mathrm{Sc} = \nu_m / D_v$, ν_m is the kinematic viscosity of the mixture of gas and vapour. δ_{T0} and δ_{M0}, defined by (3.20), can be considered as thicknesses of the thermal and diffusional boundary layers. The thickening of these layers due to the effect of the Stefan flow was described by parameters F_T and F_M (Abramzon and Sirignano 1989):

$$F_T = \delta_T / \delta_{T0}, \quad F_M = \delta_M / \delta_{M0}. \tag{3.23}$$

The following correlations were suggested in the ranges $0 \leq (B_T, B_M) \leq 20$ and $1 \leq (\mathrm{Pr}, \mathrm{Sc}) \leq 3$ (Abramzon and Sirignano 1989):

$$F_{T,M} = \left(1 + B_{T,M}\right)^{0.7} \frac{\ln\left(1 + B_{T,M}\right)}{B_{T,M}}. \tag{3.24}$$

Using film theory, we would expect that an increase in the film thicknesses, described by (3.23) and (3.24), would lead to a corresponding decrease in Nu_0 and Sh_0. The new decreased values of Nu_0 and Sh_0, called 'modified' Nusselt and Sherwood numbers in Abramzon and Sirignano (1989) (Nu^* and Sh^*), were estimated as (Abramzon and Sirignano 1989):

$$\mathrm{Nu}^* = 2 + \frac{\mathrm{Nu}_0 - 2}{F_T}, \quad \mathrm{Sh}^* = 2 + \frac{\mathrm{Sh}_0 - 2}{F_M}. \tag{3.25}$$

These parameters allow us to present the expressions for Nu and Sh as:

$$\mathrm{Nu} = \mathrm{Nu}^* \frac{\ln(1+B_T)}{B_T}, \qquad \mathrm{Sh} = \mathrm{Sh}^* \frac{\ln(1+B_M)}{B_M}. \tag{3.26}$$

The introduction of Nu^* and Sh^* allows one to present (3.15) and (3.19) as (Abramzon and Sirignano 1989):

$$\dot{m}_d = -2\pi R_d D_v \rho_{\mathrm{total}} \mathrm{Sh}^* \ln\left(1+B_M\right), \tag{3.27}$$

$$\dot{m}_d = -\frac{2\pi k_m R_d}{c_{pv}} \mathrm{Nu}^* \ln(1+B_T). \tag{3.28}$$

Note that (Abramzon and Sirignano 1989):

$$B_T = \left(1+B_M\right)^{\varphi} - 1, \tag{3.29}$$

where

$$\varphi = \left(\frac{c_{pv}}{c_{pg}}\right)\left(\frac{\mathrm{Sh}^*}{\mathrm{Nu}^*}\right)\frac{1}{\mathrm{Le}}, \tag{3.30}$$

$\mathrm{Le} = k_m/(D_v \rho_{\mathrm{total}} c_{pg})$ is the gas Lewis number but with thermal conductivity and density equal to those of a mixture of air and vapour.[1]

The evaporation process leads to the inward movement of the droplet surface (liquid/vapour interface). The effect of this movement on droplet heating was considered in a series of our papers summarised in Sect. 4.4 of Sazhin (2014).

The model based on the combination of the above-mentioned model for the gas phase and the analytical solution to the heat transfer equation for the liquid phase (Solution (3.10)) was extensively validated (Elwardany et al. 2011).

Remembering (3.14) and (3.29), one can obtain the heat rate supplied to the droplet to raise (or reduce) its temperature (internal energy) in the form:

$$\dot{q}_d = -\dot{m}_d \left[\frac{c_{pv}(T_g - T_s)}{B_T} - L(T_s)\right] = -\dot{m}_d \left[\frac{c_{pv}(T_g - T_s)}{(1+B_M)^{\varphi} - 1} - L(T_s)\right], \tag{3.31}$$

where $\dot{q}_d > 0$ when the droplet is heated; for stationary droplets: $\varphi = \left(\frac{c_{pv}}{c_{pg}}\right)\frac{1}{\mathrm{Le}}$.

An alternative approach to the calculation of $\dot{q}_d$ could be based on the analysis of the temperature distribution inside droplets, predicted by Eq. (3.10):

[1] Note that in many papers, including the most recent ones (e.g. Marti et al. 2017), it is assumed that $c_{pv} = c_{pg}$ in (3.28) and (3.30) which is obviously not correct. In Dahms and Oefelein (2016) $c_p = c_{pv}$ in the definition of Le but $c_p = c_{pg}$ in the definition of Pr, which led to ambiguity in their model.

$$\dot{q}_d = 4\pi R_d^2 k_l \left. \frac{\partial T}{\partial R} \right|_{R=R_d-0}. \quad (3.32)$$

Having substituted (3.10) into (3.32), we obtain:

$$\dot{q}_d = 4\pi R_d k_l \sum_{n=1}^{\infty} \left\{ q_n \exp\left[-\kappa_R \lambda_n^2 t\right] - \frac{\sin\lambda_n}{\|v_n\|^2 \lambda_n^2} \mu_0(0) \exp\left[-\kappa_R \lambda_n^2 t\right] - \right.$$
$$\left. - \frac{\sin\lambda_n}{\|v_n\|^2 \lambda_n^2} \int_0^t \frac{d\mu_0(\tau)}{d\tau} \exp\left[-\kappa_R \lambda_n^2 (t-\tau)\right] d\tau \right\} \left[-1 - h_0\right] \sin\lambda_n, \quad (3.33)$$

where all notations are the same as in Solution (3.10).

Once the value of $\dot{q}_d$ has been found, the evaporation rate can be found from Eq. (3.19). Remembering the definition of B_T, this equation can be rewritten as:

$$\dot{m}_d = -\frac{4\pi k_g R_d}{c_{pv}} \ln\left(1 + \frac{c_{pv}(T_g - T_s)\dot{m}_d}{L(T_s)\dot{m}_d - \dot{q}_d}\right). \quad (3.34)$$

Thus, we have two approaches to modelling the heating and evaporation of stationary droplets. The first is based on Eqs. (3.15) and (3.31) (conventional approach originally suggested in Abramzon and Sirignano 1989, Model 1), and the second is based on Eqs. (3.33) and (3.34) (Model 2). Detailed comparison between the predictions of these models was performed by Sazhin et al. (2014b). Although these predictions were qualitatively similar, there were noticeable quantitative differences between them. The reasons behind these differences are still unclear.

The models described so far are based on the assumption that droplets are perfect spheres. However, the shapes of most actually observed droplets in engineering and environmental applications are far from spherical (see Michaelides 2006; Crua et al. 2010). In most cases, the effects of non-sphericity of droplets have been investigated assuming that droplet shapes can be approximated by prolate or oblate spheroids.

The heat conduction equation inside a spheroidal body (droplet), using the Dirichlet boundary conditions, was first solved analytically more than 135 years ago (Niven 1880). This solution, however, turned out to be too complex for most practical applications. In most cases, this problem (and the related problem of mass transfer inside the body) has been investigated based on the numerical solutions to the heat transfer (and mass diffusion) equations (Jog and Hader 1997; Lima et al. 2004; He and Tafti 2017).

The problem of heat/mass transfer inside spheroidal bodies, considered in the above-mentioned papers, is complementary to the problem of heat/mass transfer from/to an ambient fluid (gas) to/from a spheroidal body, taking into account the relative velocity between the gas and the body, in the general case. The latter problem has been considered in numerous papers based on the numerical solutions to momentum and heat transfer equations in the ambient fluid (gas) in the ellipsoidal coordinate system. The analysis of Alassar (2005), Richter and Nikrityuk (2012),

Kishore and Gu (2011), Sreenivasulu et al. (2014) and Sreenivasulu and Srinivas (2015) was based on the assumption that the body surface was fixed. Juncu (2010) took into account changes in body temperature with time, while assuming that there is no temperature gradient inside the body (the thermal conductivity of the body was assumed infinitely high).

These approaches are equally applicable to solid bodies and droplets. In the case of droplets, however, apart from heating, the evaporation processes should also be taken into account in the general case. Grow (1990) was perhaps the first to solve the problem of heat and mass transfer in the vicinity of spheroidal particles assuming that their relative velocities are equal to zero, although she considered coal chars rather than droplets. One of the main limitations of this paper is that both mass and heat transfer equations were presented in the form of Laplace equations, which implies that the effects of Stefan flow from the surface of the particles were ignored. The latter effects were taken into account in the exact solutions to the mass and heat transfer equations in the gas phase around a spheroidal droplet suggested in Tonini and Cossali (2013). In that paper, it was assumed that the temperatures at all points at the surface of a droplet are the same and constant, and the droplet's shape remains spheroidal. A combined problem of spheroidal droplet heating and evaporation, similar to the one studied in Tonini and Cossali (2013), was considered in Li and Zhang (2014). As in Tonini and Cossali (2013), the authors of Li and Zhang (2014) based their analysis on the solution to the species conservation equation in the gas phase and assumed that the thermal conductivity of droplets is infinitely large. In contrast to Tonini and Cossali (2013), Li and Zhang (2014) took into account the relative velocities of droplets, assuming that the dependencies of the Nusselt and Sherwood numbers on the Reynolds and Prandtl numbers are the same as those for the spherical droplets. Also, they took into account the time dependence of droplet temperatures and sizes, although their analysis focused on oblate droplets only.

Strotos et al. (2016) presented CFD analysis of the evaporation of nearly spherical suspended droplets. They solved the Navier–Stokes, energy conservation and species transport equations; the volume-of-fluid (VOF) approach was used to capture the liquid–gas interface.

As follows from an overview of the models described above, the general problem of heating and evaporation of spheroidal droplets is far from resolved. We believe, however, that the results presented in Tonini and Cossali (2013) could be considered a starting point for solving this problem at least for slightly deformed spheroids.

The model described in Tonini and Cossali (2013) was generalised to the case of oscillating droplets under the assumption that the process can be considered quasi-steady-state (Tonini and Cossali 2014). The instantaneous and average mass and heat transfer rates over an oscillation period were functions of the oscillating frequency and amplitude. The results were compared with the predictions of the approximate model described by Mashayek (2001). The model was able to capture different evaporating mechanisms for oblate and prolate droplets.

The results of the generalisation of the model described above to the case of triaxial ellipsoidal droplets are presented in Tonini and Cossali (2016b). In this paper, a new analytical model for heat and mass transfer from deformed droplets was developed, based on the solutions to the species and energy conservation equations under steady-state conditions. It was shown that the droplet deformation enhances both the total and local mass and heat transfer. The evaporation rate from deformed droplets, having the same volume and surface area, was shown to be at a maximum for the prolate droplet and at a minimum for the oblate droplet. Purely numerical investigation of fluid flow and heat transfer from heated spheroids was conducted in Sasmal and Nirmalkar (2016).

The solutions developed in Tonini and Cossali (2013) for the gas phase surrounding a spheroidal droplet were used as boundary conditions for the solutions to heat/mass equations in the liquid phase (Zubkov et al. 2017). The temperature gradients inside and at the surface of the droplets, and the changes in their shape during the heating and evaporation process were taken into account. The effects of surface tension and droplet motion on droplet heating and evaporation were ignored. The results were applied to the analysis of an n-dodecane fuel droplet in Diesel engine-like conditions.

3.4 Hydrodynamic Models (Multi-component Droplet Heating and Evaporation)

All models for mono-component droplets discussed in the previous section are applicable to multi-component droplets. In addition to the processes considered in the previous section, however, for multi-component droplets, we need to take into account that different components evaporate at different rates, creating concentration gradients in the liquid phase. The latter leads to the liquid-phase mass diffusion of species described by the diffusion equation for the mass fractions of each component. The simplest form of this equation, when only the radial diffusion is accounted for and species diffusion coefficient D_l is constant, can be presented as (Sazhin 2014):

$$\frac{\partial Y_{li}}{\partial t} = D_l \left(\frac{\partial^2 Y_{li}}{\partial R^2} + \frac{2}{R} \frac{\partial Y_{li}}{\partial R} \right), \tag{3.35}$$

where subscripts $_l$ and $_i$ indicate liquid phase and type of species, respectively.

This equation needs to be solved subject to the boundary condition at the surface:

$$\left. \frac{\partial Y_{li}}{\partial R} \right|_{R=R_d-0} = \frac{D_v \rho_{\text{total}} \ln\left(1 + B_M\right)}{D_l \rho_l R_d} \left(Y_{li} - \varepsilon_i\right), \tag{3.36}$$

where

$$\varepsilon_i = \frac{Y_{vsi}}{\sum_i Y_{vsi}}, \tag{3.37}$$

(subscript $_v$ indicates the vapour phase) and at the centre of the droplet:

$$\left.\frac{\partial Y_{li}}{\partial R}\right|_{R=0} = 0, \tag{3.38}$$

and the relevant initial conditions. Note that Condition (3.38) can be replaced by a more general requirement that $Y_{li}(R, t)$ are twice continuously differentiable functions at $R \leq R_d$.

In the equilibrium state, the partial pressure of the ith vapour species at the surface of the droplet can be found from the equation:

$$p_{vi} = \gamma_i X_{lsi} p^*_{vi}, \tag{3.39}$$

where X_{lsi} is the molar fraction of the ith species in the liquid at the droplet surface, p^*_{vi} is the partial vapour pressure of the ith species when $X_{li} = 1$, and γ_i is the activity coefficient. If $\gamma_i = 1$ Eq. (3.39) leads to Raoult's law:

$$p_{vi} = X_{li} p^*_{vi}. \tag{3.40}$$

In many engineering applications, D_l was considered to be either infinitely small (multi-component droplets were modelled as mono-component ones) or infinitely large (perfect mixing of species). Both these simplified approaches, however, can lead to unacceptably large errors in predicted droplet temperatures and droplet evaporation times compared with the prediction of the model taking into account finite species diffusion rates inside droplets. Where species diffusion was taken into account, this was mainly performed based on the numerical solution to Eq. (3.35) (e.g. Liu et al. 2016; Yi et al. 2016a; Srivastava and Jaberi 2017). In contrast to this approach, in a series of our papers, the results of which are summarised in Sazhin (2014), a new approach to this problem based on the analytical solution to (3.35), subject to boundary conditions (3.36) and (3.38), was suggested. Assuming that R_d = const, this solution for a short time step, subject to the initial condition $Y_{li}(t = 0) = Y_{li0}(R)$, was obtained in the form (Sazhin 2014):

$$Y_{li} = \epsilon_i + \frac{1}{R}\left\{ \exp\left[D_l\left(\frac{\lambda_0}{R_d}\right)^2 t\right]\left[q_{Yi0} - Q_{Y0}\epsilon_i\right]\sinh\left(\lambda_0\frac{R}{R_d}\right)\right.$$
$$\left. + \sum_{n=1}^{\infty}\left[\exp\left[-D_l\left(\frac{\lambda_n}{R_d}\right)^2 t\right]\left[q_{Yin} - Q_{Yn}\epsilon_i\right]\sin\left(\lambda_n\frac{R}{R_d}\right)\right]\right\}, \tag{3.41}$$

where λ_0 and λ_n $(n \geq 1)$ are solutions to equations

$$\tanh\lambda = -\frac{\lambda}{h_{Y0}} \quad \text{and} \quad \tan\lambda = -\frac{\lambda}{h_{Y0}},$$

respectively, $h_{Y0} = -\left(1 + \frac{\alpha_m R_d}{D_l}\right)$,

$$Q_{Yn} = \begin{cases} -\frac{1}{||v_{Y0}||^2}\left(\frac{R_d}{\lambda_0}\right)^2 (1+h_{Y0})\sinh\lambda_0 & \text{when} \quad n=0 \\ \frac{1}{||v_{Yn}||^2}\left(\frac{R_d}{\lambda_n}\right)^2 (1+h_{Y0})\sin\lambda_n & \text{when} \quad n\geq 1 \end{cases} \tag{3.42}$$

$$q_{Yin} = \frac{1}{||v_{Yn}||^2}\int_0^{R_d} RY_{li0}(R)v_{Yn}(R)\mathrm{d}R, \quad n \geq 0, \tag{3.43}$$

$$v_{Y0}(R) = \sinh\left(\lambda_0 \frac{R}{R_d}\right), \quad v_{Yn}(R) = \sin\left(\lambda_n \frac{R}{R_d}\right), \quad n \geq 1,$$

$$||v_{Y0}||^2 = \int_0^{R_d} v_{Y0}^2(R)\mathrm{d}R = -\frac{R_d}{2}\left[1 + \frac{h_{Y0}}{h_{Y0}^2 - \lambda_n^2}\right], \tag{3.44}$$

$$||v_{Yn}||^2 = \int_0^{R_d} v_{Yn}^2(R)\mathrm{d}R = \frac{R_d}{2}\left[1 + \frac{h_{Y0}}{h_{Y0}^2 + \lambda_n^2}\right], \quad n \geq 1, \tag{3.45}$$

$Y_{lsi} = Y_{lsi}(t)$ are liquid components' mass fractions at the droplet's surface,

$$\alpha_m = \frac{|\dot{m}_d|}{4\pi\rho_l R_d^2} = \text{const.} \tag{3.46}$$

There are obvious typos in Eqs. (5.18) and (5.20) in Sazhin (2014) corrected in Sazhin (2017).

In the case of moving droplets, the distribution of mass fractions of species can be described by (3.41), but with D_l replaced by the effective diffusivity D_{eff}:

$$D_{\text{eff}} = \chi_Y D_l, \tag{3.47}$$

where the coefficient χ_Y can be approximated as:

$$\chi_Y = 1.86 + 0.86\tanh\left[2.225\log_{10}\left(\mathrm{Re}_{d(l)}\mathrm{Sc}_l/30\right)\right], \tag{3.48}$$

$\mathrm{Sc}_l = \nu_l/D_l$ is the liquid Schmidt number, ν_l is the liquid kinematic viscosity. Liquid fuel transport properties and the liquid velocity just below the droplet surface were used to calculate $\mathrm{Re}_{d(l)}$. The model based on (3.47) and (3.48) is known as the effective diffusivity (ED) model. The model, based on the assumption that species diffusivity is infinitely fast ($D_{\text{eff}} = \infty$), is referred to as the infinite diffusivity (ID) model.

As in the case of the heat transfer equation inside droplets, Solution (3.41) was generalised to the case of time-dependent droplet radii during the time step. Also, as in the case of mono-component droplets, the model based on (3.41) was validated based on the experimental results (Elwardany et al. 2011).

The model based on Eq. (3.35) or its solution (3.41) is known as the Discrete Component Model (DCM). It is typically applicable only in the case when the number of components in the droplets is small (e.g. biodiesel droplets Sazhin et al. 2014a; Qubeissi et al. 2015a) which is not the case in most automotive fuels. An alternative approach is based on the probabilistic analysis of a large number of components (e.g. Continuous Thermodynamics approach and the Distillation Curve Model; see Sazhin 2014 for details). Further developments of these models led to the Quadrature Method of Moments (QMoM) (Laurent et al. 2009) and Direct Quadrature Method of Moments (DQMoM). The latter method was further developed in Singer (2016) which led to the Direct Quadrature Method of Moments with delumping. In this family of models, a number of additional simplifying assumptions were used, including the assumption that species inside droplets mix infinitely quickly.

In contrast to the previously considered models, designed for large numbers of components, the model suggested in Sazhin et al. (2011) and Elwardany and Sazhin (2012) takes into account the diffusion of liquid species and thermal diffusion as in the classical Discrete Component Model. As in the case of the Continuous Thermodynamics approach, the new model, called quasi-discrete model, is based on the distribution function with respect to a particular property. Using a carbon number n, this function was approximated as:

$$f_m(n) = C_m(n_0, n_f)\frac{(M(n)-\gamma)^{\alpha-1}}{\beta^\alpha \Gamma(\alpha)} \exp\left[-\left(\frac{M(n)-\gamma}{\beta}\right)\right], \tag{3.49}$$

where $n_0 \leq n \leq n_f$, subscripts $_0$ and $_f$ stand for initial and final (the smallest and the largest values of n), M is the molar mass, $\Gamma(\alpha)$ is the Gamma function, α, β and γ are parameters that determine the shape of the distribution, the choice of $C_m(n_0, n_f)$ assured that $\int_{n_0}^{n_f} f_m(n)\mathrm{d}n = 1$. If fuel includes only alkanes, M (in kg/kmole) and n can be linked by the following expression:

$$M = 14n + 2. \tag{3.50}$$

As follows from the previous analysis (Sazhin 2014), the transport and thermodynamic properties of alkanes are weak functions of n. In this case, one can assume that their properties in a certain narrow range of n are close and can replace the continuous distribution (3.49) with a discrete one, consisting of N_f quasi-components with carbon numbers

$$\overline{n}_j = \frac{\int_{n_{j-1}}^{n_j} n f_m(n)\mathrm{d}n}{\int_{n_{j-1}}^{n_j} f_m(n)\mathrm{d}n}, \tag{3.51}$$

and molar fractions

$$X_j = \int_{n_{j-1}}^{n_j} f_m(n)\mathrm{d}n, \tag{3.52}$$

where j is an integer in the range $1 \leq j \leq N_f$. It was assumed that all $n_j - n_{j-1}$ are equal, i.e. all quasi-components have the same range of values of n. For the case when $N_f = 1$, the analysis of multi-component droplets is reduced to that of mono-component droplets. $\overline{n}_j$ are not integers in the general case and do not represent carbon numbers for actual components. Hence, this model is called the quasi-discrete model. These quasi-components, however, were treated as actual components in the conventional Discrete Component Model (DCM). This model is expected to be particularly useful when N_f is much less than the number of actual species in the hydrocarbon mixture. All thermodynamic and transport properties of quasi-components were determined for $n = \overline{n}_j$. For example, partial pressures of individual quasi-components were estimated as (Raoult's law is assumed to be valid):

$$p_v(\overline{n}_j) = X_{lsi}(\overline{n}_j)p^*(\overline{n}_j), \tag{3.53}$$

where X_{lsi} are the surface molar fractions of liquid quasi-components.

The main limitation of the quasi-discrete model is that it is based on the assumption that fuels consist only of alkanes, while the total molar fraction of alkanes (n-alkanes and iso-alkanes) is only about 40% of the overall composition of Diesel fuels (a similar conclusion could be drawn for gasoline fuel). Hence, the contribution of other components cannot be ignored. Also, even if we restrict our analysis to alkanes alone, it is not easy to approximate this distribution with a reasonably simple distribution function $f_m(n)$, similar to the one given by Expression (3.49). In Sazhin et al. (2014c), the quasi-discrete model was generalised to address both these problems. A realistic composition of Diesel fuels, schematically shown in Fig. 3.1, was used in the analysis presented in Sazhin et al. (2014c).

The results presented in Fig. 3.1 were simplified, taking into account that the properties of n-alkanes and iso-alkanes are rather close. Observing that the contributions of tricycloalkanes, diaromatics and phenanthrenes to Diesel fuel are rather small (less than about 1.6% for each of these components) allows us to ignore the dependence of the properties of these components on the number of carbon atoms and replace these three groups with three components, tricycloalkane, diaromatic and phenanthrene, with arbitrarily chosen carbon numbers. The molar fraction of tricycloalkanes was estimated to be 1.5647%, while the molar fractions of diaromatics and phenanthrenes were estimated to be 1.2240% and 0.6577%, respectively. Transport and thermodynamic properties of the components are summarised in Appendices 1–7 of Sazhin et al. (2014c). In the new model, the focus is shifted from the analysis of the distribution function to the direct analysis of molar fractions of the components. These are described by the matrix X_{nm}, where n refers to the number of carbon atoms and m refers to the groups (e.g. alkanes) or individual components (tricycloalkane, diaromatic and phenanthrene). The link between the values of m and the components is shown in Table 3.1.

For each m, the values of $\overline{n}_{jm}$ of quasi-components were introduced. As in the case of the original quasi-discrete model, $\overline{n}_{jm}$ are not integers in the general case. Due to the additional dimensions introduced by the subscript m, the new model is called the multi-dimensional quasi-discrete model (MDQDM). The maximal number of

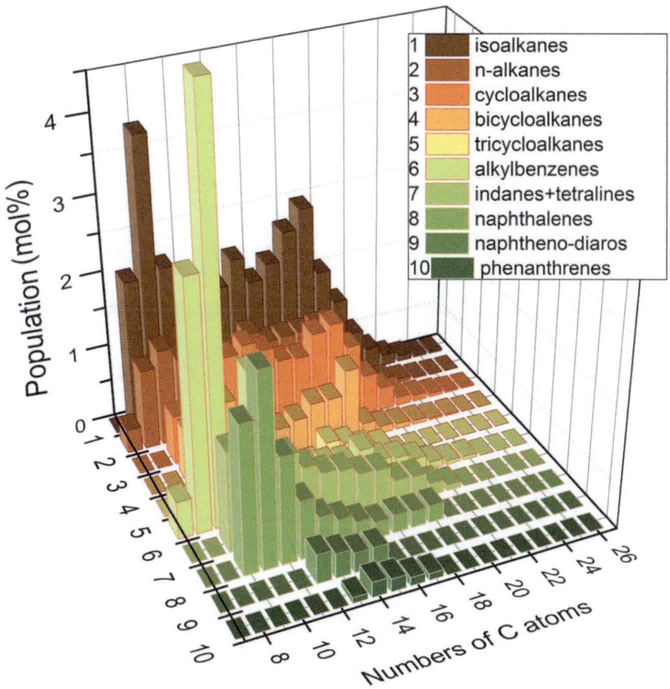

Fig. 3.1 Distribution functions of various hydrocarbons versus the numbers of carbon atoms in molecules in a representative sample of Diesel fuel. Reprinted from Fluid Phase Equilibria, Volume 356, Gun'ko et al., A quantum chemical study of the processes during the evaporation of real-life Diesel fuel droplets, pp. 146–156

Table 3.1 The relation between parameter m and groups ($m = 1$–6) and components ($m = 7$–9). Reprinted from Fuel, Volume 129, Sazhin et al., A multidimensional quasi-discrete model for the analysis of Diesel fuel droplet heating and evaporation, pp. 238–266

m	Component
1	Alkanes
2	Cycloalkanes
3	Bicycloalkanes
4	Alkylbenzenes
5	Indanes and tetralines
6	Naphthalenes
7	Tricycloalkane
8	Diaromatic
9	Phenanthrene

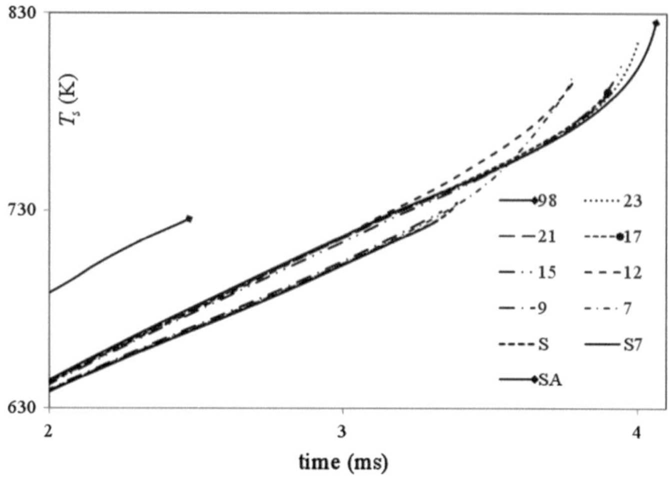

Fig. 3.2 The plots of the droplet surface temperatures T_s versus time for ten approximations of Diesel fuel composition: 98 components (indicated as (98)); 23, 21, 17, 15, 12, 9 and 7 quasi-components/components (numbers near the curves); the contributions of all groups are approximated by single quasi-components, to which the contribution of tricycloalkane is added, leading to 7 quasi-components/components (indicated as (S7)); the contribution of all 98 components is taken into account as that of a single component (indicated as (S)); the contributions of only 20 alkane components are taken into account and these are treated as a single component, with the average value of the carbon number ($C_{14.763}H_{31.526}$; indicated as (SA)). Only the final stage of droplet heating and evaporation is shown. Reprinted from Fuel, Volume 129, Sazhin et al., A multidimensional quasi-discrete model for the analysis of Diesel fuel droplet heating and evaporation, pp. 238–266

these quasi-components/components, providing the most accurate approximation of Diesel fuel, was taken to be equal to the actual number of components (98 in the case considered in Sazhin et al. (2014c)). In this case, the new model reduces to the conventional Discrete Component Model (DCM). The quasi-components in the MDQDM are treated in the same way as the quasi-components in the conventional quasi-discrete model. Also, the temperature gradient and quasi-components' diffusion inside droplets are taken into account as in the quasi-discrete model.

In Sazhin et al. (2014c), the MDQDM was applied to the analysis of heating and evaporation of a droplet with initial radius $R_{d0} = 10\,\mu\text{m}$ in air with density, temperature and pressure equal to $\rho_a = 11.9\,\text{kg/m}^3$, $T_a = 880\,\text{K}$, $p_a = 30\,\text{bar}$, respectively. All transport and thermodynamic properties for Diesel fuel and its components are given in Sazhin et al. (2014c).[2]

The plots of the droplet surface temperatures T_s and radii R_d versus time for a wide range of approximations of Diesel fuel are shown in Figs. 3.2 and 3.3.

As can be seen from Figs. 3.2 and 3.3, the approximation of 98 actual components by a single quasi-component leads to a noticeable underestimation of the droplet

[2]The results of most recent experimental and theoretical studies of Diesel fuel viscosity are presented in Kanaveli et al. (2017) and Lapuerta et al. (2017).

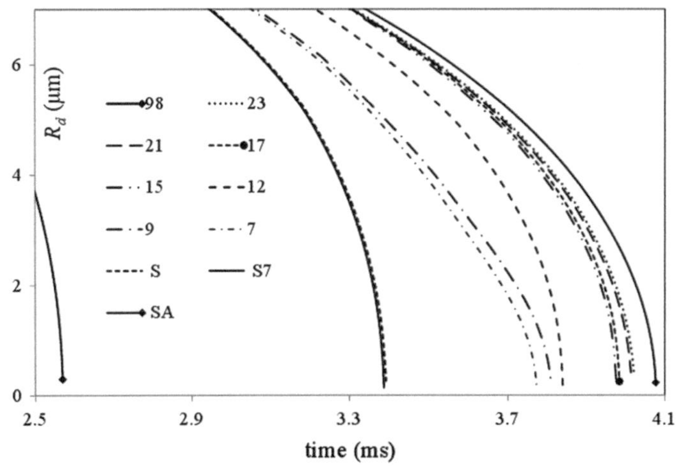

Fig. 3.3 The same as Fig. 3.2 but for the droplet radii R_d. Reprinted from Fuel, Volume 129, Sazhin et al., A multidimensional quasi-discrete model for the analysis of Diesel fuel droplet heating and evaporation, pp. 238–266

surface temperature, and an underestimation of the evaporation time by about 17%. The approximation of this fuel by a single alkane quasi-component ($C_{14.763}H_{31.526}$) (plots SA) leads to under-prediction of the evaporation time by about 37%. This is not acceptable even for qualitative analysis of the process. The plots S and S7 are almost indistinguishable. Also, plots 9 and 7 are rather close. The same applies to plots 23 and 21. This means that the contribution of diaromatics and phenanthrenes can be safely ignored in the approximation of this fuel when modelling the heating and evaporation of droplets. Both for droplet surface temperatures and radii, the accuracy of approximations improves as the number of quasi-components/components (QC/Cs) increases. In the case of 15 QC/Cs, the droplet evaporation time can be estimated with an error of about 2.5%. In the case of 21 QC/Cs, this error reduces to about 1.5%. Hence, when balancing simplicity with accuracy of the model, one can recommend the approximation of Diesel fuel with 21 QC/Cs if errors less than about 2% can be accepted. This number of QC/Cs can be reduced to 15 if errors less than about 3% can be accepted.

The application of the MDQDM to gasoline fuel droplets and a mixture of biodiesel/Diesel fuel droplets was considered in Qubeissi et al. (2015b) and Qubeissi et al. (2017).

The analysis of heating and evaporation of multi-component droplets thus far described has focused primarily on the liquid phase. It has been assumed that all vapour components in the gas phase behave as a single component. This assumption was relaxed in a number of papers, some results of which are summarised below.

In the classical Stefan–Fuchs theory, Eq. (3.15) for evaporation of mono-component droplets was derived taking into account the conservation of vapour mass flux at any point around a stationary droplet. In the case of multi-component droplets, we can impose a similar condition for all individual components in the gas phase. Following Tonini and Cossali (2015), this condition can be presented as:

$$\frac{\mathrm{d}}{\mathrm{d}R}\left(R^2\rho_{\mathrm{total}}UY_k - R^2D^{(k,m)}\rho_{\mathrm{total}}\frac{\mathrm{d}Y_k}{\mathrm{d}R}\right) = 0, \tag{3.54}$$

where subscript $_k$ refers to ambient gas ($k = 0$) or fuel vapour species ($k = 1, \ldots, n$), $R \geq R_d$ is the distance from the centre of the droplet in the gaseous phase, $D^{(k,m)}$ is the mass diffusion coefficient for the species k in the mixture, Y_k are mass fractions of species k, U is the Stefan velocity estimated as

$$U = \frac{\sum_{k=1}^{n} \dot{m}_d^{(k)}}{4\pi R^2 \rho_{\mathrm{total}}}, \tag{3.55}$$

$\dot{m}_d^{(k)}$ is the evaporation rate of species k (following Tonini and Cossali (2015) and in contrast to Eq. (3.15) we assume that $\dot{m}_d^{(k)} \geq 0$ during the evaporation process), ρ_{total} is the total density of the mixture, including ambient gas.

The analysis of Eq. (3.54) is difficult due to the fact that both ρ_{total} and $D^{(k,m)}$ are unknown functions of R. Our further analysis is based on the assumption that ρ_{total} and $D^{(k,m)}$ remain constant for all R (the assumption that ρ_{total} is constant was made when deriving Eq. (3.15)). The values of $D^{(k,m)}$ were estimated in the reference conditions as (Blanc's law):

$$D^{(k,m)} = \left(\sum_{j=0;\, j\neq k}^{n} \frac{Y_{j\,(\mathrm{ref})}}{D^{(k,j)}}\right)^{-1}, \tag{3.56}$$

where

$$Y_{j\,(\mathrm{ref})} = \frac{2Y_{j\,(\mathrm{s})} + Y_{j\,(\infty)}}{3}, \tag{3.57}$$

$Y_{j\,(\mathrm{s})}$ and $Y_{j\,(\infty)}$ are the mass fractions of species j at the surface of the droplets and in ambient gas, respectively. Expression (3.57) allows us to consider ρ_{total} under the reference conditions as well ($\rho_{\mathrm{total}} = \rho_{\mathrm{ref}}$).

Having introduced new variable $\zeta = R_d/R$, the general analytical solution to Eq. (3.54) was obtained in the form (Tonini and Cossali 2015):

$$Y_k = \alpha_k \exp\left[-\frac{\dot{m}_d^{(\mathrm{total})}}{4\pi\rho_{\mathrm{total}}R_dD^{(k,m)}}\zeta\right] + \varepsilon_k, \tag{3.58}$$

where $\dot{m}_d^{(\mathrm{total})} = \sum_{k=1}^{n} \dot{m}_d^{(k)}$,

$$\varepsilon_k = \frac{\dot{m}_d^{(k)}}{\sum_{k=1}^{n} \dot{m}_d^{(k)}} \tag{3.59}$$

is the evaporation rate of species k, α_k are unknown constants.

Recalling that $Y_k(\zeta = 0) = Y_{k\infty}$, we find that $\alpha_k = Y_{k\infty} - \varepsilon_k$. This relation for α_k allows us to rewrite Eq. (3.58) for the droplet surface ($\zeta = 1$) as:

$$Y_{ks} = \left(Y_{k\infty} - \varepsilon_k\right) \exp\left[-\frac{\dot{m}_d^{(\text{total})}}{4\pi\rho_{\text{total}} R_d D^{(k,m)}}\right] + \varepsilon_k. \tag{3.60}$$

Equation (3.58) was rearranged as (Tonini and Cossali 2015):

$$\varepsilon_k = \frac{Y_{ks} - Y_{k\infty} \exp\left[-\frac{\dot{m}_d^{(\text{total})}}{4\pi\rho_{\text{total}} R_d D^{(k,m)}}\right]}{1 - \exp\left[-\frac{\dot{m}_d^{(\text{total})}}{4\pi\rho_{\text{total}} R_d D^{(k,m)}}\right]}, \tag{3.61}$$

$$\sum_{k=1}^{n} \frac{Y_{ks} - Y_{k\infty}}{\left(1 - \exp\left[-\frac{\dot{m}_d^{(\text{total})}}{4\pi\rho_{\text{total}} R_d D^{(k,m)}}\right]\right)} = 1 = \sum_{k=1}^{n} Y_{k\infty}. \tag{3.62}$$

Non-linear equation (3.62) was used in Tonini and Cossali (2015) to calculate the total evaporation rate $\dot{m}_d^{(\text{total})}$ assuming that the values of all other parameters in this equation are known. Once the value of $\dot{m}_d^{(\text{total})}$ was obtained, the values of ε_k were calculated from Eq. (3.61).

Equation (3.54) could be formulated in terms of molar rather than mass fluxes (Tonini and Cossali 2016a). The latter equation could be solved under the assumption that the molar density of the mixture does not depend on the distance from the droplet surface. The solution to this equation would be rather similar to (3.61) and (3.62), and its explicit form was given in Tonini and Cossali (2016a). These two equations and their solutions predict slightly different evaporation rates since the conditions of constant total mass density and constant molar density of the mixture are not equivalent.

To take into account the effects of multi-component droplet movement on droplet heating and evaporation, in Tonini and Cossali (2015) (as well as in a number of other papers, e.g. Ma et al. 2016), it was assumed that there is no interaction between evaporating species. For each of these species, the Abramzon and Sirignano model (Abramzon and Sirignano 1989) was applied. The validity of this assumption is not at first evident, and this is the reason why, in many papers and books, including Sazhin (2014), the effect of relative motion between species in the gas phase has been ignored altogether.

Padoin et al. (2014), Jarvas et al. (2015), Toe et al. (2015) and Tonini and Cossali (2016a) drew attention to the fact that more accurate description of multi-component diffusion, compared with Eq. (3.54), should be based on the Maxwell–Stefan equations. Ignoring the Soret effects, diffusion due to pressure gradients and external forces, these equations can be presented as (Bird et al. 2002; Tonini and Cossali 2016a):

$$\nabla X^{(p)} = \sum_{k=0}^{n} \frac{1}{C_m D_{pk}} \left(X^{(p)} \mathbf{N}^{(k)} - X^{(k)} \mathbf{N}^{(p)} \right), \tag{3.63}$$

where $X^{(k)}$ is the molar fraction of the kth component, C_m is the molar density of the mixture, $D_{pk} = D_{kp}$ is the binary diffusion coefficient of the pth component into the kth component, $\mathbf{N}^{(p)}$ is the molar flux of the pth component, $k = 0$ refers to ambient gas.

For a multi-component spherical droplet, only the radial components of the species molar fluxes can be retained. In this case, Eq. (3.63) was presented in a similar format to that inferred from Eq. (3.54). This allowed Tonini and Cossali (2016a) to present the solution to (3.63) in a similar format to (3.61) and (3.62), but for molar fractions, assuming that the total molar density does not depend on the distance from the droplet surface. It was shown that the predictions based on Eq. (3.54) (Stefan–Fuchs equation) underestimate the total evaporation rate, especially at high ambient gas temperatures, for various droplet compositions. The largest deviation of the absolute values of the evaporation rate, predicted by the Stefan–Fuchs and Maxwell–Stefan equations, was found when none of the species mass fractions was dominant.

A new quasi-dimensional multi-component heating and evaporation model for multi-component fuel droplets was suggested in Yi et al. (2016b). In contrast to the Discrete Component Model, this model is based not on the rigorous solution to heat transfer and species diffusion equations inside droplets, but on the polynomial (quadratic) approximations of the temperature and mass fractions of species distributions inside droplets.

3.5 Kinetic and Molecular Dynamics Models

So far the modelling of droplet heating and evaporation processes has been based on the hydrodynamic approximation. Vapour at the droplet surface has been assumed to be saturated, and the evaporation has been modelled as the diffusion of vapour from the droplet surface to the ambient gas (Sazhin 2014). The limitations of this approach are well known (see Fuchs 1959). In a number of studies (see Sazhin 2014), the heating and evaporation of n-dodecane ($C_{12}H_{26}$) (an approximation for Diesel fuel) and a mixture of n-dodecane (approximating alkanes in this fuel) and p-dipropylbenzene (approximating aromatics in this fuel) droplets were studied, and a new model combining the kinetic and hydrodynamic approaches was developed.[3] In the close vicinity of droplet surfaces (10–100 molecular mean free paths), the vapour and ambient gas dynamics were investigated based on the solution to the Boltzmann equation

[3] An approximation of Diesel fuel by a mixture of n-dodecane and m-xylene was considered in Payri et al. (2015), but the implications of this approximation for kinetic modelling have not been investigated.

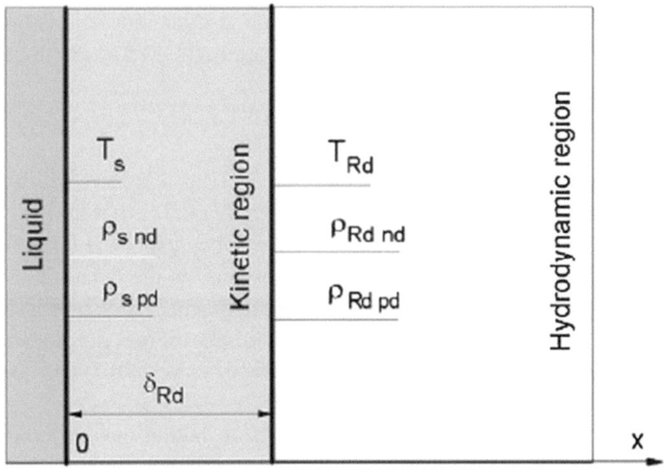

Fig. 3.4 Liquid, kinetic and hydrodynamic regions near the surface of the droplet. T_s is the droplet surface temperature, $\rho_{s\,(nd,pd)}$ are n-dodecane ($_{nd}$) and p-dipropylbenzene ($_{pd}$) vapour densities in the immediate vicinity of the droplet surface, T_{Rd} and $\rho_{Rd\,(nd,pd)}$ are the temperature and n-dodecane ($_{nd}$) and p-dipropylbenzene ($_{pd}$) vapour densities at the outer boundary of the kinetic region. Reprinted from International Journal of Heat and Mass Transfer, Volume 93, Sazhin et al., A self-consistent kinetic model for droplet heating and evaporation, pp. 1206–1217

(kinetic region), while at larger distances the study was based on the hydrodynamic equations (hydrodynamic region). The contributions of up to three components in the kinetic region (up to two components approximating Diesel fuel, and air approximated by nitrogen) were taken into account. These three regions in the vicinity of the droplet surface are schematically shown in Fig. 3.4. Inelastic collisions between molecules were taken into account using a simplified model (see Sazhin 2014 for details).

The boundary conditions between the kinetic and hydrodynamic regions were inferred based on the conservation of heat and mass fluxes at the interface between these regions. The hydrodynamic heat and mass fluxes were estimated based on the assumptions that the temperature at the outer boundary of the kinetic region is equal to the droplet surface temperature and vapour pressure at this boundary is equal to the saturated vapour pressure at a droplet surface temperature. The conservation of heat and mass fluxes at this interface allowed us to find the corrected values of temperature and vapour density (Sazhin 2014). The heat and mass fluxes in the hydrodynamic region, calculated based on these corrected values of temperature and vapour density, however, are not equal to the heat and mass fluxes in the hydrodynamic region used to find these corrected values, in the general case. This problem was addressed in Sazhin et al. (2016a), where the results of the development of a new self-consistent kinetic model for droplet heating and evaporation are described.

The solution to the Boltzmann equation in the kinetic region requires formulation of the boundary condition at the liquid/gas interface. This condition is controlled by the evaporation coefficient. The results of molecular dynamics calculations of this coefficient, using the United Atom Model, were approximated as (Sazhin 2014):

$$\beta_e(T_s) = 7 \times 10^{-6}\, T_s^2 - 9.8 \times 10^{-3}\, T_s + 3.7215. \tag{3.64}$$

where T_s is the droplet surface temperature.

One of the main limitations of the United Atom Model is that in this model, the interaction between individual molecules was described using the force field methods. The applicability of this approach is not obvious, as the dynamics of individual molecules in the vicinity of droplet surfaces are essentially quantum mechanical processes. These processes at and in the vicinity of Diesel fuel droplet surfaces are described in Gunko et al. (2013, 2014, 2015), Nasiri et al. (2015) and Sazhin et al. (2016b).

It was demonstrated that an efficient approach to taking into account quantum chemical effects on the value of the evaporation coefficients could be based on the transition state theory (TST) and quantum chemical DFT methods (Gunko et al. 2015). These were applied to several n-dodecane conformers. There was similarity between this approach and the one used previously (see Sazhin 2014). In contrast to the previous studies, however, in the analysis of Gunko et al. (2015), the TST was based on a QC DFT approach taking into account the conformerisation of n-dodecane molecules. It was demonstrated that the most accurate expression for the evaporation/condensation coefficient is the one averaged over the states of various conformers transferred between phases (Gunko et al. 2015):

$$\beta_e = \left\{ 1 - \left[\frac{\rho_g}{\rho_l} \exp \frac{\langle \Delta G_{g \to l} \rangle}{R_u T} \right]^{1/3} \right\} \exp \left\{ -0.5 \left[\left[\frac{\rho_g}{\rho_l} \exp \frac{\langle \Delta G_{g \to l} \rangle}{R_u T} \right]^{1/3} - 1 \right]^{-1} \right\}, \tag{3.65}$$

where R_u is the universal gas constant, $\rho_{g(l)}$ is the gas (liquid) density, and $\Delta G_{g \to l}$ is the change in the Gibbs free energy during the condensation process. It was assumed that the process under consideration is quasi-steady-state.

The effects of both the conformerisation and cross-conformerisation of n-dodecane molecules (CDM effects), which can contribute to the Gibbs free energies of evaporation and solvation, were taken into account.

A comparison between the results of calculations of β_e is shown in Fig. 3.5. As follows from this figure, taking into account the QC effects leads to marginal modification of the predicted β_e, except at temperatures close to the critical temperature. Thus, although the analysis of the QC effects takes into account new effects ignored in the conventional force field approach, the contribution of these effects to the values of β_e turned out to be marginal, unless temperatures were close to the critical temperature.

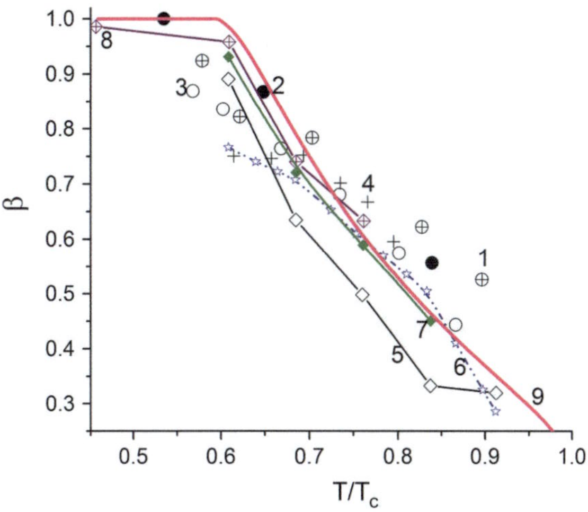

Fig. 3.5 The values of the evaporation coefficient $\beta = \beta_e$, predicted by MD, FF (symbols 1–4, curves 5–8) and Formula (3.65) (curve 9), versus normalised temperature (T/T_c, where T_c is the critical temperature). Symbols (1–4) refer to the models for structureless LJ fluids with various input parameters (Lotfi et al. 2014; Mizuguchi et al. 2010), curves 5 and 7 refer to the results obtained based on the UAM reported in Cao et al. (2011) and Xie et al. (2011), respectively, curve 6 refers to the results of calculations based on the TST model reproduced from Cao et al. (2011), curve 8 is based on the results of calculations using the model described by Mizuguchi et al. Mizuguchi et al. (2010). QC calculations were performed using DFT, ωB97X-D/cc-pVTZ and SMD/ωB97X-D/cc-pVTZ. Reprinted from Fuel, Volume 165, Sazhin et al., Quantum chemical analysis of the processes at the surfaces of Diesel fuel droplets, pp. 405–412

Acknowledgements The author is grateful to EPSRC (grants EP/K005758/1, EP/K020528/1 and EP/M002608/1) for their financial support.

References

Abramzon B, Sazhin SS (2006) Convective vaporization of fuel droplets with thermal radiation absorption. Fuel 85:32–46

Abramzon B, Sirignano WA (1989) Droplet vaporization model for spray combustion calculations. Int J Heat Mass Transf 32:1605–1618

Aggarwal SK (2014) Single droplet ignition: theoretical analyses and experimental findings. Prog Energy Combust Sci 45:79–107. https://doi.org/10.1016/j.pecs.2014.05.002. http://www.sciencedirect.com/science/article/pii/S0360128514000276

Aissa A, Abdelouahab M, Noureddine A, El Ganaoui M, Pateyron B (2015) Ranz and Marcshall correlations limits on heat flow between a sphere and its surrounding gas at high temperature. Thermal Sci 19(5):1521–1528

Alassar RS (2005) Forced convection past an oblate spheroid at low to moderate Reynolds numbers. ASME J Heat Transf 127:1062–1070

Ashna M, Rahimian MH (2017) LMB simulation of head-on collision of evaporating and burning droplets in coalescence regime. Int J Heat Mass Transf 109:520–536. https://doi.org/10.1016/j.ijheatmasstransfer.2017.01.108. http://www.sciencedirect.com/science/article/pii/S0017931016324309

Bird RB, Stewart WE, Lightfoot EN (2002) Transport phenomena. Wiley, New York

Bojko BT, DesJardin PE (2017) On the development and application of a droplet flamelet-generated manifold for use in two-phase turbulent combustion simulations. Combust Flame 183:50–65. https://doi.org/10.1016/j.combustflame.2017.04.029. http://www.sciencedirect.com/science/article/pii/S0010218017301645

Borodulin V, Letushko V, Nizovtsev M, Sterlyagov A (2017) Determination of parameters of heat and mass transfer in evaporating drops. Int J Heat Mass Transf 109:609–618. https://doi.org/10.1016/j.ijheatmasstransfer.2017.02.042. http://www.sciencedirect.com/science/article/pii/S001793101633304X

Borukhov V, Zayats G (2015) Identification of a time-dependent source term in nonlinear hyperbolic or parabolic heat equation. Int J Heat Mass Transf 91:1106–1113. https://doi.org/10.1016/j.ijheatmasstransfer.2015.07.066. http://www.sciencedirect.com/science/article/pii/S0017931015007863

Bouchenna C, Saada MA, Chikh S, Tadrist L (2017) Generalized formulation for evaporation rate and flow pattern prediction inside an evaporating pinned sessile drop. Int J Heat Mass Transf 109:482–500. https://doi.org/10.1016/j.ijheatmasstransfer.2017.01.114. http://www.sciencedirect.com/science/article/pii/S0017931016317719

Brereton GJ (2013) A discrete multicomponent temperature-dependent model for the evaporation of spherical droplets. Int J Heat Mass Transf 60:512–522

Cao B-Y, Xie J-F, Sazhin SS (2011) Molecular dynamics study on evaporation and condensation of n-dodecane at liquid-vapor phase equilibria. AIP J Chem Phys 134(16):164309

Chen X, Wang X, Chen PG, Liu Q (2017) Thermal effects of substrate on Marangoni flow in droplet evaporation: response surface and sensitivity analysis. Int J Heat Mass Transf 113:354–365. https://doi.org/10.1016/j.ijheatmasstransfer.2017.05.076. http://www.sciencedirect.com/science/article/pii/S0017931016341448

Clift R, Grace JR, Weber ME (1978) Bubbles, drops and particles. Academic Press, New York

Crua C, Shoba T, Heikal M, Gold M, Higham C (2010) High-speed microscopic imaging of the initial stage of diesel spray formation and primary breakup. In: SAE International 2010-01-2247, pp 1–10

Dahms RN, Oefelein JC (2016) The significance of drop non-sphericity in sprays. Int J Multiphase Flow 86:67–85. https://doi.org/10.1016/j.ijmultiphaseflow.2016.07.010. http://www.sciencedirect.com/science/article/pii/S030193221530104X

Elwardany AE, Sazhin SS (2012) A quasi-discrete model for droplet heating and evaporation: application to Diesel and gasoline fuels. Fuel 97:685–694

Elwardany AE, Gusev IG, Castanet G, Lemoine F, Sazhin SS (2011) Mono- and multi-component droplet cooling/heating and evaporation: comparative analysis of numerical models. Atomization Sprays 21:907–931

Feng S, Cui X, Li A (2016) Fast and efficient analysis of transient nonlinear heat conduction problems using combined approximations (CA) method. Int J Heat Mass Transf 97:638–644. https://doi.org/10.1016/j.ijheatmasstransfer.2016.02.061. http://www.sciencedirect.com/science/article/pii/S0017931015316550

Fuchs NA (1959) Evaporation and droplet growth in gaseous media. Pergamon Press, London

Gouesbet G, Grehan G (2011) Generalized Lorenz-Mie theories. Springer, Berlin

Grow DA (1990) Heat and mass transfer to an elliptical particle. Combust Flame 80:209–213

Gunko VM, Nasiri R, Sazhin SS, Lemoine F, Grisch F (2013) A quantum chemical study of the processes during the evaporation of real-life diesel fuel droplets. Fluid Phase Equilib 356:146–156

Gunko VM, Nasiri R, Sazhin SS (2014) A study of the evaporation and condensation of n-alkane clusters and nanodroplets using quantum chemical methods. Fluid Phase Equilibria 366:99–107

Gunko VM, Nasiri R, Sazhin SS (2015) Effects of the surroundings and conformerisation of n-dodecane molecules on evaporation/condensation processes. J Chem Phys 142(3):034502

He L, Tafti DK (2017) Heat transfer in an assembly of ellipsoidal particles at low to moderate Reynolds numbers. Int J Heat Mass Transf 114:324–336. https://doi.org/10.1016/j.ijheatmasstransfer.2017.06.068. http://www.sciencedirect.com/science/article/pii/S0017931017312942

Heywood JB (1988) Internal combustion engines fundamentals. McGraw-Hill, New York

Hristov J (2015) An approximate analytical (integral-balance) solution to a non-linear heat diffusion equation. Thermal Sci 19(2):723–733

Jarvas G, Kontos J, Hancsok J, Dallos A (2015) Modeling ethanol-blended gasoline droplet evaporation using COSMO-RS theory and computation fluid dynamics. Int J Heat Mass Transf 84:1019–1029. https://doi.org/10.1016/j.ijheatmasstransfer.2014.12.046. http://www.sciencedirect.com/science/article/pii/S0017931014011545

Jog MA, Hader MA (1997) Transient heat transfer to a spheroidal liquid drop suspended in an electric field. Int J Heat Fluid Flow 18:411–418

Juncu G (2010) Unsteady heat transfer from an oblate/prolate spheroid. Int J Heat Mass Transf 53:3483–3494

Kanaveli I-P, Atzemi M, Lois E (2017) Predicting the viscosity of diesel/biodiesel blends. Fuel 199:248–263. https://doi.org/10.1016/j.fuel.2017.02.077. http://www.sciencedirect.com/science/article/pii/S0016236117302272

Khayat R, deBruyn J, Niknami M, Stranges D, Khorasany R (2015) Non-Fourier effects in macro- and micro-scale non-isothermal flow of liquids and gases review. Int J Thermal Sci 97:163–177. https://doi.org/10.1016/j.ijthermalsci.2015.06.007. http://www.sciencedirect.com/science/article/pii/S1290072915001726

Kishore N, Gu S (2011) Momentum and heat transfer phenomena of spheroid particles at moderate Reynolds and Prandtl numbers. Int J Heat Mass Transf 55:2595–2601

Lapuerta M, Rodrguez-Fernndez J, Fernndez-Rodrguez D, Patio-Camino R (2017) Modeling viscosity of butanol and ethanol blends with diesel and biodiesel fuels. Fuel 199:332–338. https://doi.org/10.1016/j.fuel.2017.02.101. http://www.sciencedirect.com/science/article/pii/S0016236117302521

Laurent C, Lavergne G, Villedieu P (2009) Continuous thermodynamics for droplet vaporization: comparison between Gamma-PDF model and QMoM. Comptes Rendus Mecanique 337(6):449–457. https://doi.org/10.1016/j.crme.2009.06.004. http://www.sciencedirect.com/science/article/pii/S1631072109000680

Li S-N, Cao B-Y (2016) On defects of Taylor series approximation in heat conduction models. Int J Heat Mass Transf 98:824–832. https://doi.org/10.1016/j.ijheatmasstransfer.2016.03.067. http://www.sciencedirect.com/science/article/pii/S001793101531749X

Lima DR, Farias SN, Lima GB (2004) Mass transport in spheroids using the Galerkin method. Braz J Chem Eng 21:667–680

Liu L, Liu Y, Mi M, Wang Z, Jiang L (2016) Evaporation of a bicomponent droplet during depressurization. Int J Heat Mass Transf 100:615–626. https://doi.org/10.1016/j.ijheatmasstransfer.2016.05.007. http://www.sciencedirect.com/science/article/pii/S0017931015314897

Li J, Zhang J (2014) A theoretical study of the spheroidal droplet evaporation in forced convection. Phys Lett A 378(47):3537–3543. https://doi.org/10.1016/j.physleta.2014.10.020. http://www.sciencedirect.com/science/article/pii/S037596011401038X

Lotfi A, Vrabec J, Fischer J (2014) Evaporation from a free liquid surface. Int J Heat Mass Transf 73:303–317

Marti F, Martinez O, Mazo D, Garman J, Dunn-Rankin D (2017) Evaporation of a droplet larger than the Kolmogorov length scale immersed in a relative mean flow. Int J Multiphase Flow 88:63–68. https://doi.org/10.1016/j.ijmultiphaseflow.2016.09.019. http://www.sciencedirect.com/science/article/pii/S0301932215300604

Mashayek F (2001) Dynamics of evaporating drops. Part II: Free oscillations. Int. J Heat Mass Transf 44(8):1527–1541

Ma X, Zhang F, Han K, Song G (2016) Numerical modeling of acetone-butanol-ethanol and diesel blends droplet evaporation process. Fuel 174:206–215. https://doi.org/10.1016/j.fuel.2016.01.091. http://www.sciencedirect.com/science/article/pii/S0016236116001125

Michaelides E (2006) Particles, bubbles and drops. World Scietific, New Jersey

Mierzwiczak M, Chen W, Fu W-J (2015) The singular boundary method for steady-state non-linear heat conduction problem with temperature-dependent thermal conductivity. Int J Heat Mass Transf 91:205–217. https://doi.org/10.1016/j.ijheatmasstransfer.2015.07.051. http://www.sciencedirect.com/science/article/pii/S0017931015007711

Mizuguchi H, Nagayama G, Tsuruta T (2010) Molecular dynamics study on evaporation coefficient of biodiesel fuel. In: Seventh international conference on flow dynamics, p 386

Moore TJ, Jones MR (2015) Solving nonlinear heat transfer problems using variation of parameters. Int J Thermal Sci 93:29–35. https://doi.org/10.1016/j.ijthermalsci.2015.02.002. http://www.sciencedirect.com/science/article/pii/S1290072915000538

Nasiri R, Gunko VM, Sazhin SS (2015) The effects of internal molecular dynamics on the evaporation/condensation of n-dodecane. Theoret Chem Acc 134:N83

Niven C (1880) On the conduction of heat in ellipsoids of revolution. Philos Trans R Soc Lond 171:117–151

Olguin H, Gutheil E (2014) Influence of evaporation on spray flamelet structures. Combust Flame 161(4):987–996. https://doi.org/10.1016/j.combustflame.2013.10.010. http://www.sciencedirect.com/science/article/pii/S0010218013003842

Padoin N, Toe ATD, Rangel LP, Ropelato K, Soares C (2014) Heat and mass transfer modeling for multicomponent multiphase flow with CFD. Int J Heat Mass Transf 73:239–249. https://doi.org/10.1016/j.ijheatmasstransfer.2014.01.075. http://www.sciencedirect.com/science/article/pii/S0017931014001161

Payri R, Viera JP, Pei Y, Som S (2015) Experimental and numerical study of lift-off length and ignition delay of a two-component diesel surrogate. Fuel 158:957–967. https://doi.org/10.1016/j.fuel.2014.11.072. http://www.sciencedirect.com/science/article/pii/S0016236114011764

Qi H, Guo X (2014) Transient fractional heat conduction with generalized Cattaneo model. Int J Heat Mass Transf 76:535–539. https://doi.org/10.1016/j.ijheatmasstransfer.2013.12.086. http://www.sciencedirect.com/science/article/pii/S0017931014003858

Qubeissi MA, Sazhin SS, Crua C, Turner J, Heikal MR (2015a) Modelling of biodiesel fuel droplet heating and evaporation: effects of fuel composition. Fuel 154:308–318. https://doi.org/10.1016/j.fuel.2015.03.051. http://www.sciencedirect.com/science/article/pii/S0016236115003476

Qubeissi MA, Sazhin SS, Turner J, Begg S, Crua C, Heikal M (2015b) Modelling of gasoline fuel droplets heating and evaporation. Fuel 159:373–384

Qubeissi MA, Sazhin SS, Elwardany AE (2017) Modelling of blended Diesel and biodiesel fuel droplet heating and evaporation. Fuel 187:349–355. https://doi.org/10.1016/j.fuel.2016.09.060. http://www.sciencedirect.com/science/article/pii/S0016236116309176

Rahman M, Saghir M (2014) Thermodiffusion or Soret effect: historical review. Int J Heat Mass Transf 73:693–705. https://doi.org/10.1016/j.ijheatmasstransfer.2014.02.057. http://www.sciencedirect.com/science/article/pii/S0017931014001859

Richter A, Nikrityuk PA (2012) Drag forces and heat transfer coefficients for spherical, cuboidal and ellipsoidal particles in cross flow at sub-critical Reynolds numbers. Int J Heat Mass Transf 55:1343–1354

Rybdylova O, Al Qubeissi M, Braun M, Crua C, Manin J, Pickett LM, de Sercey G, Sazhina EM, Sazhin SS, Heikal M (2016) A model for droplet heating and its implementation into ANSYS Fluent. Int Commun Heat Mass Transf 76:265–270

Sasmal C, Nirmalkar N (2016) Momentum and heat transfer characteristics from heated spheroids in water based nanofluids. Int J Heat Mass Transf 96:582–601. https://doi.org/10.1016/j.ijheatmasstransfer.2016.01.054. http://www.sciencedirect.com/science/article/pii/S0017931015307225

Sazhin SS, Al Qubeissi M, Kolodnytska R, Elwardany A, Nasiri R, Heikal M (2014a) Modelling of biodiesel fuel droplet heating and evaporation. Fuel 115:559–572

Sazhin SS, Al Qubeissi M, Xie J-F (2014b) Two approaches to modelling the heating of evaporating droplets. Int Commun Heat Mass Transf 57:353–356

Sazhin SS, Al Qubeissi M, Nasiri R, Gunko VM, Elwardany AE, Lemoine F, Grisch F, Heikal MR (2014c) A multi-dimensional quasi-discrete model for the analysis of Diesel fuel droplet heating and evaporation. Fuel 129:238–266

Sazhin SS, Shishkova IN, Al Qubeissi M (2016a) A self-consistent kinetic model for droplet heating and evaporation. Int J Heat Mass Transf 93:1206–1217

Sazhin SS, Gunko VM, Nasiri R (2016b) Quantum-chemical analysis of the processes at the surfaces of diesel fuel droplets. Fuel 165:405–412

Sazhin SS (2017) Modelling of fuel droplet heating and evaporation: recent results and unsolved problems. Fuel 196:69–101. https://doi.org/10.1016/j.fuel.2017.01.048. http://www.sciencedirect.com/science/article/pii/S0016236117300583

Sazhin SS (2006) Advanced models of fuel droplet heating and evaporation. Prog Energy Combust Sci 32:162–214

Sazhin SS (2014) Droplets and sprays. Springer, Heidelberg

Sazhin SS, Elwardany A, Sazhina EM, Heikal MR (2011) A quasi-discrete model for heating and evaporation of complex multicomponent hydrocarbon fuel droplets. Int J Heat Mass Transf 54:4325–4332

Singer SL (2016) Direct quadrature method of moments with delumping for modeling multicomponent droplet vaporization. Int J Heat Mass Transf 103:940–954. https://doi.org/10.1016/j.ijheatmasstransfer.2016.07.067. http://www.sciencedirect.com/science/article/pii/S0017931016309450

Sirignano WA (2014) Advances in droplet array combustion theory and modeling. Prog Energy Combust Sci 42:54–86. https://doi.org/10.1016/j.pecs.2014.01.002. http://www.sciencedirect.com/science/article/pii/S0360128514000033

Snegirev AY (2013) Transient temperature gradient in a single-component vaporizing droplet. Int J Heat Mass Transf 65:80–94

Snegirev A, Talalov VA, Tsoi AS, Sazhin SS, Crua C (2012) Advancement in turbulent spray modeling: the effect of internal temperature gradient in droplets. In: Proceedings of International Symposium on Advances in Computational Heat Transfer (16 July Bath, 2012, UK) CHT12-MP09

Sreenivasulu B, Srinivas B (2015) Mixed convection heat transfer from a spheroid to a Newtonian fluid. Int J Thermal Sci 87:1–18. https://doi.org/10.1016/j.ijthermalsci.2014.08.002. http://www.sciencedirect.com/science/article/pii/S1290072914002270

Sreenivasulu B, Srinivas B, Ramesh K (2014) Forced convection heat transfer from a spheroid to a power law fluid. Int J Heat Mass Transf 70:71–80. https://doi.org/10.1016/j.ijheatmasstransfer.2013.10.065. http://www.sciencedirect.com/science/article/pii/S0017931013009253

Srivastava S, Jaberi F (2017) Large eddy simulations of complex multicomponent diesel fuels in high temperature and pressure turbulent flows. Int J Heat Mass Transf 104:819–834. https://doi.org/10.1016/j.ijheatmasstransfer.2016.07.011. http://www.sciencedirect.com/science/article/pii/S0017931016304392

Strotos G, Malgarinos I, Nikolopoulos N, Gavaises M (2016) Predicting the evaporation rate of stationary droplets with the VOF methodology for a wide range of ambient temperature conditions. Int J Thermal Sci 109:253–262. https://doi.org/10.1016/j.ijthermalsci.2016.06.022. http://www.sciencedirect.com/science/article/pii/S1290072915302738

Subramanian VR, Diwakar VD, Tapriyal D (2005) Efficient macro-micro scale coupled modeling of batteries. J Elecrochem Soc 152(10):A2002–A2008

Toe ATOD, Padoin N, Ropelato K, Soares C (2015) Cross diffusion effects in the interfacial mass and heat transfer of multicomponent droplets. Int J Heat Mass Transf 85:830–840. https://doi.org/10.1016/j.ijheatmasstransfer.2015.01.131. http://www.sciencedirect.com/science/article/pii/S0017931015001490

Tonini S, Cossali GE (2012) An analytical model of liquid drop evaporation in gaseous environment. Int J Thermal Sci 57:45–53

Tonini S, Cossali GE (2013) An exact solution of the mass transport equations for spheroidal evaporating drops. Int J Heat Mass Transf 60:236–240

Tonini S, Cossali G (2014) An evaporation model for oscillating spheroidal drops. Int Commun Heat Mass Transf 51:18–24. https://doi.org/10.1016/j.icheatmasstransfer.2013.12.001. http://www.sciencedirect.com/science/article/pii/S0735193313002388

Tonini S, Cossali GE (2015) A novel formulation of multi-component drop evaporation models for spray applications. Int J Thermal Sci 89:245–253. https://doi.org/10.1016/j.ijthermalsci.2014.10.016. http://www.sciencedirect.com/science/article/pii/S1290072914003007

Tonini S, Cossali G (2016a) A multi-component drop evaporation model based on analytical solution of Stefan-Maxwell equations. Int J Heat Mass Transf 92:184–189. https://doi.org/10.1016/j.ijheatmasstransfer.2015.08.014. http://www.sciencedirect.com/science/article/pii/S0017931015301551

Tonini S, Cossali G (2016b) One-dimensional analytical approach to modelling evaporation and heating of deformed drops. Int J Heat Mass Transf 97:301–307. https://doi.org/10.1016/j.ijheatmasstransfer.2016.02.004. http://www.sciencedirect.com/science/article/pii/S0017931015317415

Wang C, Xu R, Song Y, Jiang P (2017) Study on water droplet flash evaporation in vacuum spray cooling. Int J Heat Mass Transf 112:279–288. https://doi.org/10.1016/j.ijheatmasstransfer.2017.04.111. http://www.sciencedirect.com/science/article/pii/S0017931017303319

Will J, Kruyt N, Venner C (2017) An experimental study of forced convective heat transfer from smooth, solid spheres. Int J Heat Mass Transf 109:1059–1067. https://doi.org/10.1016/j.ijheatmasstransfer.2017.02.018. http://www.sciencedirect.com/science/article/pii/S0017931016331763

Wittig K, Nikrityuk P, Richter A (2017) Drag coefficient and Nusselt number for porous particles under laminar flow conditions. Int J Heat Mass Transf 112:1005–1016. https://doi.org/10.1016/j.ijheatmasstransfer.2017.05.035. http://www.sciencedirect.com/science/article/pii/S0017931017305604

Xiao H, Zhao L, Li Z, Wei M, Guo G (2016) Development of a simplified model for droplet vaporization. Thermal Sci 20(1):337–345

Xie J-F, Sazhin SS, Cao B-Y (2011) Molecular dynamics study of the processes in the vicinity of the n-dodecane vapour/liquid interface. Phys Fluids 23(11):112104

Yadav AK, Chowdhury A, Srivastava A (2017) Interferometric investigation of methanol droplet combustion in varying oxygen environments under normal gravity. Int J Heat Mass Transf 111:871–883. https://doi.org/10.1016/j.ijheatmasstransfer.2017.03.125. http://www.sciencedirect.com/science/article/pii/S0017931016341424

Yi P, Long W, Feng L, Wang W, Liu C (2016a) An experimental and numerical study of the evaporation and pyrolysis characteristics of lubricating oil droplets in the natural gas engine conditions. Int J Heat Mass Transf 103:646–660. https://doi.org/10.1016/j.ijheatmasstransfer.2016.07.084. http://www.sciencedirect.com/science/article/pii/S0017931016309991

Yi P, Long W, Jia M, Tian J, Li B (2016b) Development of a quasi-dimensional vaporization model for multi-component fuels focusing on forced convection and high temperature conditions. Int J Heat Mass Transf 97:130–145. https://doi.org/10.1016/j.ijheatmasstransfer.2016.01.075. http://www.sciencedirect.com/science/article/pii/S0017931015308760

Yin C (2015) Modelling of heating and evaporation of n-heptane droplets: towards a generic model for fuel droplet/particle conversion. Fuel 141:64–73. https://doi.org/10.1016/j.fuel.2014.10.031. http://www.sciencedirect.com/science/article/pii/S0016236114010266

Zhang L, Shang X (2015) Analytical solution to non-Fourier heat conduction as a laser beam irradiating on local surface of a semi-infinite medium. Int J Heat Mass Transf 85:772–780. https://doi.org/10.1016/j.ijheatmasstransfer.2015.02.024. http://www.sciencedirect.com/science/article/pii/S0017931015001878

Zhukovsky K (2016a) Exact solution of Guyer-Krumhansl type heat equation by operational method. Int J Heat Mass Transf 96:132–144. https://doi.org/10.1016/j.ijheatmasstransfer.2016.01.005. http://www.sciencedirect.com/science/article/pii/S0017931015315611

Zhukovsky K (2016b) Violation of the maximum principle and negative solutions for pulse propagation in Guyer-Krumhansl model. Int J Heat Mass Transf 98:523–529. https://doi.org/10.1016/j.ijheatmasstransfer.2016.03.021. http://www.sciencedirect.com/science/article/pii/S0017931016301971

Zubkov VS, Cossali GE, Tonini S, Rybdylova O, Crua C, Heikal M, Sazhin SS (2017) Mathematical modelling of heating and evaporation of a spheroidal droplet. Int J Heat Mass Transf Part B 108:2181–2190. https://doi.org/10.1016/j.ijheatmasstransfer.2016.12.074. http://www.sciencedirect.com/science/article/pii/S001793101633438X

Chapter 4
Combustion of Multi-component Fuel Droplets

Srinibas Karmakar, S. K. Som and D. Chaitanya Kumar Rao

4.1 Introduction and Background

The evaporation and combustion of liquid fuel droplets play a significant role in various fields of science and technology, such as power and process industries, chemistry, medicine, and environmental processes. The research work in the area of droplet evaporation is going on over the last few decades. A phenomenal progress has already taken place in the theory of evaporation and combustion of liquid fuel droplets and sprays. The theoretical progress includes the development of physical models and computer codes capable of solving model equations. The experimental advances, on the other hand, involve the development of improved instruments for calibrating the model predictions and for investigations of several complex phenomena occurring in the process of droplet combustion. A large number of investigations carried out in the field have contributed well to the scientific understanding of the subject. A liquid droplet (in the context of evaporation/combustion study) can be classified into three categories based on its constituents. The first one is a pure component droplet which comprises a single component. The second one is a multi-component droplet which consists of two or more liquids (miscible or immiscible), and the third one is a nanofuel droplet where solid nanoparticles are suspended in the liquid droplet. The understanding of physical processes and their consequences in the combustion of pure component fuel droplet

S. Karmakar (✉) · D. C. K. Rao
Department of Aerospace Engineering, Indian Institute of Technology,
Kharagpur 721302, West Bengal, India
e-mail: skarmakar@aero.iitkgp.ernet.in

S. K. Som
Department of Mechanical Engineering, Indian Institute of Technology,
Kharagpur 721302, West Bengal, India

S. Basu et al. (eds.), *Droplets and Sprays*, Energy, Environment, and Sustainability,
https://doi.org/10.1007/978-981-10-7449-3_4

is well established. Therefore, in the present chapter, more attention has been paid to the discussion on different underlying features during combustion of multi-component and nanofuel droplets. A brief review of numerical and experimental investigations carried out in these fields is included.

The characteristics of droplet combustion are controlled by the physical processes associated with heating and evaporation of the liquid droplet. Fundamentally, the droplet evaporation process constitutes two events occurring in tandem. One is the detachment of liquid molecules from droplet surface, and the other is the diffusion of vapor molecules into ambient gas phase. The classical approach of predicting droplet evaporation rate is based on a purely diffusion controlled model where the details of molecular detachment at droplet surface are ignored, and the droplet surface is assumed to be in thermodynamic equilibrium with the adjacent vapor at saturated state. Accordingly, the evaporation rate is predicted to be the same as the rate of mass diffusion from droplet surface to ambient gas phase. The model is also known as hydrodynamic model. On the other hand, some of the recently used predictive tools are based on kinetic model and molecular dynamics model which take into account of non-equilibrium state at droplet surface along with the details of molecular detachment at the surface either by incorporating an empirical evaporation coefficient or by the analysis of molecular dynamics at liquid–gas interface.

The physical processes in a pure component droplet combustion are relatively simple as compared to those occurring in case of combustion of multi-component fuel droplet and nanofuel droplet. A review of pioneering and classical works in the field has been well documented (Chigier 1976, 1977, 1983; Faeth 1977, 1983, 1987; Law 1982; Sirignano 1983, 1988, 1990). A comprehensive review of the latest developments on droplet heating and evaporation is also available in the form of a self-sufficient text (Sazhin 2006, 2017). Attention in earlier investigations has been paid mostly on the various aspects of droplet heating and transport processes, predictive models for transport coefficients, and rate of vaporization from liquid droplets. The focus was made on the influences of gas-phase flow field and the induced circulatory flow in the liquid phase on droplet evaporation rate, and shape and size of flame surrounding the droplet.

Most of the transportation fuels are actually multi-component in nature; however, many of them have been modeled as the closely resembled pure fuels for simplified analysis. Although due to the increasing demand for combustion efficiency and stringent emission requirements, multi-component commercial fuels have become very important. For example, blends of gasoline and ethanol have been used widely in many countries. The widely varying physical and chemical properties of the liquid constituents in the multi-component fuel blends influence the performance of the combustion process in the engines. To understand this multi-component effect on the combustion process of droplet, Law (1982) proposed to consider three factors: (i) relative concentrations and volatilities of the constituents, (ii) miscibility, and (iii) intensity of internal circulation. The phenomena which are majorly observed in evaporation/combustion of multi-component liquid fuels are puffing or minor ejection, micro-explosion or major ejection, and abrupt explosion. Depending on the properties of the constituents, only puffing may be observed in certain combination of fuel blends whereas both puffing and micro-explosion may be witnessed in other combinations. These puffing and micro-explosion can lead to droplet disintegration which is termed as secondary atomization. This secondary atomization is often considered to be advantageous from the combustion point view; however, the controlling parameters leading to puffing/micro-explosion and hence secondary atomization are not well understood. This section of the chapter tries to recollect the existing propositions and make a logical sequence for better understating of the underlying mechanism.

Nanofuel or nanofluid fuel is a new class of fuel in which energetic nanoparticles are dispersed in conventional hydrocarbon fuel. Nanofuel is a broad class of fuel where base fuel can be either liquid fuel or solid fuel. However, in the present context, the discussion of nanofuel is restricted by considering the liquid fuel as a base fuel. It is important to note here that nanofuels generally contain a few nanoscaled materials which are fundamentally different from the slurry fuels that were studied much earlier (Roy Choudhury 1992). Recent advancement in nanotechnology has generated renewed interest in processing and utilization of nanofuels. Nanoparticles being smaller than micron-sized particles can have some positive attributes such as higher specific surface area, higher reactivity, and potential to store energy at the surface. Due to these characteristics, nanoparticles may often be useful in shortening ignition delays, reduction in burn times, and better combustion than their micron-size counter parts (Gan and Qiao 2012). Nanofuels are broadly multi-component fuels; however, the multi-component feature is even more complex due to the presence of solid particles as one of the constituents. This section of the chapter mainly emphasizes the evaporation/combustion of nanofuels barring the studies involved in investigation of combustion characteristics of nanofuels in engine. To develop a fundamental understanding of evaporation/combustion of nanofuels, it is necessary to understand the following: characteristics of nanoparticles (size, morphology, energy content, etc.), preparation of nanofuel, effect of surfactant or surface functionalization on the stability of nanofuel, physicochemical properties of nanofuels, how these properties affect the evaporation/combustion characteristics.

4.2 Numerical Studies on Droplet Evaporation and Combustion

A liquid droplet undergoes the process of heating and evaporation simultaneously when it is exposed to a high-temperature gas. The heating of droplet, in turn, enhances the rate of evaporation characterized by the detachment of liquid molecules from the droplet surface and the subsequent diffusion of vapor molecules into surrounding gas phase.

In a classical hydrodynamic model, the phenomenon of molecular detachment from liquid to gas phase at the surface is ignored. The evaporation rate is determined by the rate of diffusion of vapor molecules in gas phase in consideration of droplet surface to be in thermodynamic equilibrium. The phenomenon of liquid detachment at the surface is taken care of by kinetic model or molecular dynamics (MD) model. In case of reactive liquid component(s) of the droplet, combustion reaction takes place in gas phase with the appearance of a flame front at the surface of stoichiometric composition. Most of the modeling investigations consider the combustion reactions to be instantaneous and hence the flame is infinitely thin. The liquid droplet is finally heated up by the flame front. Therefore, the evaporation rate and hence the burning rate of a liquid fuel droplet depend on (i) flame temperature, (ii) shape and size of the flame front, and (iii) location relative to the droplet surface. The gas-phase flow field has a profound influence on the above-said characteristic parameters influencing the droplet burning rate.

The flow in gas phase around a liquid droplet may arise due to (i) forced flow by external agency and (ii) buoyancy-induced flow because of temperature and species concentration gradients in the gas phase. However, for small droplets, the buoyancy-induced flow will be less pronounced compared to the forced flow. The nature of gas flow past a vaporizing droplet and of induced flow in the liquid phase within the droplet has been investigated by several researchers (Harper and Moore 1968; Prakash and Sirignano 1978, 1980; Tong and Sirignano 1982; Sadhal 1983; Sundararajan and Ayyaswamy 1984; Rangel and Fernandez-Pello 1984; Gogos et al. 1986; Dwyer and Sanders 1988; Dwyer 1988, 1989; Abramzon and Sirignano 1989; Ayyaswamy et al. 1990; Dash et al. 1991; Dash and Som 1991a, b; Curtis and Farrell 1992; Chiang et al. 1992; Michaelides et al. 1992; Biswal et al. 1999) during last few decades. The major assumptions in almost all flow models are, namely (i) the droplet remains spherical because of low values of Weber number encountered in situations of interest and (ii) the gas-phase flow field in absence of any inherent fluctuation in flow parameters is quasi-steady since characteristic time for development of gas-phase flow field is two to three orders of magnitude less than the droplet lifetime. However, the assumption fails near or above the critical pressure of gas phase. The interesting features in the flow of gas over a vaporizing droplet are, namely (i) the large non-uniform outward radial flow at droplet surface due to evaporation which is known as Stefan flow, (ii) separation near wake region, and (iii) induction of circulatory flow in the droplet phase. The typical values of droplet Reynolds number encountered in practice usually conform to a laminar flow of gas around the droplet. However, the expansion of gaseous combustion products in near vicinity of the droplet causes turbulence in local gas phase over and above the inherent free stream turbulence.

The influence of buoyancy in gas-phase flow field around large drops depends upon the relative magnitudes of inertia and buoyancy forces. The buoyancy force arises due to both temperature and species concentration gradients in gas phase. A typical Froude number, based on drop diameter and free stream velocity relative to droplet, is usually considered as the criterion for the influence of buoyancy-driven flow over forced flow. The droplet-phase flow field depicts a

Hills-type spherical vortex core surrounded by a viscous boundary layer and internal wake. In forced flow, a typical envelope flame surrounds the droplet which turns to a wake flame at higher free stream velocity. In buoyancy-driven flow, an envelope flame appears for aiding flows, while a hat-shaped flame appears in case of opposing flows.

The rates of heat transfer and mass transfer for vaporizing droplet are usually provided in terms of Nusselt number (Nu) and Sherwood number (Sh) correlations, respectively. A large number of such empirical and numerical correlations are available in the literature. These are usually expressed in a general form as

$$Nu = \frac{(2 + K\mathrm{Re}^{1/2}\,\mathrm{Pr}^{1/3})}{(1+B)^n} \tag{4.1}$$

$$Sh = \frac{(2 + KSc^{1/2}\,\mathrm{Pr}^{1/3})}{(1+B)^n} \tag{4.2}$$

where Re, Pr, Sc, and B are, respectively, the Reynolds number, Prandtl number, Schmidt number, and transfer number.

The transfer number B is defined as

$$B = \frac{Cp_g(T_g - T_s)}{\Delta H_V} \tag{4.3}$$

where, Cp_g, T_g, T_s, and ΔH_V are, respectively, the specific heat of gas phase, gas-phase free stream temperature, surface temperature of droplet, and the enthalpy of vaporization of liquid at droplet surface temperature. The different relations for various values of K and n have been provided. The accuracy of a given relation depends upon the range of $\mathrm{Pe_h}$ (heat transfer Peclet number = $\mathrm{Re} \cdot \mathrm{Pr}$) and $\mathrm{Pe_m}$ (mass transfer Peclet number = $\mathrm{Re} \cdot \mathrm{Sc}$) for which the relationship is prescribed. The most widely used values of the parameters supported by both experiments and theory are K = 0.60, n = 1.

The above relations given by Eqs. (4.1) and (4.2) can be used for both steady-state and unsteady-state (during droplet heat up) droplet evaporation. The use of the relations (Eqs. 4.1 and 4.2) during an unsteady state of evaporation requires a modified value of enthalpy of vaporization for the liquid in consideration of heat flux to the drop (Dash et al. 1991). A widely popular empirical equation for mass transfer in this regard is due to Renksizbulut et al. (1991).

$$Sh(1 + B_m)^{0.7} = 2 + 0.87\mathrm{Re}_m^{1/2} Sc^{1/3} \tag{4.4}$$

The validity of the relation lies in a wide range of $10 < \mathrm{Re_m} < 2000$. The subscript m represents that Reynolds number and transfer number are based on mean values of physical properties.

A multi-component droplet consists of more than one liquid component (miscible or immiscible). The vaporization of different components takes place from the droplet surface at different rates depending upon the respective vapor pressure at surface temperature. Therefore, the more volatile liquid evaporates faster thus creating concentration gradient in the liquid phase leading to a liquid-phase mass diffusion within the droplet.

Most of the studies based on hydrodynamic (purely diffusion controlled) model ignore the phenomena like nucleation, bubble growth, and subsequent puffing or micro-explosion while modeling the droplet heating and evaporation. The general framework of the model is usually based on the solution of conservation equations of heat, mass, and momentum in both gas and liquid phases through coupling at the interface with appropriate boundary conditions. The usual assumptions comprise the sphericity of droplet, quasi-steady fields of flow, temperature and concentration in gas phase, negligible Dufour and Soret effect, and the interface is at thermodynamic equilibrium. Simplified analyses, in the absence of any convection effect (for a small droplet in a stagnant medium), assume spherical symmetry. However, in case of buoyancy-induced flow or forced flow in gas phase, spherical–symmetric assumption fails, and one has to take into consideration both radial and azimuthal space coordinates as independent variables. A general formulation for a bicomponent liquid droplet with spherical symmetry is reproduced below for a ready reference.

Gas phase:

Continuity:

$$v_r = v_S \frac{R^2}{r^2} \tag{4.5}$$

where, v_S is the uniform Stefan flow velocity at droplet surface, and R is the instantaneous droplet radius.

Energy conservation:

$$v_r \frac{dT}{dr} = \alpha^v \frac{1}{r^2} \frac{d}{dr}\left(r^2 \frac{dT}{dr}\right) \tag{4.6}$$

where, α^v is the thermal diffusivity of gas phase, and T is the temperature.

Species conservation:

$$v_r \frac{dC_1^v}{dr} = D_{1a} \frac{1}{r^2} \frac{d}{dr}\left(r^2 \frac{dC_1^v}{dr}\right) \tag{4.7}$$

$$v_r \frac{dC_2^v}{dr} = D_{2a} \frac{1}{r^2} \frac{d}{dr}\left(r^2 \frac{dC_2^v}{dr}\right) \tag{4.8}$$

where C is the mass fraction. Subscripts represent the species while the superscript v represents the vapor phase. The term D_{1a} represents the diffusion coefficient of vapor component 1 in air in presence of vapor component 2, while D_{2a} represents the diffusion coefficient of vapor component 2 in air in presence of vapor component 1.

In case of a spatially constant thermal diffusivity (α^v) and mass diffusivities (D_{1a}, D_{2a}), Eqs. (4.6), (4.7), and (4.8) can be solved analytically with appropriate boundary conditions as

$$\frac{T - T_S}{T_\infty - T_S} = \frac{\exp\left(-\frac{v_S R^2}{r\alpha^v}\right) - \exp\left(-\frac{v_S R}{\alpha^v}\right)}{1 - \exp\left(-\frac{v_S R}{\alpha^v}\right)} \tag{4.9}$$

$$\frac{C_1^v - C_{1S}^v}{C_{1\infty}^v - C_{1S}^v} = \frac{\exp\left(-\frac{v_S R^2}{rD_{1a}}\right) - \exp\left(-\frac{v_S R}{D_{1a}}\right)}{1 - \exp\left(-\frac{v_S R}{D_{1a}}\right)} \tag{4.10}$$

$$\frac{C_2^v - C_{2S}^v}{C_{2\infty}^v - C_{2S}^v} = \frac{\exp\left(-\frac{v_S R^2}{rD_{2a}}\right) - \exp\left(-\frac{v_S R}{D_{2a}}\right)}{1 - \exp\left(-\frac{v_S R}{D_{2a}}\right)} \tag{4.11}$$

The second subscripts ∞ and S represent the free stream and droplet surface, respectively.

The diffusion coefficient D_{1a} and D_{2a} are evaluated in terms of their individual binary diffusion coefficients from a generic relation (Gavhane et al. 2016).

$$D = \mathfrak{D}_{ij}\left[1 + \frac{C_k\{(M_k/M_j)\mathfrak{D}_{ik} - \mathfrak{D}_{ij}\}}{C_i\mathfrak{D}_{jk} + C_j\mathfrak{D}_{ik} + C_k\mathfrak{D}_{ij}}\right] \tag{4.12}$$

where $\mathfrak{D}_{ij}$ is the binary diffusion coefficient of i in j in a two-component system of i in j, and D_{ij} is the diffusion coefficient of i in j in presence of a third component k.

Droplet phase:

The energy balance is written in consideration of the droplet to be of uniform temperature during the process of its heating up as

$$\frac{4}{3}\pi R^3 \rho^l C_p^l \frac{dT^l}{dt} = k^v 4\pi R^2 \left(\frac{dT}{dr}\right)_{r=R} - \dot{m}_{1e} 4\pi R^2 h_{fg1} - \dot{m}_{2e} 4\pi R^2 h_{fg2} \tag{4.13}$$

Species continuity equation:

$$\frac{\partial C_1^l}{\partial t} = D_{12}\frac{1}{r^2}\frac{\partial}{\partial r}\left(r^2 \frac{\partial C_1^l}{\partial r}\right) \tag{4.14}$$

The most important feature in this context is the determination of liquid mass fraction of a species at droplet surface which provides one of the interface boundary conditions for the solution of Eq. (4.14). The instantaneous mass fraction of a species in liquid phase at droplet surface is found out from the consideration of bulk mass fraction of the species in liquid phase (Bhattacharya et al. 1996).

The vapor mass fraction of a component at droplet surface is determined in consideration of thermodynamic equilibrium at droplet surface and the mixture of air and liquid vapors to be an ideal gas as

$$C_{i,s}^{v} = (p_{vi}M_i) / \sum_{i=1}^{n} p_{vi}M_i \tag{4.15}$$

In consideration of liquid droplet to be an ideal mixture, the vapor pressure (the partial pressure of vapor) of a component at droplet surface is found out from Raoult's law as

$$p_{vi} = p_{vi}^{\circ} X_{i,S}^{l} \tag{4.16}$$

where p_{vi}° is the vapor pressure of species i as a pure component at the same temperature and $X_{i,S}^{l}$ is the mole fraction of component i in its liquid phase at droplet surface.

Mass evaporation rate is predicted to be the same as the rate of mass transfer of the liquid vapor from the droplet surface to its incipient gas phase as

$$\dot{m}_{ie} = -\rho^{v} D_{ia} \left(\frac{dC_i^{v}}{dr} \right)_{r=R} + \rho^{v} v_S C_{iS}^{v} \tag{4.17}$$

$$\dot{m} = \sum_{i=1}^{n} \dot{m}_{ie} = \rho^{v} v_S \tag{4.18}$$

The Stefan flow velocity v_S which can be written in consideration of droplet surface being impermeable to non-evaporating species, as

$$v_S = \frac{-\sum_{i=1}^{n} D_{ia} \left(\frac{dC_{ia}}{dr} \right)_{r=R}}{1 - \sum_{i=1}^{n} C_{is}} \tag{4.19}$$

The modeling investigations reveal three important features as follows:

1. The more volatile component evaporates at a much faster rate, yet remains inside the core and tends to diffuse toward the surface because of concentration gradients created by the prior vaporization. The concentration of more volatile component at the surface reduces to a negligible value within the period of droplet transient heating.
2. The steady-state droplet temperature exceeds the boiling of lighter component at the existing state in droplet core.

3. The droplet temperature and evaporation histories are controlled by the heavier component while the ignition characteristics are determined by the lighter component.

In theoretical modeling, there exist two limiting cases, (a) the rapid regression or zero diffusivity limit and (b) the rapid mixing (uniform concentration) or infinite-diffusivity limit. In a low-temperature environment, the gas-phase diffusion is small and the droplet lifetime is long so that the liquid-phase diffusion can be considered to be a fast one and concentrations in the liquid may be assumed to be uniform, and hence, the limit (b) can be applied. A condition of sequential vaporization of the components depending upon their volatilities can take place similar to distillation. The volatility is usually characterized by normal boiling point of the component. In combustion systems, droplet vaporization takes place in a high-temperature ambient where gas-phase diffusion is very high, while the liquid-phase diffusion is relatively quite low. Therefore, limit (b) does not hold good while limit (a) may hold good in some cases.

The presence of liquid vapors in free stream increases the droplet lifetime. The increase is more profound due to the presence of less volatile liquid vapor alone as compared to that due to the presence of more volatile liquid vapor alone (Gavhane et al. 2016). This influence is more prominent at lower ambient temperature. The widely popular empirical relation for single component droplet due to Renksizbulut (Eq. 4.4) can be used in case of multi-component droplet with reasonably fair accuracy provided, logical and appropriate modifications of mass transfer number B_m and mass transfer coefficient are made in consideration of all constituent vapor concentrations at both droplet surface and free stream along with mass fluxes of the component vapors. The interested readers may see the reference (Gavhane et al. 2016).

In realistic situations, the droplet may consist of a vast number of components. The above-stated approach may not be satisfactory under the situations. An alternative approach based on continuous thermodynamics (Tamim and Hallett 1995; Zhu and Reitz 2001) has been developed for the purpose. This refers to a well-mixed droplet model (infinite-diffusivity limit) and is based on the introduction of a distribution function. The type of function to be chosen depends upon the nature of fuel droplet. Sometimes a simple function approximates well for many practically important fuels. The main distinguishing and attractive feature of this approach of modeling multi-component droplet heating and evaporation is that few parameters defining the distribution function characterize the composition of fuel instead of mole fractions of a large number of individual species. The readers may see the references (Tamim and Hallett 1995; Zhu and Reitz 2001) for detailed analyses of the model stated above.

The models described so far was implicitly based on the assumption that the gas–liquid interface is at thermodynamic equilibrium, and both the liquid and gas phases can be treated as continuum. This is valid only at moderately low ambient gas pressure and for droplets which are not too small. Under this situation, the velocity distribution of molecules and its random motion at the interface do not influence the number of molecules per unit volume and their average properties. Hence, the physical properties of both the phases remain uninfluenced by molecular motion and their distribution. However, such consideration becomes questionable at high ambient pressure and when droplet diameter is too small. Under such circumstances, the physical properties of both gas and liquid phases depend not only on the average molecular velocity but on its statistical distribution (expressed by appropriate distribution functions). The kinetic model and molecular dynamic simulation have been brought forward under such situations.

A parameter known as Knudsen number, K_n (defined by the ratio of the molecular mean free path to the characteristic length of the system) is considered as the criterion for deviation from the continuum. In case of droplet evaporation, the characteristic length scale is the droplet diameter. The assumption of continuum holds good when $K_n \leq 0.01$. On the other extreme, when $K_n > 10$, flow becomes free molecular in nature and the assumption of continuum fails. The range of K_n spanning from 0.01 to 0.1 is known as slip regime, where no slip boundary condition becomes invalid although continuum conservation equations can still be used for describing the flow fields. In the context of droplet evaporation, the gas phase can be regarded as the continuum at several mean free path away from the droplet surface. The range $0.1 < K_n < 10$ refers to transition regime.

The kinetic models have been employed to study the liquid droplet evaporation since the pioneering works of Hertz (1882), Knudsen (1915), Bond and Struchtrup (2004). The model is centered on the fact that there exists a Knudsen layer at the droplet surface which takes care of kinetic effect through molecular collision considering the departure from continuum and thermodynamic equilibrium in the particular region. The gas phase surrounding the Knudsen layer is referred as hydrodynamic region. The heat transfer in the Knudsen layer is assumed to be negligible. In addition, it is also assumed that pure gas-phase diffusion removes the vapor at the interface of Knudsen layer and hydrodynamic region. It has been recognized that the thickness of 10 mean free path for the Knudsen layer matched well with the hydrodynamic solution at the interface (Sazhin and Shishkova 2009). Considering both incoming and outgoing molecular fluxes at the droplet surface to be Maxwellian, the evaporation mass flux (Kryukov et al. 2004; Labuntsov and Kryukov 1979) can be written as

$$\dot{m}_{e,kine} = 0.6\sqrt{2R_gT_s}\left[\left(1 - 2\sqrt{\pi}\frac{\dot{m}_{e,kine}}{\rho_s\sqrt{2R_gT_s}}\frac{1-\sigma}{\sigma}\right)\rho_s - \rho_{Rd}\right] \times \sqrt{\rho_{Rd}\Big/\left(1 - 2\sqrt{\pi}\frac{\dot{m}_{e,kine}}{\rho_s\sqrt{2R_gT_s}}\frac{1-\sigma}{\sigma}\right)\rho_s} \tag{4.20}$$

where ρ_{Rd} is the density of water vapor, T_s is the temperature of interface, and σ is the evaporation or condensation coefficient.

The density of water vapor is calculated by equating the kinetic mass flux with diffusive mass flux at the outer boundary of Knudsen layer. Kryukov et al. (2004)

performed a comparative study on evaporation of diesel fuel droplets at high pressure (30 atm) using diffusion controlled hydrodynamic approach and kinetic approach. Considering the evaporation of fuel droplet into its vapor, it was concluded that kinetic model predicts relatively longer evaporation time as well as higher droplet temperature compared to the predictions from purely diffusion controlled hydrodynamic model. It was suggested that the kinetic effects should be considered for modeling the evaporation of diesel fuel droplets under realistic conditions. However, their predictions from both the models were not compared with empirical findings as a standard of calibration. Sazhin et al. (2007) studied the evaporation of droplets into a background gas using the kinetic model and made similar predictions as stated by Kryukov et al. (2004). Further details of advanced models corresponding to fuel droplet evaporation are available in a review paper by Sazhin (2006).

A comparison of purely diffusion controlled, kinetic and empirical models suggests that kinetic model usually predicts a longer lifetime when compared to diffusion controlled model at high ambient vapor concentration in free stream (Pati et al. 2011). This deviation becomes more prominent for smaller diameter droplets and higher ambient temperature. In addition, the predictions from kinetic model become close to empirical ones for $\sigma = 0.5$. The kinetic model becomes the rate predictive model during the evaporation of droplet in a vapor-rich ambient conditions at high temperature. Hence, for accurately determining the droplet lifetime, kinetic model is superior to the purely diffusion controlled model.

The reliability of kinetic model depends on the accuracy of the value of evaporation or condensation coefficient, σ used in Eq. (4.20). The values of σ are usually obtained from experiments and are not known for many fuels. The values for water are used for this purpose, and the value of σ for a particular situation is being tuned by matching the model result with experiment. Therefore, the accurate and ultimate approach should be the determination of σ based on the analysis of molecular dynamics on the liquid–gas interface. This leads to a new paradigm of modeling droplet evaporation based on molecular dynamics (MD) simulation.

The interaction between the molecules are usually taken care of by Lennard-Jones 12-6 potential (Consolini et al. 2003; Yang and Pan 2005; Xu et al. 2004; Anisimov et al. 1999; Walther and Koumoutsakos 2001), and the path of particles is determined by Newton's second law where the interactive forces between the molecules are determined from the analytical expression of Lennard-Jones potential. A major limitation of the model is the consideration of a large number of molecules which is computation intensive. The actual number of molecules to be analyzed depends upon available computing power. A relatively detailed discussion on both kinetic and MD models is available in the literature (Sazhin et al. 2007).

There are relatively much less number of modeling studies pertaining to droplet heating and evaporation that take care of nucleation, bubble growth, and subsequent droplet breakup resulting in puffing and micro-explosion in case of a multi-component droplet comprising constituents with widely varying volatility. The nucleation is usually homogeneous in nature. For practical fuels, heterogeneous nucleation is possible because of the existence of dissolved gases and impurities. However, it is difficult to quantify the associated impurities. Moreover, the temperature at which heterogeneous nucleation happens is bounded by the superheat limit and the saturation temperature (Blander and Katz 1975; Gerum et al. 1979).

In almost all the modeling works, the bubble generation is described by a homogeneous nucleation theory, and its subsequent growth is described by a modified Rayleigh equation (Robinson and Judd 2004) as

$$\rho_l\left[R_b\frac{d^2R_b}{dt^2}+1.5\left(\frac{dR_b}{dt}\right)^2\right]=P_g-P_l-\frac{2\sigma}{R_b} \tag{4.21}$$

where R_b is the bubble radius, ρ_l is the density of liquid droplet, and σ is the surface tension. P_g is the gas pressure in the bubble, and P_l is the liquid droplet pressure.

Some studies add viscous terms into the Rayleigh equation to account for the viscous effect during bubble expansion as

$$\rho_l\left[R_i\frac{d^2R_i}{dt^2}+1.5\left(\frac{dR_i}{dt}\right)^2\right]=P_{gi}-P_l-\frac{2\sigma}{R_i}-\frac{4\mu}{R_i}\frac{dR_i}{dt}-\frac{4\kappa}{R_i^2}\frac{dR_i}{dt} \tag{4.22}$$

where μ and κ are liquid viscosity and surface viscosity, respectively.

However, Robinson and Judd (2004) showed that viscous effect is negligible compared to the other terms. The onset of micro-explosion depends on the ambient pressure and temperature, initial droplet size, the composition and types of liquid fuel constituting the droplet. Zeng and Lee (2007) studied the breakup process by the analysis of linear instability. They reported that optimum composition and high ambient pressure favor micro-explosion. However, extremely high pressure suppresses micro-explosion since the volatility difference between the constituents decreases with pressure.

The influence of ambient temperature on onset of micro-explosion may be contrasting in nature. This depends on the race between the enhancement of droplet heat up and its evaporation with temperature. For constituent liquids with lower volatility, the heat up process may dominate and thus the normalized onset radius (NOR: the square of the ratio of droplet radius at the onset of micro-explosion to its initial radius) increases with ambient temperature. But it is opposite for liquid droplets with higher volatility where the evaporation becomes stronger with ambient temperature and dominate over the droplet heating up resulting in lower NOR. An increase in ambient temperature from 1000 K to 2300 K, the NOR is decreased by about 40% for a 20% butanol–80% biodiesel droplet (Shen et al. 2010). Micro-explosion may not take place below a threshold value of ambient temperature. No micro-explosion is observed during the lifetime of a 20% butanol–80% biodiesel droplet below an ambient temperature of 1000 K (Shen et al. 2010). This is evident due to the fact that at low ambient temperature, the heat transfer rate will not be rapid enough such that the droplet reaches the superheat limit.

Micro-explosion of an oxygenate diesel blend is possible under typical diesel engine environments, and the effect is stronger for droplets with larger sizes or a near 50/50 composition. Shen et al. (2010) modeled the breakup process from a surface energy approach by imposing minimal surface energy on the system. Sauter mean radius (SMR) was determined based on the simulated results of bubble growth, droplet characteristics at the onset of micro-explosion, and predictions from the breakup model along with a generic probability distribution function. They also reported the existence of an optimal droplet size for the onset of micro-explosion, and it was concluded that micro-explosion is possible in biofuel and diesel blends under engine operating conditions (Shen et al. 2010). The Sauter mean radius of secondary droplets is estimated to be 30–40% of initial droplet radius for droplets with initial radius less than 50 μm. The interested readers may see the cited references for more detailed discussion with mathematical analyses in this area.

There exists an optimal value of initial droplet size (for given composition and nature of constituent fuels) and an optimal value of initial droplet composition (for given initial size) for the maximum value of NOR (early nucleation). This can also be attributed to the physical race between the droplet heat up and its evaporation with droplet size and liquid composition. The optimal value of initial droplet radius for a 50% ethanol–50% tetradecane is 25 μm at ambient pressure and temperature of 40 atm and 825 K, respectively, as reported in the literature (Shen et al. 2010). Micro-explosion is not observed during the droplet lifetime with an initial radius less than 10 μm (Shen et al. 2010). The optimal composition for ethanol–tetradecane droplet is approximately 50-50, while for an ethanol–biodiesel droplet, the optimal composition is 30% ethanol and 70% biodiesel.

In case of emulsions, the growth of a vapor bubble, originated at the oil/water interface, results in puffing. The breakup of vapor bubble as well as the water sub-droplet dynamics subsequent to puffing determines the oil droplet breakup outcome (Shinjo et al. 2014). The physics of puffing and micro-explosion phenomena can be investigated using high-fidelity interface-capturing simulation (Shinjo et al. 2014). For water-in-oil emulsions, size and location of the water sub-droplet control the dynamics of parent droplet after the occurrence of puffing event. Figure 4.1 represents the puffing and after-puffing dynamics of a water-in-oil emulsion droplet for a three-dimensional case. As seen in the figure, puffing creates a hole in the parent droplet resulting in the expulsion of vapor. Since the liquid wall is thin near the surface of parent droplet, the bubble pushes this particular side strongly toward outside. The subsequent rupture of liquid wall results in the expulsion of water vapor along with liquid fragments (Fig. 4.1(b): $t = 2.5$ μs). In addition, at the edge of liquid–gas interface near the water/oil boundary, the edge becomes unstable due to Landau–Darrieus (LD) instability. A small water sub-droplet results in rapid detachment and the subsequent oil droplet breakup is limited, whereas when a sub-droplet is large and initially located near the center of parent droplet, the droplet breakup is more intense (Shinjo et al. 2014). When the micro-explosion is initiated by the simultaneous growth and breakup of multiple individual vapor bubbles, each breakup is confined and initially independent;

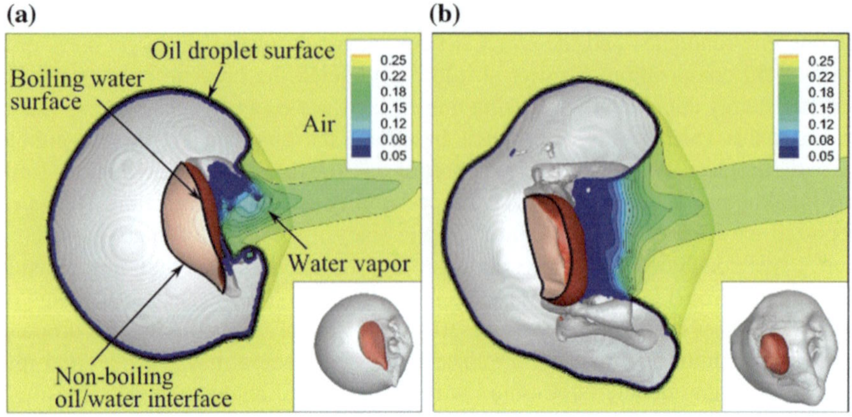

Fig. 4.1 Puffing and post-puffing dynamics of a water-in-oil emulsion droplet with 4.3% water sub-droplet volume fraction. **a** $t = 1.5$ μs and **b** $t = 2.5$ μs. The gray surface indicates the oil droplet surface. The darker gray and lighter gray (red and pink, respectively) surfaces show the water sub-droplet's boiling surface and inert non-boiling interface attached to the liquid oil, respectively. The contours on the plane that is perpendicular to the z-axis and contains the CG of the oil droplet show the oxygen mass fraction. The darkest gray and lightest gray (blue and yellow, respectively) correspond to water vapor and air, respectively (Shinjo et al. 2014)

however, their mutual interactions are probable at a later stage, which can increase the degree of breakup.

Figure 4.2 shows a 3D case with multiple explosions. Initially, each breakup proceeds independently as local puffing. Subsequently, mutual interactions between individual puffing starts to occur, and the entire droplet breaks apart, whereas a single puffing event does not have the potential to break apart the entire parent oil droplet (Shinjo et al. 2014).

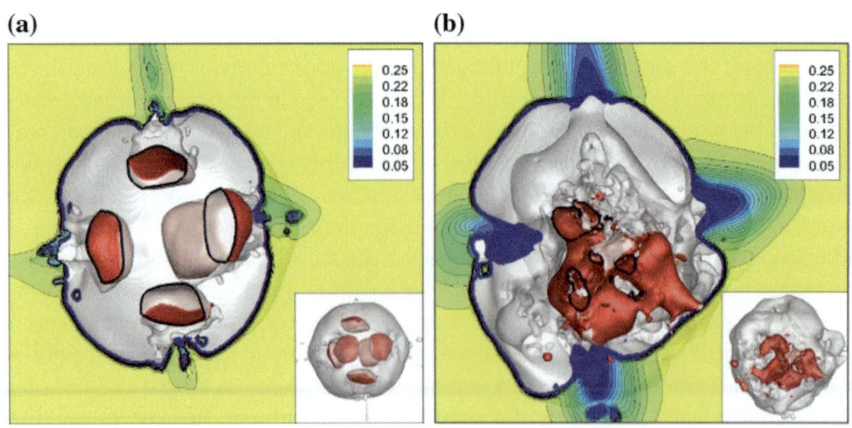

Fig. 4.2 Micro-explosion induced by multiple puffing. **a** $t = 1.0$ μs and **b** $t = 3.0$ μs (Shinjo et al. 2014)

Although our current understanding of the physics of puffing/micro-explosion is limited, improvement in the micro-explosion models has been of great interest. Further numerical investigations on the occurrence of instabilities in burning miscible and emulsion droplets can improve our understanding of the disruptive phenomena. Further, the underlying physics can be helpful in regulating puffing/micro-explosion for enhancing the spray combustion in realistic conditions.

4.3 Experimental Studies on Multi-component Droplet Combustion

There exists a substantial interest in the utilization of multi-component or emulsion fuels such as alcohol/oil mixtures and water/oil emulsions. The vast difference in the physical and chemical properties of the components of these fuels leads to combustion characteristics that are rather different compared to pure fuels. The temporal variations of the relative gasification rates, concentration, and spatiotemporal distribution of the particular fuel components within the multi-component droplet are primarily controlled by the following three factors: (a) the relative concentrations and volatilities of the liquid components, (b) the miscibility of the liquid components, and (c) the intensity of motion (internal circulation) within the droplet (Law 1982). The miscibility of the components affects the phase change, and thus the surface vapor pressure. Internal circulation, on the other hand, influences the rate with which the liquid constituents are transported to the surface for the gasification to take place (Law 1982). It is important to understand that a volatile liquid cannot undergo gasification unless it is exposed at the droplet surface, and this exposure can be achieved through either surface regression or diffusion and internal circulation. Compared to liquid-phase thermal diffusion and surface regression, the liquid-phase mass diffusion is an extremely slow process (Law 1982). Hence, it is reasonable to expect that the liquid constituent will be trapped inside the core of the droplet during most of the droplet lifetime.

In the case of a bicomponent fuel with wide volatilities, both volatility differential and diffusional resistance are important factors due to which a droplet exhibits three-staged behavior (Law 2010). Figure 4.3 represents the $d_s^2(t)$ plot of a heptane/hexadecane (70/30 v/v) droplet where the first stage involves the preferential gasification of the more volatile component (heptane). The second stage is represented by the flat portion of the $d_s^2(t)$ plot, indicating transitional droplet heating as the less volatile component begins to participate in the gasification process actively. The third stage of the $d_s^2(t)$ plot shows a linear behavior, demonstrating the possible attainment of diffusion-limited quasi-steady state.

While the combustion of multi-component droplets is diffusion controlled, the diffusional resistance is not sufficiently strong to yield the quasi-steady behavior, especially for mixtures with large volatility differential. The fact that the third stage

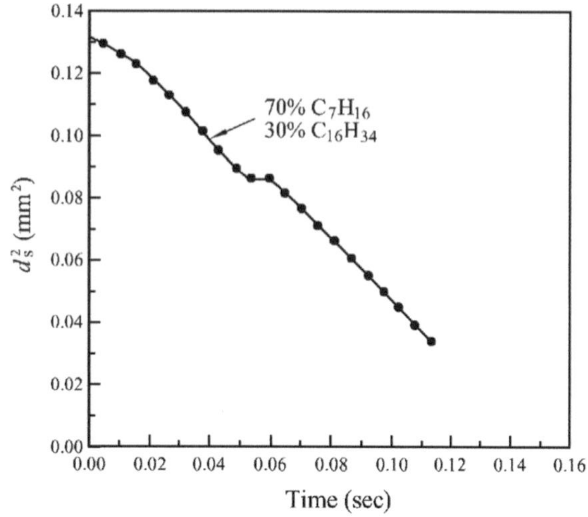

Fig. 4.3 Experimental d_s^2 plot demonstrating a three-stage behavior during the gasification of a heptane–hexadecane droplet with widely varying volatilities (Law 2010)

of Fig. 4.3 is linear because once vigorous burning is established, the burning rates of different alkanes are quite close to each other. Even though batch distillation is not the exact gasification mechanism for multi-component droplets, it is nevertheless a good approximation under two situations. The first is during the early part of the droplet lifetime when the more volatile component in the surface layer is preferentially gasified and the second circumstance is for relatively slow vaporization such that additional time is available for liquid-phase mass diffusion to be effective.

4.3.1 Review of Experimental Approaches in Droplet Combustion

Different experimental approaches are adopted to study the combustion of fuel droplets under various operating conditions. Based on the intrusive nature of the experimental systems, the approaches can be broadly classified as (1) intrusive approaches and (2) non-intrusive or low-intrusive approaches. An intrusive approach is referred to a technique where the foreign objects (such as thermocouple) interact with the droplet during the experimentation. However, these techniques have the potential to evaluate certain inner characteristics of the droplet such as the spatiotemporal variation of the temperature inside the droplet. The most common intrusive approach is suspended droplet technique. Low-intrusive techniques include freely falling technique and Leidenfrost technique. A brief review of the above-mentioned techniques is discussed as follows.

Suspended droplet technique is an appropriate example of the intrusive approach in which a single droplet is suspended on a thin quartz/silicon carbide fiber or a thin

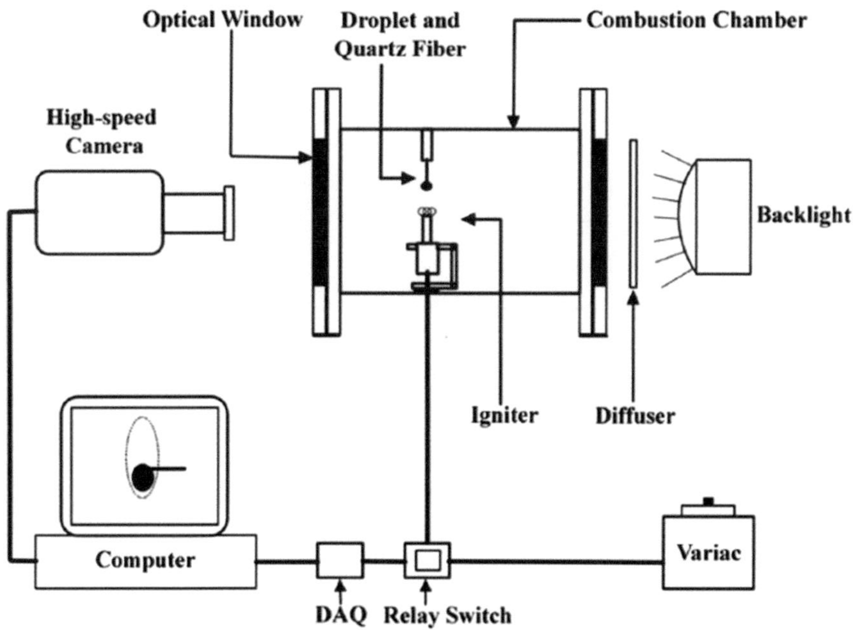

Fig. 4.4 Schematic of experimental setup of droplet suspended on a quartz fiber (Rao et al. 2017)

thermocouple wire. The suspended technique using a filament to support the fuel droplet was employed in the studies conducted even in early 50s (Godsave 1953; Goldsmith and Penner 1954; Kobayasi 1955). Figure 4.4 shows a representative schematic of the experimental setup where the droplet is suspended on quartz fiber. The suspending technique is fairly straightforward to set up and easy to perform the experiment; especially performing cinematography of the stationary droplet is relatively easy. However, the intrusion of the fiber or thermocouple into the droplet is a matter of concern in this technique particularly, the distortion of droplet from spherical shape, heat transfer to and from the fiber (more important when thermocouple in use). Some of these limitations are discussed elsewhere (Law 2010). Though the fiber or thermocouple wire acts as a foreign body to the droplet, their influences on droplet burning behavior are often neglected especially when the fiber diameter is less than 100 μm. The suspended droplet is either exposed to high-temperature furnace or it is ignited by a heated loop of nichrome wire or glow plug. The droplet burning process is then recorded by a high frame rate and high-resolution camera.

The captured videos are processed using image processing software to obtain data on droplet regression rate, flame and sooting characteristics, and various phenomena occurring inside the droplet such as vapor growth and puffing/micro-explosion features in multi-component droplets.

In the case of non-intrusive or low-intrusive approaches, the interaction between the droplet and the measurement system is considered to be negligible or very low.

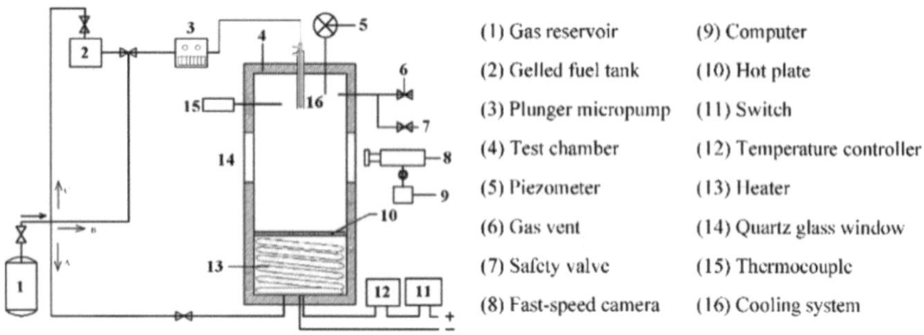

Fig. 4.5 Representative experimental setup of freely falling droplet (Liu et al. 2012)

Examples of such approaches are freely falling droplet or drop tube, and Leidenfrost approach. Figure 4.5 shows a typical experimental setup of combustion under freely falling conditions. Limitation of these non-intrusive approaches is that extractions of data often rely on the extrapolation from the operating conditions or on a visual observation of the experiment (Mura et al. 2014). However, there are several advantages of the freely falling droplet technique such as ability to study a small sized droplet, no interference from the suspending fiber, and the capability of using volatile fuels. Since the droplet does not remain stationary in this approach, it is comparatively difficult to obtain detailed information of the combustion process especially through photography (Law and Law 1982). In all the droplet combustion experiments conducted under normal gravity condition, the effect of buoyancy is prevalent which in turn affects the droplet burning rate and flame shape. The techniques which are employed to minimize the effect of buoyancy are (i) droplet combustion under micro-gravity or gravity-free environment and (ii) droplet combustion under low pressure. The key reason of using these techniques is to maintain the spherical symmetry of the droplet. Examples of such techniques are available in the literature such as droplet combustion in gravity-free environments (Kumagai and Isoda 1957; Okajima and Kumagai 1975; Dietrich et al. 1996) and droplet combustion in a low-pressure environment (Law et al. 1980a). Some relevant discussions in this regard are available in the book by Law (2010).

4.3.2 *Disruptive Phenomena in Multi-component Miscible Droplets*

The combustion of multi-component droplets with wide volatility differential can undergo sudden fragmentation, which is usually referred as micro-explosion. The basic mechanism responsible for this micro-explosion event for miscible multi-component mixtures is the diffusional entrapment of the volatile components in the droplet's inner core. The droplet interior has a relatively higher concentration of the more volatile component while the droplet temperature attains a high value since it is controlled by the more abundant, lower volatile component at the surface. The liquid constituent in the droplet interior can be heated beyond the local boiling point and thereby possess a substantial amount of superheat. There is a maximum limit on the extent of superheat a liquid can accumulate. If the droplet temperature is sufficiently high such that this limit is reached, then the liquid element will homogeneously nucleate and gasify, leading to intense internal pressure buildup and subsequent fragmentation of the droplet. Experimentally, micro-explosions have been frequently observed (Lasheras et al. 1980; Wang et al. 1984; Wang and Law 1985; Lasheras et al. 1981). It has been well established that the micro-explosion event can occur only if the volatilities of the components are sufficiently different and their initial concentrations lie within an optimum range. The reason being that micro-explosion requires the non-volatile components to drive up the droplet temperature and the volatile components to facilitate internal nucleation. It is also recognized that the occurrence of micro-explosion is facilitated with increasing pressure (Wang and Law 1985). These three distinctive properties have all been verified experimentally. In particular, it has been verified that the optimum composition of a bicomponent mixture for enhanced micro-explosion event is around 50-50%.

The superheating of the more volatile liquid results in bubble nucleation, which can be either heterogeneous (in emulsion) or homogeneous (in miscible fuels). The subsequent explosive growth of the vapor bubble causes fragmentation of liquid droplet into small secondary droplets. Complete disintegration of the droplet is termed as micro-explosion whereas partial fragmentation of a droplet is termed as puffing. Figure 4.6a and b illustrates a schematic diagram of puffing and micro-explosion. Initially, the pressure inside the vapor bubble is high compared to the ambient liquid pressure. The tiny bubbles coalesce with relatively larger bubble due to the internal circulation inside the droplet. This leads to the formation of a bigger bubble. Since the nucleation sites are dependent on the proportion of higher

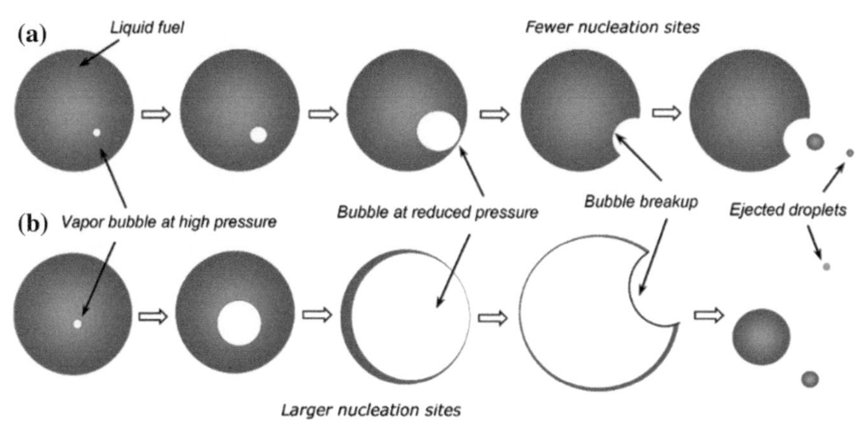

Fig. 4.6 Schematic diagram of **a** puffing and **b** micro-explosion (Rao et al. 2017)

volatile component, a lower proportion of the higher volatile component results in fewer nucleation sites and hence less coalescence of tiny bubbles. Therefore, the bubble cannot grow further and breaks apart resulting in the ejection of small secondary droplets. This breakup of a relatively smaller bubble is referred to puffing. A higher proportion of more volatile component initiates a significant number of nucleation sites, which is associated with the coalescence of tiny bubbles. This results in the formation of a bigger bubble, and its subsequent breakup leads to micro-explosion. In essence, the volatility differential, the relative concentration of constituents in multi-component fuel droplet, as well as the size and location of the bubble at the pre-breakup instant, controls the intensity of the droplet fragmentation (Wang et al. 1984; Wang and Law 1985; Lasheras et al. 1981).

There are a number of studies available on puffing and micro-explosion phenomena during combustion of multi-component miscible fuel droplets. Many experimental investigations were conducted using freely falling droplet approach during early 80s in which micro-explosion phenomenon was reported (Lasheras et al. 1980, 1981; Wang et al. 1984, 1985). The experiments were performed on binary n-paraffin mixtures to investigate the disruptive burning characteristics. It was observed that a minimum difference between the normal boiling points of the components as well as a certain initial concentration of the more volatile component must exist for disruption phenomena to occur. Moreover, the disruptive burning was found to be the result of homogeneous bubble nucleation within the interior of the droplet, which is followed by the growth and breakup of the vapor bubble. Niioka et al. (1986) performed an experimental investigation on the combustion and micro-explosion behavior of miscible fuel droplets under high pressure. The tested mixtures consist of n-heptane (more volatile component) and n-hexadecane (less volatile component). In accordance with the previous study (Wang and Law 1985), the droplet exhibits a three-staged combustion behavior in d^2 versus time plots; however, this staged burning diminishes with the increase in pressure.

The occurrence of micro-explosion is stochastic and therefore it cannot be predicted in a deterministic way by the classical criterion using the superheat limit and droplet temperature (Mikami et al. 1998). The occurrence probability of micro-explosion is rather controlled by the ratio of the liquid-phase lifetime to the nucleation time and depends largely on the droplet diameter (Mikami et al. 1998). Based on the quasi-steady assumption in droplet combustion, the occurrence of micro-explosion is expressed using the following criterion:

$$T_{LS} \leq T_d \quad (4.23)$$

where T_{LS} is referred as the superheat limit for the initial concentration in the droplet and T_d is the calculated droplet temperature in the quasi-steady combustion (Law and Law 1982). Figure 4.7 represents a typical example of these temperatures for an n-heptane/n-hexadecane mixtures.

However, there are studies available in literature in which micro-explosions were observed even when the criteria were not satisfied (Lasheras et al. 1980), and the

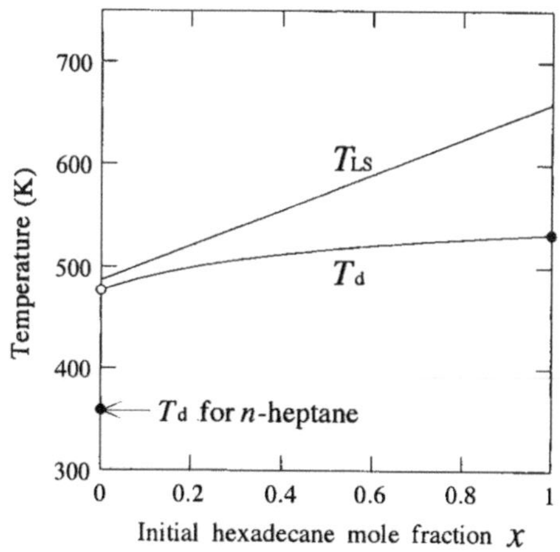

Fig. 4.7 Limit of superheat T_{LS} and calculated droplet temperature T_d in quasi-steady combustion for n-heptane/n-hexadecane mixtures at 0.1 MPa as functions of the initial hexadecane mole fraction x in the droplet (Mikami et al. 1998)

micro-explosions were not observed even though the above criteria were fulfilled (Mikami et al. 1998; Yang et al. 1990). Therefore, the micro-explosion phenomena are considered as stochastic in nature. The occurrence probability of micro-explosion (P_{ME}) evaluates the stochastic nature of micro-explosion (Mikami et al. 1998).

$$P_{ME} = \frac{n_{ME}}{N_0} \tag{4.24}$$

Here, n_{ME} is defined as the number of droplets that show micro-explosion, and N_0 is defined as the sample droplet number. Figure 4.8 shows a typical variation of P_{ME} with number of droplet sample (N_0) for n-hexane/n-hexadecane droplets with an initial concentration of n-hexadecane as 0.40 (mole basis). It can be observed from the figure that as the number of droplet sample (N_0) increases, the uncertainty

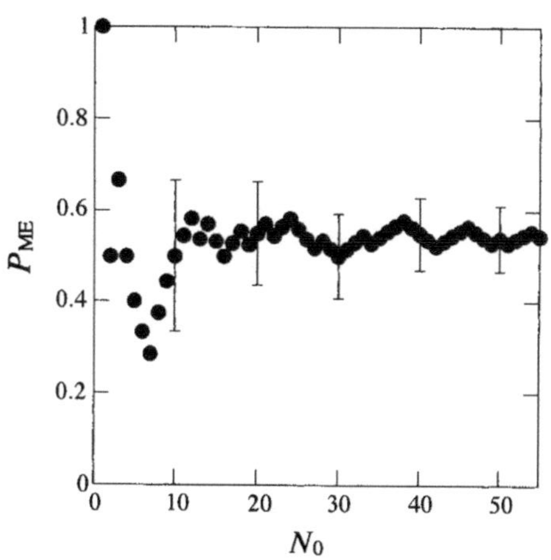

Fig. 4.8 Variation of occurrence probability of micro-explosion (P_{ME}) with sample droplet number N_0 for hexane/hexadecane droplets (Mikami et al. 1998)

of probability parameter (PME) decreases. In fact, beyond the number of droplet sample 20, the uncertainty associated with PME remains almost constant. Therefore, it is generally suggested to consider large value of number of droplet samples in order to get a reliable value of the probability of occurrence of micro-explosion phenomena.

The combustion of practical miscible fuels such as butanol/Jet A-1 and acetone-butanol-ethanol (A-B-E)/Jet A-1 fuel droplets also undergo puffing and micro-explosion (Rao et al. 2017). The sequence of processes which lead to micro-explosion are onset of nucleation, growth of vapor bubble, and subsequent breakup of the bubble as well as parent droplet. The onset of nucleation can be characterized by normalized squared onset diameter (NOD), which is the square of the ratio of droplet diameter at nucleation to the initial droplet diameter. When the volatility differential among the fuel components is smaller, nucleation of vapor bubble is delayed. Similarly, higher volatility differential among the fuel components results is earlier nucleation (Rao et al. 2017). The NOD value can be determined by carefully observing the first appearance of the bubble visible from the high-speed images (Rao et al. 2017). The growth of vapor bubble and its subsequent breakup can also be visualized through high-speed images. A typical sequence of pictures of bubble growth and micro-explosion of A-B-E droplet is shown in Fig. 4.9, where 0 ms represents the time when the bubble first appears. As seen in the figure, the vapor bubble, which has grown from the nucleus, is located nearly symmetric within the droplet. The bright area at the center of the droplet is due to optical effects.

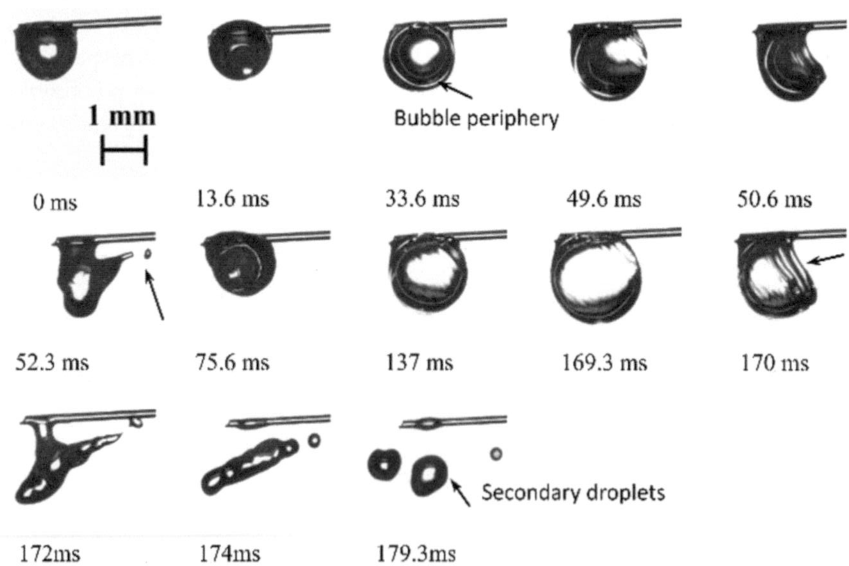

Fig. 4.9 Sequence of images of bubble growth and micro-explosion of a typical A-B-E droplet (Rao et al. 2017)

A similar scattering of light highlights the bubble periphery which appears like a bright ring. The vapor bubble starts to grow until it breaks (50.6 ms), resulting in the ejection of secondary droplets. The nucleation occurs again, and as the bubble grows, a portion of its relatively smooth surface bulges out of the droplet (Fig. 4.9). The occurrence of puffing (52.3 ms) creates turbulence inside the droplet resulting in the formation of more nucleation sites and in turn leading to the growth of a bigger bubble (Shinjo et al. 2014; Rao et al. 2017). As the bubble grows and droplet continues to evaporate, the peripheral liquid sheet becomes thinner and at the same time is subjected to an internal pressure higher than that of the ambiance. This causes the tearing of the liquid sheet into small droplets at a later stage (at 179.3 ms in Fig. 4.9).

The characteristics of disruptive phenomena can also be represented through the temporal evolution of droplet diameters. Figure 4.10 shows a comparison of the evolution of droplet diameters for pure Jet A-1, pure butanol, and A-B-E. The droplets of Jet A-1 and pure butanol show a smooth and continuous temporal regression of diameter due to evaporation without any disruption by bubble growth and breakup. On the other hand, two prominent spikes characterize the regression profile of A-B-E. The sudden increase in droplet diameter followed by its immediate rapid decrease is due to the expansion and subsequent breakup of the bubble. The first spike corresponds to the bubble growth leading to puffing phenomenon while the second spike corresponds to the micro-explosion phenomenon. The spikes represent the maximum droplet diameter at which the bubble breaks apart. It is noticeable that the time from bubble generation to the final breakup is very short (of the order of 1/100 of the average droplet lifetime). The drop size after the breakup is significantly smaller, and thus, the vaporization is greatly enhanced.

Micro-explosion results in the ejection of multiple droplets with both larger and smaller diameters; however, puffing produces relatively smaller diameter droplets compared to micro-explosion (Rao et al. 2017). Moreover, the growth of vapor

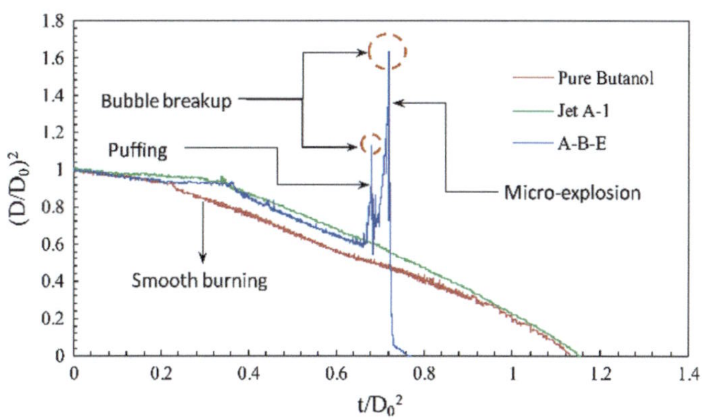

Fig. 4.10 Comparison between the temporal evolutions of droplet diameter of typical Jet A-1, pure butanol, and A-B-E droplets (Rao et al. 2017)

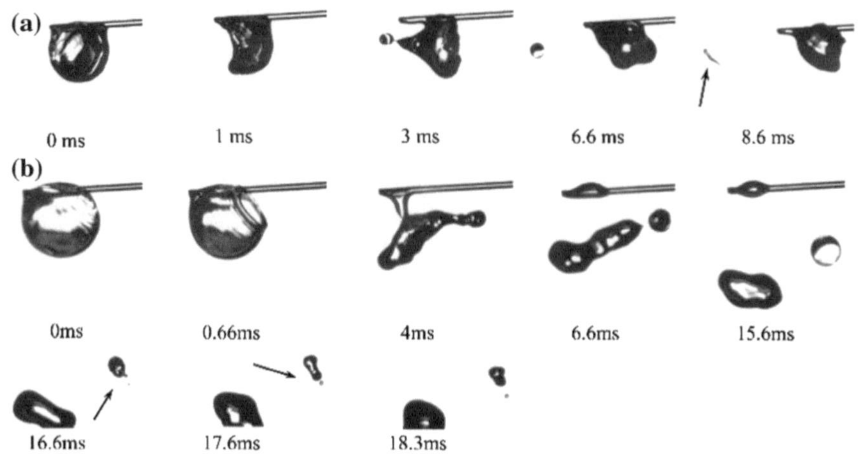

Fig. 4.11 Sequence of images of puffing/micro-explosions in secondary droplets

bubble in the secondary droplets leads to further fragmentation of the secondary droplet. Some sequences of images of multiple puffing/micro-explosions are shown in Fig. 4.11. Here, 0 ms represents the onset of micro-explosion in parent droplet. After the occurrence of first puffing or micro-explosion, a bubble starts to grow in the separated droplet and subsequently breaks apart leading to the ejection of secondary droplets. Due to the presence of a substantial proportion of volatile component in the secondary droplets, the bubble forms and starts to grow again in those droplets leading to second micro-explosion (Rao et al. 2017).

4.3.3 *Disruptive Phenomena in Emulsion Droplets*

In contrast to a miscible mixture (single liquid phase), emulsions are multi-phase mixtures consisting of an immiscible lower boiling point constituent (usually water) dispersed in higher boiling point base fuel (usually heavier hydrocarbons) along with or without a trace amount of chemical surfactant for stabilization (Wang and Law 1985; Lasheras et al. 1979; Randolph and Law 1986; Chung and Kim 1990). The emulsions are generally categorized into two types: water-in-oil emulsion and oil-in-water emulsion. The most commonly studied emulsions in practical combustion laboratories are water-in-oil emulsions, in which a number of fine water micro-droplets are dispersed in the continuous phase of base fuel. Water-in-oil emulsions have been extensively investigated in a wide range of combustors such as gas turbine engines, diesel engines, furnaces, and boilers. Since the participation of water in the evaporation/combustion is expected to lower the droplet temperature, a reduction in NOx emissions has been reported in many investigations.

Contrary to a miscible mixture (where components can freely diffuse), diffusion of the dispersed phase water micro-droplets in the continuous oil phase is negligible. However, similar to the miscible mixtures, the droplet temperature is limited by the higher boiling point component (oil), which is much greater than that of lower boiling point component (water). The embedded water micro-droplets can then be heated to its limit of superheat resulting in violent micro-explosion. Micro-explosion of emulsions has been observed to occur more frequently and with greater intensity as compared to droplets of miscible mixtures. This is because, in a miscible blend, the tendency of the more volatile component to nucleate is diluted by the less volatile component since they are mixed at the molecular level. In addition, continuous and rapid bubble growth requires a correspondingly rapid rate of supply of the volatile component to the nucleation site which, however, is not favored in the presence of strong diffusional resistance. In the case of water-in-oil emulsions, the water micro-droplets are themselves quite large (order of a few microns for macro-emulsions). Therefore, they contain sufficient mass for immediate and continuous conversion into vapor to facilitate micro-explosion.

One of the initial studies on combustion of emulsion droplets include the investigation of single droplets of heavy oil and water-in-oil emulsions by gravimetric as well as photographic techniques (Jacques et al. 1977). The effect of water content (up to 15%) on the combustion characteristics of emulsions was studied. In addition, the asphaltene content of the base oil was varied in the investigation. The emulsion droplet was observed to exhibit swelling and disruption during the pre-ignition period. The increase in water content of the emulsion was found to have minimal effect on the rate constant, which can be attributed to the increased pre-ignition mass loss caused by the disruptive phenomenon, which further indicates a reduction in the actual burning rate constant. The ignition delay during the combustion of droplet increases with an increase in water content due to the lowered droplet temperature.

Theoretical as well as experimental investigation on burning water-in-oil emulsion droplets indicates that the relative volatilities and concentration of the components and intensity of internal circulation play a major role during the combustion of emulsions (Law et al. 1980b). The fact that water and oil do not mix is one of the important reasons of different combustion characteristics of emulsions from those of miscible multi-component fuel droplets. In particular, the existence of internal circulation inhibits micro-explosion in water-in-oil emulsion droplets. The experimental investigation on the combustion of water/n-octane emulsion droplets at different ambient pressures with and without forced convection was performed by Law et al. (1980b). The suspended, spark ignition, droplet technique and cine-microphotography were employed to determine the time history of the droplet size. The burning rate constants (K) as a function of the initial water content, ambient oxidizer concentration, and the intensity of external convection are shown in Fig. 4.12. It can be seen that the burning rate constant initially decreases with an increase in water content (for low water content) but subsequently levels off at a higher proportion of water. In the case of higher convective intensities, the leveling off also seems to occur relatively earlier (Law et al. 1980b).

An experimental investigation on the micro-explosive behavior of emulsified hydrocarbons droplets has also been performed under weightless conditions, which

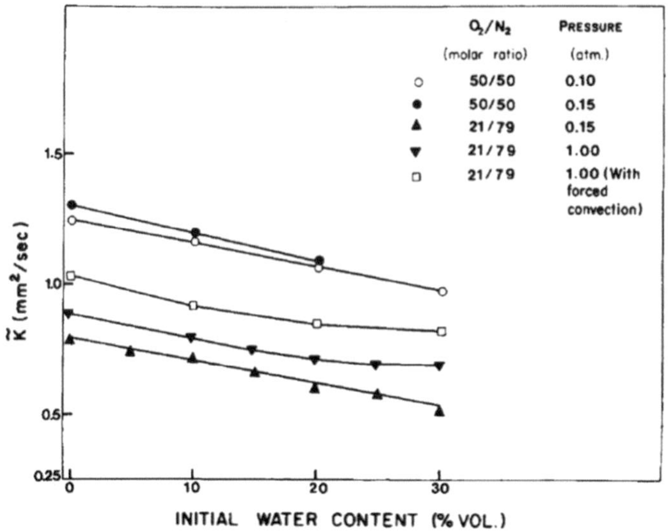

Fig. 4.12 Burning rate constants as functions of initial water content, ambient oxidizer concentration, and the intensity of external convection (Law et al.1980b)

were then compared with the combustion behavior of free-falling emulsions under the influence of gravity (Kimura et al. 1986). In weightless conditions (zero gravity, no buoyancy), no violent micro-explosion occurs with regard to n-tridecane, n-tetradecane, n-pentadecane, gas–oil, and JP-4 emulsions; however, micro-explosions occur in free-falling combustion in oxygen-enriched atmospheres under one gravity. Under zero gravity conditions, both O/W and W/O emulsions result in "spray-like" dispersion of fine droplets (10–20 μm). Similarly, in a typical emulsion droplet (of size 1.2–2.8 mm) composed of JP-4, N_2H_4, and surfactant, micro-explosion does not take place. However, in burning C_9–C_{12} n-paraffin emulsions, mild micro-explosion occurs which results in the expulsion of relatively large diameter droplets in the order of 100 μm. In case of n-pentane, n-heptane, and n-octane emulsions, disruption does not occur except the flashing in the end. In contrast to zero gravity conditions, free-falling droplets of n-paraffin hydrocarbons having more than 13 carbon atoms show micro-explosion under the effect of normal gravitation, while C_9–C_{12} n-paraffin shows mild micro-atomization. In case of freely falling droplets, water in emulsion is drained from the dispersed state since the heat comes back from the flame zone and forced to coalesce by convection in the presence of buoyancy. However, the rate of water coalescence under zero gravity conditions is considered to be very slow, resulting in spray-like micro-explosion. A similar investigation was performed on the combustion characteristics of n-heptane, n-decane, n-hexadecane, and iso-octane emulsion droplets with different amount of water in freely falling conditions (Wang and Chen 1996). The effect of water emulsification on the ignition, extinction, and micro-explosion characteristics of the droplet was studied. It was reported that while the water emulsification prolongs the ignition delay, the effect is minimal for sufficiently volatile fuels. For the fuels which are relatively less volatile than water, micro-explosion is possible resulting in violent fragmentation or continuous mass ejection. Moreover, the probability of onset and intensity of micro-explosion are further enhanced with increasing water content (up to 30%).

The emulsification of hydrocarbons with water reduces the droplet flame luminosity as well as soot formation in a low convection environment (Jackson and Avedisian 1998).

In case of water/heptane emulsions (10/90 v/v), the burning process occurs in three stages. Stage 1 represents burning of heptane, stage 2 indicates burning of both heptane and water, and finally, surfactant controls the burning process in stage 3 (Fig. 4.13). In the final stage of combustion, the transition to surfactant-dominated vaporization results in micro-explosions. Even though the boiling points of heptane and water are almost identical, micro-explosions are observed. The reason for the disruptive nature is attributed to surfactant enrichment at the droplet surface during the combustion process which would raise the droplet temperature above that required for nucleation of vapor bubble within the emulsion.

Apart from emulsions, micro-explosion phenomenon is also observed in compound droplets (Chen and Lin 2011). Unlike an emulsion, a compound droplet consists of a water core and a fuel shell, which originates from the phase separation of emulsions (Segawa et al. 2000; Kadota and Yamasaki 2002; Kadota et al. 2007). The vaporization rate, ignition delay time, and occurrence of micro-explosion of the compound droplets are primarily affected by three experiment parameters: droplet diameter, water content, and environmental oxygen concentration (Chen and Lin

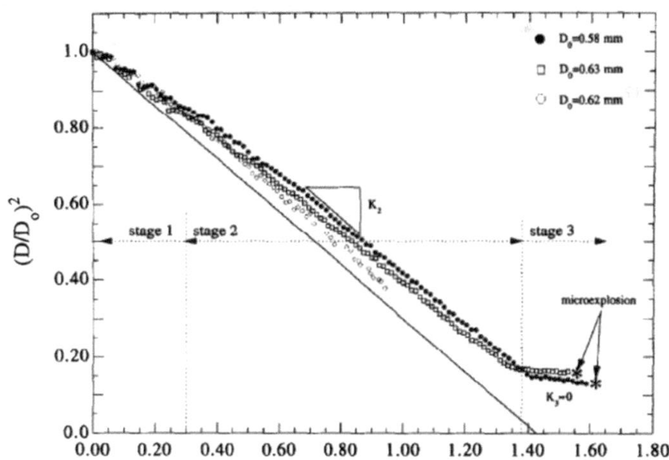

Fig. 4.13 Variation of normalized droplet diameter with a scaled time for 10% water emulsion droplets. Solid line represents a prediction from the numerical analysis (Jackson and Avedisian 1998)

2011). The ignition delay time increases with increasing water content, and the vaporization rate increases with increasing ambient oxygen concentration. The most influential parameter for micro-explosion in compound droplets is droplet diameter, followed by ambient oxygen concentration while the water content has the least effect on micro-explosions.

As discussed in the experimental techniques section, another low-intrusive approach for studying micro-explosions in emulsion droplet is Leidenfrost burning (Mura et al. 2010, 2012, 2014; Kadota et al. 2007). In this approach, a droplet is placed upon a heated surface which then levitates on its vapor, avoiding the contact between the sample and the heated surface. Due to the surface tension, the droplet retains the spherical shape. The Leidenfrost effect and the mechanisms responsible for the phenomenon are widely known (Leidenfrost 1966; Eberhart et al. 1975; Avedisian and Fatehi 1988). Statistical analyses such as Weibull distribution are used to obtain the characteristics of the micro-explosion during the Leidenfrost burning of an emulsion droplet (Kadota et al. 2007). Weibull analysis is employed to obtain the distribution function of the waiting time for the onset of micro-explosion as well as to derive the mathematical expression for the rate of micro-explosion as a function of water content and emulsion temperature. The waiting time for the inception of micro-explosion decreases with the increase in the boiling point of the base fuel, initial water content, ambient pressure, and test surface temperature. Similarly, the rate of micro-explosion increases with the increase in the normal boiling point of the base fuel.

The temperature of micro-explosion is also related to the distribution of the dispersed water droplets inside the parent droplet. As the Sauter mean diameter of water droplets is reduced, the temperature of micro-explosion rises to a very high level, highlighting a significant degree of metastability of the liquid water. It has also been indicated that if the size of water droplets is large, the coalescence phenomenon becomes predominant whereas coalescence becomes less important with finer distribution (Mura et al. 2012).

4.4 Introduction to Combustion of Nanofuel Droplet

Nanofluid fuels or popularly termed as nanofuels are new class of fuels in which energetic nanoparticles are dispersed in liquid fuels. The nanofuels generally contain low amount nanoparticles and are fundamentally different from slurries in which the micron-sized particles are loaded at significantly higher loading densities (40–80%). Combustion of slurry fuel droplets has been extensively studied. Most of these investigations involve combustion of droplets with micron-sized particles such as boron (Antaki and Williams 1987; Takahashi et al. 1989), aluminum (Turns et al. 1987; Wong and Turns 1987; Byun et al. 1999), carbon, and a blend of aluminum and carbon (Szekely and Faeth 1982; Sakai and Saito 1983; Wong and Turns 1989; Lee and Law 1991). A detailed review on the combustion of slurry fuel droplet is reported in the literature (Roy Choudhury 1992). Key challenges observed in burning of slurry fuels having high particle loading is that combustion of large agglomerates requires a significantly longer time than that of individual particles. Recent advancement in nanotechnologies has opened up the possibility to use these energetic particles at their nanometer size ranges. There are several positive attributes of nanoparticles which have driven interest toward the development of the nanofuel such as reduced ignition delay, high specific surface area, possibility of more complete combustion than micron-size counterparts.

In spite of potential benefits of nanoparticles as additives, very limited studies are available on combustion of nanofuels. One such study includes an investigation on the hot-plate ignition probability of nanoparticle-laden diesel droplets in the range of temperatures from 688 to 768 °C (Tyagi et al. 2008). The ignition probability for Al- and Al_2O_3-laden droplets of two different sizes (15 and 50 nm) is considerably higher than that of pure diesel. Moreover, ignition probability is not influenced by either the type of nano-additives or particle size. One of the commonly performed investigation includes combustion of nano particle laden droplet suspended on a fiber. Gan et al. (2011) performed an experimental investigation on combustion of ethanol and n-decane droplets containing aluminum particles of diameters 80, 5, and 25 μm with a particle loading of 10% by weight.

The experimental procedure for combustion study of nanofuel droplet is similar to that of pure component or multi-component droplet. Additional step followed for nanofuel is the preparation of nanofuel. Generally, nanofuel is prepared by mixing the particles in liquid fuel and then sonicated the mixture using a probe-type or bath-type sonicator. During sonication, the alternating high-pressure and low-pressure cycles are generated due to which the attracting force between individual particles becomes mitigated and hence the chance of agglomeration is reduced. In many occasions, surfactant is also added to promote chemical stabilization. The choice of surfactant and its concentration in nanofuel is suspension depends on the combination of nanoparticles and liquid selected. For example, Sorbitan Oleate ($C_{24}H_{44}O_6$) is a common surfactant used to increase dispersion stability of metal nanoparticles in n-alkanes, and it is usually utilized in the emulsion of water and fuel (Kadota and Yamasaki 2002; Gan and Qiao 2011). Surfactant also has an influence on the combustion characteristics of nanofuel droplet. Gan et al. (2011) also used Sorbitan Oleate as surfactant in their study, and the maximum concentration of surfactant was kept as 2.5%. Figure 4.14 shows a typical burning sequence of n-decane/nano-Al droplet. Five distinct stages observed during the combustion process are: (1) pre-heating and ignition stage, (2) classical droplet combustion stage, (3) micro-explosion stage, (4) surfactant flame stage, and (5) aluminum droplet flame stage. In the first stage, evaporation occurs on the droplet surface, and the droplet is surrounded by the vapor cloud. The evaporation decreases the droplet size; however, the decrease in the droplet size is minimal in this stage. Temperature of the droplet increases after the ignition of droplet. The droplet diameter regression and temperature data for n-decane/nano-Al droplet case are shown in Fig. 4.15. The liquid components inside droplet (n-decane and Sorbitan Oleate) follow the characteristics of a typical multi-component droplet

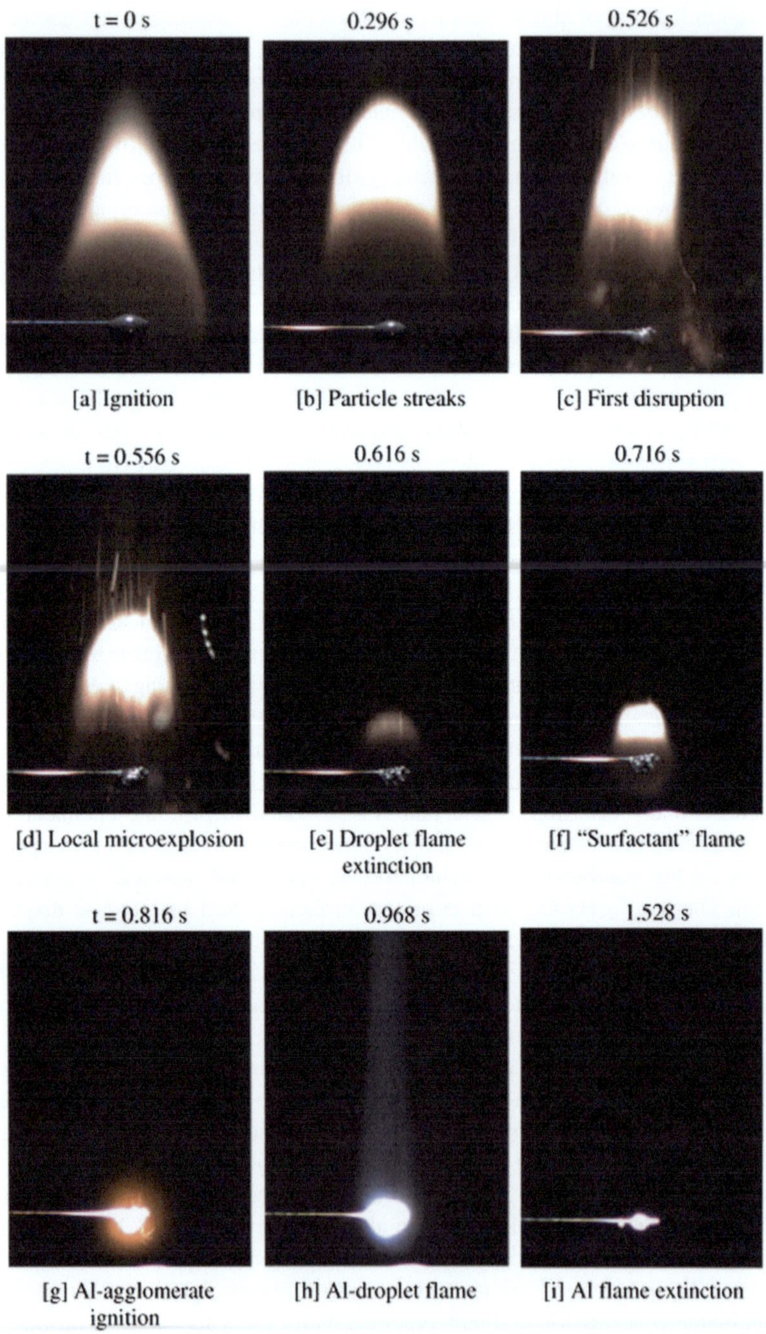

Fig. 4.14 Typical burning sequence of a n-decane droplet loaded with 10% nano-aluminum (particle size 80 nm) and 2.5% surfactant (Gan and Qiao 2011)

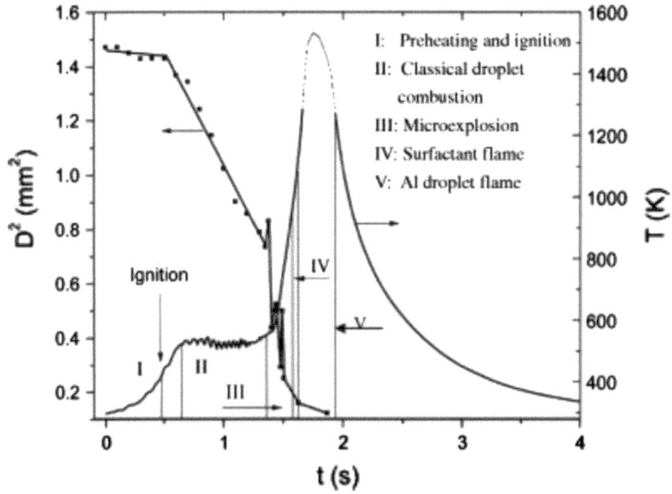

Fig. 4.15 Droplet diameter regression and temperature histories during combustion of a stabilized n-decane/nano-Al droplet (Gan and Qiao 2011)

because of the significant difference in the boiling points (n-decane being more volatile than Sorbitan Oleate).

The second stage is characterized by steady vaporization and burning of the droplet similar to the classical burning feature of single component droplet. In this stage, an envelope flame surrounds the droplet. The droplet diameter regresses steadily with time approximately following the classical d^2 law (Fig. 4.15). The third stage, i.e., micro-explosion stage is characterized by the strong shape oscillations of the droplet, which is caused due to the formation of bubbles and subsequently their growth inside the droplet and fragmentation of parent droplet into smaller secondary droplets. The micro-explosion phenomenon in combustion of nanofuel droplet is slightly different than that observed in combustion of miscible fuel droplet or emulsion droplet. In nanofuel droplet, the effect of nanoparticles in terms of particle diffusion, aggregation dynamics, and chemical reaction can influence the characteristics of micro-explosion in nanofuel droplet. This will be discussed separately in the later part of the chapter. After the micro-explosion stage, the envelope flame extinguishes, and another envelope diffusion flame surrounds the large agglomerate which is considered as "surfactant flame." This surfactant flame is formed due the burning of surfactant or its pyrolyzed products. The surfactant flame is observed in all particle loadings and surfactant concentration regardless of base fuel or particle size. Interestingly, the surfactant flame has not been observed in earlier studies on combustion of slurry fuel droplets. The last distinct stage is combustion of aluminum agglomerate. Agglomerate ignites once the surfactant flame extinguishes and a bright white glow is visible (Fig. 4.14h) due to burning of aluminum. Apart from the studies on aluminum nanoparticles, the burning characteristics of droplet with dilute and dense suspensions comprising iron and boron nanoparticles have also been studied in detail (Gan et al. 2012). For dilute suspensions (< 1% loading), both the droplet and particles burn simultaneously; however, in case of dense nano-suspensions, most of the particles burn as large agglomerate at a later part of burning when all the liquid fuel is consumed.

Gan et al. (2011) compared the burning behaviors of droplets with nano-suspension (80 nm Al particles) and micron-suspensions (5 μm Al particles) in order to understand the particle agglomeration within a burning droplet and its influence on the overall combustion characteristics of the droplet. In contrast to nanoparticle-laden droplets, the burning process of micron-suspension cases consisted only three distinctive stages: the pre-heating and ignition, classical combustion, and micro-explosion. In addition, the micro-explosion behavior is different for the two cases. Micro-explosion in micron-suspension occurs with much stronger intensity and comparatively at later part of burning. In case of nano-suspensions, agglomeration residue of burnt aluminum was found on the fiber, whereas very low amount of residue was found on the fiber and most of the residue fell on the chamber wall for micro-suspensions. The difference in micro-explosion observed for these two categories of suspension is most likely due the difference in agglomeration during the combustion of droplet since very low amount of agglomerated residue of burnt aluminum is formed in case of nano-suspensions. There are many theories proposed by researchers to explain the micro-explosion behavior of slurry fuel droplets. For example, Takahashi et al. (1989) conducted a study on combustion of slurry fuel consisting of JP-10 and boron, and a three-step mechanism was proposed, which includes d^2 law combustion, shell formation, and disruption. Byun et al. (1999) proposed an additional stage (pressure buildup stage) in addition to the above proposition. In essence, these mechanisms mainly emphasize the formation of aggregate shell inside the droplet during the early stage of burning. The aggregate shell formed inside the droplet usually traps the liquid fuel inside. A pressure increase will follow. The shell temperature can exceed the normal boiling point of the trapped liquid fuel, reaching the surfactant pyrolysis temperature. Meanwhile, the pyrolysis renders the shell texture less permeable to the fuel vapor, causing significant pressure buildup within the fuel droplet. Wong et al. (1989) proposed that the formation of impermeable shell might be promoted by pyrolyzed surfactant. Vapor pressure trapped inside the shell may increase due to the impermeable nature of the shell. Once the pressure reaches a sufficiently high value, the shell ruptures causing disintegration of the droplet. The particle size can have an influence on the structure, strength, and permeability of this aggregate shell. Gan et al. (2011) suggested that the aggregate shell formed inside the droplet is densely packed and impermeable for the case of micron-suspension, whereas in nano-suspension the shell is considered to be porous and permeable.

A nanofuel droplet can exhibit minor or major disruptive events based on the strength, structure, and porosity of the aggregate shell and vapor pressure buildup. To investigate and quantify these disruptive events, combustion of bicomponent (ethanol–water) droplets containing various concentrations (0.5%, 1% in the dilute range and 5%, 7.5% in the dense range) of nano particles were studied in detail (Miglani and Basu 2015). The secondary atomization can be characterized by identifying various modes in case dilute and dense concentrations. For lower particle loading (0.5 and 1%), the disruptive behavior is characterized by low-intensity atomization modes which cause small-scale distortion of the flame. On the contrary, for higher particle loading (5 and 7.5%), the disruptive behavior is governed by the high-intensity bubble ejections along with severe perturbations in the envelope flame. Five distinct modes of secondary breakup have been identified with a help of two parameters, (i) ejection impact parameter (α) and (ii) droplet void fraction (ϕ) (Miglani and Basu 2015). The modes of atomization classified based on the severity of the ejection are: (1) needle-type ligament ejections ($\alpha_{local} << 0.01$); $\phi \approx 0$; high kinetic energy), (2) needle ejection with tip-base breakup ($\alpha_{local} << 0.1$), $\phi \approx 0$, (3) low momentum needle-type ligaments with only tip breakup ($0.1 < \alpha_{local} < 0.3$); $0.1 < \phi < 0.2$, (4) low momentum, thick ligaments (αlocal > 0.3); $\phi > 0.3$, and (5) localized catastrophic fragmentation with multiple ligament formation ($\alpha_{local} \rightarrow 1$); $\phi \rightarrow 1$. First three modes are considered to be low-intensity modes and are present in the entire droplet lifecycle for lower particle loadings (0.5% and 1%), whereas last two modes being high-intensity modes usually occur in the cases of higher particles loadings (5 and 7.5%). Jump from low-intensity modes to high-intensity modes generally occurs when the particle loading increases. The low-intensity breakup modes induce mild shape oscillations whereas the high-intensity modes result in intense volumetric shape oscillation which results in axis-switching. In addition, distinct regime of single major bubble is identified to be evaporative Darrieus–Landau instability during the pre-breakup phase (Miglani et al. 2014). A detail discussion on these secondary modes, influence of these modes on the surface undulation of the droplets, and the typical time scales of these modes is available in the review article by Basu and Miglani (2016).

Tanvir et al. (2014) conducted combustion experiment with droplet stream of ethanol with and without addition of Al nanoparticles (1–5% by weight). They analyzed the thermophysical properties of the nanofluid fuels such as conductivity, viscosity, and surface tension. It was observed that there was significant increase in the thermal conductivity whereas there was no appreciable change in viscosity and surface tension with particle addition up to 5 wt%. Aluminum nanoparticles seemed to enhance the droplet burning rate significantly as high as 140% for 5% particle loading. Another study was conducted by Javed et al. 2015 to investigate the effects of different concentrations of Al nanoparticles on the autoignition and combustion characteristics of heptane-based nanofuel fuel droplets at various ambient temperatures (from 600 °C to 850 °C). They observed that nano-Al loaded heptane droplet did not follow classical d^2-law combustion; rather, the nanofuel droplets exhibited disruptive burning (bubble formation, expansion, and micro-explosions) irrespective of nano-Al concentration. In the low ambient temperature range (600–700 °C), there was no appreciable difference in the average gasification rate between pure heptane and heptane/nano-Al droplets; however, it was observed to be significantly higher for nano-Al-loaded heptane droplets than that of pure heptane droplets in the higher temperature range (750–850 °C). There was almost no residue left on the fiber after the combustion of nano-Al/heptane droplet got completed, and no separate aluminum flame was observed in this case—contrary to the observations of

Gan et al. (2011). Strategies which can enhance the burning rate by introducing micro-explosions of higher intensity and minimize the particle aggregation by better dispersion using appropriate surfactant and/or surface capping of the particles are going to be very useful for making nanofuel as next-generation viable fuel (Sundaram et al. 2017).

4.5 Conclusion

The evaporation and combustion of liquid fuel droplets have widespread applications in power and process industries, chemistry, medicine, and environmental processes. The research work in the area of droplet evaporation is going on over the last few decades. A remarkable progress has already taken place in the theory of evaporation and combustion of liquid fuel droplets and sprays. The theoretical progress includes the development of physical models and computer codes capable of solving model equations. The experimental progress, on the other hand, involves the development of improved instruments for calibrating the model predictions and for investigations of several complex phenomena occurring in the process of droplet combustion. A large number of works carried out in the field have contributed well to the scientific understanding of the subject. The present chapter discusses different underlying features of combustion of multi-component and nanofuel droplets. A brief review of numerical and experimental investigations carried out in these fields is included in this chapter.

References

Abramzon B, Sirignano WA (1989) Droplet vaporization model for spray combustion calculations. Int J Heat Mass Transf 32(9):1605–1618

Anisimov SI, Dunikov DO, Zhakhovskii VV, Malyshenko SP (1999) Properties of a liquid–gas interface at high-rate evaporation. J Chem Phys 110(17):8722–8729

Antaki P, Williams FA (1987) Observations on the combustion of boron slurry droplets in air. Combust Flame 67(1):1–8

Avedisian CT, Fatehi M (1988) An experimental study of the Leidenfrost evaporation characteristics of emulsified liquid droplets. Int J Heat Mass Transf 31(8):1587–1603

Ayyaswamy PS, Sadhal SS, Huang LJ (1990) Effect of internal circulation on the transport to a moving drop. Int Commun Heat Mass Transf 17(6):689–702

Basu S, Miglani A (2016) Combustion and heat transfer characteristics of nanofluid fuel droplets: a short review. Int J Heat Mass Transf 96:482–503

Bhattacharya P, Ghosal S, Som SK (1996) Evaporation of multicomponent liquid fuel droplets. Int J Energy Res 20(5):385–398

Biswal LD, Datta A, Som SK (1999) Transport coefficients and life history of a vaporising liquid fuel droplet subject to retardation in a convective ambience. Int J Heat Fluid Flow 20(1):68–73

Blander M, Katz JL (1975) Bubble nucleation in liquids. AIChE J 21(5):833–848

Bond M, Struchtrup H (2004) Mean evaporation and condensation coefficients based on energy dependent condensation probability. Phys Rev E Stat Nonlinear Soft Matter Phys 70(6):1–21

Byun DY, Baek SW, Cho JH (1999) Microexplosion of aluminum slurry droplets. Int J Heat Mass Transf 42(24):4475–4486

Chen C-K, Lin T-H (2011) Burning of water-in-dodecane compound drops. Atomization Sprays 21(10):867–881

Chiang CH, Raju MS, Sirignano WA (1992) Numerical analysis of convecting, vaporizing fuel droplet with variable properties. Int J Heat Mass Transf 35(5):1307–1324

Chigier NA (1976) The atomization and burning of liquid fuel sprays. Prog Energy Combust Sci 2 (2):97–114

Chigier NA (1977) Instrumentation techniques for studying heterogeneous combustion. Prog Energy Combust Sci 3(3):175–189

Chigier N (1983) Group combustion models and laser diagnostic methods in sprays: a review. Combust Flame 51:127–139

Chung SH, Kim JS (1990) An experiment on vaporization and microexplosion of emulsion fuel droplets on a hot surface. Symp (Int) Combust 23(1):1431–1435

Consolini L, Aggarwal SK, Murad S (2003) A molecular dynamics simulation of droplet evaporation. Int J Heat Mass Transf 46(17):3179–3188

Curtis EW, Farrell PV (1992) A numerical study of high-pressure droplet vaporization. Combust Flame 90(2):85–102

Dash SK, Som SK (1991a) Ignition and combustion of liquid fuel droplet in a convective medium. J Energy Res Technol 113(3):167–170

Dash SK, Som SK (1991b) Transport processes and associated irreversibilities in droplet combustion in a convective medium. Int J Energy Res 15(7):603–619

Dash SK, Sengupta SP, Som SK (1991) Transport processes and associated irreversibilities in droplet evaporation. J Thermophys Heat Transf 5(3):366–373

Dietrich DL, Haggard JB, Dryer FL, Nayagam V, Shaw BD, Williams FA (1996) Droplet combustion experiments in spacelab. Symp (Int) Combust 26(1):1201–1207

Dwyer HA (1988) Calculations of unsteady reacting droplet flows. Symp (Int) Combust 22 (1):1923–1929

Dwyer HA (1989) Calculations of droplet dynamics in high temperature environments. Prog Energy Combust Sci 15(2):131–158

Dwyer HA, Sanders BR (1988) A detailed study of burning fuel droplets. Symp (Int) Combust 21 (1):633–639

Eberhart JG, Kremsner W, Blander M (1975) Metastability limits of superheated liquids: bubble nucleation temperatures of hydrocarbons and their mixtures. J Colloid Interface Sci 50(2):369–378

Faeth GM (1977) Current status of droplet and liquid combustion. Prog Energy Combust Sci 3 (4):191–224

Faeth GM (1983) Evaporation and combustion of sprays. Prog Energy Combust Sci 9(1–2):1–76

Faeth GM (1987) Mixing, transport and combustion in sprays. Prog Energy Combust Sci 13 (4):293–345

Gan Y, Qiao L (2011) Combustion characteristics of fuel droplets with addition of nano and micron-sized aluminum particles. Combust Flame 158(2):354–368

Gan Y, Qiao L (2012) Radiation-enhanced evaporation of ethanol fuel containing suspended metal nanoparticles. Int J Heat Mass Transf 55(21–22):5777–5782

Gan Y, Lim YS, Qiao L (2012) Combustion of nanofluid fuels with the addition of boron and iron particles at dilute and dense concentrations. Combust Flame 159(4):1732–1740

Gavhane S, Pati S, Som SK (2016) Evaporation of multicomponent liquid fuel droplets: influences of component composition in droplet and vapor concentration in free stream ambience. Int J Therm Sci 105:83–95

Gerum E, Straub J, Grigull U (1979) Superheating in nucleate boiling calculated by the heterogeneous nucleation theory. Int J Heat Mass Transf 22(4):517–524

Godsave GAE (1953) Studies of the combustion of drops in a fuel spray–the burning of single drops of fuel. Symp (Int) Combust 4(1):818–830

Gogos G, Sadhal SS, Ayyaswamy PS, Sundararajan T (1986) Thin-flame theory for the combustion of a moving liquid drop: effects due to variable density. J Fluid Mech 171:121–144

Goldsmith M, Penner SS (1954) On the burning of single drops of fuel in an oxidizing atmosphere. J Jet Propul 24(4):245–251

Harper JF, Moore DW (1968) The motion of a spherical liquid drop at high Reynolds number. J Fluid Mech 32(2):367–391

Hertz H (1882) Ueber die verdunstung der flussigkeiten, insbesondere des quecksilbers, im luftleeren raume. Ann Phys 253(10):177–193

Jackson GS, Avedisian CT (1998) Combustion of unsupported water-in-n-heptane emulsion droplets in a convection-free environment. Int J Heat Mass Transf 41(16):2503–2515

Jacques MT, Jordan JB, Williams A, Hadley-Coates L (1977) The combustion of water-in-oil emulsions and the influence of asphaltene content. Symp (Int) Combust 16(1):307–319

Javed I, Baek SW, Waheed K (2015) Autoignition and combustion characteristics of heptane droplets with the addition of aluminum nanoparticles at elevated temperatures. Combust Flame 162(1):191–206

Kadota T, Yamasaki H (2002) Recent advances in the combustion of water fuel emulsion. Prog Energy Combust Sci 28(5):385–404

Kadota T, Tanaka H, Segawa D, Nakaya S, Yamasaki H (2007) Microexplosion of an emulsion droplet during Leidenfrost burning. Proc Combust Inst 31(2):2125–2131

Kimura M, Ihara H, Okajima S, Iwama A (1986) Combustion behaviors of emulsified hydrocarbons and JP-4/N2H4 droplets at weightless and free falling conditions. Combust Sci Technol 44(5–6):289–306

Knudsen M (1915) Die maximale verdampfungsgeschwindigkeit des quecksilbers. Ann Phys 352 (13):697–708

Kobayasi K (1955) An experimental study on the combustion of a fuel droplet. Symp (Int) Combust 5(1):141–148

Kryukov AP, Levashov VY, Sazhin SS (2004) Evaporation of diesel fuel droplets: kinetic versus hydrodynamic models. Int J Heat Mass Transf 47(12–13):2541–2549

Kumagai S, Isoda H (1957) Combustion of fuel droplets in a falling chamber. Symp (Int) Combust 6(1):726–731

Labuntsov DA, Kryukov AP (1979) Analysis of intensive evaporation and condensation. Int J Heat Mass Transf 22(7):989–1002

Lasheras JC, Fernandez-Pello AC, Dryer FL (1979) Initial observations on the free droplet combustion characteristics of water-in-fuel emulsions. Combust Sci Technol 21(1–2):1–14

Lasheras JC, Fernandez-Pello AC, Dryer FL (1980) Experimental observations on the disruptive combustion of free droplets of multicomponent fuels. Combust Sci Technol 22(5–6):195–209

Lasheras JC, Fernandez-Pello AC, Dryer FL (1981) On the disruptive burning of free droplets of alcohol/n-paraffin solutions and emulsions. Symp (Int) Combust 18(1):293–305

Law CK (1982) Recent advances in droplet vaporization and combustion. Prog Energy Combust Sci 8(3):171–201

Law CK (2010) Combustion physics. Cambridge university press

Law CK, Law HK (1982) A d2-law for multicomponent droplet vaporization and combustion. AIAA J 20(4):522–527

Law CK, Chung SH, Srinivasan N (1980a) Gas-phase quasi-steadiness and fuel vapor accumulation effects in droplet burning. Combust Flame 38:173–198

Law CK, Lee CH, Srinivasan N (1980b) Combustion characteristics of water-in-oil emulsion droplets. Combust Flame 37:125–143

Lee A, Law CK (1991) Gasification and shell characteristics in slurry droplet burning. Combust Flame 85(1–2):77–93

Leidenfrost JG (1966) On the fixation of water in diverse fire. Int J Heat Mass Transf 9(11):1153–1166

Liu Z, Hu X, He Z, Wu J (2012) Experimental study on the combustion and microexplosion of freely falling gelled unsymmetrical dimethylhydrazine (UDMH) fuel droplets. Energies 5 (8):3126–3136

Michaelides EE, Liang L, Lasek A (1992) The effect of turbulence on the phase change of droplets and particles under non-equilibrium conditions. Int J Heat Mass Transf 35(9):2069–2076

Miglani A, Basu S (2015) Effect of particle concentration on shape deformation and secondary atomization characteristics of a burning nanotitania dispersion droplet. J Heat Transf 137:1–8

Miglani A, Basu S, Kumar R (2014) Insight into instabilities in burning droplets. Phys Fluids 26:32101

Mikami M, Yagi T, Kojima N (1998) Occurrence probability of microexplosion in droplet combustion of miscible binary fuels. Symp (Int) Combust 27(2):1933–1941

Mura E, Josset C, Loubar K, Huchet G (2010) Effect of dispersed water droplet size in microexplosion phenomenon for water in Oil emulsion. Atomization Sprays 20(9):791–799

Mura E, Massoli P, Josset C, Loubar K, Bellettre J (2012) Study of the micro-explosion temperature of water in oil emulsion droplets during the Leidenfrost effect. Exp Thermal Fluid Sci 43:63–70

Mura E, Calabria R, Califano V, Massoli P, Bellettre J (2014) Emulsion droplet micro-explosion: analysis of two experimental approaches. Exp Thermal Fluid Sci 56:69–74

Niioka T, Sato J (1986) Combustion and microexplosion behavior of miscible fuel droplets under high pressure. Symp (Int) Combust 21(1):625–631

Okajima S, Kumagai S (1975) Further investigations of combustion of free droplets in a freely falling chamber including moving droplets. Symp (Int) Combust 15(1):401–417

Pati S, Chakraborty S, Som SK (2011) Influence of ambient vapor concentration on droplet evaporation in a perspective of comparison between diffusion controlled model and kinetic model. Int J Heat Mass Transf 54(21–22):4580–4584

Prakash S, Sirignano WA (1978) Liquid fuel droplet heating with internal circulation. Int J Heat Mass Transf 21(7):885–895

Prakash S, Sirignano WA (1980) Theory of convective droplet vaporization with unsteady heat transfer in the circulating liquid phase. Int J Heat Mass Transf 23(3):253–268

Randolph AL, Law CK (1986) Time-resolved gasification and sooting characteristics of droplets of alcohol/oil blends and water/oil emulsions. Symp (Int) Combust 21(1):1125–1131

Rangel RH, Fernandez-Pello AC (1984) Mixed convective droplet combustion with internal circulation. Combust Sci Technol 42(1–2):47–65

Rao DCK, Karmakar S, Som SK (2017) Puffing and micro-explosion behavior in combustion of butanol/jet A-1 and acetone-butanol-ethanol (A-B-E)/jet A-1 fuel droplets. Combust Sci Technol 189(10):1796–1812

Renksizbulut M, Nafziger R, Li X (1991) A mass transfer correlation for droplet evaporation in high-temperature flows. Chem Eng Sci 46(9):2351–2358

Robinson AJ, Judd RL (2004) The dynamics of spherical bubble growth. Int J Heat Mass Transf 47(23):5101–5113

Roy Choudhury P (1992) Slurry fuels. Prog Energy Combust Sci 18(5):409–427

Sadhal SS (1983) Flow past a liquid drop with a large non-uniform radial velocity. J Fluid Mech 133:65–81

Sakai T, Saito M (1983) Single-droplet combustion of coal slurry fuels. Combust Flame 51:141–154

Sazhin SS (2006) Advanced models of fuel droplet heating and evaporation. Prog Energy Combust Sci 32(2):162–214

Sazhin SS (2017) Modelling of fuel droplet heating and evaporation: Recent results and unsolved problems. Fuel 196:69–101

Sazhin SS, Shishkova IN (2009) Kinetic algorithm for modeling the droplet evaporation process in the presence of heat flux and background gas. Atomization Sprays 19(5):473–489

Sazhin SS, Shishkova IN, Kryukov AP, Levashov VY, Heikal MR (2007) Evaporation of droplets into a background gas: kinetic modelling. Int J Heat Mass Transf 50(13–14):2675–2691

Segawa D, Yamasaki H, Kadota T, Tanaka H, Enomoto H, Tsue M (2000) Water-coalescence in an oil-in-water emulsion droplet burning under microgravity. Proc Combust Inst 28(1):985–990

Shen C, Cheng WL, Wang K, Lee CF (2010) Estimating the secondary droplet size distribution after micro-explosion of bio-fuel droplets. In: ILASS-Americas Proceedings of the institute for liquid atomization and spray systems—North and South America, vol. 22, Cincinnati

Shinjo J, Xia J, Ganippa LC, Megaritis A (2014) Physics of puffing and microexplosion of emulsion fuel droplets. Phys Fluids 26(10):103302

Sirignano WA (1983) Fuel droplet vaporization and spray combustion theory. Prog Energy Combust Sci 9(4):291–322

Sirignano WA (1988) An integrated approach to spray combustion model development. Combust Sci Technol 58(1–3):231–251

Sirignano WA (1990) Fluid dynamics and transport of droplets and sprays. Cambridge University Press

Sundaram D, Yang V, Yetter RA (2017) Metal-based nanoenergetic materials: synthesis, properties, and applications. Prog Energy Combust Sci 61:293–365

Sundararajan T, Ayyaswamy PS (1984) Hydrodynamics and heat transfer associated with condensation on a moving drop: solutions for intermediate Reynolds numbers. J Fluid Mech 149:33–58

Szekely GA, Faeth GM (1982) Combustion properties of carbon slurry drops. AIAA J 20(3):422–429

Takahashi F, Heilweil IJ, Dryer FL (1989) Disruptive burning mechanism of free slurry droplets. Combust Sci Technol 65(1):151–165

Tamim J, Hallett WLH (1995) A continuous thermodynamics model for multicomponent droplet vaporization. Chem Eng Sci 50(18):2933–2942

Tanvir S, Qiao L (2014) Effect of addition of energetic nanoparticles on droplet-burning rate of liquid fuels. J Propul Power 31(1):408–415

Tong AY, Sirignano WA (1982) Analytical solution for diffusion and circulation in a vaporizing droplet. Symp (Int) Combust 19(1):1007–1020

Turns SR, Wong SC, Ryba E (1987) Combustion of aluminum-based slurry agglomerates. Combust Sci Technol 54(1–6):299–318

Tyagi H, Phelan PE, Prasher R, Peck R, Lee T, Pacheco JR, Arentzen P (2008) Increased hot-plate ignition probability for nanoparticle-laden diesel fuel. Nano Lett 8(5):1410–1416

Walther JH, Koumoutsakos P (2001) Molecular dynamics simulation of nanodroplet evaporation. J Heat Transf 123(4):741–748

Wang CH, Chen JT (1996) An experimental investigation of the burning characteristics of water-oil emulsions. Intern Commun Heat Mass Transf 23(6):823–834

Wang CH, Law CK (1985) Microexplosion of fuel droplets under high pressure. Combust Flame 59(1):53–62

Wang CH, Liu XQ, Law CK (1984) Combustion and microexplosion of freely falling multicomponent droplets. Combust Flame 56(2):175–197

Wong SC, Turns SR (1987) Ignition of aluminum slurry droplets. Combust Sci Technol 52(4–6):221–242

Wong SC, Turns SR (1989) Disruptive burning of aluminum/carbon slurry droplets. Combust Sci Technol 66(1–3):75–92

Xu X, Cheng C, Chowdhury IH (2004) Molecular dynamics study of phase change mechanisms during femtosecond laser ablation. J Heat Transfer 126(5):727–734

Yang TH, Pan C (2005) Molecular dynamics simulation of a thin water layer evaporation and evaporation coefficient. Int J Heat Mass Transf 48(17):3516–3526

Yang JC, Jackson GS, Avedisian CT (1990) Combustion of unsupported methanol/dodecanol mixture droplets at low gravity. Symp (Int) Combust 23(1):1619–1625

Zeng Y, Lee CF (2007) Modeling droplet breakup processes under micro-explosion conditions. Proc Combust Inst 31(2):2185–2193

Zhu GS, Reitz RD (2001) A model for high-pressure vaporization of droplets of complex liquid mixtures using continuous thermodynamics. Int J Heat Mass Transf 45(3):495–507

Part III
Atomization Principles and Injection Strategies

Chapter 5
On Primary Atomization in Propulsive Device Fuel Injectors—A Short Review

Kuppuraj Rajamanickam, Achintya Mukhopadhyay and Saptarshi Basu

Abstract This chapter provides a brief review of primary atomization mechanisms in spray nozzle relevant to the propulsive devices. Attention which focused on experimental efforts has been made in understanding the primary atomization. Primary atomization involved in two widely used class of nozzles namely, pressure jet and twin-fluid (air-assist) atomizer is explicitly considered.

Keywords Primary atomization · Liquid jet/sheet breakup · Near-nozzle phenomena · Pressure jet atomizer · Air-assist atomizer

5.1 Introduction

The basic understanding of spray formation from spray nozzle is of importance in many industrial applications (Lefebvre 1988). In general, the spray formation is perceived in two forms, namely primary and secondary atomizations. Though resultant droplet size in the spray is dictated by secondary atomization, the size scales of the droplets subjected to secondary atomization itself are governed by primary atomization. The mechanism of primary atomization is determined by several coupled parameters like injection pressure, nozzle design, ambient gas density, and velocity.

The crucial and complicated nature makes primary atomization as an inevitable phenomenon in liquid spraying process. Classical studies reported primary atomization as an instability driven phenomenon. For instance, as soon as the liquid

K. Rajamanickam · S. Basu (✉)
Department of Mechanical Engineering, Indian Institute of Science, Bangalore 560012, India
e-mail: sbasu@mecheng.iisc.ernet.in

A. Mukhopadhyay
Department of Mechanical Engineering, Jadavpur University, Kolkata 700032, West Bengal, India
e-mail: achintya.mukho@gmail.com

S. Basu et al. (eds.), *Droplets and Sprays*, Energy, Environment, and Sustainability,
https://doi.org/10.1007/978-981-10-7449-3_5

jet leaves from the nozzle, it is subjected to some disturbances as a result wavy pattern is observed at the liquid–air interface. These waves continue to grow in time and space, ultimately results in the breakup of a liquid jet into the droplets. The size of the resultant droplet from this breakup event is shown to be comparable with a length scale of the waves (Savart 1833; Rayleigh 1878). This observation clearly indicates that it is necessary to identify the dominant wavelength at which the liquid jet/sheet is likely to breakup. The two widely used approaches include linear stability analysis (LSA) and experimental investigation of the spray (preferably time-resolved imaging) (Ashgriz 2011).

Here, LSA is basically a theoretical approach which involves solving a dispersion relation which governing the stability of an interface for an applied disturbance, whereas experiments mostly involve shadow imaging of the spray and first order estimates like breakup length, spray spread (cone angle) can be retrieved from the images. In addition, researchers have proposed several other experimental techniques to delineate the quantitative information about primary atomization. The main idea of this paper is to briefly present the various experimental investigations attempted to understand the primary atomization processes.

Though correlations are available for final droplet size distribution in the spray as a function of injection conditions, a more detailed fundamental understanding of primary atomization is needed to construct the models.

In Sect. 5.1, the importance of experimental investigation over primary atomization in understanding the spray formation is briefly discussed. In subsequent sections, the various aspects of primary atomization mechanisms occur in two widely used atomizer configurations are presented. Several non-dimensional numbers proposed by researchers in relevant to primary atomization also provided.

5.2 Primary Atomization

In spray formation process, primary atomization represents the first disintegration of ligaments/droplets from the continuous column of liquid jet/sheet. As soon as the liquid leaves the spray nozzle, it is subjected to various forms of aerodynamic instability and subsequently breakup into fragments and droplets (see Figs. 5.1 and 5.3). These droplets may undergo further breakup and it is termed as secondary atomization. Hence, any spray formation involves four major processes as follows:

- Formation of liquid sheet/jet (green-colored box in Fig. 5.1)
- Breakup of liquid sheet/jet and formation of ligaments (marked with yellow circles in Fig. 5.1)
- Breakup of ligament and formation of parent droplet (red-colored box in Fig. 5.1)
- Breakup of parent droplet into several daughter droplets

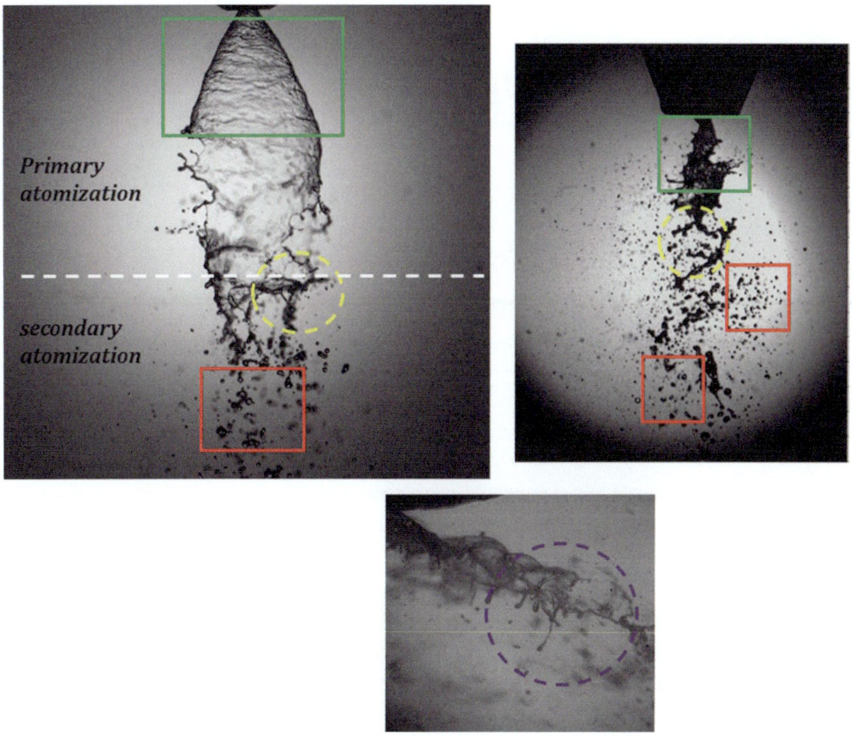

Fig. 5.1 Stages involved in atomization process

Primary atomization can be represented by various non-dimensional numbers; among these, the most widely used parameter is Weber number (We), which gives the quick estimate to identify when liquid jet/sheet becomes likely to breakup

$$We \sim \frac{\rho_a U_R^2 D}{\sigma} \quad (5.1)$$

where ρ_a is air density (kg/m^3); U_R is relative velocity between liquid jet and ambient; D is liquid jet diameter; σ is surface tension (N/m). The condition pertains to $We \gg 1$ is considered as favorable situation for liquid jet breakup. The other important non-dimensional numbers will be shown in subsequent sections.

Nowadays, various techniques range from pressure atomizing to ultrasonic have been used in spray industry to produce sprays. Among, these two techniques namely pressure-atomizing and air-assist (twin-fluid) nozzles remain popular till date. The fundamental atomization mechanism involved in these techniques is illustrated in Fig. 5.2. In pressure-atomizing nozzles, the liquid will be usually injected at very high injection pressure into the quiescent environment. At the nozzle exit, the applied injection pressure is converted into kinetic energy, which

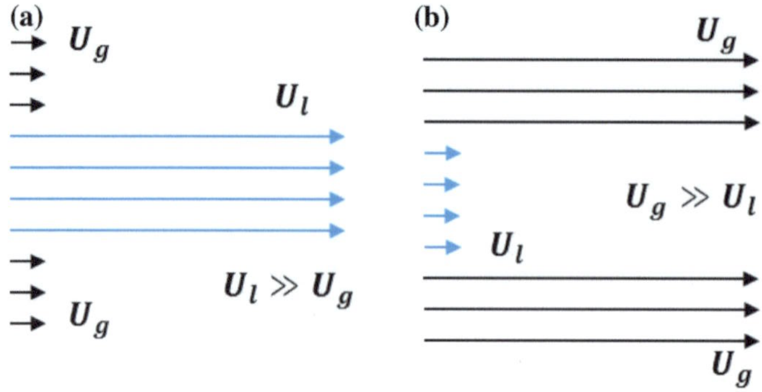

Fig. 5.2 Typical gas- and liquid-phase velocity profile in pressure-atomizing and air-assist atomizers

acts as an energy source for primary atomization. On the other hand, in the air-assist nozzle, the liquid is usually exposed to high-velocity air (~100 m/s). The high momentum associated with gas phase disrupts the liquid sheet.

Though in both the cases, the instability waves at the interface are initiated due to velocity difference (i.e., shear), and the form of shear (axial or azimuthal) is determined by coflowing gas phase. For example, in some atomizers swirling flow is used, in this case, shear will be predominantly in azimuthal direction rather than axial component.

In pressure-atomizing nozzle, the primary atomization is solely determined by fuel injector itself, whereas in air-assist atomizers, primary atomization is determined by the form of gas-phase instabilities induced at liquid–gas interface.

Another, an important feature is to identify the mechanism of primary breakup, the form of jet disruption determines the spatial distribution of the droplets. For the given nozzle, the primary breakup mechanism may vary as function of its operating conditions. In coaxial atomizer, at very low air-flow rate, the liquid jet exhibits orderly mannered breakup (Fig. 5.3a), whereas at very high flow rate, the presence of various size scale of turbulent eddies in the air flow makes the chaotic breakup of the liquid jet (Fig. 5.3c). Hence, it is necessary to identify the primary breakup mechanism for the given nozzle at various operating conditions.

The detailed investigation in primary atomization not only yields the primary droplet size, it also helps to perceive qualitative and quantitative information in near-nozzle breakup dynamics. The near-nozzle breakup is much important, particularly in air-assist atomizers, because this is the zone where momentum exchange happens between gas and liquid phase. Hence, the knowledge gained in primary atomization zone can be used to optimize the nozzle design. Further, it is also mandatory to identify the dominant instability mechanisms and its frequency signatures in coaxial atomizers, because this information is mandatory to ensure the stable operation in liquid-fueled combustors like gas turbines, rocket engines.

The mode of primary atomization is governed by two major parameters namely, whether the liquid is injected in form of liquid sheet or continuous jet and energy

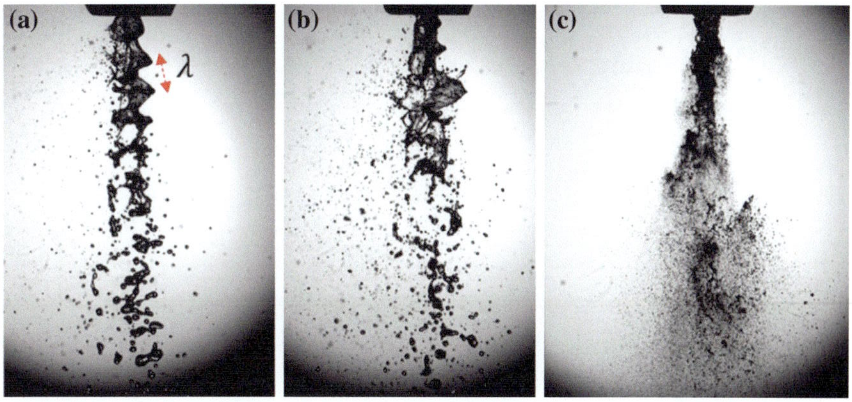

Fig. 5.3 Visualization of liquid jet breakup in coaxial air-assist atomizer: **a** $U_g = 5$ m/s; **b** $U_g = 10$ m/s; $U_g = 60$ m/s

source utilized for atomization (i.e., pressure or air assist). For instance, the pressure jet-based atomizer is widely used in diesel engines. Most of the oil burners utilize pressure swirl atomizer (simplex nozzle) to deliver the hollow-cone spray in the form of swirling liquid sheet. Likewise, coaxial atomizer exhibits widespread fanning in rocket engines, gas turbine combustors, etc. In each of these cases, the mode of atomization is completely different and needs to be investigated thoroughly to understand the degree of atomization.

Primary atomization can be quantified by a number of parameters, of these, in experimental perspective, most of the researchers explicitly considered length scale of the wavelength (see Fig. 5.3a) formed at interface, primary droplet size, breakup, or intact length of the liquid sheet/jet.

5.3 Primary Atomization in Pressure-Atomizing Nozzles

In this chapter, an experimental investigation carried out on primary atomization in straight jet and hollow-cone liquid-sheet atomizer is briefly discussed. The various correlations and its characteristic features in primary atomization also presented.

5.3.1 *Pressure Jet Atomizer*

Straight jet pressure atomizer is one of the most widely used spray nozzles in diesel engines, jet cleaning applications. In this atomizer, the liquid velocity at the nozzle exit acts as a main governing parameter in the breakup of the liquid jet. Researchers used jet stability curve which is drawn as a function of liquid injection velocity (U_L)

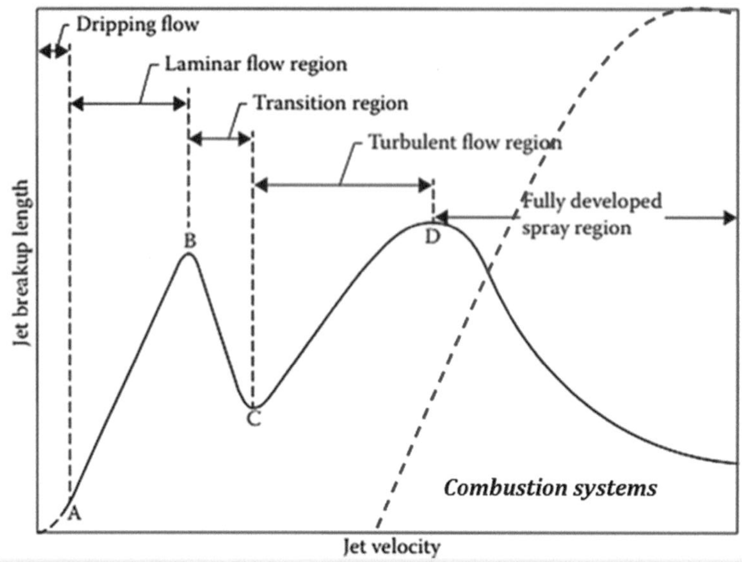

Fig. 5.4 Jet stability curve (adapted from Lefebvre and McDonell 2017)

as a criterion to explain the mode of atomization. The graph (Fig. 5.4) clearly depicts that initial increase in breakup length (L_b) with U_L in the laminar regime, sudden depression in the L_b is shown for the turbulent regime. The turbulent regime is the condition where the spray is realized in the spray nozzle and most of the spray nozzles are usually operated at this condition only (particularly combustion systems). The corresponding velocity in this regime is approximately in the range of ≥ 200 m/s and the waves formed at the liquid jet surface are extremely small in size, which ultimately yields very small-sized droplets.

5.3.2 *Experimental Approaches in Primary Atomization*

It should be noted that at these high-velocity conditions, the wave formation over the jet is no longer exhibits well-defined wavy pattern as observed in Rayleigh breakup regime. The high level of turbulence associated with these jets makes diagnostics as a highly complicated task in the near-nozzle breakup zone. Moreover, the presence of highly dense liquid core in this region (see Fig. 5.5) possesses additional challenges for imaging the primary atomization process. The classical shadowgraph system which employs LED-based stroboscopic lamp may fail to yield a crisp picture of the liquid core. Further, the high velocity of the liquid core results in streak formation in the images. This can be overcome by employing the laser-based illumination system, since laser light source can offer extremely low pulse duration, this can be utilized to get the time frozen snapshot of the spray.

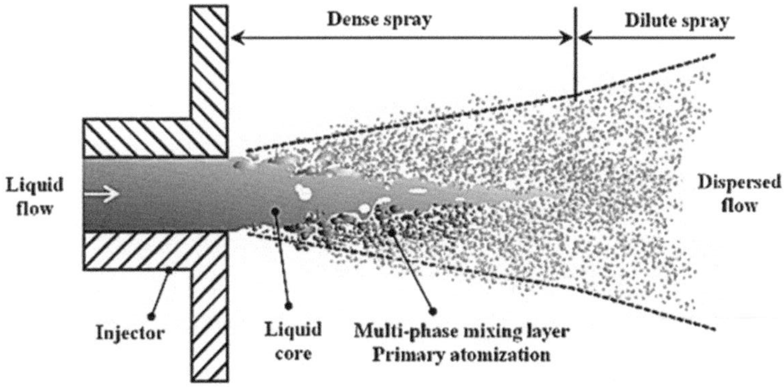

Fig. 5.5 Typical structure of high-pressure diesel spray (reprinted with permission from Faeth 1996)

The spatial ensemble technique like laser-based planar imaging also widely attempted in sprays. It involves passing a thin sheet of laser across the defined section of the spray. The scattered light by droplets is usually imaged with camera, spatially resolved image yields access to quantities like penetration length, cone angle. Even the quantities like droplet size and velocity can also acquire from planar measurements with arrangements like Particle image velocimetry (PIV) and Interferometric Mie imaging (IMI). Again, the dense and two-phase nature of the spray in near nozzle (Fig. 5.5) results in multiple scattering, which essentially makes this technique as inefficient in near-nozzle spray characterization. Attempts have been made to filter these multiple scattering in sprays using spatially modulated laser sheet (Kristensson 2012). Later, this technique received greater attention and the same is known as Structured Laser Illumination Planar Imaging (SLIPI) Fig. 5.6, clearly, depicts the potential of SLIPI technique in imaging the dense region of spray. In particular, the liquid core region can be imaged properly with this system. Berrocal et al. (2012) employed SLIPI technique to image the diesel engine sprays, their results elucidated the presence of large-sized droplets in the near-nozzle region (due to the liquid core), which is failed to detectable in conventional planar imaging.

Researchers also successfully utilized X-ray absorption system to acquire the images of the spray in high-pressure injector (MacPhee et al. 2002; Lai al. 2011). Moon et al. (2010) implemented X-ray-based phase contrast imaging to diagnose the primary breakup characteristics in double-hole diesel injector. The images of the spray acquired (2.0 after ms start of injection) at very close to the nozzle exit shows the existence of swirl-like vortex structure (see Fig. 5.6). The X-ray-based measurements also yield the information about the length scales of the waves formed at the jet surface. This is acquired from the intensity profile measured at the wavy surface of the sheet.

The wavelength information suggests the increasing trend of vortical structures strength with injection pressure. From this, it is concluded that, at very high

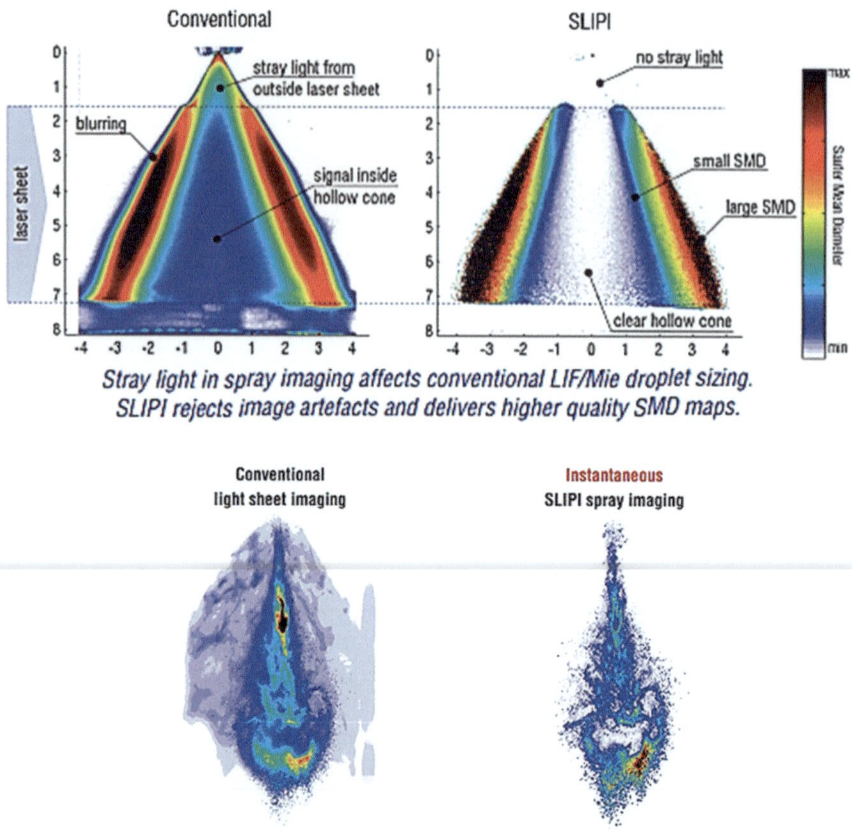

Fig. 5.6 Comparison between conventional and SLIPI-based planar laser imaging of sprays (reprinted with permission from SLIPI 2017)

injection pressure, the sharp decrease in wavelength (Fig. 5.7) and increased strength of wave amplitude result in the rapid breakup of the liquid jet. More details on modern experimental techniques and their limitations/difficulties in spray measurements can be found in Bachalo (2000) and Berrocal (2006) (Fig. 5.8).

MacPhee et al. (2002) investigated the high-pressure sprays under realistic conditions using high-speed X-ray-based optical imaging system. The high-resolution images acquired in the near-nozzle region reveal the occurrence of shock wave. The trailing edge speed of the fuel spray in near vicinity is found to be ~600 m/s and the same is reduced to ~180 m/s at 20 mm downstream of the nozzle. The fluctuations in the mass distribution in the spray edges are corroborated with the presence of shock waves. Later, the Schlieren experiments carried out by Wang et al. (2017) also showed the presence of shock waves in the high injection pressure sprays (see Fig. 5.9).

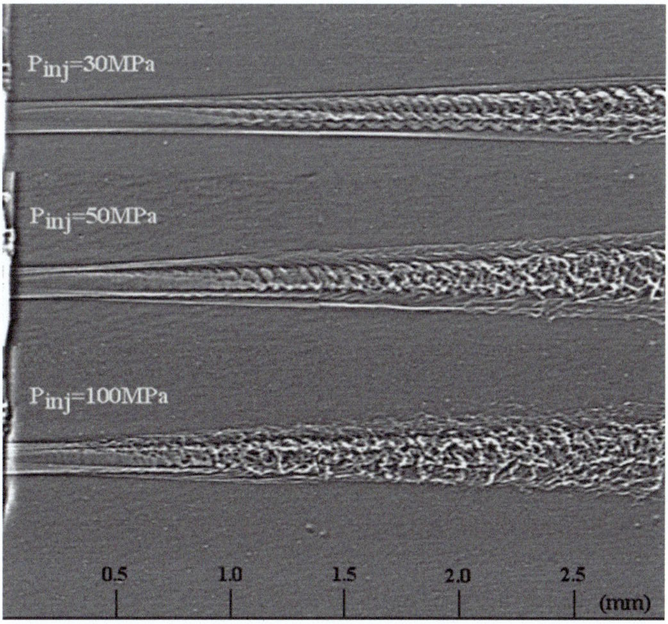

Fig. 5.7 Near-nozzle spray structures in diesel fuel injector acquired using X-ray-based phase contrast imaging (adapted from Moon et al. 2010)

Fig. 5.8 Illustration of wave formation over the liquid jet surface as a function of injection pressure (adapted from Moon et al. 2010)

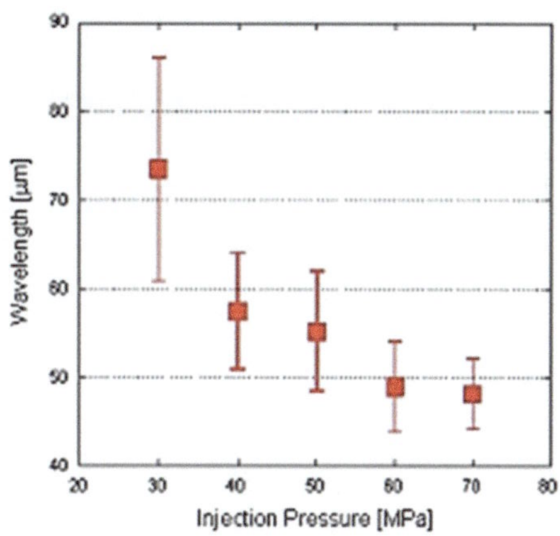

Later, the usage of ballistic imaging (Galland et al. 1995; Wang et al. 1991) in the dense sprays shows its capability in acquiring better spatial resolution images as compared to image quality offered by X-ray technique. Due to its special features over the past decade, ballistic imaging received greater attention among the researchers. This technique relies on photons which are coming out of the sprays, once after the spray is illuminated with very short pulse duration laser. From the

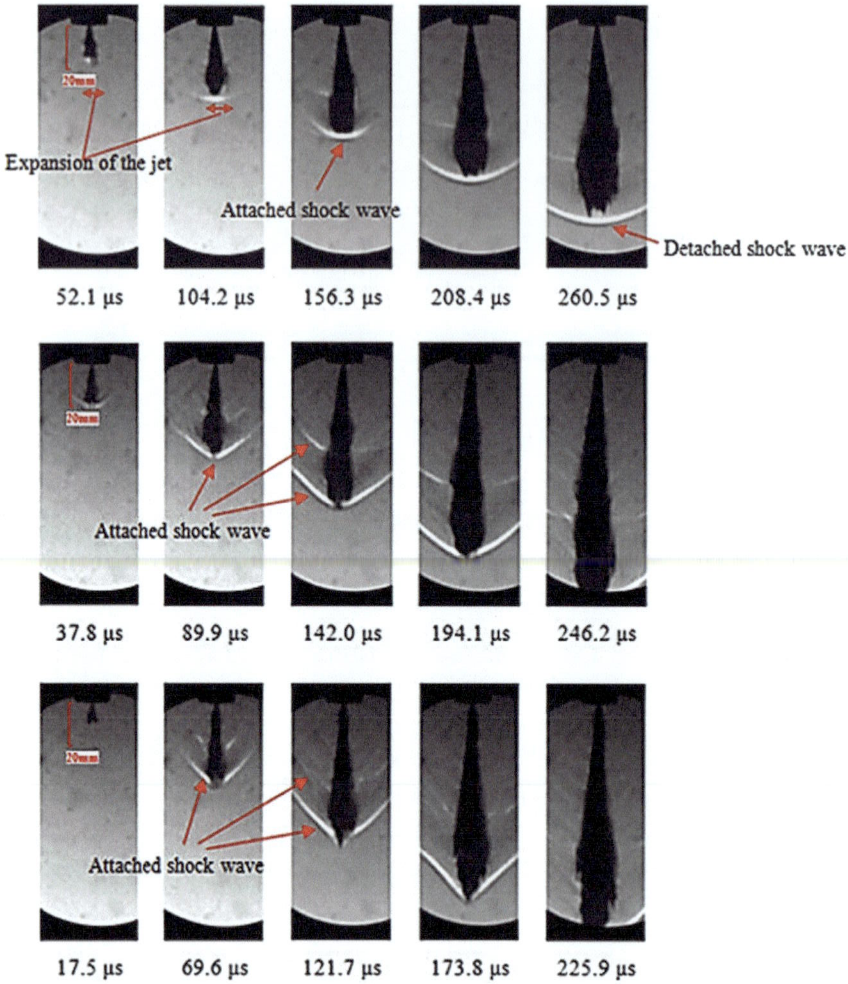

Fig. 5.9 Occurrence of shockwave in high injection pressure sprays (reprinted with permission from Wang et al. 2017)

incident light, a small portion of photons exits directly without any significant scattering and the arrival time of these photons is very less as compared to other scattered photons. These unscattered photons are termed as ballistic photons; later, same is used to reconstructing the spray images. More the ballistic photons, higher the reconstructed image resolution, since the ballistic photons directly represent the spray structure (Linne et al. 2009).

This technique adopts special filter named optical Kerr effect (OKE) gate to segregate the diffused photons from the ballistic photons. Linne et al. (2006) successfully implemented this technique to investigate the near-nozzle spray which is emanated from 155-μm orifice single-hole diesel injector at 1000 MPa. The scheme of the optical system utilized is schematically shown in Fig. 5.9, for brevity detailed optical specifications and working are discussed in this article.

The typical spray structure captured in the near nozzle (~3.5 mm) yields several information about the spray. For instance, the white spot (shown as a void in Fig. 5.10a) in the image indicates the presence of gas vapor on the edge of the spray. The dark blackish region in the center indicates the dominance of liquid phase. This information helps to delineate, how homogeneously the fuel and air mass fraction distributed in the given spatial region. The images captured at different time instants after the start of injection reveal the presence of roll-up structures and detachment of ligaments from the central core region (Fig. 5.10b). In addition, the periodic structures presented at the spray edge also reported.

Several researchers (Driscoll et al. 2003; Araneo et al. 1999; Goldsworthy et al. 2011) implemented planar laser measurements (LIF, PIV) to quantify the flow

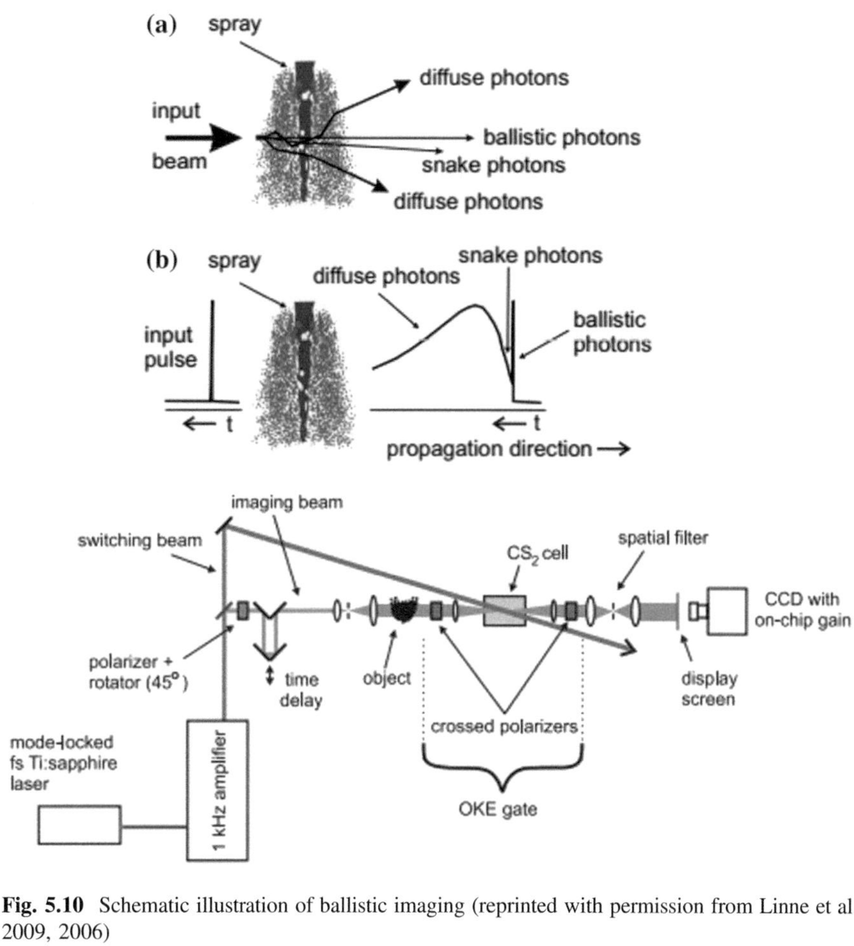

Fig. 5.10 Schematic illustration of ballistic imaging (reprinted with permission from Linne et al. 2009, 2006)

structure formed in the near nozzle. Since these measurements yield the detailed flow field around the nozzle, it can help to identify the zone where maximum turbulence in the spray distributed. It also helps to corroborate the how the flow field influences the spatial distribution of droplet and hence mixing field. For instance, Cao et al. (2000) carried out LIF and PIV measurements using rhodamine dye to visualize the dynamics of internal structures in the spray. The obtained flow field shows the presence of 'branch-like structures,' later, the heterogeneous distribution of interpreted as the cause of these structures.

Accumulation/clustering of the droplet in the region of low-vorticity zone also reported. Further, the quantification is given for surrounding air entrainment in the spray as a function of droplet velocity.

5.3.3 *Pressure Swirl Atomizer*

Like straight jet pressure atomizer, pressure swirl atomizer (also known as a simplex nozzle) is one of the widely used nozzles in combustion applications. The liquid is fed into the series of tangential slots present inside the swirl number, before exiting the orifice, liquid has got sufficient swirling motion inside the nozzle. Upon exiting, the liquid comes like thin liquid sheet whose thickness is very small compared to the nozzle orifice size itself. This is due to the fact the swirling motion induces the formation of air core inside the orifice. Since the given volume of liquid is dispensing in the form of a thin liquid sheet (rather than a straight jet), finer droplets can be achieved at much lesser injection pressure as compared to straight jet pressure atomizer. The usual operating pressure of the simplex nozzle falls in the range of 1–5 MPa. Most of the pressure swirl atomizers are of hollow-cone type.

The classical theory on the mechanism of primary atomization and breakup of the liquid sheet is addressed by several researchers (Squire 1953; Dombrowski and Johns 1963; Fraser et al. 1962). The breakup process (i.e., drop formation) in liquid sheet involves two stages, namely, ligament formation from sheet breakup and the further breakup of ligaments into droplets.

The resultant ligament and droplet diameter are shown to be a function of the length scale of the wave formed over the surface.

$$D_L = \left(\frac{2}{\pi}\lambda t_s\right)^{0.5} \tag{5.2}$$

where t_s is liquid sheet thickness (mm); λ is wavelength (mm). In general, it is shown that the reduction in wavelength scale with increase in sheet velocity (i.e., injection pressure).

Saha et al. (2012) provided experimental values on wavelength observed for the different working condition of the atomizer. High-resolution shadow images are used for this purpose, the obtained result shows good agreement with the existing theory for maximum wavelength.

Presence of two forms of instability, namely short and long-wave instability are present in the primary breakup region. From Fig. 5.11, it is perceived that short waves are attenuated higher Weber numbers (i.e., higher injection pressure), because sheet experiences higher aerodynamic force in this regime due to increased injection velocity. Later, the diameter of the ligaments and resultant droplets formed in the primary zone is shown to be scaled with wavelength (Figs. 5.12 and 5.13).

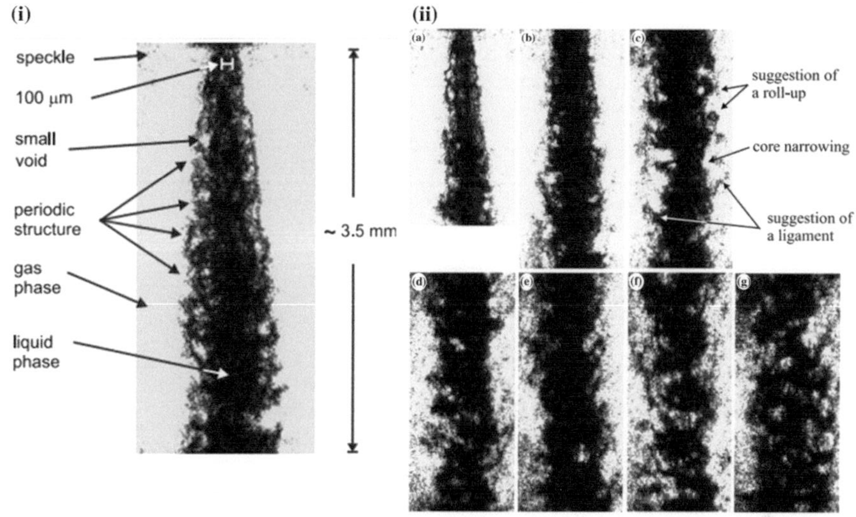

Fig. 5.11 Typical near-nozzle diesel fuel spray structure acquired at 1000 MPa (reprinted with permission from Linne et al. 2006)

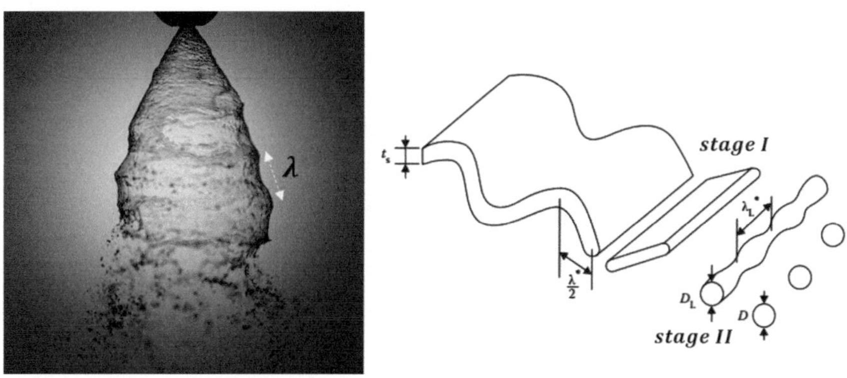

Fig. 5.12 Wave formation over the surface of liquid sheet (adapted from Lefebvre and McDonell 2017)

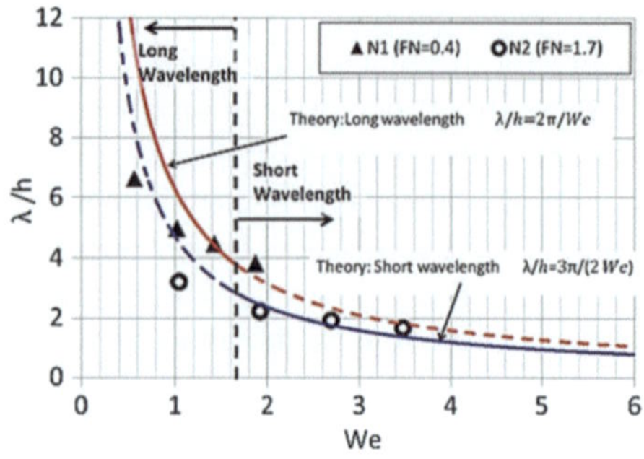

Fig. 5.13 Variation of wavelength as a function of liquid injection pressure (reprinted with permission from Saha et al. 2012)

The theoretical expression also is given for ligament diameter in both regimes (i.e., short and long wave)

$$D_L = \sqrt{\frac{16t_s}{We}} \tag{5.3}$$

$$D_L = \sqrt{\frac{24t_s}{We}} \tag{5.4}$$

5.4 Primary Atomization in Twin-Fluid Atomizer

The combined action of two fluids (gas and liquids) in the atomization process makes twin-fluid atomization as fundamentally different and highly challenging task. The review by Lasheras and Hopfinger (2000) highlighted the complexities associated with quantifying the near-nozzle breakup of liquid jet exposed high-speed air. In pressure-atomizing nozzles, the atomization is mostly influenced by nozzle design (orifice size, injection pressure), whereas in twin-fluid nozzles, atomization is determined by several parameters. The design includes both internal

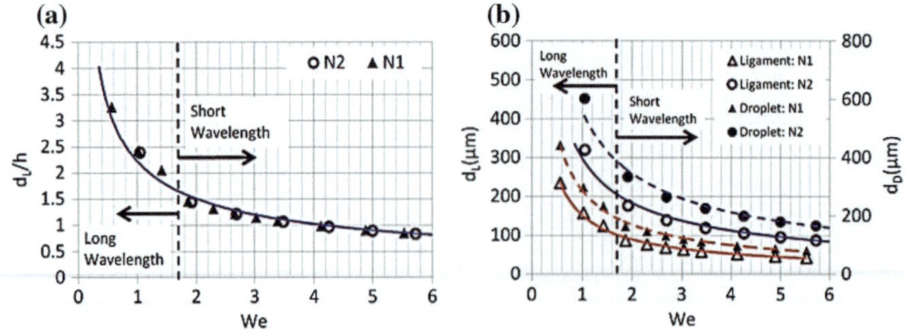

Fig. 5.14 Variation of primary ligament diameter as function of injection pressure (reprinted with permission from Saha et al. 2012)

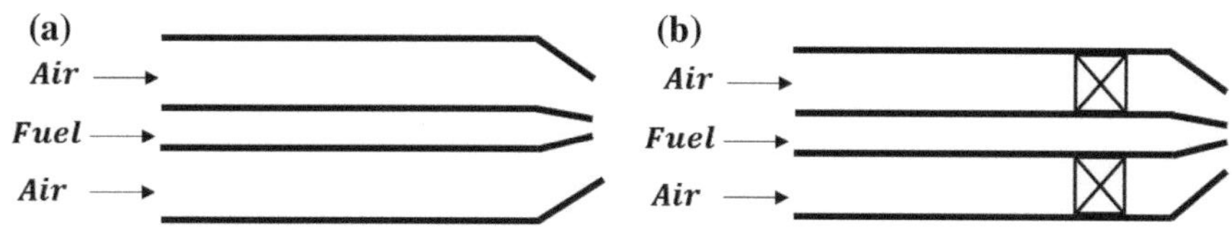

Fig. 5.15 Schematic of twin-fluid atomizers: **a** plain coaxial air-assist atomizer; **b** coaxial swirl air-assist atomizer

and external air mixing configuration, in this chapter primary atomization mechanism pertains to only external mixing is considered. The atomization mechanism may vary as a function of airflow scheme, i.e., axial round jet (Fig. 5.14a) or swirling air jet (Fig. 5.14b). Both configurations are widely used in several engineering applications (Fig. 5.15).

Further, in some atomizers liquid is discharged in the conical/flat sheet instead of round column of the jet. In addition to the Weber number, momentum flux ratio between two phases (gas and liquid) is considered as main governing parameter in air-assist atomizers (Lefebvre 1990)

$$MR = \frac{\rho_a U_a^2}{\rho_l U_l^2} \tag{5.5}$$

where, ρ_a, ρ_l and U_a, U_l is air, liquid density (kg/m3) and velocity (m/s), respectively. Like injection pressure, in twin-fluid nozzles most of the investigations are reported as function of coflowing gas velocity (i.e., *MR*). Hence, the resultant droplet size (SMD) is found to be scaled as $SMD\, \alpha\, U_g^{-n}$, where n varies from 0.7 to 1.5 (Lasheras et al. 1998).

5.4.1 Plain Coaxial Air-Assisted Atomizer

The breakup and atomization of a liquid jet in high-speed coaxial gas flow has been widely investigated by many researchers. When, MR $\gg 1$, the transfer of kinetic energy from fast moving air to liquid causes the explosive breakup of liquid jet/sheet, this process is also known as Airblast atomization. Coaxial atomization is viewed as the result of shear instability arises at the interface due to the discontinuity in the velocity of two parallel moving fluids (Lin et al. 1990; von Helmholtz 1868). The length scales of the unstable wave predicted by Raynal (1997) through experiments show inconsistency with the predictions made from KH instability analysis. His experiments showed the formation of velocity boundary layer at the nozzle surface yields continuous velocity profile at the nozzle exit (Fig. 5.16).

Flow visualization (shadowgraphy) studies carried out in the coaxial atomizer at different MR values suggests various regimes of a jet breakup. From Fig. 5.17, it is clear that for the given liquid Reynolds number (Re_l), increase in air velocity causes the breakup to shift from axisymmetric to shear, fiber-type atomization (Fig. 5.17).

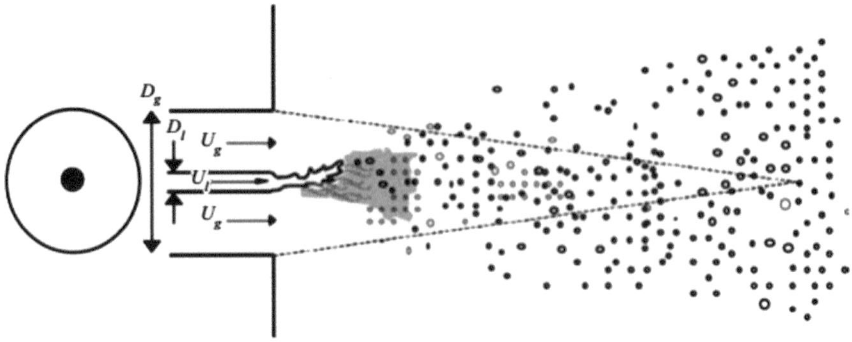

Fig. 5.16 Schematic illustration of liquid jet breakup by high-speed coaxial gas stream (reprinted with permission from Varga et al. 2003)

At very high air-flow rates, the dominance of shear at the interface brings chaotic nature in the breakup of the liquid jet (see Fig. 5.16f–h). At low flow rates MR < 1, length scale of intact portion of liquid core (L_b) is shown to be function of liquid injection velocity (i.e., Re_l), whereas for high air-flow rates, L_b is found to be scaled with value of MR

$$L_b/D = \frac{6}{\sqrt{MR}}\left|1 - \frac{U_l}{U_a}\right|^{-1} \tag{5.6}$$

The empirical correlations proposed by other researchers (Rehab et al. 1997; Hiroyasu 1991; Chigier and Reitz 1996) also exhibit consistence with the results reported through experiments (Lasheras and Hopfinger 2000). The expression 5.6 is proposed by balancing the dynamic pressure across the interface between two phases.

Later, the detailed experimental and theoretical investigation carried out by Matas et al. (2011), Lasheras et al. (1998) and Marmottant and Villermaux (2004) noticed the presence of 'δ_g' in primary instability of two-phase mixing layer (see Fig. 5.18). Moreover, the mechanism of atomization and stability of interface is found be great dependent function of δ_g. Marmottant and Villermaux (2004) proposed two forms of analysis namely, zero and nonzero vorticity layer limit. The zero-vorticity layer assumption essentially yields conventional KH instability treatment of an interface, which assumes discontinuity in the velocity profile.

On the other hand, the inclusion of vorticity layer thickness closely predicts the unstable nature of the interface, particularly for higher MR values. Further, the unstable wavelength arises from the primary instability of liquid–gas interface is shown as function of gas vorticity layer thickness δ_g

$$\lambda_p = \delta_g\{\rho_l/\rho_a\}^{1/2} \tag{5.7}$$

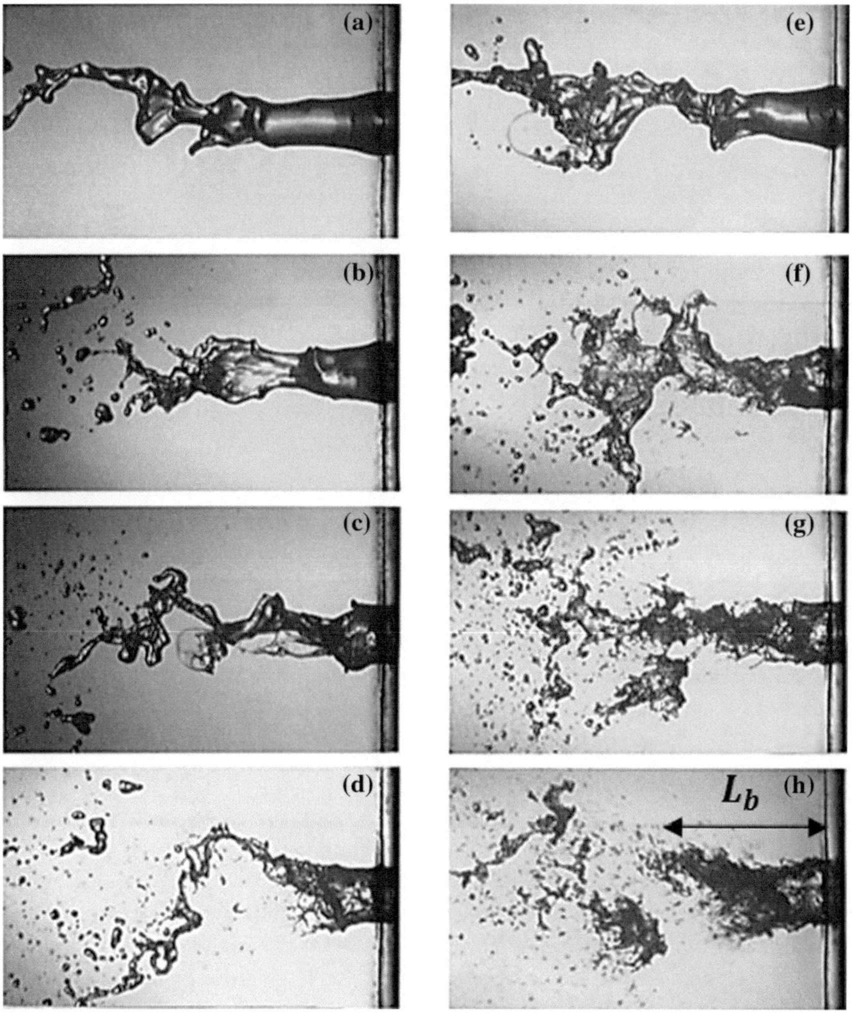

Fig. 5.17 Qualitative visualization of primary atomization of liquid jet in coaxial air-assist atomizer (reprinted with permission from Lasheras et al. 1998)

The shedding frequency of the liquid jet subjected to high MR values is found toe scales with $f_t \sim U_i/\lambda_p$, where U_i is interfacial velocity, it can be calculated from the expression proposed by Dimotakis (1986) in two-dimensional shear layer entrainment experiment

$$U_i \sim \frac{\sqrt{\rho_l}U_l + \sqrt{\rho_g}U_g}{\sqrt{\rho_l} + \sqrt{\rho_g}} \tag{5.8}$$

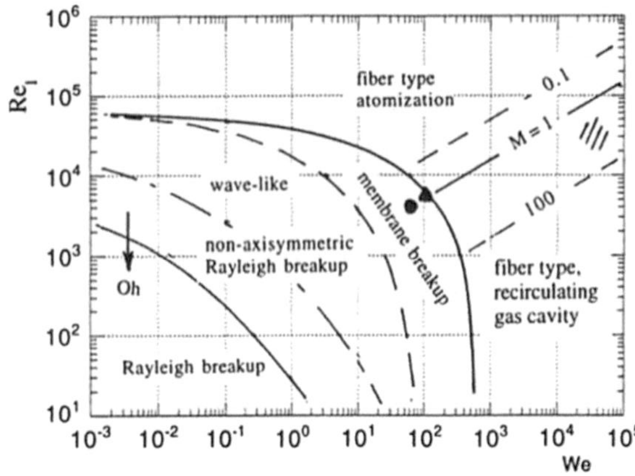

Fig. 5.18 Breakup regime map for coaxial atomizer (reprinted with permission from Hopfinger 1998)

The theoretical frequency f_t has been cross-validated by experimental investigations in several class coaxial air-assist geometries (Matas et al. 2015; Fuster et al. 2013).

Whenever the strength of the interfacial waves exceeds the restoring surface tension force, the surface of the liquid jet/sheet tends to peel off and gives rise to ligaments. The size scale of the ligaments formed during much higher air velocity is shown to be comparable with wavelength arises during primary instability

$$D_L = \frac{1}{4}\left(\frac{d_l}{(MR)^{1/2}}\right) * \lambda_p / X_b \tag{5.9}$$

Here, d_l is liquid jet diameter (m); X_b—breakup length.

Smoke flow visualization (Fig. 5.19) carried out in two-phase mixing layers clearly depicts the dominance of gas-phase Kelvin–Helmholtz waves in governing the primary atomization process (Jerome et al. 2013). Though coflow air momentum is predominantly in the horizontal direction, the large angle ejection of droplets from the liquid film is reported (Fig. 5.20).

It has been shown that, the series of ligament ejection from the surface of the liquid film is shown to be synchronized with the shedding frequency of the recirculation vortex formed at the interface (Fig. 5.19b).

5.4.2 *Swirl Coaxial Air-Assist Atomization*

Liquid fuel injection in coaxial swirling air-flow field becomes common practice in combustors to enhance the efficiency and reduce the flame height (Lefebvre 1998). Like plain air-assist atomizer, here also spray formation is governed by coaxial air flow. However, three-dimensional nature of swirl flows gives rise to several

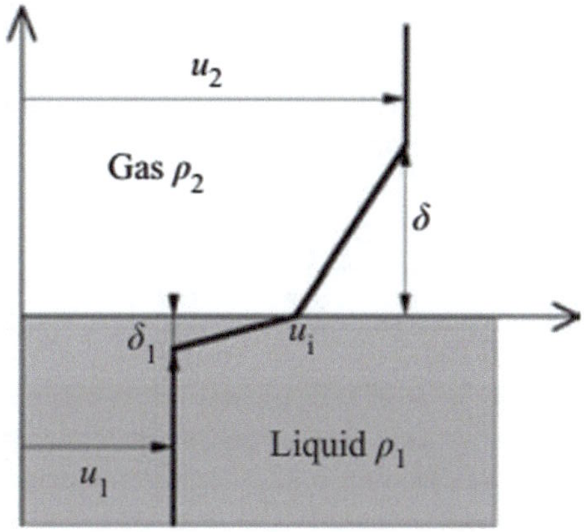

Fig. 5.19 Illustration of vorticity/shear layer formation near the interface (reprinted with permission from Marmottant and Villermaux 2004)

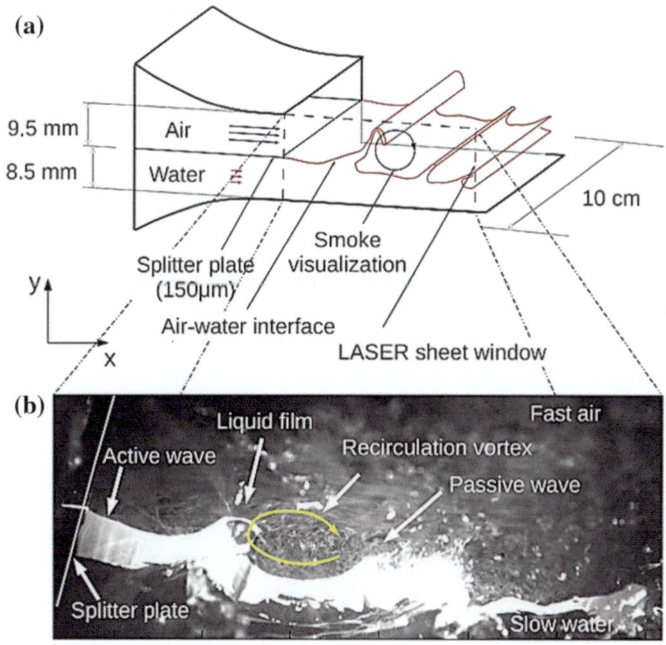

Fig. 5.20 Schematic illustration of wave formation at two-phase mixing layers in coaxial air-assist atomizer (reprinted with permission from Jerome et al. 2013)

complicated breakup mechanisms (Panchagnula et al. 1996). For example, swirl flow exhibits shear in axial as well as the azimuthal direction which eventually leads spanwise and streamwise breakup of liquid jet/sheet (Rajamanickam and Basu 2017).

In general, the swirling strength of the coflowing air is quantified by non-dimensional number named Swirl number (S). It represents the ratio between fluxes of tangential to axial momentum. It is well known that the flow exhibits central toroidal recirculation (CTRZ) in the position close to injector while swirl number attains a value of ≥ 0.6. The typical flow field observed across swirler for $S \sim 0.81$ is shown in Fig. 5.19, the flow reversal induced by vortex bubble breakdown (VBB) results in the formation of two counter-rotating eddies in the near field (Fig. 5.19a). Further, this zone is identified as the region of higher vorticity magnitude (see Fig. 5.19b).

Rajamanickam and Basu (2017) employed time-resolved measurements to delineate the spatiotemporal process involved with the liquid sheet injected into the strong coaxial swirling flow field (Fig. 5.21). They implemented proper orthogonal decomposition (POD) analysis to identify the most energetic structure involved in primary atomization of the liquid sheet. The results suggest the dominance of centrifugally unstable mode in the momentum transfer process; this mode has been identified as one of the inherent structures present in the strong swirling jet (Gallaire and Chomaz 2003) (Fig. 5.22).

The trend of the primary breakup length scales observed for liquid sheet breakup in swirling flows is illustrated in Fig. 5.21. The liquid sheet exhibits uniform breakup in the absence of coaxial swirling air (Fig. 5.21a); however, oscillation is observed in the magnitude of length scales in the temporal space while the liquid sheet is interacting with swirling air flow (Fig. 5.23).

For example, at t = 0.0565 s, the magnitude of X_{bl} is higher than X_{br}, whereas for the same flow rate (MR = 6100), at t = 0.0577 s, the value of X_{br} is more than X_{bl}.

The instantaneous images acquired from high-speed shadowgraphy experiments reveal different breakup mechanism even for fixed air-flow rate. Later, this is found

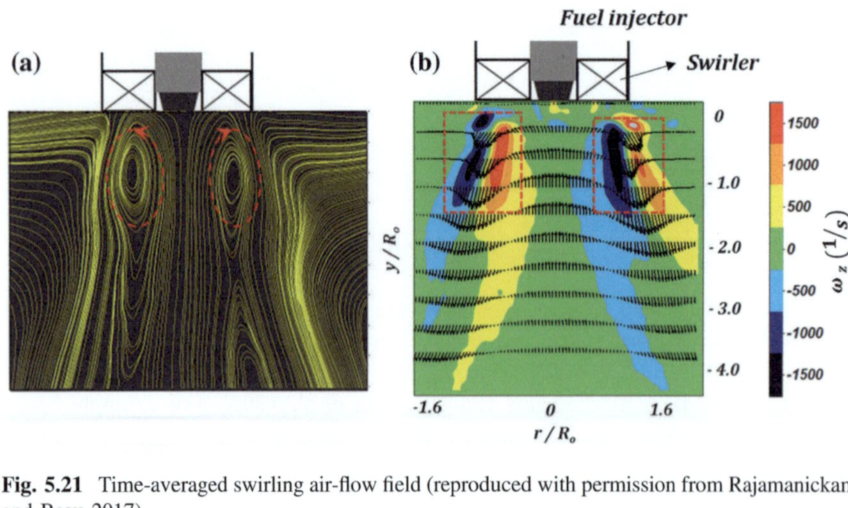

Fig. 5.21 Time-averaged swirling air-flow field (reproduced with permission from Rajamanickam and Basu 2017)

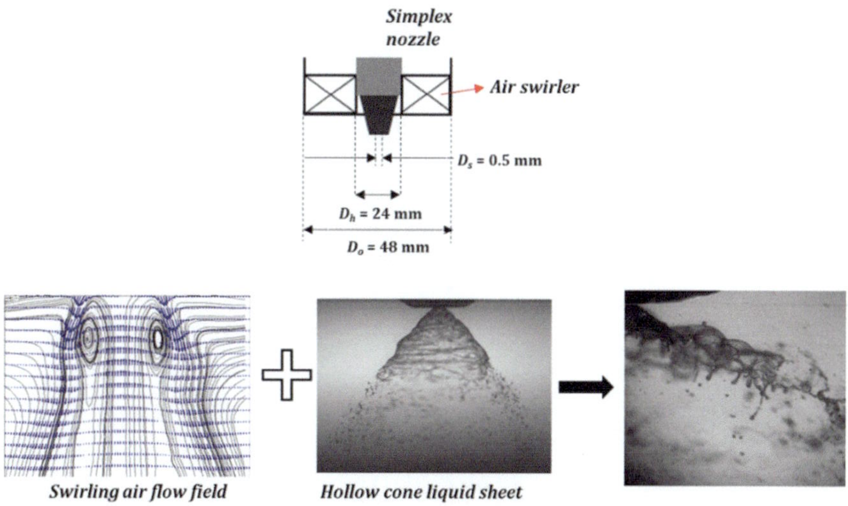

Fig. 5.22 Geometrical details of coaxial swirl atomizer assembly (reprinted with permission from Rajamanickam and Basu 2017)

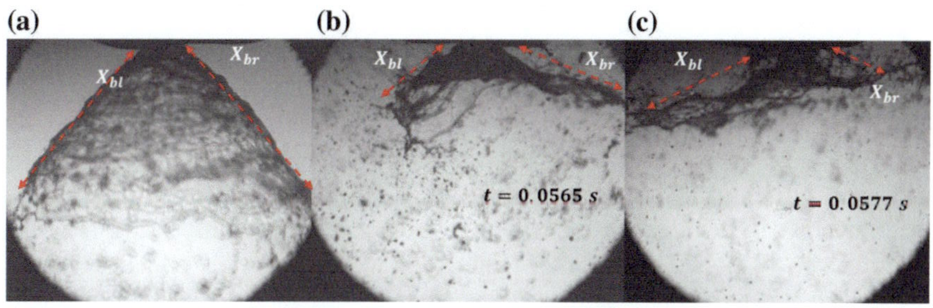

Fig. 5.23 Instantaneous visualization of liquid sheet breakup in swirling flow; **a** MR = 0; **b** MR = 6100; MR = 6100

to be the cause of dynamic nature (roll-up time) of eddies present in the flow field. To delineate this effect, modified Weber number is proposed in terms of circulation (see Eq. 5.9)

$$\overline{We_{\overline{\Gamma}}} \sim \frac{\rho_a^2 \left(\frac{\overline{\Gamma}}{2\pi \overline{r}_{cc}}\right)^2}{\left(\frac{\sigma}{t_f}\right)} \quad (5.10)$$

where $\overline{\Gamma}$ is time-averaged circulation strength (m^2/s) obtained in the region where vortices show maximum strength (shown as red dotted lines in Fig. 5.19b); r_{cc} is vortex core radius (m); σ is surface tension (N/m); t_f is liquid sheet thickness (m).

It is known that in highly turbulent swirling flows, dynamic evolution of vortices results in instantaneous fluctuations in circulation strength Γ', which ultimately brings fluctuations in the Weber number $\left(We'_{\Gamma'}\right)$ in temporal domain. As a result, liquid sheet perceives varying aerodynamic force across the surface of the liquid sheet. Further, FFT implemented over the instantaneous breakup length scales yields the frequency value close to POD temporal mode pertains to mode 1. This quantifies the dominance of gas-phase instabilities in breakup dynamics of liquid sheet. More detailed insights on swirl spray interaction in near-nozzle region can be found in Rajamanickam and Basu (2017).

The proposed theoretical expression for breakup length based on maximum wavelength shows close resemblance with the experimental value.

$$X_b \approx 0.1 \left\{ \frac{1}{\sqrt{\frac{\rho_a}{\rho_l}} \frac{(U_a - U_i)}{\delta_g}} \right\} f_{KH} * \delta_g \sqrt{\frac{\rho_l}{\rho_a}} \tag{5.11}$$

where f_{KH}—Kelvin–Helmholtz frequency (Hz).

5.5 Summary and Conclusions

The detailed experimental investigations pertaining to the primary atomization of liquid jet/sheet in relevance to pressure-atomizing nozzle and twin-fluid injector have been discussed. The attempts made and difficulties associated with the experimental characterization of near-nozzle breakup also briefly presented. It is clear that in both the cases, it is mandatory to quantify the primary breakup mechanism for a complete understanding of spray formation process. The length scale of the wave formed at the liquid–air interface is found to be a main governing parameter in primary atomization.

Acknowledgements Financial support from DST Swarnajayanti Fellowship scheme and NCCRD is great fully acknowledged.

References

Araneo L, Coghe A, Brunello G, Cossali GE (1999) Experimental investigation of gas density effects on diesel spray penetration and entrainment. SAE Technical Paper

Ashgriz N (2011) Handbook of atomization and sprays: theory and applications. Springer Science & Business Media

Bachalo W (2000) Spray diagnostics for the twenty-first century. Atomization and sprays vol 10, no 3–5

Berrocal E (2006) Multiple scattering of light in optical diagnostics of dense sprays and other complex turbid media

Berrocal E, Kristensson E, Hottenbach P, Aldén M, Grünefeld G (2012) Quantitative imaging of a non-combusting diesel spray using structured laser illumination planar imaging. Appl phys b 109(4):683–694

Cao Z-M, Nishino K, Mizuno S, Torii K (2000) PIV measurement of internal structure of diesel fuel spray. Exp Fluids 29:S211–S219

Chigier N, Reitz RD (1996) Regimes of jet breakup and breakup mechanisms- Physical aspects. In: Recent advances in spray combustion: Spray atomization and drop burning phenomena, vol 1, pp 109–135

Dimotakis PE (1986) Two-dimensional shear-layer entrainment. AIAA J 24(11):1791–1796

Dombrowski N, Johns WR (1963) The aerodynamic instability and disintegration of viscous liquid sheets. Chem Eng Sci 18(3):203–214

Driscoll KD, Sick V, Gray C (2003) Simultaneous air/fuel-phase PIV measurements in a dense fuel spray. Exp Fluids 35(1):112–115

Faeth GM (1996) Spray combustion phenomena. In: Symposium (international) on combustion, vol 26, pp 1593–1612

Fraser RP, Eisenklam P, Dombrowski N, Hasson D (1962) Drop formation from rapidly moving liquid sheets. AIChE J 8(5):672–680

Fuster D et al (2013) Instability regimes in the primary breakup region of planar coflowing sheets. J Fluid Mech 736:150–176

Gallaire F, Chomaz J-M (2003) Instability mechanisms in swirling flows. Phys Fluids (1994-present) 15(9):2622–2639

Galland PA et al (1995) Time-resolved optical imaging of jet sprays and droplets in highly scattering medium. American Society of Mechanical Engineers, New York, NY (United States)

Goldsworthy L, Bong C, Brandner PA (2011) Measurements of diesel spray dynamics and the influence of fuel viscosity using PIV and shadowgraphy. Atomization and sprays vol 21, no 2

Hiroyasu H (1991) Break-up length of a liquid jet and internal flow in a nozzle. In: Proceedings ICLASS-91, pp 275–282

Hopfinger EJ (1998) Liquid jet instability and atomization in a coaxial gas stream. In: Advances in Turbulence VII. Springer, pp 69–78

Jerome JJS, Marty S, Matas J-P, Zaleski S, Hoepffner J (2013) Vortices catapult droplets in atomization. Phys Fluids (1994-present) 25(11):112109

Kristensson E (2012) Structured laser illumination planar imaging SLIPI applications for spray diagnostics

Lai M-C et al (2011) Characterization of the near-field spray and internal flow of single-hole and multi-hole sac nozzles using phase contrast X-ray imaging and CFD. SAE Int J Engines 4 (2011-1-681):703–719

Lasheras JC, Hopfinger EJ (2000) Liquid jet instability and atomization in a coaxial gas stream. Annu Rev Fluid Mech 32(1):275–308

Lasheras JC, Villermaux E, Hopfinger EJ (1998) Break-up and atomization of a round water jet by a high-speed annular air jet. J Fluid Mech 357:351–379

Lasheras JC, Villermaux E, Hopfinger EJ (1998) Break-up and atomization of a round water jet by a high-speed annular air jet. J Fluid Mech 357:351–379

Lefebvre A (1988) Atomization and sprays, vol. 1040. CRC press

Lefebvre AH (1990) Energy considerations in twin-fluid atomization. In ASME 1990 international gas turbine and aeroengine congress and exposition, p V003T06A002–V003T06A002

Lefebvre AH (1998) Gas turbine combustion. CRC press

Lefebvre AH, McDonell VG (2017) Atomization and sprays. CRC press

Lin SP, Lian ZW, Creighton BJ (1990) Absolute and convective instability of a liquid sheet. J Fluid Mech 220:673–689

Linne M, Paciaroni M, Hall T, Parker T (2006) Ballistic imaging of the near field in a diesel spray. Exp Fluids 40(6):836–846

Linne MA, Paciaroni M, Berrocal E, Sedarsky D (2009) Ballistic imaging of liquid breakup processes in dense sprays. Proc Combust Inst 32(2):2147–2161

MacPhee AG et al (2002) X-ray imaging of shock waves generated by high-pressure fuel sprays. Science 295(5558):1261–1263

Marmottant P, Villermaux E (2004) On spray formation. J Fluid Mech 498:73–111

Matas J-P, Marty S, Cartellier A (2011) Experimental and analytical study of the shear instability of a gas-liquid mixing layer. Phys Fluids (1994-present) 23(9):094112

Matas J-P, Marty S, Dem MS, Cartellier A (2015) Influence of gas turbulence on the instability of an air-water mixing layer. Phys Rev Lett 115(7):074501

Moon S et al (2010) Ultrafast X-ray phase-contrast imaging of high-speed fuel sprays from a two-hole diesel nozzle. In: ILASS Americas, 22nd annual conference on liquid atomization and spray systems

Panchagnula MV, Sojka PE, Santangelo PJ (1996) On the three-dimensional instability of a swirling, annular, inviscid liquid sheet subject to unequal gas velocities. Phys Fluids (1994-present) 8(12):3300–3312

Rajamanickam K, Basu S (2017) Insights into the dynamics of spray–swirl interactions. J Fluid Mech 810:82–126

Rayleigh L (1878) On the instability of jets. Proc Lond Math Soc 1(1):4–13

Raynal L, Villermaux E, Lasheras JC, Hopfinger EJ (1997) Primary instability in liquid-gas shear layers. In Symposium on turbulent shear flows, 11th, Grenoble, France, pp 27–1

Rehab H, Villermaux E, Hopfinger EJ (1997) Flow regimes of large-velocity-ratio coaxial jets. J Fluid Mech 345:357–381

Saha A, Lee JD, Basu S, Kumar R (2012) Breakup and coalescence characteristics of a hollow cone swirling spray. Phys Fluids 24(12):124103

Savart F (1833) Mémoire sur le choc d'une veine liquide lancée contre un plan circulaire. Ann chim 54(56):1833

Squire HB (1953) Investigation of the instability of a moving liquid film. Br J Appl Phys 4(6):167

Structure laser illumination planar imaging (SLIPI) (2017). http://www.lavision.de/de/products/spraymaster/slipi/index.php. Accessed 20 Jul 2017

Varga CM, Lasheras JC, Hopfinger EJ (2003) Initial breakup of a small-diameter liquid jet by a high-speed gas stream. J Fluid Mech 497:405–434

von Helmholtz HLF (1868) Discontinuirliche Flüssigkeits-Bewegungen

Wang L, Ho PP et al (1991) Ballistic 2-D imaging through scattering walls using an ultrafast optical Kerr gate. Science 253(5021):769

Wang Y, Yu Y, Li G, Jia T (2017) Experimental investigation on the characteristics of supersonic fuel spray and configurations of induced shock waves. Sci Rep 7

Chapter 6
A Comprehensive Model for Estimation of Spray Characteristics

Souvick Chatterjee, Achintya Mukhopadhyay and Swarnendu Sen

Abstract Spray characteristics play an important role in determining efficiency of gas turbine or rocket combustors. The breakup of liquid jet is a complex nonlinear process governed by the fundamentals of well-known instabilities like Rayleigh instability and Kelvin–Helmholtz instability. Current availability of powerful computing tools has made computational fluid dynamics (CFD) along with other analytical and numerical techniques a viable tool for the design of combustors which reduced the requirement of expensive experimental studies in the design process. In this chapter, we demonstrate a system which incorporates a liquid jet that breaks up in the presence of a strong swirling field and sandwiched between two swirling air flow streams. Our work includes computational fluid dynamics, analytical technique (linear stability), and statistical tool (entropy formulation) to model the spray characteristics in the form of breakup length and droplet distribution from nozzle geometry and inlet kinematic conditions. The internal hydrodynamics of the nozzle is modeled in commercial software named Ansys. The output of this simulation in the form of flow kinematics is used to estimate the growth rate of instability associated with atomization using a linear stability analysis. The breakup length which is a function of this growth rate is found to closely match with experimental values. A statistical method known as maximum entropy formulation (MEF) is further used with inputs as the breakup length and a mean diameter value (obtained from linear stability analysis) to estimate the droplet diameter distribution. Thus, a comprehensive model is described in this chapter which is a useful prediction tool for spray characteristics and hence is a significant contribution to the spray and droplet community.

S. Chatterjee (✉) · A. Mukhopadhyay · S. Sen
Department of Mechanical Engineering, Jadavpur University, Kolkata, West Bengal, India
e-mail: souvickchat@gmail.com

A. Mukhopadhyay
e-mail: achintya.mukho@gmail.com

S. Basu et al. (eds.), *Droplets and Sprays*, Energy, Environment, and Sustainability,
https://doi.org/10.1007/978-981-10-7449-3_6

6.1 Introduction

Smaller droplets with fine atomization are desired in most combustion systems like diesel, spark ignition, gas turbine engines for high volumetric heat release rate leading to a wider burning range and lower exhaust concentration of pollutant emission (Lefebvre 1984). Current availability of powerful computing tools has made computational fluid dynamics (CFD) along with other analytical and numerical techniques a viable tool for the design of combustors which reduced the requirement of expensive experimental studies in the design process. In this work, we include computational fluid dynamics, analytical technique (linear stability), and statistical tool (entropy formulation) to model the spray characteristics in the form of breakup length and droplet distribution from nozzle geometry and inlet kinematic conditions.

In the process of atomization, a liquid jet or sheet is disintegrated by kinetic energy of the liquid itself or by exposure to high-velocity gas. The mechanism of breakup is inherently complex with the interaction of multiple forces. As has been summarized by Tharakan et al. (2013), in liquid rocket combustors, there exist two types of injectors: impinging jet and coaxial swirl. While in the former, two or more liquid jets impact to form a flat liquid sheet, the latter operates by imparting momentum to the fuel with or without co-flowing gas resulting in an axisymmetric liquid sheet. These injectors are similar to the pressure swirl injectors in gas turbines. For subcritical operation, atomization is governed by similar factors in both liquid rocket and gas turbine injectors.

Although the internal flow dynamics are quite different in different type of atomizers, in general, atomization is a two-step procedure: primary and secondary atomization. In primary atomization, the liquid jet/sheet breaks up into unstable ligaments as a result of the aerodynamic interaction between liquid and surrounding gas, and subsequently, the ligaments further breakdown into smaller droplets. Droplets thus formed interact with the surrounding fluid and undergo further breakup leading to formation of smaller droplets. This process is known as secondary atomization. Secondary atomization strongly depends on the flow characteristics of the combustor, is closely related to phenomena like turbulent transport of droplets, and is not considered in the present work. There exists established literature relating to the ligament and droplet diameters as shown by Senecal et al. (1999). However, determination of ligament diameter is dependent on the type of atomizer such that for airblast atomizers, it is not dependent on sheet thickness (Schmidt et al. 2003) but is directly proportional to square root of sheet thickness for pressure swirl atomization (Senecal et al. 1999).

Different types of nozzles produce three canonical geometrical shapes of liquid: jet, planar sheet, and annular sheet (Lin 2003). While planar and cylindrical sheets are widely studied, annular sheets still remain relatively unexplored and the current work is focused on such liquid sheets. Experimental studies have been performed by many researchers over the years on annular sheets. Motivated by scientific interest in the fluid motion of a hollow jet and by the potential utility of a method

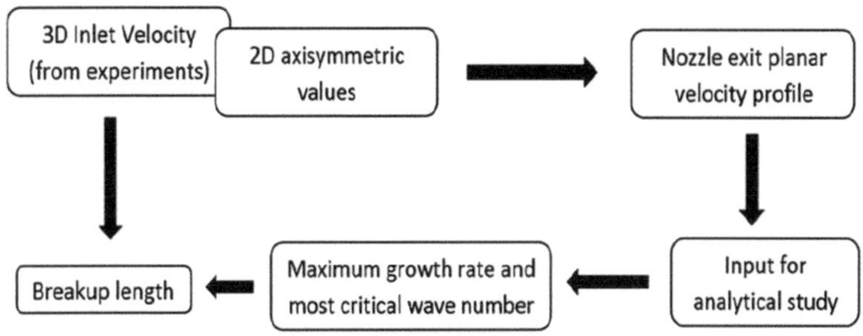

Fig. 6.1 Flowchart of the comprehensive model

for the mass production of rigid shells of high quality, Kendall performed experiments on annular jet and studied their stability (Kendall 1986). Previous works include experiments on four airblast geometry as reported by Lavergne et al. (1993). Another noteworthy work of experiments on annular jets using PDPA is the doctoral dissertation of Shen (1997). The authors of this paper have also reported experimental studies on hybrid atomizers using different post-processing techniques and geometric modularity against which the current comprehensive model results are validated.

A comprehensive spray model is very useful for an exhaustive understanding of the atomization process. Tharakan et al. (2013) provided an excellent review of existing comprehensive models and their shortcomings. As has been mentioned by them, the basic steps in the modeling are: (i) modeling of internal hydrodynamics of the injector leading to prediction of liquid sheet velocity at nozzle exit, (ii) modeling of breakup with inputs from internal hydrodynamics model which predicts breakup length and mean droplet diameter, and (iii) prediction of droplet size and velocity distribution using predicted breakup length and diameter from previous step. The sequential steps involved in the model are shown graphically in Fig. 6.1. A comprehensive model incorporating all the three modules is rare in literature and is addressed in this work. The first step is addressed by a computational approach, an analytical method is used to model the second segment, and the third step is modeled by a statistical tool.

Sub-comprehensive models using two of the above modules have been shown in the literature. For pressure swirl atomizers, Liao et al. (2000) and Ibrahim and Jog (2007) incorporated the first two modules in their analysis. A stability analysis combined with entropy formulation has been shown by Nath et al. (2011) and Movahednejad et al. (2011). This chapter summarizes all the three modules and compares the breakup length as obtained from the comprehensive model with in-house experiments.

6.2 Design of Atomizer

The hybrid atomizer in the current study has a swirling liquid sheet sandwiched between an inner and an outer swirling air stream. All the three fluids are fed tangentially into the atomizer thus creating a high swirling flow field. Noteworthy, both liquid and air streams possess tangential velocity in the same direction. Both the air streams are fed using a compressor, whereas the liquid is recycled with the aid of a gear pump. The flow rates are monitored using rotameters in all the three streams. Kerosene is used as fuel in the current study. Figure 6.2 shows a dimensional schematic of the atomizer along with a photograph of the actual injector highlighting the nozzle exit plane.

As can be seen from Fig. 6.2c, backlighting is used to illuminate the spray. The diffused backlighting system included 2 linear halogen lamps of 1000 W each. Images were captured at 234 frames per second (fps) with a pixel resolution of 180 × 430. With a proper calibration with a linear scale, such an image covered a region of interest of 15 cm × 35 cm. As can be seen from Fig. 6.2d, the outer profile of the sheet breaks up after a certain axial location. The portion of the liquid stream prior to breakup of outer profile is termed as sheet and the downstream section is the spray region. We define this critical axial distance as the breakup length (L_b). The maximum width of the sheet prior to breakup is denoted as the sheet width (SW). Hence, length and width of the green box denoting the sheet

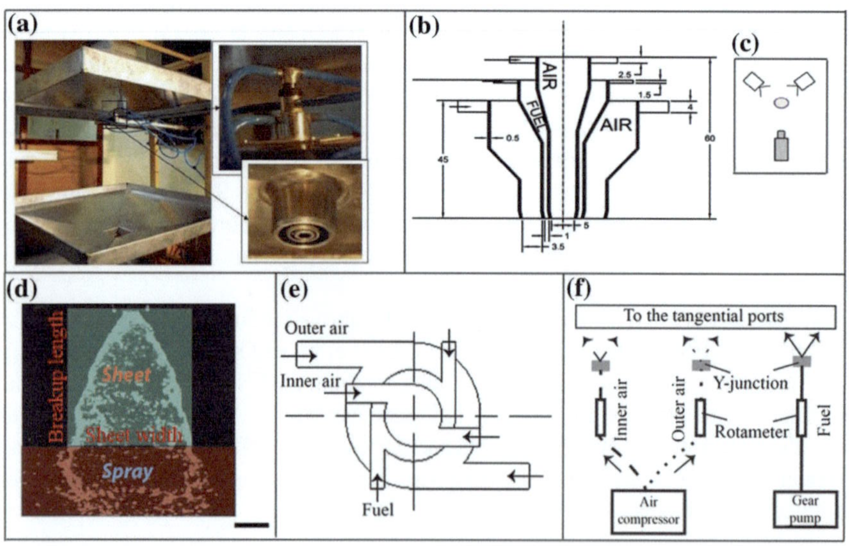

Fig. 6.2 **a** Experimental setup. **b** Schematic of atomizer with dimensions in mm. **c** Imaging technique with backlight and high-speed camera. **d** Snapshot of a spray showing the sheet and spray region, Scale bar: 3 mm. **e** Top view of atomizer showing the tangential entry ports. **f** Schematic of the setup showing the fuel and air pathways

region in Fig. 6.2d denote the breakup length (L_b) and sheet width (SW), respectively. The spray is characterized by the non-dimensional numbers Reynolds number (Re) and Weber Number (We) for the gas streams (inner/outer) and liquid stream, respectively, such that:

$$\mathrm{Re}_{g,j} = \frac{\rho_g U_{g,j} D_j}{\mu_g} \quad \mathrm{j = in, ou;}\ We_l = \frac{\rho_l U_l^2 D_l}{\sigma_l}$$

A detailed parametric study of different sheet profiles for a range of these kinematic parameters is shown in Chatterjee et al. (2015). Figure 6.3 reveals interesting features about the spray structure at different values of the flow hydrodynamic parameters. Five different regimes have been found each showing a different spray structure. A low liquid and air flow typically forms a jet in case of which neither the effect of pressure swirl nor airblast is seen. A high air flow along with liquid with lower Weber Number typically represents an airblast atomizer where early atomization close to the nozzle is observed as shown by the rectangular symbols. However, owing to the presence of high swirl in the flow field, a high liquid flow widens the spray creating a conical sheet. But the aerodynamic force destabilizes this sheet, creating perforations which are prominent in the presence of high air flow.

This gives rise to the perforated conical sheet at high liquid Weber Numbers. Perforated sheets have also been reported by Park et al. (2004) in the breakup of planar liquid sheet from an airblast nozzle. This kind of spray is similar to the "tulip" stage mentioned by Lefebvre (1989), but such stages are known to be observed for pressure swirl atomizers where there occurs formation of air core as a consequence of the swirling motion of the liquid. Thus, the motion of the inner air core and the liquid is not independent of each other. On the other hand, in the present atomizer, the three streams are separated by walls, and thus, their motions within the injector are independent of each other. Hence, the perforated conical sheet formed at high liquid and low air flow combination in the present study is

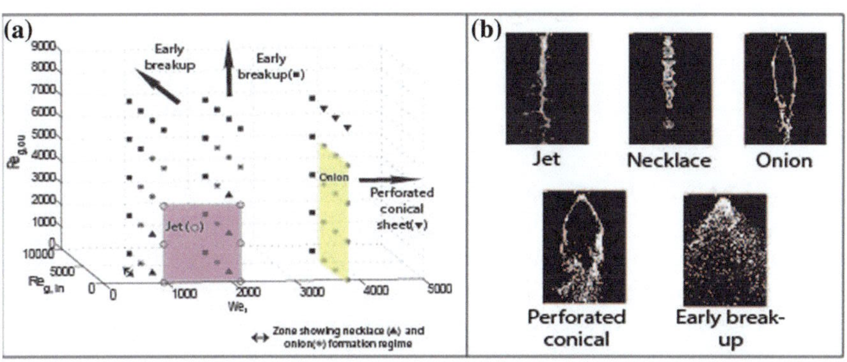

Fig. 6.3 **a** Different regimes of sheet breakup. **b** Characteristic image of each regime

distinct from a conventional pressure swirl atomizer. Apart from these three spray structures formed at boundaries of flow regimes, two other types of sprays are formed in the zone marked in Fig. 6.3a: a less popular necklace mode and a predominantly occurring onion mode. The onion-type sheet profile is common in literature where the sheet bulges outward after exiting from the nozzle and converges further downstream showing a relatively higher breakup length. We found this at a pretty broad spectrum of flows as is shown in the regime diagram (Fig. 6.3a). This type of profile leads to a perforated conical sheet if the liquid flow is increased and to an early breakup regime when air flows are increased. Hence, this structure looks like a stable sheet at an intermediate regime of flows. The necklace formation is a unique observation found at a small regime of low liquid flow and moderate air flow where the sheet modulates into segments of smaller structures and creates the pattern of a necklace. With the knowledge of these sheet structures, a comprehensive model is developed which is compared with the experimental results.

6.3 Model Description

The model constitutes three segments which are described in the following three subsections.

6.3.1 *Internal Hydrodynamics: Computational Fluid Dynamics*

The simulations to understand flow pattern inside the nozzle are done in a commercial package named **ANSYS Fluent**®. The primary equations that need to be solved throughout the domain are Reynolds averaged continuity and momentum equations denoted as:

$$\frac{\partial \rho}{\partial t} + \nabla \cdot (\rho \bar{u}) = 0$$

$$\frac{\partial (\rho \bar{u})}{\partial t} + \nabla \cdot (\rho \bar{u}\bar{u}) = -\nabla p + \nabla \cdot (\mu(\nabla \bar{u} + \nabla \bar{u}^T)) + \rho g$$

where ρ, u, and p are density, velocity, and pressure, respectively.

Navier–Stokes equation is coupled with Volume of Fluid (VOF) surface tracking technique on a fixed Eulerian structured mesh to take care of the fluid–air interaction. In the VOF model, a single set of equations is shared by the fluids, and the volume fraction of each of the fluid is tracked throughout the domain. Water has been used as surrogate fuel in the study and is regarded as the primary fluid for the

VOF studies. The volume fraction of one fluid in the cell is denoted as $\alpha = 0$ for an empty cell; $\alpha = 1$ for a full cell; and $0 < \alpha < 1$ when a cell contains the interface between the first and second fluids. The full domain is initialized with air at time, $t = 0$. The position of a free surface between two immiscible phases is tracked in VOF model by an additional advection equation for the additional phase:

$$\frac{\partial \alpha_i}{\partial t} + u \cdot \nabla \alpha_i = 0$$

The pressure-velocity coupling has been modeled using a SIMPLE algorithm. While the momentum equations are discretized by first-order upwind scheme, Geo-Reconstruct scheme is used for the volume fraction. The geometric reconstruction scheme represents the interface between fluids using a piecewise-linear approach. It assumes that the interface between two fluids has a linear slope within each cell which is used for calculation of advection of fluid through the cell faces.

Turbulence in the momentum equation is modeled by standard k-ϵ model. Though researchers have reported to observe some shortcomings in the ability of this standard model, still there is no conclusive information available in the literature regarding suitable modification of the same (Mohammadi and Pironneau 1994). Also models like algebraic stress model (ASM) and Reynolds stress model (RSM) are either too complex or do not perform significantly better in predicting the strong swirling flow (Datta and Som 2000).

At the exit plane, the second axial derivatives of the variables are equated to zero to ensure smooth transition of the flow. The boundary condition at the wall is considered to satisfy the no-slip condition. Also, an initial solution of the simulation revealed large values of y^+, defined as a non-dimensional wall distance such that $y^+ = \frac{u_* y}{\nu}$, where u_* is the friction velocity at the nearest wall, y is the distance to the nearest wall, and ν is the local kinematic viscosity of the fluid. Hence, y^+ adaption was used to refine the height of the cells along the wall boundary which brought the y^+ value within desirable range ($30 < y^+ < 60$) (King 2010).

As all practical atomizers are three dimensional, first a 3D model is developed for our computations. But a 3D transient model proved computationally very expensive, and hence, a 2D axisymmetric model is also investigated. While in the 3D model a tangential inlet geometry provides the required swirl velocity to the flow, an axisymmetric swirl model is used for the two-dimensional case as explained later. Two different such axisymmetric models were found to exist in the literature (a) by conserving angular momentum, total mass flow rate, and kinetic energy of the fluid at the inlet ports (Ibrahim 2006) and (b) by conserving fluid flow rate and angular momentum at the inlet ports (Datta and Som 2000). The additional conservation of kinetic energy in the former formulation results into a shorter inlet width as compared to the latter where the fluid enters the injector through the entire length of the port. In the context of pressure swirl atomizers, such simplification to a 2D axisymmetric model was found to give excellent results as shown by Ibrahim et al. (2006). In method (a), the axisymmetric assumption requires determination of equivalent velocities since it requires a radial and tangential velocity as inlet boundary condition as compared to the 3D geometry inlet via tangential ports. Similar to the conversion done by Ibrahim (2006), the radial and tangential velocities at the inlet are calculated by matching the angular momentum, total mass flow rate, and the kinetic energy of the fluid at the inlet ports. The inlet tangential and radial velocities of the swirl chamber are given by $W_{inlet} = \frac{Q}{A_p}\frac{(D_s - D_p)}{D_s}$ and $V_{inlet} = \sqrt{(\frac{Q}{A_p})^2 - W_{inlet}^2}$, respectively, where Q is the net fluid intake through the tangential ports. Using V_{inlet}, the inlet width d_h can be calculated such that the net fluid flow rate remains conserved. This leads to the equation $\pi D_s d_h V_{inlet} = Q$ from which the width of the annular zone, d_h , through which the fluid enters the atomizers in the 2D configuration is obtained.

In method (b), flow rate and angular momentum are conserved. Let A_{tube} be cross-sectional area of the tube through which flow is occurring. Hence, for inner air, liquid, and outer air streams, respectively:

$$A_{tube,in} = \frac{\pi}{4} d_i^2$$

$$A_{tube,l} = \frac{\pi}{4} (D_{ol}^2 - D_{il}^2)$$

$$A_{tube,oa} = \frac{\pi}{4} (D_{oa}^2 - D_{ia}^2)$$

where d_i is the diameter of the inner air tube, and D_o and D_i are outer and inner diameters of the annular tubes. Now, with flow rate conserved (say Q), the axial velocity in each of the streams will be given by:

$$V_{axial} = \frac{Q}{A_{tube}}$$

Now, if A_p be the port area through which the fluid enters the atomizer, angular momentum (M_θ) is given by:

$$M_\theta = \dot{m} \frac{Q}{A_p}$$

such that

$$M_\theta = \int_{R_i}^{R_o} 2\pi r dr u \rho V_\theta$$

The limit R_i and R_o varies for the three different streams, and a linear swirl velocity profile is assumed for all the streams.

Inner air

Swirl Velocity Profile: $V_\theta = V_{\theta,w}\frac{r}{R}; \mathrm{R_i} = 0; R_o = \frac{d_i}{2}$.

Hence, $M_\theta = \frac{2\pi\rho u V_{\theta,w}}{R}\frac{R^3}{3}$.

Using the above two expression of angular momentum, $V_{\theta,w}$ is calculated which provided the velocity profile of the tangential velocity. Thus, both the axial and tangential velocities are known. Similar methodology is used for the other fluid streams such that the boundary conditions only vary. The varying boundary conditions for the liquid and outer air are shown below:

Liquid

$$V_\theta = V_{\theta,w}\frac{r - R_i}{R_o - R_i};\ R_i = \frac{D_{il}}{2};\ R_o = \frac{D_{ol}}{2}$$

Outer air

$$R_o = \frac{D_{ol}}{2};\ R_i = \frac{D_{ia}}{2};\ R_i = \frac{D_{ia}}{2}$$

Thus, an axial velocity and a linear swirl velocity profile are obtained for all the three fluid streams which are used as the inlet velocity conditions for the simulations.

As can be seen from Table 6.1, the error in the cone angle and peak exit swirl velocity value using the axisymmetric assumption are within acceptable ranges when Method A is used in the 2D case. Method B is seen to overestimate the swirling nature of the spray such that a higher peak swirl velocity and cone angle are obtained. Hence, as is used in Method A, conservation of kinetic energy is important to simulate the 2D axisymmetric configuration. Summing this with the

Table 6.1 Comparison between 3D and 2D axisymmetric results

Parameter	Inlet liquid tangential velocity (m/s)	3D case	2D case Method A	2D case Method B	Error % Method A	Error % Method B
Cone Angle	150	42.9	43.2	51.5	0.699	20.05
	200	43.7	45	54	2.975	23.57
	250	44.2	44.5	55.4	0.679	25.34
	300	45.2	46	56	1.77	23.9
	350	46.4	17.1	56,3	1.509	21.34
3D peak exit swirl velocity	150	245.9	243.6	312.56	0.944	27.11
	200	301.9	298.5	389.47	1.139	29
	250	350.2	348.6	446.74	0.459	27.57
	300	398.1	395.9	467.16	0.631	17.26
	350	452.5	454.8	548.9	0.508	21.3

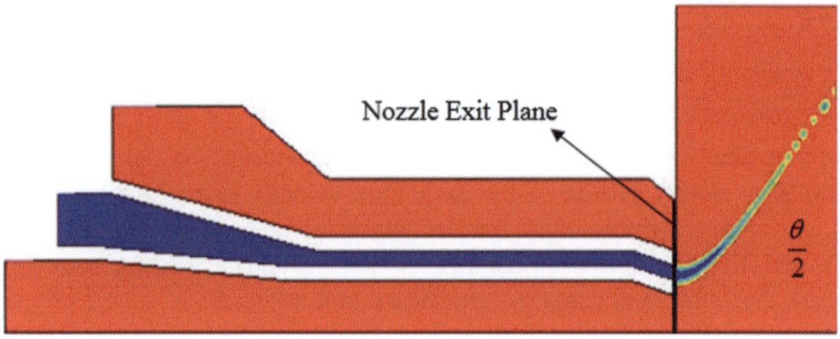

Fig. 6.4 Typical phase diagram showing the cone angle and nozzle exit plane

advantage of saving computational power and time, the 2D axisymmetric assumption using Method A proves to be a fruitful substitute for a parametric study and is discussed subsequently.

The output parameters analyzed from the simulations are the cone angle and the velocity profiles at the nozzle exit plane. Both from the 2D and the 3D simulations, we obtain a phase contour as shown in Fig. 6.4. While for the 2D case output is directly obtained from the simulations, in the 3D case, an axisymmetric plane at the middle of the geometry is chosen wherein the phase contour is studied. The phase contours directly provide a measure of the spray cone angle, θ as shown in Fig. 6.4.

The multiphase volume of fluid (VOF) method formulation is checked for mesh independence and also validated with published experimental results. Two parameters namely spray cone angle and maximum swirl magnitude at the exit plane are monitored for varying node numbers, keeping the kinematic conditions constant as shown in Table 6.2. The results became independent of spatial resolution for nodes more than 2.5e5, for the 2D axisymmetric study, and for 3e5 nodes in the 3D configuration. The bold setting in Table 6.2 is used for our study.

The experimental results of Jeon et al. (2011) have been used to validate the mathematical model presented earlier. Their experiments involved a recession zone inside the atomizer, and hence, we modeled the same using VOF in Fluent. In this 2D axisymmetric model, the central gas core is fed axially, whereas the liquid is injected radically and hence possess a radial and a tangential component. For different liquid axial velocity (U_{la}) and momentum flux ratio (M), a measurement of gas velocity to liquid axial velocity, the evolution of spray cone angle according to

Table 6.2 Mesh independent study for 2D axisymmetric configuration

Number of nodes	Spray cone angle	Max. swirl velocity at exit plane
22478	42.84	234
86072	42.87	241.8
115905	42.86	243.32
256457	**42.86**	**243.31**
367191	42.86	243.31

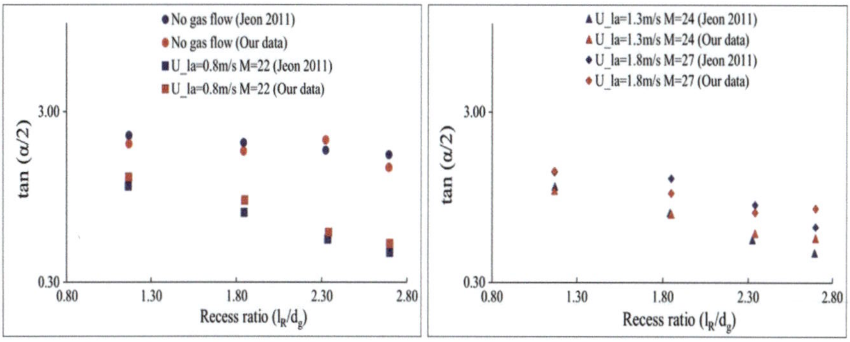

Fig. 6.5 Validation of numerical model with literature

recess length are studied. As can be seen from Fig. 6.5, the similar trend of decrease of cone angle with increase in recess length is observed. In addition, the cone angle values are in excellent quantitative agreement with the experimental results (Jeon et al. 2011).

6.3.2 *Linear Stability Analysis*

The second module for the comprehensive model is a linear stability analysis to estimate the breakup length and a mean droplet diameter size. This analysis uses output of the simulation results, described in the previous section, as input conditions for the axial and swirl velocities of the liquid and air streams. The analysis is based on the assumption of inviscid flow and the presence of free swirl which makes the flow irrotational. Such assumptions have also been used by Ibrahim and Jog (2008) for annular sheets and Li (1994) for planar sheets. Thus, velocity potential can be defined for all the three streams, such that the mass conservation equation becomes $\nabla^2\phi=0$, where ϕ is the velocity potential. Details of the methodology are presented in Chatterjee et al. (2013). Considering a potential flow, perturbation is applied to the liquid and air potentials. The unperturbed interface η_0 is deformed, and temporal evolution of the interface is estimated. The interfaces must satisfy the kinematic and dynamic boundary conditions. According to the kinematic boundary condition, the radial velocities at the interface of the fluids in contact are equal at all times. Hence, the general kinematic boundary condition is $v_j=\frac{D\eta}{Dt}$, where j denotes the phase (liquid/air). The dynamic boundary condition requires that the forces on two sides of the interface balance each other. For inviscid fluids, this implies that the difference in liquid and gas pressure across the interface is balanced by surface tension. Thus, at each interface, the force balance can be expressed as $p_l-p_g=\sigma\kappa$. For such surface, the curvature κ is defined as $\kappa=\nabla\cdot\hat{n}$, where $\hat{n}$ is the unit normal of the liquid–gas interface directed toward the gas phase.

The potential formulations are expanded along with the appropriate kinematic and dynamic boundary conditions in Taylor series as suggested by Jazzayeri and Li (2000) such that details of the expressions obtained are shown in Chatterjee et al. (2013). The following initial conditions are used:

$$\eta_{i1}(x,\theta,0) = \cos(kx+n\theta); \eta_{i1,t}(x,\theta,0) = 0$$
$$\eta_{o1}(x,\theta,0) = \cos(kx+n\theta); \eta_{o1,t}(x,\theta,0) = 0$$

In temporal stability, a small disturbance is applied at all points on the interface initially. Subsequently, stability is decided on the basis of whether the deformation grows or decays in time. The velocity potentials are obtained from the first-order equations by using the method of Laplace Transform along with some involved algebra. The theoretical formulation along with solution of dispersion equation is done using ***Mathematica***. The dimensionless dispersion equation in Laplace domain is obtained which is of fourth order.

For temporal instability, the wave frequencies are complex for given real axial wave numbers. The two solutions with positive real parts of complex wave frequency represent two families of unstable solutions, while the other two solutions with negative real parts of complex wave frequency correspond to stable solutions which are not concerned in this work. Also, by inverse Laplace Transform, a relation is obtained between the amplitudes of the disturbances at the two interfaces, which is generally complex. Thus, the two interfacial waves differ not only in the magnitude but also in the phase angle. A close to zero phase difference represents the para-sinuous mode while a value closer to π corresponds to the para-varicose mode which are similar to the sinuous and varicose modes in case of planar sheets. The model is validated with established results. As can be seen in Fig. 6.6, excellent agreement is found for the variation of growth rate with wave number between our results and that of Shen and Li (1996) for the same input conditions. Consistent

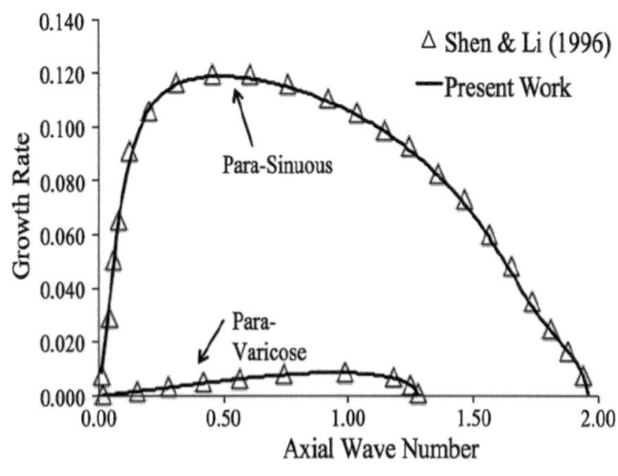

Fig. 6.6 Variation of growth rate for para-sinuous ad para-varicose modes

with literature, para-sinuous mode showed a significantly higher growth rate than the para-varicose mode.

In this work, we formulate a sheet breakup length based on the analogy of the prediction of breakup length of cylindrical liquid jet (Reitz and Bracco 1982). If the surface disturbance amplitude has reached a value of η_b at breakup with initially being η_0 at nozzle exit, a breakup time τ can be calculated as:

$$\eta_b = \eta_0 \exp(\Omega_s \tau)$$
$$\Rightarrow \tau = \frac{1}{\Omega_s} \ln\left(\frac{\eta_b}{\eta_0}\right)$$

where Ω_s is the maximum growth rate obtained from the linear stability analysis. Thus, the sheet breakup length is given by:

$$L_b = V\tau = \frac{V}{\Omega_s} \ln\left(\frac{\eta_b}{\eta_0}\right)$$

where V is the axial velocity. Based on experimental measurements, Dombrowski and Hooper (1962) suggested a value of 12 for the term for planar sheets, and it has been used by Senecal et al. (1999) and Moon et al. (2010). However, Fu et al. (2010) measured this parameter from their images of a conical sheet and measured it to be 2.5. An analysis of experiments carried out in-house shows this value in the range of 1.5–2.5. Hence, in this work, using nozzle geometry and hydrodynamic parameters of Carvalho and Heitor (1998), breakup length is estimated using range of values of $\frac{\eta_b}{\eta_0}$ and compared with experimental breakup lengths. Figure 6.7 shows this comparison. As can be seen at the air flow regimes discussed in this work, experimental breakup lengths are well comparable to analytical ones with $\ln\left(\frac{\eta_b}{\eta_0}\right) = 2.5$. This seems consistent with experimental observations where it was seen conical sheets are predominantly seen at relatively lower inner air after which the sheet tends to breakup close to the nozzle.

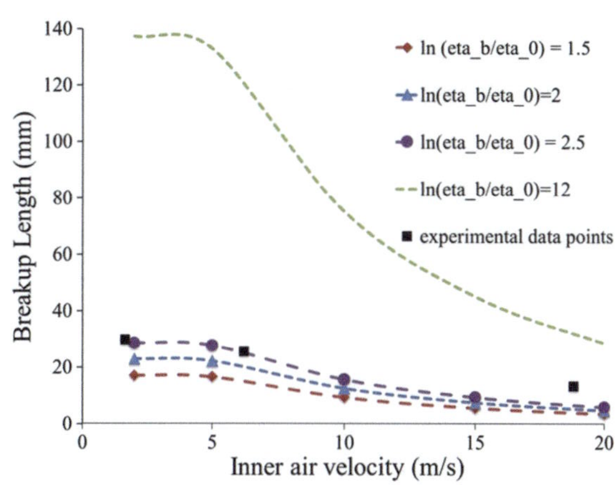

Fig. 6.7 Comparison of breakup length with experimental data for different reported

The ability of linear stability to predict the temporal frequency of interface oscillation was also investigated. From LSA, for a particular hydrodynamic configuration, first the imaginary part of the frequency is identified corresponding to the maximum growth rate. Since, this frequency maps with the maximum growth rate, it can be considered as the frequency prior to sheet breakup as is measured in the experimental study of Wahono et al. (2008). The frequency measured is compared with that of experimentally obtained by Wahono et al. (2008), hence the same nozzle geometry and hydrodynamic parameters are used.

Also, for proper comparison, the angular frequency is converted to a linear frequency using length and timescales. The non-dimensional Strouhal number is calculated given by $St = \frac{fh_l}{V}$ where f is the linear frequency, V is liquid axial velocity, and h_l is liquid sheet thickness. The comparison between Strouhal number thus computed, and those estimated by Wahono et al. (2008) is shown in Table 6.3. The corresponding inner air (u_i), water (u_w), and outer air (u_o) velocities for the cases are shown in the table. As can be seen, the Strouhal numbers closely match except a few cases (C4–C8). However, these cases have been shown in Wahono et al. (2008) to have a high outer air momentum ratio. Thus, this increased difference at these cases can be attributed to the presence of higher instability in the system. The results of the experimental investigation mentioned earlier clearly showed role of air in increasing instability which looks consistent with this frequency estimation. Thus, a nonlinear analysis is required to obtain proper estimates under these conditions.

6.3.3 *Comparison with In-house Experiments*

We did not have provision for measurement of pressure gradient across the injector. The rotameters used in the experiments provided flow rates which correspond to a particular inlet pressure for all the streams. The nozzle is designed to eject spray into atmosphere. As part of the comprehensive study, the absolute pressure drop of the liquid across the injector was estimated from the CFD analysis to be ~1.26–5 Pa when the inlet flow rate was *2 LPM*. The nozzle exit axial and swirl velocity profiles obtained from CFD analysis for different outer air corresponding to an inner flow of *10 LPM* ($Re_{g,in} = 2777$) are shown in Fig. 6.8a and b. Flow conditions are similar to the ones used in the experimental study. The increased outer air at the nozzle inlet leads to a higher axial and swirl velocity at the exit atomizer which increases instability in the sheet as can be seen from Fig. 6.8c. Inner air velocity profile, particularly swirl velocity profile is not affected significantly by change in outer air flow rate. An increased outer air flow creates higher maximum growth rate

Table 6.3 Comparison of Strouhal number with established experimental results

Case number (Shen and Li 1996)	Velocity conditions (m/s)	Strouhal number (Shen and Li 1996)	Strouhal number (from our LSA)
CI	$u_i = 3.2$; $u_w = 1$; $u_o = 8.0$	0.0244	0.02045
C2	$u_i = 2.8$; $u_w = 1$; $u_o = 11.1$	0.0366	0.024
C3	$u_i = 9.1$; $u_w = 1$; $u_o = 12.1$	0.0406	0.0294
C4	$u_i = 10.1$; $u_w = 1$; $u_o = 16.4$	0.0488	0.0354
C5	$u_i = 4.5$; $u_w = 1$; $u_o = 16.8$	0.0528	0.0312
C6	$u_i = 2.7$; $u_w = 1$; $u_o = 17.5$	0.0528	0.0313
C7	$u_i = 4.5$; $u_w = 1$; $u_o = 18.0$	0.0528	0.0318
C8	$u_i = 14.3$; $u_w = 2$; $u_o = 14.9$	0.0528	0.0364
C9	$u_i = 3.2$; $u_w = 2$; $u_o = 8.1$	0.0064	0.0146
C10	$u_i = 8.5$; $u_w = 2$; $u_o = 12.7$	0.0163	0.0237
C11	$u_i = 14.5$; $u_w = 2$; $u_o = 14.5$	0.0223	0.0293
C12	$u_i = 5.2$; $u_w = 2$; $u_o = 16.1$	0.0244	0.0249
C13	$u_i = 2.6$; $u_w = 2$; $u_o = 17.3$	0.0284	0.0251
C14	$u_i = 9.7$; $u_w = 2$; $u_o = 15.9$	0.0284	0.0272

and larger critical wave number except for the case of a very low flow in the outer air channel as compared to when there is no outer air. A slight decrease in maximum growth rate is observed for $Re_{g,ou} = 1736$ (*10 LPM*) relative to for $Re_{g,ou} = 0$. This results in a small increase in breakup length at this intermediate $Re_{g,ou}$ as shown in Fig. 6.8d, and this trend is consistent with our observation from the experimental data. A higher growth rate at increased outer air corresponds to an early breakup configuration with very low breakup lengths.

Similar observations have been made for inner air *20 LPM* ($Reg_{g,in} = 5555$) in Fig. 6.9 where the exit velocity profiles are shown only for a characteristic case of $Re_{g,ou} = 5208$ (*20 LPM*) since for other outer air flows, trends are observed similar to Fig. 6.8a and b.

At a higher inner air flow of 30 LPM ($\mathrm{Re}_{\mathrm{g,in}} = 8333$), liquid sheets breakup early as was observed from the experiments. The growth rate curves for different

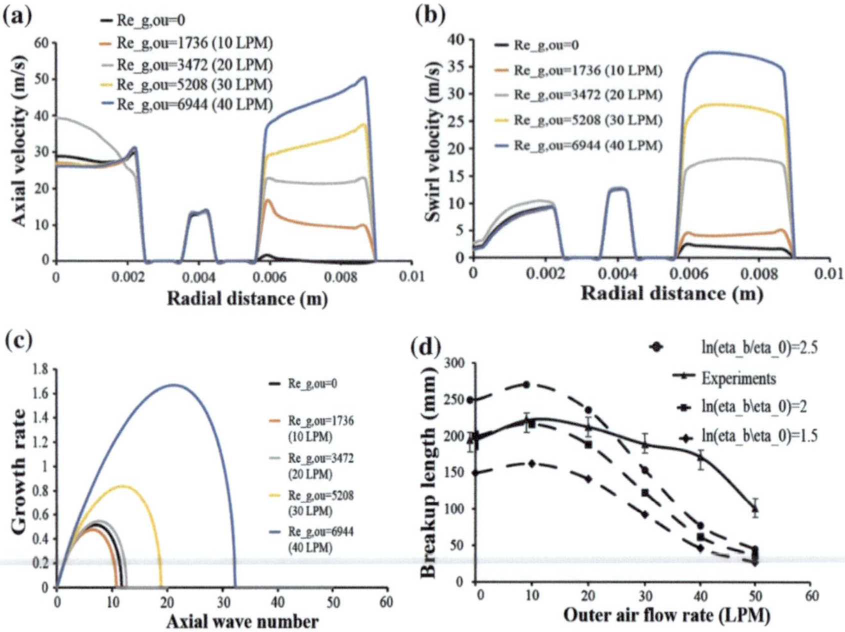

Fig. 6.8 Nozzle exit velocity profiles, growth rate curve, and comparison with experimental breakup lengths for liquid flow 2 LPM and inner air 10 LPM. **a** and **b** Axial and swirl velocity profile at nozzle exit. **c** Growth rate curve for different outer air configuration. **d** Breakup length estimated from maximum growth rate and compared with experimentally measured results

outer air flows collapse at this high inner air leading to the same maximum growth rate and hence breakup lengths for all $Re_{g,ou}$ as seen in Fig. 6.10. Also, as compared to the low and moderate inner air flows, lower breakup length is observed for a 30 LPM inner air flow from both the comprehensive model and experiments. The experiments also show a much smaller variation in breakup length, which is consistent with the predictions. Also, at this inner air flow, a $\ln\frac{\eta_b}{\eta_0}$ equals 12 shows closer match to the experimental values. Hence, with increase in air flow rates, higher values of this parameter are required for comparisons with experiments. The value of $\ln\frac{\eta_b}{\eta_0} = 12$ has been used by multiple researchers over the years (Dombrowski and Hooper 1962; Moon et al. 2010). The fact that this value is showing good comparisons with experimental breakup lengths at high air flows for a conical sheet is a significant contribution to existing knowledge on atomization.

Growth rate curves and axial velocity profiles at the nozzle exit plane for the interesting sheet configurations discussed earlier (necklace, onion, and early breakup) are shown in Fig. 6.11. As can be seen, the early breakup regime is the most favorable condition for breakup with increased growth rate and maximum critical wave number. While the necklace has a higher growth rate compared to the onion mode, the critical wave number is comparable for both these cases.

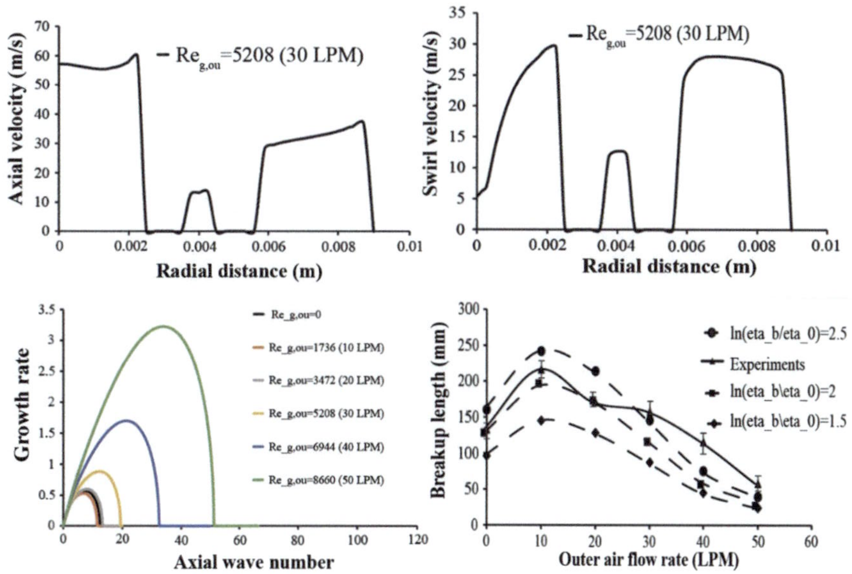

Fig. 6.9 Nozzle exit velocity profiles, growth rate curve, and comparison with experimental breakup lengths for liquid flow 2 LPM and inner air 10 LPM

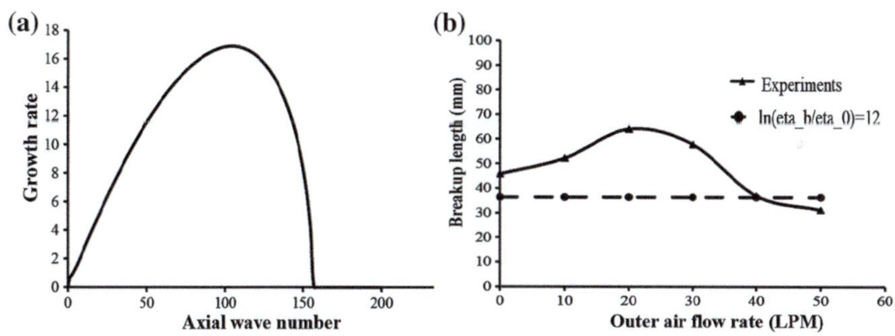

Fig. 6.10 **a** Growth rate variation with axial wave number and **b** Comparison with experimental breakup lengths for liquid flow 2 LPM and outer air 30 LPM

The ligament diameter and a mean droplet diameter can also be estimated from the results of the linear stability analysis using established empirical relations. Senecal et al. (1999) showed different ligament diameters for short and long waves in the case of pressure swirl atomizers. But, for both of those, the ligament diameter varies as $d_L = \sqrt{\frac{h}{k_S}}$ where h is the sheet thickness and k_s is the most critical wave number. For airblast atomizers, as shown by Schmidt et al. (1999), the ligament diameter is independent of sheet thickness. However, for both pressure swirl and

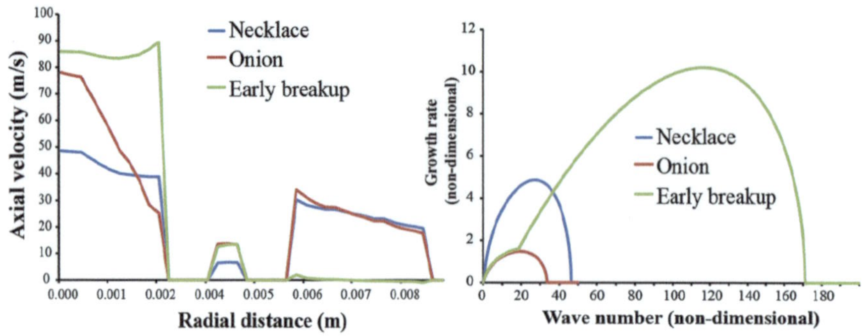

Fig. 6.11 Axial velocity profile and growth rate curves for necklace, onion, and early breakup regime

airblast atomizers, the ligament diameter is related to a mean droplet diameter by the same relation given by:

$$d_D = 1.88 d_L (1 + 3Oh)^{\frac{1}{6}}$$

where $Oh = \frac{\mu}{(\rho \sigma d_L)^{\frac{1}{2}}}$ is the Ohneshorge number.

6.3.4 *Maximum Entropy Formulation*

The maximum entropy formulation (MEF) is based on the information entropy theory and is used for prediction of size and velocity distribution. According to this theory, the most probable droplet size and velocity distributions maximize information entropy or Shannon entropy of the system subjected to the constraint conditions. For a liquid spray, the size and velocity spectra are discretized into finite number of classes. A joint probability function $P_{i,j}$ is defined to express the probability of a droplet to be in diameter class '*i*' and velocity class '*j*' such that:

$$P_{i,j} = \frac{\dot{n_{i,j}}}{\dot{N}}.$$

where $\dot{n}_{i,j}$ is the number of drops per unit time having diameter d_i and velocity v_j, and $\dot{N} = \sum_{i=1}^{m} \sum_{j=1}^{k} \dot{n_{i,j}}$ represents the total number of drops per unit time in the spray in the entire diameter range of $i = 1(1)m$ and velocity range of $j = 1(1)k$. The Shannon entropy for the joint probability distribution is given by Sovani et al. (1999):

$$S = -\sum_{i=1}^{m}\sum_{j=1}^{k} P_{i,j}\ln(P_{i,j})$$

The formulation discussed in this work is similar to the one used by Nath et al. for planar sheets (Nath et al. 2011). A detailed algorithm of the formulation is presented in the doctoral thesis of the same author (Nath 2011). An estimate of the breakup length and mean diameter is used in this formulation as obtained from the stability analysis shown in the previous section.

6.4 Constraint Conditions

The constraint conditions for maximizing the Shannon entropy in a liquid sheet breakup problem are:

- ***Normalization condition***: $\sum_{i=1}^{m}\sum_{j=1}^{k} P_{i,j} = 1$
- ***Mass balance***: This considers conservation of mass during the breakup process such that evaporation effects are ignored. So, the mass mean diameter of sheet (d_m) would be related to the mass mean diameter of all droplets (d_i) in the spray by the relation: $\sum_{i=1}^{m}\sum_{j=1}^{k} \dot{n_{i,j}}\left(\frac{\pi}{6}d_i^3\rho_l\right) = \dot{N}\left(\frac{\pi}{6}d_m^3\rho_l\right)$. In terms of probability, this implies $\sum_{i=1}^{m}\sum_{j=1}^{k} P_{i,j}D_i^3 = 1$ where D_i represents the non-dimensional droplet diameter such that $D_i = \frac{d_i}{d_m}$.
- ***Momentum balance***: The liquid sheet is in contact with air streams in either direction. Momentum is transferred from air to liquid on both inner and outer surface over the breakup length leading to the conservation of momentum given by:
  Momentum of liquid at atomizer exit + Momentum transfer to liquid from inner air due to drag + Momentum transfer to liquid from outer air due to drag = Total momentum of generated droplets

Mathematically, this corresponds to (as a time rate equation):

$$\dot{m}_l u_l + \left(\frac{1}{2}\rho_g(u_{g,i}-u_l)(|u_{g,i}-u_l|)\pi d_i' l C_{f,i}\right) + \left(\frac{1}{2}\rho_g(u_{g,o}-u_l)(|u_{g,o}-u_l)\pi d_o l C_{f,o}\right)$$
$$= \sum_{i=1}^{m}\sum_{j=1}^{k} \dot{n_{i,j}}\left(\frac{\pi}{6}d_i^3\rho_l\right)u_j$$

where l and g are subscripts for liquid and gas, respectively, 'l' is the liquid breakup length, 'd_i'' and 'd_o' are inner and outer diameter of the liquid sheet; C_f is the drag coefficient on liquid sheet by air and equals $\frac{1.328}{\sqrt{Re_g}}$. The air Re is estimated from nozzle exit velocity conditions. Using liquid velocity u_l and inner diameter of liquid sheet d_i' as velocity and length scales (same as used in the linear stability analysis), in terms of probability distribution, the above equation becomes:

$$\sum_{i=1}^{m}\sum_{j=1}^{k} P_{i,j} D_i^3 V_j = 1 + \frac{2\rho L}{(D_o^2-1)}\left(C_{f,i}(U_i-1)(|U_i-1|) + D_o C_{f,o}(U_o-1)(|U_o-1|)\right)$$

where $L = \frac{l}{d_i'}$ is the non-dimensional breakup length, $V_j = \frac{u_j}{u_l}$ is the non-dimensional velocity distribution of the droplets, $\rho = \frac{\rho_g}{\rho_l}$ is the air to liquid density ratio; and U_i and U_o are the non-dimensional inner and outer air velocities.

- ***Energy balance***: This considers conservation of energy between the annular liquid sheet and that of droplets formed in the process. Hence, similar to the momentum equation, this requires:
  Kinetic Energy of liquid at atomizer exit + Energy transferred from inner air to liquid + Energy transferred from outer air to liquid = Total droplet kinetic energy + Total droplet surface energy

  Mathematically as a time rate equation, this is represented as:

$$\begin{aligned}&\frac{1}{2}\dot{m}u_l^2 + \left(\frac{1}{2}\rho_g(u_{g,i}-u_l)|u_{g,i}-u_l|\pi d_i' l C_{f,i}(u_{g,i}-u_l)\right)\\&+\left(\frac{1}{2}\rho_g(u_{g,o}-u_l)|u_{g,o}-u_l|\pi d_o l C_{f,o}(u_{g,o}-u_l)\right)\\&=\sum_{i=1}^{m}\sum_{j=1}^{k}\dot{n}_{i,j}\frac{1}{2}\left(\frac{\pi}{6}d_i^3\rho_l\right)u_j^2 + \sum_{i=1}^{m}\sum_{j=1}^{k}\dot{n}_{i,j}\,\sigma\pi d_i^2\end{aligned}$$

  In non-dimensional terms and probability distribution, this equation becomes:

$$\begin{aligned}&\sum_{i=1}^{m}\sum_{j=1}^{k} P_{i,j}D_i^3V_j^2 + B\sum_{i=1}^{m}\sum_{j=1}^{k} P_{i,j}D_i^2\\&=1+\frac{4\rho L}{D_o^2-1}\left(C_{f,i}(U_i-1)^2|U_i-1| + C_{f,o}D_o(U_o-1)^2|U_o-1|\right)\end{aligned}$$

where $B = \frac{12}{\frac{\rho_l u_l^2 d_m}{\sigma}}$, with σ is the surface tension of liquid with respect to air.

The primary objective of the MEF formulation is maximization of Shannon entropy given by Shannon (1948) under the constraint equations which can be re-written as:

$$Z = \sum_{i=1}^{m}\sum_{j=1}^{k} P_{i,j} - S_1 = 0$$

$$Y = \sum_{i=1}^{m}\sum_{j=1}^{k} P_{i,j}D_i^3 - S_2 = 0$$

$$X = \sum_{i=1}^{m}\sum_{j=1}^{k} P_{i,j}D_i^3 V_j - S_3 = 0$$

$$W = \sum_{i=1}^{m}\sum_{j=1}^{k} P_{i,j}D_i^3 V_j^2 + B\sum_{i=1}^{m}\sum_{j=1}^{k} P_{i,j}D_i^2 - S_4 = 0$$

where $S_1 = S_2 = 1$

$$S_3 = 1 + \frac{2\rho L}{(D_o^2 - 1)}(C_{f,i}(U_i - 1)^2 + D_o C_{f,o}(U_o - 1)^2)\text{and}$$
$$S_4 = 1 + \frac{4\rho L}{D_o^2 - 1}(C_{f,i}(U_i - 1)^3 + C_{f,o}D_o(U_o - 1)^3)$$

The value of distribution function $P_{i,j}$ which maximizes Shannon entropy $\boldsymbol{S}$ is given by:

$$\frac{\partial S}{\partial P_{i,j}} + \lambda_1\frac{\partial Z}{\partial P_{i,j}} + \lambda_2\frac{\partial Y}{\partial P_{i,j}} + \lambda_3\frac{\partial X}{\partial P_{i,j}} + \lambda_4\frac{\partial W}{\partial P_{i,j}} = 0$$

where λ_1, λ_2, λ_3, and λ_4 are the Lagrange multipliers used for optimization of the system.

Calculating the partial derivatives, the probability distribution $P_{i,j}$ is solved which becomes:

$$P_{i,j} = \exp[-(1 + \lambda_1 + \lambda_2 D_i^3 + \lambda_3 D_i^3 V_j + \lambda_4(D_i^3 V_j^2 + BD_i^2))]$$

The expression for $P_{i,j}$ is subsequently used in the constraint equations to obtain four equations for the four unknown Lagrange multipliers λ_1, λ_2, λ_3, and λ_4. These equations are solved simultaneously using a Newton–Raphson scheme till convergence. The solution procedure for solving this is similar to the one used by Nath (2011). The converged solutions of the Lagrange multipliers are used to obtain the distribution function which is subsequently used to calculate the droplet size diameter distribution.

The droplet size distribution which is a product of the probability distribution function with the diameter values is validated for an annular liquid sheet instability study published in the literature. As can be seen from Fig. 6.12, a good validation particularly for the tail of the distribution is obtained both with experiments as well as previous MEF formulations.

Using the comprehensive methodology flowchart described earlier, the droplet diameter distribution is obtained for a pressure swirl and an airblast configuration based on the mean diameter and breakup length obtained from the linear stability

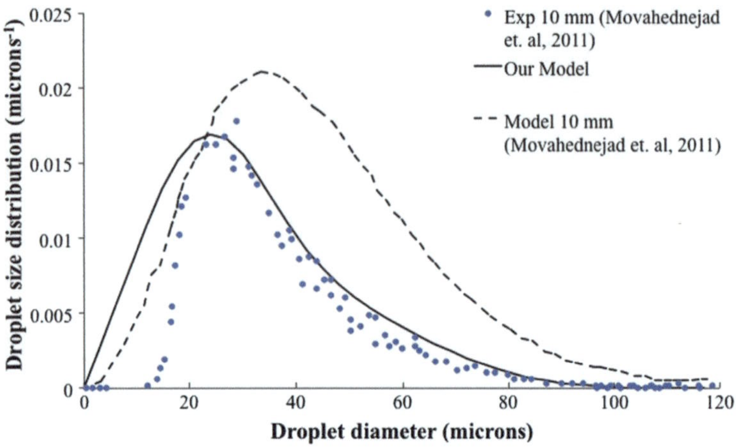

Fig. 6.12 Validation of our MEF formulation with previous literature

analysis. While for a pressure swirl atomizer, there is high liquid inlet swirl and very low air velocities; reverse inlet conditions exist for the airblast scenario. Also, as mentioned earlier, the ligament diameter is dependent on sheet thickness for pressure swirl atomizers, but for airblast, it is independent of the sheet thickness. Figure 6.13 shows the droplet diameter distribution for a typical pressure swirl and airblast atomizer as obtained using the comprehensive analysis. The width of the histogram obtained for the airblast atomizer is smaller as compared to that of the pressure swirl one which signifies the increased probability of obtaining a relatively more uniform distribution of droplets using the airblast atomizer. Thus, the droplet

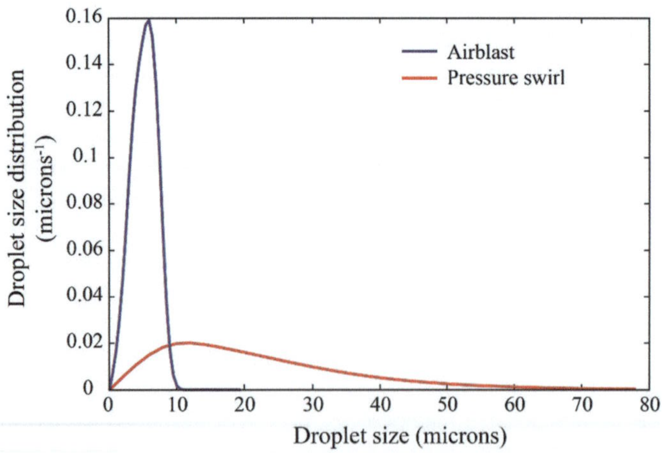

Fig. 6.13 Droplet diameter distribution obtained using MEF for a typical pressure swirl and an airblast atomizer

distribution obtained does not use any experimental data but is final outcome of a series of modeling using different methodology which feed information to each other.

6.5 Conclusions

The book chapter shows a three-step comprehensive analysis of spray modeling which has not been reported in literature till date. The three-step formulation couples linear stability analysis and internal hydrodynamics and adds maximum entropy formulation to that. The maximum growth rate and the most critical wave number corresponding to maximum growth rate are obtained using linear stability analysis for the different sheet profiles found in experiments done in-house. Consistent with experimental results, the early breakup mode of sheet is most unstable with higher growth rate. The breakup lengths are obtained using established empirical relations from the maximum growth rate. The breakup length estimated from this model showed good comparisons with the experimentally obtained breakup lengths. The temporal frequency as obtained from the stability analysis is also compared with reported frequencies obtained from experiments on annular nozzle. In terms of estimating frequencies as well, the stability analysis showed good comparisons especially at low and moderate air flows. While for moderate air flows, good agreement with experimental breakup lengths are obtained in the region of $1.5 \leq \ln\left(\frac{\eta_b}{\eta_0}\right) \leq 2.5$, at higher air flows, $\ln\left(\frac{\eta_b}{\eta_0}\right) = 12$ shows a better agreement.

Further, a statistical methodology of maximum entropy formulation is shown in this chapter for annular sheets to predict the droplet size distribution. This methodology maximizes Shannon entropy using the constraints in the form of normalization and mass, momentum, and energy balance. This statistical formulation is validated with previous experimental studies on annular sheets. As part of the comprehensive analysis, the droplet diameter distribution of a typical pressure swirl and an airblast atomizer is determined. The airblast atomizer shows a distribution with lower width and also shifted slightly toward left signifying the utility of the same in producing finer and more uniform spray.

References

Carvalho IS, Heitor MV (1998) Liquid film break-up in a model of a prefilming airblast nozzle. Exp Fluids 24(5–6):408–415

Chatterjee S, Samanta S, Mukhopadhyay A, Sen S (2013a) Effect of confinement and prefilming on the stability of an annular liquid sheet. At Sprays 23(5):401–418

Chatterjee S, Samanta S, Mukhopadhyay A, Sen S (2013) Effect of confinement and prefilming on the stability of an annular liquid sheet. At Sprays

Chatterjee S, Das M, Mukhopadhyay A, Sen S (2015) Experimental investigation of breakup of annular liquid sheet in a hybrid atomizer. J Propul Power 31(5):1232–1241

Datta A, Som SK (2000) Numerical prediction of air core diameter, coefficient of discharge and spray cone angle of a swirl spray pressure nozzle. Int J Heat Fluid Flow 21(4):412–419

Dombrowski N, Hooper PC (1962) The effect of ambient density on drop formation in sprays. Chem Eng Sci 17(4):291–305

Fu Q-F, Yang L-J, Qu Y-Y, Gu B (2010) Linear stability analysis of a conical liquid sheet. J Propul Power 26(5):955–968

Ibrahim A (2006) Comprehensive study of internal flow field and linear and nonlinear instability of an annular liquid sheet emanating from an atomizer. University of Cincinnati

Ibrahim AA, Jog MA (2007) Nonlinear breakup model for a liquid sheet emanating from a pressure-swirl atomizer. J Eng Gas Turbines Power 129(4):945–953

Ibrahim AA, Jog MA (2008) Nonlinear instability of an annular liquid sheet exposed to gas flow. Int J Multiph Flow 34(7):647–664

Ibrahim AA, Jog MA, Jeng SM (2006) Effect of liquid swirl-velocity profile on the instability of a swirling annular liquid sheet. At Sprays 16(3):237–263

Jazayeri SA, Li X (2000) Nonlinear instability of plane liquid sheets. J Fluid Mech 406:281–308

Jeon J, Hong M, Han Y-M, Lee SY (2011) Experimental study on spray characteristics of gas-centered swirl coaxial injectors. J Fluids Eng 133(12):121303–121307

Kendall JM (1986) Experiments on annular liquid jet instability and on the formation of liquid shells. Phys Fluids 29(7):2086–2094

King R (2010) Active flow control II: papers contributed to the conference "Active Flow Control II 2010", Berlin, Germany, 26–28 May 2010. Springer, Berlin, Heidelberg

Lavergne G, Trichet P, Hebrard P, Biscos Y (1993) Liquid sheet disintegration and atomization process on a simplified airblast atomizer. J Eng Gas Turbines Power-Trans ASME 115(3):461–466

Lefebvre AH (1984) Fuel effects on gas turbine combustion-liner temperature, pattern factor, and pollutant emissions. J Aircr 21(11):887–898

Lefebvre AH (1989) Atomization and sprays. Hemisphere Pub. Corp.

Li X (1994) On the instability of plane liquid sheets in two gas streams of unequal velocities. Acta Mech 106(3):137–156

Liao Y, Benjamin MA, Jeng SM, Jog MA (2000) Instability of an annular liquid sheet surrounded by swirling airstreams. AIAA J 38:453–460

Lin SP (2003) Breakup of liquid sheets and jets. Cambridge University Press

Mohammadi B, Pironneau O (1994) Analysis of the K-epsilon turbulence model. Wiley

Moon Y, Kim D, Yoon Y (2010) Improved spray model for viscous annular sheets in a swirl injector. J Propul Power 26(2):267–279

Movahednejad E, Ommi F, Hosseinalipour SM, Chen CP, Mahdavi SA (2011) Application of maximum entropy method for droplet size distribution prediction using instability analysis of liquid sheet. Heat Mass Transf 47(12):1591–1600

Nath S (2011) Modeling of spray formation by disintegration of liquid sheet. Jadavpur University, India

Nath S, Datta A, Mukhopadhyay A, Sen S, Tharakan TJ (2011) Prediction of size and velocity distributions in sprays formed by the breakup of planar liquid sheets using maximum entropy formalism. At Sprays 21(6):483–501

Park J, Huh KY, Li XG, Renksizbulut M (2004) Experimental investigation on cellular breakup of a planar liquid sheet from an air-blast nozzle. Phys Fluids 16(3):625–632

Reitz RD, Bracco FV (1982) Mechanism of Atomization of a Liquid Jet. Phys Fluids 25 (10):1730–1742

Schmidt DP, Chiappetta LM, Goldin GM, Madabhushi RK (2003) Transient multidimensional modeling of air-blast atomizers. Atomization Spray 13(4):373–394

Senecal PK, Schmidt DP, Nouar I, Rutland CJ, Reitz RD, Corradini ML (1999) Modeling high-speed viscous liquid sheet atomization. Int J Multiph Flow 25(6–7):1073–1097

Shannon CE (1948) A mathematical theory of communication. Bell Syst Tech J 27(3):379–423

Shen J (1997) Formation and characteristics of sprays from annular viscous liquid jet breakup. University of Victoria

Shen J, Li X (1996) Instability of an annular viscous liquid jet. Acta Mech 114(1):167–183

Sovani SD, Sojka PE, Sivathanu YR (1999) Prediction of drop size distributions from first principles: the influence of fluctuations in relative velocity and liquid physical properties. At Sprays 9(2):133–152

Tharakan T, Mukhopadhyay A, Datta A, Jog M (2013) Trends in comprehensive modeling of spray formation. Int J Spray Combust Dyn 5(2):123–180

Wahono S, Honnery D, Soria J, Ghojel J (2008) High-speed visualisation of primary break-up of an annular liquid sheet. Exp Fluids 44(3):451–459

Chapter 7
Modeling of Flash Boiling Phenomenon in Internal and Near-Nozzle Flow of Fuel Injectors

Kaushik Saha, Michele Battistoni and Sibendu Som

Abstract Detailed analysis of the internal and the near-nozzle flow of fuel injectors is a necessity for a comprehensive understanding of any internal combustion engine performance. For gasoline direct injection engines, under part-load conditions, the in-cylinder pressure can be subatmospheric when the high-temperature fuel is injected, resulting in flash boiling. Detailed experimental characterization of such complex phenomena is extremely difficult. Three-dimensional computational fluid dynamics (CFD) simulations provide key insights into the flash boiling phenomena. The Spray G injector from Engine Combustion Network (ECN) has been considered for this study, which has eight counter-bored holes. Homogeneous relaxation model is used to capture the rate of phase change. Standard and RNG $k - \epsilon$ turbulence models have been employed for modeling turbulence effects. Based on apriori thermodynamic estimates, three types of thermodynamic conditions have been explored: non-flashing, moderate flashing, and intense flashing. Numerical analyses showed that with more flashing the spray plumes grow wider due to the volume expansion of the rapidly forming fuel vapor. Mainly single-component fuel is studied in this work. Iso-octane is considered as the gasoline surrogate for this study. Binary component blends of isooctane and ethanol were also tested for blended fuel flashing predictions using the existing numerical setup. After careful estimation of blended fuel saturation properties, the simulations indicated that blended fuels can be more volatile than the individual components and thus exhibit more flashing compared to the cases with single-component fuels.

K. Saha (✉) · S. Som
Argonne National Laboratory, 9700 S Cass Avenue, Lemont, IL 606439, USA
e-mail: ksaha@anl.gov

S. Som
e-mail: ssom@anl.gov

M. Battistoni
University of Perugia, Perugia, Italy
e-mail: michele.battistoni@unipg.it

S. Basu et al. (eds.), *Droplets and Sprays*, Energy, Environment, and Sustainability,
https://doi.org/10.1007/978-981-10-7449-3_7

Nomenculture

CFL	Courant–Friedrichs–Lewy number
D	Mass diffusivity (m^2/s)
Ja	Jakob number
m_{vap}, m_{liq}	Mass of vapor and liquid phases in a cell (kg)
p	Local pressure (Pa)
$P_{ch}, P_{chamber}$	Chamber pressure (kPa)
p_{crit}	Critical pressure (Pa)
P_{inj}	Injection pressure (MPa)
P_{sat}, p_{sat}	Saturation pressure (Pa)
R_p	Pressure ratio
S_m	Source term in species transport equations
t	Time (s)
T_{fuel}	Fuel temperature (K)
T_{sat}	Saturation temperature (K)
u	Local cell velocity (m/s)
u_i	Advecting mean velocity (m/s)
u_j	Advected mean velocity (m/s)
V	Volume of a cell (m^3)
x	Local cell vapor quality
$\bar{x}$	Local cell equilibrium quality
Y_m	Mass fraction of mth species

Greek

α	Void fraction (vapor and non-condensable gases)
ϵ	Turbulence dissipation rate (m^2/s^3)
θ, θ_0	Equilibrium timescale and empirical time constant (s)
$\rho, \rho_v, \rho_l, \rho_g$	Density of mixture, vapor, liquid, and gas (kg/m^3)

7.1 Introduction

Gasoline direct injection (GDI) engines have been preferred to port fuel injection (PFI) engines for almost a decade. Experimental studies have shown that GDI offers better fuel economy and superior performance when compared to PFI Harada et al. (1997); Iwamoto et al. (1997). Developing a comprehensive understanding of spray formation in GDI systems is warranted for further improvements in GDI engine combustion. The GDI sprays are affected by multiple factors including "flash boiling." Flash boiling occurs in modern GDI engines at throttled conditions, and it significantly affects spray formation Weber and Leick (2014). Throttled operating conditions create low in-cylinder chamber pressures ($\approx$30 kPa). Consequently, when heated gasoline-type fuel (at around 363 K) is injected into such in-cylinder environments, liquid fuel is subjected to a superheated condition (the ambient pressure is lower than the saturation pressure corresponding to the fuel temperature). Flashing causes rapid bulk conversion of liquid fuel to gaseous vapor, leading to volume expansion and thus affecting spray width and penetration. Therefore, flashing is either advantageous or disadvantageous, depending on chamber design and injection timing as well as injector orientation. As a result, flashing for GDI engines is becoming a relevant topic of research. Flashing is usually perceived as a non-equilibrium phenomenon since heat transfer is non-negligible and requires a finite timescale for the phase change to occur Neroorkar (2011).

Some experimental studies have been undertaken to investigate the flash boiling effects on GDI spray patterns. Vanderwege and Hochgreb (1998) resorted to planar laser-induced fluorescence (PLIF) and Mie scattering on a pressure-swirl injection system. They observed the spray transformation from hollow cone to solid cone, and the degree of superheat was increased. The authors inferred through image processing that the droplet diameter, due to flashing, reduced by 40%. Using bubble-point calculation, they also recommended that superheating by roughly 20 K is required for the flash boiling to noticeably affect the spray atomization. Weber and Leick Weber and Leick (2014) used mainly high-speed Mie scattering and high-speed shadowgraphy for their analysis. Additionally, droplet velocities were estimated by a shadow particle image velocimetry technique. Reduction in individual spray plume length and increment in the initial cone angle due to flashing were observed. It was noticed that during intense flashing two diametrically opposite spray plumes might interact with each other. In a recent experimental study with a GDI injector similar to Spray G (2014), Montanaro and Allocca (2015) used a Delphi Valve Covered Orifice (VCO) injector with isooctane fuel. Variation of the injection pressure, the fuel temperature, and the chamber pressure was carried out, keeping the chamber temperature constant at 296 K. They observed that subatmospheric chamber pressure and elevated fuel temperatures caused substantial flashing.

Considerable work has been done in the literature in terms of numerical studies on the flash boiling phenomenon. Kawano et al. (2004) considered bubble growth, bubble perturbation, and bubble breakup to model flashing sprays. Zeng and Lee (2001) followed a similar approach to study n-pentane sprays subjected to different degrees of superheat. They inferred that more flashing is caused by an increment in superheat, leading to smaller drop sizes (~50% reduction in SMD, when flashing is considered in the model), wider sprays, and even enhanced plume interactions. They also deduced that for lower degrees of superheat, aerodynamic forces dominate, while flashing takes over at higher degrees of superheat. Apart from bubble-based models, studies using homogeneous relaxation model (HRM) Bilicki and Kestin (1990), Downar-Zapolski et al. (1996), Gopalakrishnan and Schmidt (2008) was reported, which is based on empirical approach. HRM provides the timescale for the local condition to reach the thermal equilibrium.

Theoretically, HRM lies in between the two thermodynamic limits for two-phase models: the homogeneous equilibrium model (HEM) and the homogeneous frozen model (HFM) Downar-Zapolski et al. (1996), Brennen (1995). For HEM, the two phases are assumed to be mixed homogeneously with the heat transfer occurring spontaneously, which is not feasible in a real-world scenario. The other extreme,

HFM, assumes heat transfer takes infinitely long time. Thus, HRM reasonably captures underlying physics of the practical two-phase flow scenarios. The vapor–liquid equilibrium properties are necessary for HRM implementation and have been estimated for pure as well as blended fuels, for GDI applications Neroorkar (2011), Gopalakrishnan and Schmidt (2008). Recently, Moulai et al. (2015) and Saha et al. (2016a) performed numerical simulations of isooctane flash boiling in a multi-hole GDI fuel injector (Spray G) under conditions specified in the Engine Combustion Network (ECN) (2014). In these studies, the outlet chamber is mainly filled with isooctane. They explored different flashing and non-flashing conditions in their work. The role of the model parameters of HRM and their impact on the extent of flashing were also reported by Saha et al. (2017) very recently.

It is evident that study of internal and near-nozzle flashing is essential for in-depth understanding of GDI systems. In this chapter, the effect of operating conditions and turbulence models will be shown using a single-component gasoline surrogate (isooctane). The operating conditions and fuel injector considered in this study are based on the baseline Spray G condition (2014) and some parametric variations of that baseline condition. For blended fuel, varying proportions of isooctane and ethanol have been used and tested under mild to moderate flashing condition. The chapter is organized in the following way: First, the numerical setup and the governing equations will be described. Next, the simulation results indicating the effect of different boundary conditions on near-nozzle flow patterns and their thermodynamic implications and classifications of non-flashing, moderate flashing, and intense flashing will be presented. This will be followed by the effect of turbulence models and finally, simulation results from blended fuel flash boiling will be shown.

7.2 Model Formulation

In this study, it is considered that isooctane (physical surrogate for gasoline) is flowing through a multi-hole GDI nozzle, denoted as the Spray G injector in the ECN (2014). The conditions studied in this work are summarized in Table 7.1. From prior work by our group Saha et al. (2016a) and other research groups Moulai et al. (2015), it has been shown that the Spray G condition is supposed to be non-flashing since the chamber pressure is higher than the saturation pressure corresponding to the fuel temperature. However, surface vaporization of the liquid jets is expected owing to convective heating from the hot environment. In the case of Spray G2, flashing is feasible because the chamber condition renders the fuel to be superheated. For Spray G2, the chamber is kept at room temperature; therefore, convective vaporization due to the ambience should be very low. The Spray G3 condition is also considered, which has a higher fuel temperature and is subjected to a chamber pressure of one atmosphere. Thus, the degree of superheat is much higher indicating a possibility of intense flash boiling. The chamber temperature under the Spray G3 condition is also at room temperature; hence, convective vaporization of liquid jets due to the ambience should be very low.

Table 7.1 Cases studied

Parameters	Spray G	Spray G2	Spray G3
Inj. press. (MPa)	20	20	20
Chamber press. (kPa)	600	53	100
Chamber temp. (K)	573	293	293
Fuel temp. (K)	363	363	413
Chamber fluid	N_2	N_2	N_2
Degree of superheat, ΔT (K)	N/A	12.34	40.68
Pressure ratio, R_P	0.13	1.48	2.83
Thermodynamic state	Non-flashing, vaporizing	Moderate flashing	Intense flashing

$\Delta T = T_{\text{Fuel}} - T_{\text{sat}}(P_{\text{chamber}})$; $R_P = P_{\text{sat}}/P_{\text{chamber}}$

For testing the effect of pure and blended fuels: isooctane, ethanol, and different blends (Ethanol 0.3 and 0.85 volume fraction, i.e., isooctane 0.7 and 0.15 volume fraction, respectively) have been used. Typically, blended fuels while vaporizing or flashing should vary in composition; hence, the properties may change. It would be more realistic to consider such variations in run time. However, as an initial attempt, it is assumed that the blended fuels while flashing will undergo the phase change such that the relative proportion of the individual components will remain same during the simulation. Such approaches have been shown to be reasonable in the past Neroorkar (2011). However, there is a scope for modification in this regard and will be the subject of future studies.

The REFPROP database (2013) has been used to create property database for blends. The saturation properties are estimated based on bubble-point calculation through REFPROP. The physical properties such as liquid viscosity, liquid thermal conductivity, and liquid-specific heat for the blend are obtained through mixing rules (mass fraction based). The vapor thermal conductivity and vapor viscosity of the blend are assumed to be same as those of air since these values are approximately in the same order for gaseous species and orders of magnitude lower than the corresponding liquid phases.

7.2.1 Nozzle Geometry and Computational Domain

The Spray G nozzle has eight counter-bored holes with the nominal diameters of the holes and the counter-bores as 165 μm and 388 μm, respectively. The length-to-diameter ratios of the nozzle holes and counter-bores are around unity. There are five dimples (reducing the flow passage) on the injector wall, making the upstream flow passages leading to the holes asymmetric. Hence, it is imperative to consider the full geometry to achieve meaningful predictions. The nominal geometry of the Spray G

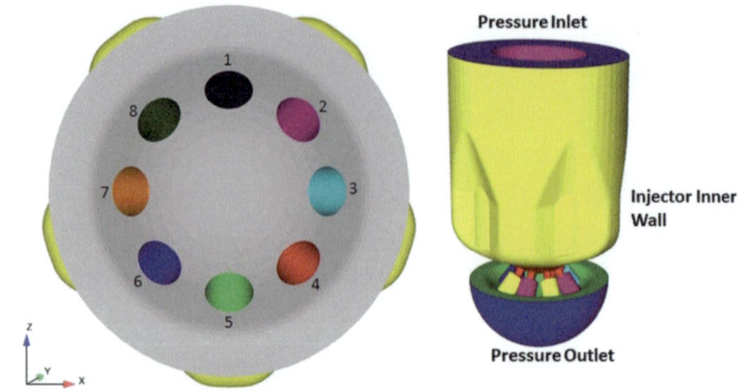

Fig. 7.1 (Left) Layout of the eight-stepped holes of the Spray G injector (bottom view) and (Right) the computational domain considered in this study

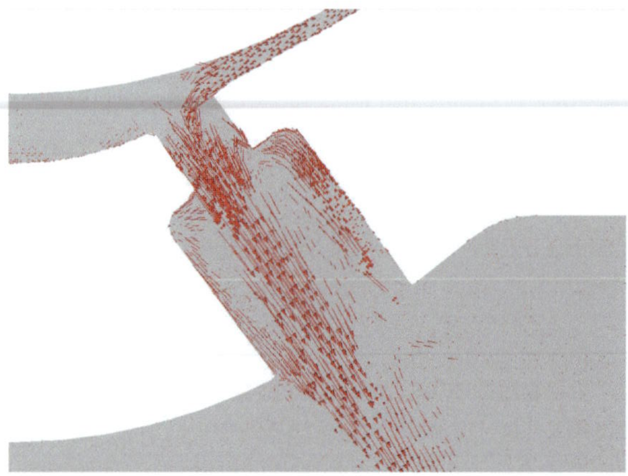

Fig. 7.2 Backflow velocity vectors in the counter-bore region for the Spray G case on a 2D cut-plane

nozzle was available through the ECN (2014). Figure 7.1 shows the eight counter-bored holes and the five dimples. A hemispherical outlet domain with a diameter of 9 mm is used. A previous study Saha et al. (2016a) indicated that results using a 9-mm outlet domain could be reasonable for internal and near-nozzle flow analysis. Past studies with this injector demonstrated that the chamber gas backflows into the counter-bore Moulai et al. (2015), Saha et al. (2016b). The stepped or counter-bored hole, with the backflow velocity vectors, is shown in Fig. 7.2.

For simulating the blended fuel flash boiling three-dimensional simulations with a sector domain (1/8th) has been used to save computational time, since the focus is on the effect of fuel physical properties. Symmetric boundary condition has been used at the two side-faces of the domain. Simulations are performed using the CONVERGE code Richards et al. (2015), which uses a cut cell technique to generate the mesh automatically during the run time. In order to capture the sharp gradients in the flow variables, fixed embedding has been used near the walls, inside the holes and in the near-nozzle regions, and inside the chamber. In the current study, results using 140 μm as the base grid (maximum grid size) and 17.5 μm minimum grid size have been presented because this grid resolution has been shown to provide reasonable estimates with affordable wall-clock times based on our previous studies Saha et al. (2016b).

7.2.2 Governing Equations

The mixture multiphase model has been used by solving the governing equations of mass, momentum, species and energy conservation, additional scalar equations for turbulence. Compressibility of both the liquid and vapor phases has been taken into account. The governing equations adopted for this study are available in the literature Saha et al. (2017), Richards et al. (2015).

The species considered are liquid fuel, vapor fuel, N_2, and O_2. The densities of all the gaseous species have been estimated using the ideal gas equation. α_m and Y_m represent the volume fraction and mass fraction of the individual species, respectively. The individual species concentrations are determined from the species conservation equations:

$$\frac{\partial \rho Y_m}{\partial t} + \frac{\partial u_j \rho Y_m}{\partial x_j} = \frac{\partial}{\partial x_j}\left(\rho D \frac{\partial Y_m}{\partial x_j}\right) + S_m \tag{7.1}$$

where S_m represents the source or sink term. There are no source/sink terms for non-condensable gases (N_2 and O_2). S_m in the vapor species transport equation provides the source term for phase change. The relevant details for estimating S_m in phase change scenarios are provided in the "Homogeneous Relaxation Model" subsection.

The transport equations and model constant for the standard and re-normalization group (RNG) $k - \epsilon$ models are not stated here, which can be obtained from the literature Saha et al. (2017), Richards et al. (2015). The details about the numerical implementation and material properties are available in prior publications Saha et al. (2016a), Saha et al. (2017).

7.2.2.1 Homogeneous Relaxation Model

S_m in the fuel species conservation equations is solved using the HRM formulation. The vapor quality is mathematically represented as $x = \frac{Y_v}{Y_v + Y_l} = \frac{m_{\text{vap}}}{m_{\text{vap}} + m_{\text{liq}}}$. The rate of change of local vapor quality, $(\frac{Dx}{Dt})$, provides the estimate of S_m. HRM enables one to express $\frac{Dx}{Dt}$ as:

$$\frac{Dx}{Dt} = \frac{\bar{x} - x}{\theta} \tag{7.2}$$

The timescale θ is calculated as

$$\theta = \theta_0 \alpha^{-0.54} \psi^{-1.76} \tag{7.3}$$

where

$$\theta_0 = 3.84 \times 10^{-7};\ \psi = \frac{|p_{\text{sat}} - p|}{p_{crit} - p_{sat}};\ \alpha = \alpha_v + \alpha_{N_2} + \alpha_{O_2}$$

These specific values of the parameters have been known to be effective in previous studies Neroorkar (2011), Downar-Zapolski et al. (1996). When the local pressure is lower than the saturation pressure, the liquid will tend to vaporize and the local quality will tend to approach the equilibrium quality and the corresponding timescale will be estimated by the timescale θ. Therefore, the flashing is triggered in the computation when local pressure suffers depressurization and drops below the saturation pressure. HRM is capable of predicting phase changes from liquid to vapor and vice versa, whenever the condition is thermodynamically feasible. Naturally, in a high-temperature environment, such as the Spray G case, HRM can predict vapor formation due to convective heating of the liquid jets. Since HRM was developed/verified only for flashing and cavitation problems, vaporization predictions by HRM, due to convective heating may not be reasonable.

Non-condensable gases have been included in α despite the fact that gases were not considered in the original HRM formulation. It is our understanding that in a given cell, when the liquid volume fraction is very low compared to the gaseous species, considering only the fuel vapor volume fraction may not provide a proper physical representation of the thermodynamic condition in that cell. Moreover, the work done by Downar-Zapolski et al. (1996) indicated that the relaxation time exponentially decreases to very small values with increments in the void fraction. Once the void fraction exceeds 10%, the relaxation timescale continues to decrease asymptotically (cf. Fig. 7.3). This indicates that mathematically, the inclusion of gases in the void fraction should not hamper prediction of the relaxation timescale. Additionally, under flash boiling conditions, the gases do not need to provide energy for phase change. The superheated liquid provides the necessary enthalpy for phase change under flash boiling conditions. For the flashing cases Spray G2 and Spray G3, the chamber temperatures were specifically kept low to ensure that there is negligible phase change due to convective heating by the ambience. The relation between S_m and $\frac{Dx}{Dt}$ has been explained by Battistoni et al. (2015). If x^0 is the current local vapor quality of a computational cell and x^1 is the corresponding value for the next time-step, and the time-step size is Δt, then x^1 can be calculated as $x^1 = \bar{x} - \left(\bar{x} - x^0\right) e^{-\Delta t/\theta}$. Hence,

$$S_m = \frac{(x^1 - x^0)(m_{\text{vap}} + m_{\text{liq}})}{V \Delta t} = (x^1 - x^0)\rho \frac{(Y_v + Y_l)}{\Delta t} \tag{7.4}$$

where V is the volume of the cell and m_{vap} and m_{liq} are the masses of the vapor and liquid phases, respectively, in that cell.

7.3 Results and Discussions

In this section, first the effect of thermodynamically different boundary conditions will be demonstrated through HRM predictions. After that, results from turbulence model variation effects will be provided followed by simulation outcomes of blended fuel flashing.

7.3.1 Boundary and Operating Conditions

Previous studies Saha et al. (2016a, b) assessed the predictions qualitatively as well as quantitatively and proved the credibility of the numerical predictions using the current setup. Hence, those findings are not shown here for the sake of brevity. The thermodynamic conditions reported here are Spray G, Spray G2, and Spray G3 as shown in Table 7.1. The expected outcomes for these three conditions have been already stated earlier through thermodynamic assessment. Figure 7.3 presents the vapor mass fraction contours at 0.1 ms. The predicted results are in accordance with prior estimation based on thermodynamic analysis. The Spray G calculation indicates that the peripheries of the liquid jets are vaporizing because of convective heating in the high-temperature chamber. Spray G2 is mildly flashing since the degree of superheat is not very high. The degree of superheat is very high for Spray G3 and hence there is abundant vapor formation, i.e., intense flashing in the chamber and significant plume-to-plume interaction, as shown in the simulation result. Detailed experimental findings with quantified information on vapor concentration are still not available in the literature. At this stage, it is not feasible to provide a comparison of predictions of vapor concentration with experimental data. However, the prior thermodynamic calculations, along with observations by Weber and Leick Weber and Leick (2014), point to the fact that when the pressure ratio ($R_P = P_{\text{sat}}/P_{\text{chamber}}$) exceeds two, there is the potential for aggressive flash boiling. Weber and Leick Weber and Leick (2014) also observed that under intense flash boiling conditions, it is feasible for two opposing spray plumes to interact with each other (cf. Fig. 7.8). The vapor fuel mass fraction contour for Spray G3 (in Fig. 7.3) shows that the HRM captures aggressive flash boiling resulting in plume merging and collapsing.

7.3.2 Turbulence Models

Typically, for all the cases, the standard $k - \epsilon$ turbulence model has been used. To understand the role of the turbulence model, the RNG $k - \epsilon$ turbulence model has also been used for the Spray G condition in Table 7.1. Figure 7.4 shows the effect of turbulence model variation in terms of vapor mass fraction. Considerable differences

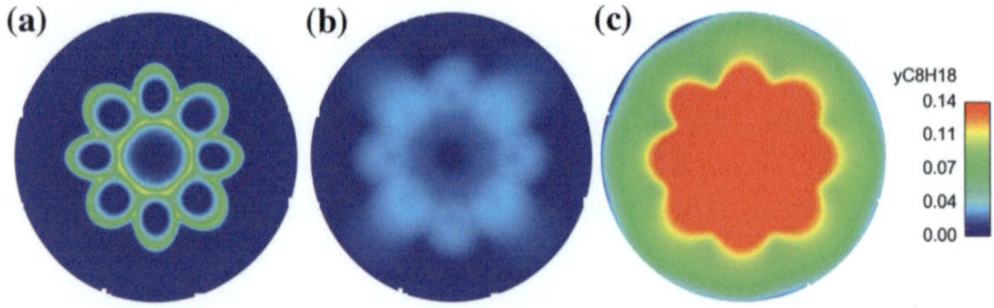

Fig. 7.3 Effect of operating and boundary conditions on two-phase flow through Spray G injector under G, G2, and G3 conditions. Vapor mass fraction contours are shown through a horizontal cut-plane 2 mm from the injector tip

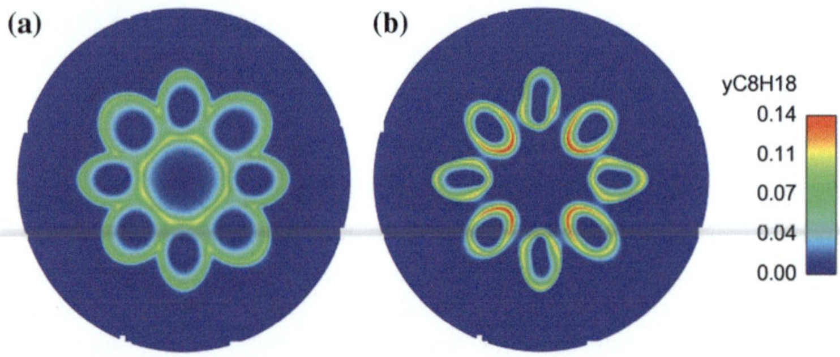

Fig. 7.4 Effect of changing the RANS turbulence model on the fuel vapor distribution on a horizontal plane (**a** Standard and **b** RNG), 2mm downstream of the injector tip for the Spray G injector under the non-flashing condition from Table 7.1

are observed, especially the lower "vapor dispersion" for the RNG $k-\epsilon$ turbulence model, which has been shown previously by our work Saha et al. (2017).

In order to understand the reason for the differences between the standard and RNG $k-\epsilon$ models, it is worthwhile to look at the turbulent viscosity distributions in the chamber with the two different turbulence models. We have verified that for RANS-based turbulence simulations, at reasonably fine mesh resolutions, turbulent viscosity is dominant compared to molecular viscosity and numerical viscosity. The differences in the viscosity contours are noteworthy, as seen in Fig. 7.5. Higher viscosity predictions with the standard $k-\epsilon$ model imply more turbulence-induced mixing, which can contribute to higher vapor dispersion, as seen in Fig. 7.4. Unlike the standard $k-\epsilon$ model, the RNG $k-\epsilon$ model accounts for low, medium, and high strain rates in the flows Saha et al. (2017). In our prior publication Saha et al. (2016b), the effect of increasing the outlet domain was investigated. The issue of vapor buildup near the outlet boundary could be mitigated to a great extent, even when the standard $k-\epsilon$ model was used. However, the larger domain would mean more cells and hence the need for more computational resources. Hence, the RNG $k-\epsilon$ turbulence model should be a reasonable choice for these problems with in-nozzle and near-nozzle analyses. Although the results in Fig. 7.3 are obtained with

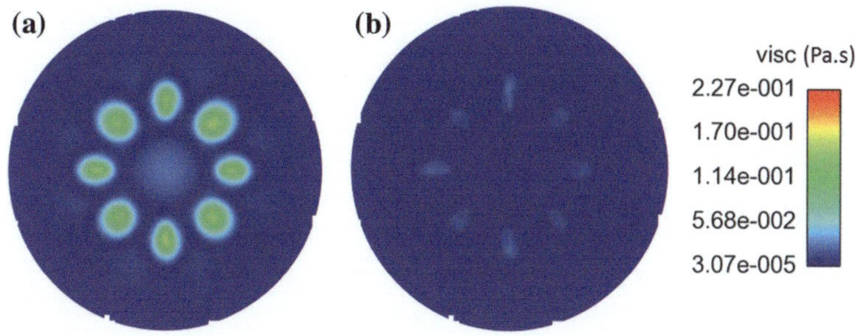

Fig. 7.5 Effect of changing the RANS turbulence model on the turbulent viscosity distribution on a horizontal plane (**a** Standard and **b** RNG), 2 mm downstream of the injector tip for the Spray G injector under the non-flashing condition from Table 7.1

the standard $k - \epsilon$ model, the qualitative findings for effects of different operating conditions are still valid.

7.3.3 Blended Fuels

It is important to estimate the blend saturation properties and compare them with those of the individual components, especially since there is significant interest in USA and worldwide to run sprak ignition engines with ethanol blends. Previous work Neroorkar (2011) indicates that the blend, depending on its composition and binary thermodynamic properties, could be more volatile than the individual components. Using REFPROP (2013), the saturation properties of the different blends were estimated. Figure 7.6 shows the blend saturation properties at different fuel temperatures for different varying proportions of binary isooctane and ethanol blends. It is evident that for certain compositions, the blend is more volatile than the individual components for a wide range of temperatures. The blends which have saturation properties different than the constituent species tend to have an azeotropic point with specific proportions of the constituent species. At azeotropic point, the vapor mixture has the same proportion of the constituents as the liquid mixture Fedali et al. (2014). The saturation pressure of such blends could be higher or lower than those of the individual species. Isooctane and ethanol blends, as already shown, have higher saturation pressures and hence can be considered positive azeotropes Boublikova and Lu (1969). It should also be noted that ethanol-octane blends could be unstable at low fluid temperatures and hence REFPROP estimates would be reasonable for high fuel temperature blends of ethanol and isooctane. The fuel temperature considered in this study is 363 K. Therefore, the simulation results using REFPROP (2013) estimates for blend properties could be considered reasonable.

Next, simulation results for the other single-component fuel (e.g., ethanol) and blended fuel (isooctane and ethanol) under flashing conditions in Eulerian frame-

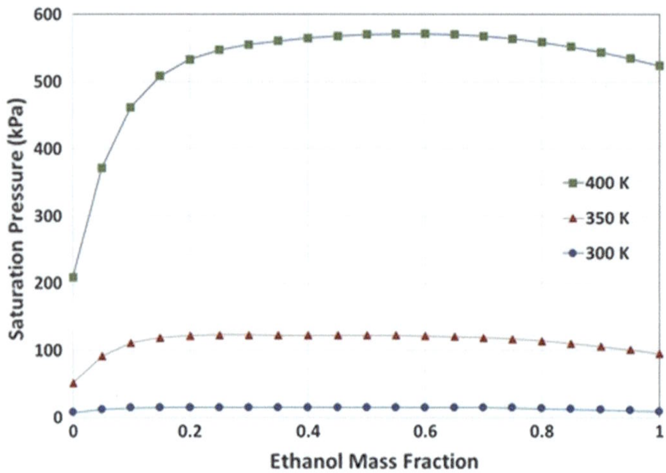

Fig. 7.6 Blend saturation pressure values estimated using REFPROP v9.1 (2013) at different fuel temperatures for varying proportions of blends (iso-octane and ethanol)

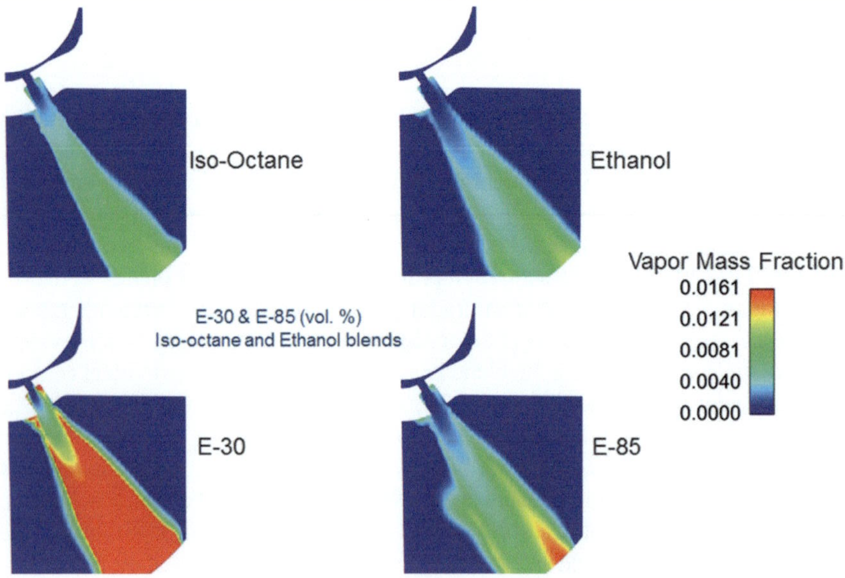

Fig. 7.7 Vapor mass fraction contours of pure and blended fuels at steady-state conditions

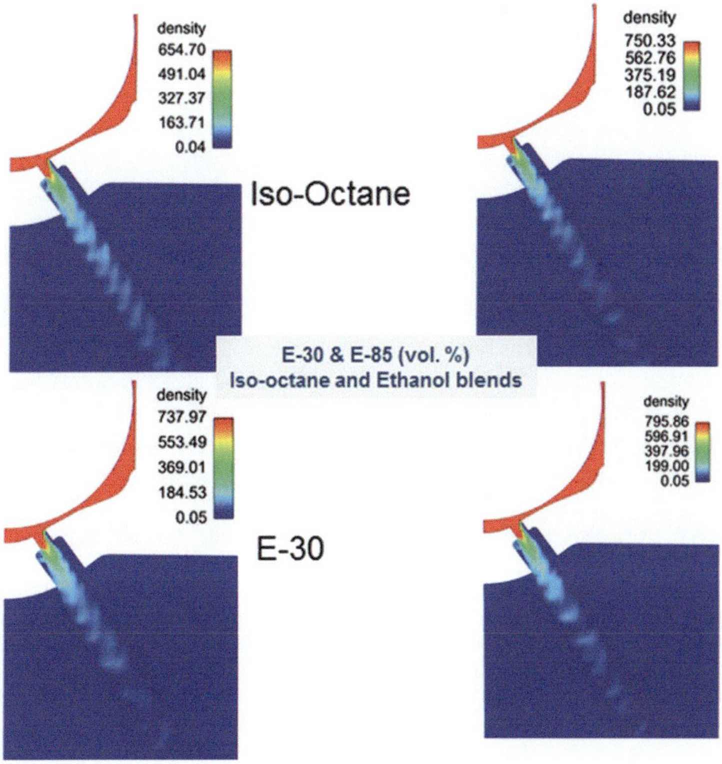

Fig. 7.8 Density contours of pure and blended fuels at steady-state conditions

work (in-nozzle and near-nozzle) are presented. Figure 7.7. shows the zoomed views of sac, orifice, and near-nozzle region on 2D cut-planes and compares the vapor mass fraction contours after reaching steady state for a flashing condition ($P_{chamber} = 53$ kPa and $T_{fuel} = 363$ kPa). The fuel temperature and chamber pressure are the determining parameters for ensuring a flashing condition for the sprays and the above-mentioned conditions ensure that fuel will be superheated in the chamber and hence undergo flash boiling. It was evident from Fig. 7.6 that E-30 (ethanol 30% by volume; isooctane 70% by volume) is more volatile than the individual components. E-85 being dominantly ethanol has the saturation properties closer to ethanol. The relatively higher concentration of vapor being formed for E-30 indicates that this specific blend will have a higher tendency to flash. The density contours for the four different cases, two pure and two blends, are shown in Fig. 7.8. The blends are expected to have densities in between the individual components. However, for E-85 the density exceeds that of ethanol. The bulk moduli of the blends were estimated based on the relative proportions of the individual components that can lead to a higher than expected density of the blended fuel.

7.4 Summary and Concluding Remarks

A numerical study has been carried out for a typical GDI fuel injector which is an extension of several years o research in our group. The HRM is shown to be capable of predicting non-flashing yet vaporizing, moderately flashing, and intensely flashing scenarios, in accordance with the calculations of the pressure ratio and the degree of superheat. Aggravated flashing leads to enhanced plume interaction and plume collapse. The RNG $k - \epsilon$ turbulence model can be suitable for the GDI spray applications because it can minimize the physical vapor buildup in small outlet domains. Earlier work indicated that larger domains were needed with standard $k - \epsilon$ model. Therefore, RNG $k - \epsilon$ model turns out to be an economic, yet physically correct turbulence modeling approach for these calculations. Blended fuel flashing has been explored, and blends of isooctane and ethanol are more volatile than individual constituent species. As a result, the blended fuels experience enhanced flashing for a condition under which the single-component fuels experience mild to moderate flashing.

Acknowledgements UChicago Argonne, LLC, Operator of Argonne National Laboratory (Argonne), a US Department of Energy Office of Science laboratory, is operated under Contract No. DE-AC02-06CH11357. The US Government retains for itself, and others acting on its behalf, a paid-up nonexclusive, irrevocable worldwide license in said article to reproduce, prepare derivative works, distribute copies to the public, and perform publicly and display publicly, by or on behalf of the Government. This research was partially funded by the US Department of Energy (DOE) Office of Vehicle Technologies, Office of Energy Efficiency and Renewable Energy under Contract No. DE-AC02-06CH11357. The authors wish to thank Gurpreet Singh and Leo Breton at DOE, for their support. The authors would like to thank Ronald Grover at General Motors and Lyle Pickett at Sandia National Laboratory for providing the nominal geometry of the Spray G nozzle. The authors want to express their gratitude to Yanheng Li and Eric Pomraning at Convergent Sciences Inc. for providing a better understanding of the volume-averaging method for VOF simulations. The authors are grateful to LCRC computing resources at Argonne National Laboratory.

References

Battistoni M, Duke DJ, Swantek AB, Tilocco FZ, Powell CF, Som S (2015) Effects of non-condensable gas on cavitating nozzles. Atomization Sprays 25(6):453483

Bilicki Z, Kestin J (1990) Physical aspects of the relaxation model in two-phase flow. Proc Roy Soc Lond A 428:379397

Boublikova L, Lu BC-Y (1969) Isothermal vapourliquid equilibria for the ethanoln-octane system. J Appl Chem 19:8992

Brennen CE (1995) Cavitation and bubble dynamics. Oxford University Press

Downar-Zapolski P, Bilicki Z, Bolle L, Franco J (1996) The non-equilibrium relaxation model for one-dimensional flashing liquid flow. Int J Multiph Flow 22(3):473483

Fedali S, Madani H, Bougriou C (2014) Modeling of the thermodynamic properties of the mixtures: prediction of the position of azeotropes for binary mixtures. Fluid Phase Equilib 379:120–127

Gopalakrishnan S, Schmidt D (2008) A computational study of flashing flow in fuel injector nozzles. SAE Technical Paper 2008-01-0141

Harada J, Tomita T, Mizuno H, Zenichiro M, Ito Y (1997) Development of direct injection gasoline engine. SAE Technical Paper 970540

Iwamoto Y, Noma K, Nakayama O, Yamauchi T, Ando H (1997) Development of gasoline direct injection engine. SAE Technical Paper 970541

Kawano D, Goto Y, Odaka M, Senda J (2004) Modeling atomization and vaporization processes of flash-boiling spray. SAE Technical Paper 2004-01-0534

Montanaro A, Allocca L (2015) Flash boiling evidences of a multi-hole GDI spray under engine conditions by Mie-scattering measurements. JSAE 2015-01-1945

Moulai M, Grover R, Parrish S, Schmidt D (2015) Internal and near-nozzle flow in a multi-hole gasoline injector under flashing and non-flashing conditions. SAE Technical Paper 2015-01-0944

Neroorkar KD (2011) Modeling of flash boiling flows in injectors with gasoline-ethanol fuel blends. PhD thesis, University of Massachusetts-Amherst

Reference fluid thermodynamic and transport properties-REFPROP v9.1 (2013)

Richards KJ, Senecal PK, Pomraning E (2015) CONVERGE v2.2 Documentation. Convergent Sciences Inc

Saha K, Battistoni M, Som S (2017) Investigation of homogeneous relaxation model parameters and implications on GDI spray formation. Atomization Sprays 27(4):345–365

Saha K, Som S, Battistoni M, Li Y, Quan S, Senecal PK (2016a) Modeling of internal and near-nozzle flow for a GDI fuel injector. ASME J Energy Resour Technol 138(5):(052208)111

Saha K, Som S, Battistoni M, Li Y, Pomraning E, Senecal PK (2016b) Numerical investigation of two-phase flow evolution of in- and near-nozzle regions of a gasoline direct injection engine during needle transients. SAE Int J Engines 9(2)

Spray G ECN gasoline spray combustion, Sandia National Laboratory. http://www.sandia.gov/ecn/index.php. Accessed from Dec 2014

Vanderwege BA, Hochgreb S (1998) The effect of fuel volatility on sprays from high-pressure swirl injectors. Proc Combust Inst 27:1865–1871

Weber D, Leick P (2014) Structure and velocity field of individual plumes of flashing gasoline direct injection sprays. In: 26th annual conference on liquid atomization and spray systems ILASS-Europe 2014

Zeng Y, Lee CF (2001) An atomization model for flash boiling sprays. Combust Sci Technol 169:4567

Chapter 8
Novel Fuel Injection Systems for High-Speed Combustors

Kuppuraj Rajamanickam, Swapneel Roy and Saptarshi Basu

Abstract A spray nozzle which can produce extremely small droplet size (≤ 10 μm) finds relevance in drying, liquid-fueled engines, etc. In addition to the droplet size, its momentum too has equal importance in high-speed combustion systems like scramjets, pulse detonation engines for rapid mixing. Many of the nozzles which adopt standalone techniques (pressure, air-assist) fail to produce finer size droplets. Efforts have been made to hybridize two or more techniques to achieve the finer atomization. For instance, standard air-assist atomizer can be combined with effervescent/ultrasonic means to achieve further reduction in droplet size. This chapter presents the comprehensive aspects of such type of hybrid atomizers. Features such as mode of operation, benefits, shortcomings, areas of application are discussed in greater details.

Keywords Hybrid atomizers · Modes of atomization · Finer droplet size
Scramjet engines · Air-assist atomization · Effervescent sprays
Spray drying

8.1 Introduction

Over the past two decades, high-speed combustors have been the subject of intense research among air breathing propulsion engines (Segal 2009). Efficient and clean combustion become two major driving factors in modern day combustion systems. However, for very high-speed flow, the combustor presents difficult challenges particularly with fuel–air mixing process. Ferri (1973) highlighted the tightly coupled nature of fluid dynamics and chemical reaction time scales in supersonic combustors.

Unlike gas-fueled combustor, the additional droplet evaporation time makes liquid-phase combustion a highly challenging task. In liquid-fueled combustors,

K. Rajamanickam · S. Roy · S. Basu (✉)
Department of Mechanical Engineering, Indian Institute of Science, Bangalore, India
e-mail: sbasu@mecheng.iisc.ernet.in

S. Basu et al. (eds.), *Droplets and Sprays*, Energy, Environment, and Sustainability,
https://doi.org/10.1007/978-981-10-7449-3_8

fuel injector (spray nozzle) forms the building block in determining the efficient operation of the system (Bayvel 1993). The characteristic features of spray nozzles like droplet size and velocity have significant influences in homogeneous mixture formation inside the combustor. Particularly, in high-speed combustors (scramjet engines; flow speed ~1–2 km/s), extremely low residence time of flow demands rapid mixing of fuel–air (Swithenbank 1967). In these combustors, spray nozzle should deliver extremely fine-sized droplets (≤ 15 μm) at very high momentum (i.e., velocity), to promote droplet penetration in high-speed flows along with curtailed evaporation time. Efforts have been made to accelerate the fuel–air mixing process in high-speed flows (Bogdanoff 1994; Menon 1989; Gutmark et al. 1995). The attempts include both from the perspective of achieving lower droplet evaporation time and usage of special structures (struts, ramps) to promote turbulence mixing. The additional challenges in the high-speed combustor include flame stabilization, minimization of friction/drag and shock wave, momentum losses incurred during fuel injection into high-speed flows (Bogdanoff 1994).

Though existing designs like ultrasonic, effervescent, flash boiling, electrospray atomizers offer extremely small-sized droplets, the offset against these atomizers is low throughput and velocity of the drops (Lefebvre 1988). Several attempts have been made to hybridize two or more atomization techniques to bridge the shortfall associated with each of these techniques. For instance, conventional pressure atomizing nozzle can be mated with electrostatic technique to enhance droplet dispersion and atomization processes.

In this article, initially, various design challenges and typical atomizer requirements in high-speed combustors are briefly discussed. Subsequently, some of the novel techniques and recent developments in atomizers for high-speed combustors are presented.

8.2 Fuel Injector Needs and Challenges in High-Speed Combustors

In general, fuel atomization is achieved in atomizers by inducing momentum difference between the liquid and surrounding gas. In pressure atomizers, liquid jet/sheet is injected at very high pressure into the quiescent ambient (Fig. 8.1a), whereas, in twin-fluid atomizers, the slow-moving liquid is injected into high-speed air stream (Fig. 8.1b). Essentially in both the cases, the shear/momentum difference between the two phases (gas and liquid) forms the major source of energy required for atomization (see Eq. 1)

$$MR = \frac{\rho_g U_g^2}{\rho_l U_l^2} \tag{1}$$

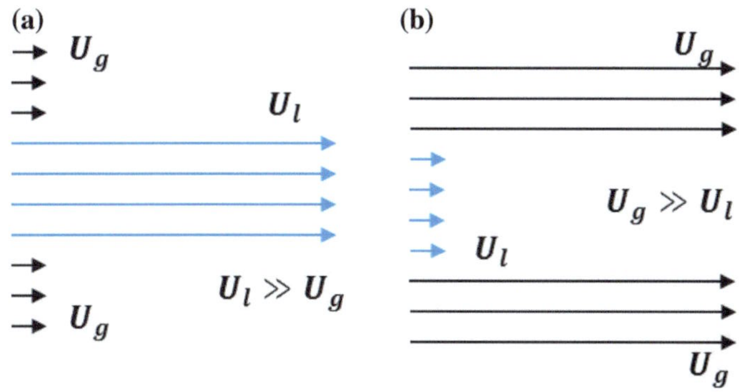

Fig. 8.1 Typical gas- and liquid-phase velocity profile: **a** pressure atomizing nozzle; **b** air-assist nozzle

Here, ρ_g, ρ_l is gas and liquid phase density (kg/m^3) and U_g, U_l is gas and liquid phase velocity (m/s). Pressure atomizing nozzles are widely used in diesel engines. Modern day engines utilize high-pressure pumps to pressurize the liquid as high as ~1600 bar (Guerrassi and Dupraz 1998; Stumpp and Ricco 1996). This yields high-velocity fuel jet at the injector exit. The ejected fuel jet is subjected to complex instabilities; the waves formed over the liquid surface results in disintegration of the jet into tiny ligaments which further undergo disintegration into daughter droplets.

The steady (continuous flow) nature of combustion in aero-propulsive devices motivated the engine designers to use twin-fluid atomizers (Lefebvre 1998). Here, fuel is continuously injected into the fast-moving air inside the combustor. The degree of atomization (particularly primary) is governed by velocity difference between the two phases. But, unlike pressure atomizing nozzles, the final spray is not solely dependent on the atomizer itself; the turbulence associated with the air field leads to additional effects like secondary atomization and spatial clustering of droplets (Lasheras et al. 1998). Particularly, these effects are more pronounced in high-speed flows (Ma $\geq$ 1). Further, the surrounding air flow pattern also plays a major role in spray formation. For instance, aero gas turbine engines utilize swirling air flows to stabilize the flame, whereas most of the high-speed combustors (rocket engines, scramjet) utilize normal axial round jets.

For high Mach number flows (Ma > 1), the injected droplets should have sufficient penetration momentum to overcome the stagnation pressure close to injector tip (Segal 2009). For efficient combustion, the liquid-fueled combustor should satisfy the following criterion

$$\tau_{evap} + \tau_{mixing} + \tau_{chemical} < \tau_{residence} \quad (2)$$

where $\tau_{evap}, \tau_{mixing}, \tau_{chemical}, \tau_{residence}$ are time scales associated with evaporation, mixing, reaction, and residence, respectively. Here, τ_{evap} is solely determined by the size of droplets delivered by spray nozzle. It should be noted that lower the droplet size lesser is the evaporation time scale and vice versa. Mixing time scale $\left(\tau_{mixing}\right)$ is linked to the turbulence level associated with the air flow and injected droplets. It is perceived that the mixing can be enhanced by modifying the turbulent intensity in the flows. Hence, in addition to droplet size and velocity, spray orientation is identified as the major determinant in flame stabilization and combustion efficiency. The spray orientation is mainly governed by the scheme of injection, i.e., parallel or perpendicular (jet in crossflow) with respect to the free-stream flow. Both orientations have been extensively investigated, and brief comparison of the same is presented in (Bogdanoff 1994; Seiner et al. 2001). Further, several aspects like pulsed injection, inclusion of lobes in injectors, multipoint injections have also been implemented to enhance mixing. The additional accessories mounted to the fuel injector should not form any obstacles to the free-stream flow, so as to avoid any drag/friction losses.

The tightly coupled nature of fluid mechanics and chemical reaction brings an additional challenge in injector design for high-speed combustors. To minimize the time scales associated with τ_{evap} and τ_{mixing}, it is mandatory to have the fuel injector deliver ultrafine droplets at high velocity.

8.3 Fuel Injection System in High-Speed Combustors

The basic understanding of spray atomization and mixing is of prime importance in high-speed flows to optimize the system. Studies revealed complexities such as shock waves, flow separation, wall-bounded effects, recirculation vortices in high-speed flow injection systems. As mentioned earlier, in high-speed combustors, fuel atomizer and its injection scheme play a major role. Hence in this chapter, the widely adopted fuel injection systems are presented.

8.3.1 *Transverse Fuel Injection System*

Transverse injection of fuel into the supersonic flow (jet in crossflow) is one of the widely adopted techniques in high-speed combustors (Segal 2009). Here, usually, the fuel jet is injected at the sonic condition. The basic flow interaction mechanism is illustrated in Fig. 8.2. The shear layer induced between injected and incoming sonic jet leads to the formation of 3D bow shock upstream of the jet (see Fig. 8.2b), which ultimately results in separation of the boundary layer at a location close to the injector. Further, the injected jet penetration length is shown to be a direct function of momentum flux ratio (J) (Funk and Orth 1967; Billig and Schetz 1966).

Experimental and numerical studies by Lee et al. (1992), Ben-Yakar et al. (2006), Yuan et al. (1999) pointed out the evolution of vortices in the near field of the reactants. Though momentum flux ratio (J) determined the near-field dynamics,

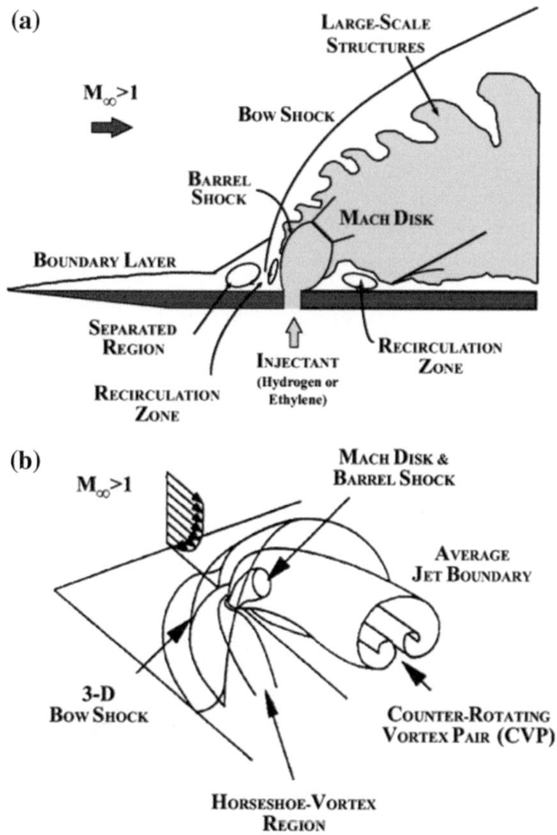

Fig. 8.2 Illustration of flow pattern formed during transverse injection of fuel into supersonic combustor (reprinted with permission from Ben-Yakar et al. 2006)

the far-field mixing is shown to be a function of injected jet velocity. High-speed diagnostics (PLIF and Schlieren) performed by Ben-Yakar et al. (2006) for the same J between hydrogen and ethylene jet reveal different mixing characteristics. Here, hydrogen (injected at very high velocity) shows lower penetration length with the presence of large-scale eddies over long distances (Fig. 8.3a). Opposite is observed for ethylene jet (Fig. 8.3b). This contradictory observation indicates the degree of complexities in identifying the fundamental relationships among the governing parameters in high-speed combustors.

8.3.2 *Parallel Injection Systems*

Though the detached shock layer in the upstream of the injected jet opens up potential avenues for transverse/normal fuel injection, the pressure loss and hot spots near the wall demand an alternate form of injection. To overcome the above-mentioned shortcomings, the parallel fuel injection scheme is investigated by

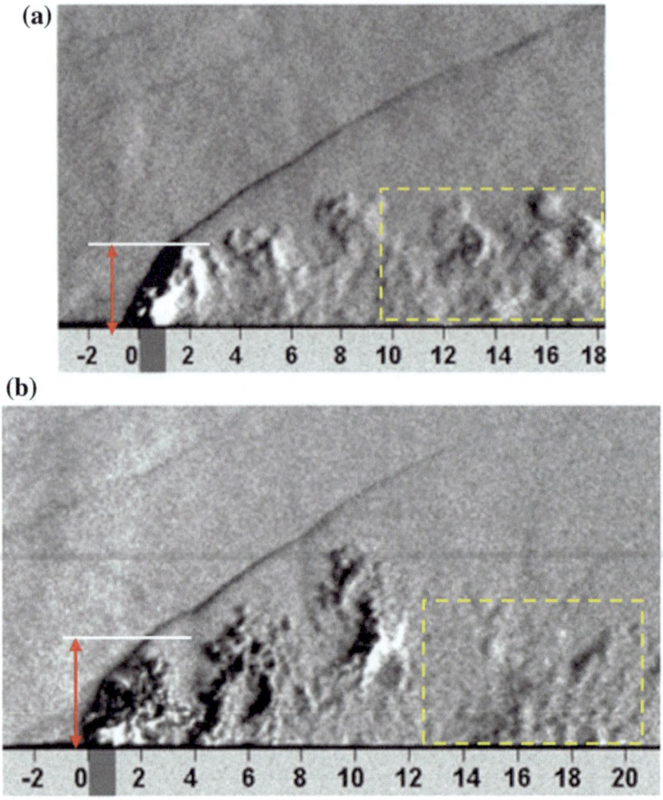

Fig. 8.3 Jet penetration in supersonic crossflow (reproduced with permission from Ben-Yakar et al. 2006)

several researchers (Northam et al. 1989; Masuya et al. 1995; Sunami et al. 2005). But the earlier parallel injection suffers from poor combustion efficiency due to extremely poor mixing in high-speed flows. The axial vortices induced by shear layer growth between two coaxial fluid streams become the major detrimental parameter for mixing (see Fig. 8.4). The induced streamwise vortices engulf the fluid from both the phases and form the interface. The mixing takes place at this interface, and it is enhanced with corrugations in interfacial area.

Mixing enhancement in the turbulent shear layer is observed with the addition of streamwise vortices in the flow direction (Sunami and Scheel 2002; Riggins and Vitt 1995). In view of this, different parallel injection configurations are tested. Among those, ramp-based injectors are investigated in greater details (Rogers et al. 1998). The widely used ramp configurations are compression and expansion types (see Fig. 8.5). The supersonic flow over the ramp is shown to induce shock and expansion waves, which results in the formation of additional vortices in the flow due to generation of baroclinic torque. The engine test carried out with ramps

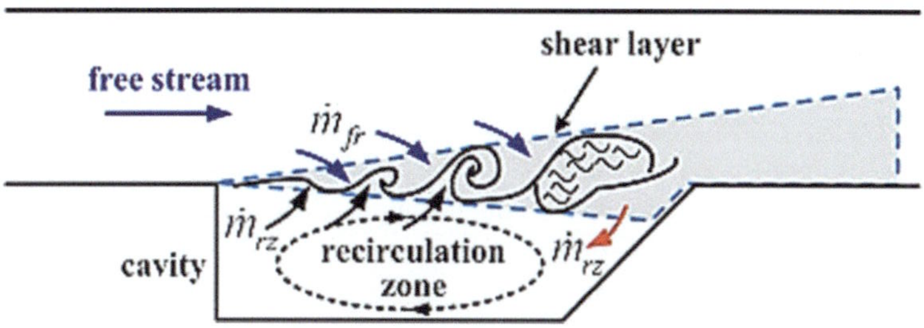

Fig. 8.4 Illustration of mixing layer formation in coaxial jets (reprinted with permission from Wang et al. 2017)

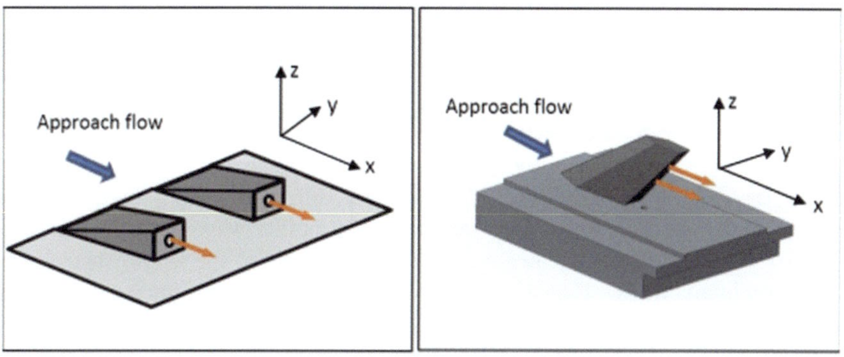

Fig. 8.5 Schematic of strut and ramp injectors used in supersonic combustors (adapted from Cabell et al. 2014)

proved to be an ideal candidate for improving the combustion efficiency in supersonic flows. Particularly, compression ramp injector is shown to exhibit best autoignition of the fuel in comparison to expansion ramp injector. The design variables like ramp height, angle, and flow area are found to be major deciding parameters in the performance of the ramp injectors. It is also recognized that the generation of strong axial vorticity beyond certain limit results in damping of transverse vortices, which leads to adverse effects toward mixing (Leibovich 1984).

Other methods like insertion of tabs, backward facing step, cavity injection, lobe mixers, counterflow mixers are also attempted in high-speed flows to enhance the mixing efficiency (Seiner et al. 2001). Additionally, the external forcing of a jet using Helmholtz resonators, acoustic excitation, piezoelectric actuators are also investigated.

8.4 Conventional Standalone Atomizers

Mixing (τ_{mixing}) and evaporation (τ_{evap}) time scales are crucial in successful operation of high-speed combustors. In the previous sections, the various techniques attempted to enhance the mixing in high Mach number flows are briefly discussed. However, in liquid-fueled combustor, the additional challenge lies in minimizing the evaporation time scale (τ_{evap}) by injecting fine-sized droplets into the free-stream flow. Subsequently, various types of atomizers which yield low droplet size are presented.

8.4.1 *Coaxial Atomizers*

Coaxial atomizers are one of the most used fuel injectors in steady-flow combustion systems like gas turbine and rocket engines. It falls under twin-fluid atomizer family, which means atomization of liquid is assisted by another fluid. In this atomizer, fuel is injected at very low velocity (~1–3 m/s) into high-speed air flow (~100–150 m/s) (Lefebvre 1990). The velocity difference across the gas–liquid interface induces shear, which ultimately drives instability waves over the interface. The size of the resultant ligament/droplet is comparable to the length scale of the most dominant wave. Hence, coflowing air acts as the primary source of atomization in coaxial atomizers (Fig. 8.6).

In combustors, some portion of the entry airflow itself diverted inside the atomizer passage to act as coaxial air flow. This added advantage makes coaxial atomizer as the best candidate in gas turbine engines. Since the droplets are entrained by high momentum gas jet, this offers strong penetration for the fuel spray (Engelbert et al. 1995; Mayer 1994). Based on the requirements of coaxial airflow

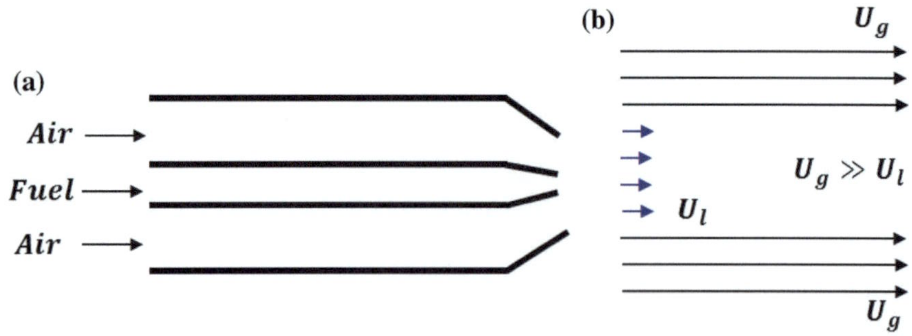

Fig. 8.6 **a** Schematic of air-assist atomizer; **b** typical gas–liquid phase velocity profile in air-assist atomizer

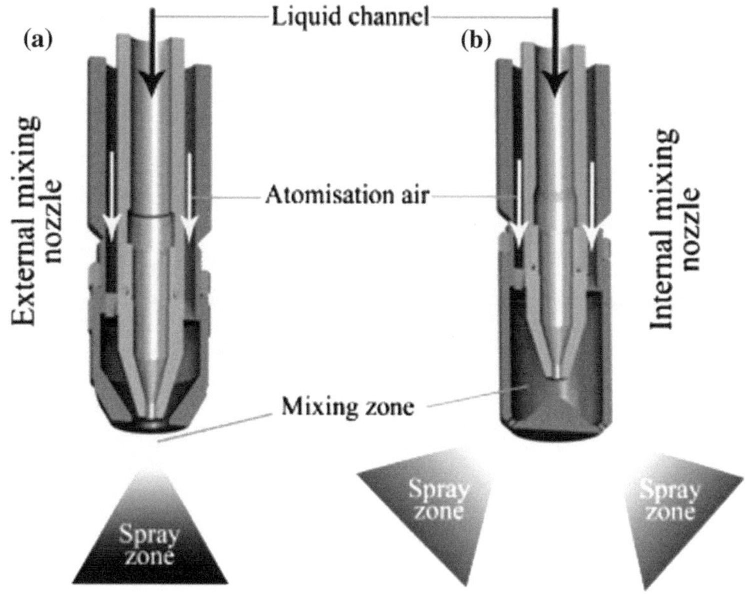

Fig. 8.7 **a** Internal mixing air-assist atomizer; **b** external mixing air-assist atomizer (reproduced with permission from Hede et al. 2008)

pattern, air–liquid meeting point, these atomizers are categorized into different types (Fig. 8.7a, b). For instance, in gas turbine engines, the air flows through a swirler before it interacts with liquid. Further, in some injectors, liquid will be delivered in form of sheet rather than round column of the jet. The two widely used twin-fluid atomizer in a gas turbine is schematically shown in Fig. 8.8. In the prefilming airblast atomizer, fuel is discharged in the form of slow-moving film into a very high-speed air flow (Jasuja 1979; El-Shanawany and Lefebvre 1978). The relative velocity difference between the two phases results in shear, which eventually causes the breakup of liquid sheet.

The main influencing parameter in determining the final droplet size is found to be momentum ratio (Eq. 1) or air to liquid ratio (ALR) (Lefebvre 1992; Rajamanickam and Basu 2017). Studies have reported the decreasing trend of droplet SMD with increasing level of coaxial air (i.e., air velocity; see Fig. 8.9). The detailed correlation involving air and liquid velocity, fluid properties like surface tension, liquid and gas phase density for the different twin-fluid atomizer configurations are listed in Lefebvre (1980). Since many of the atomizers fitted in combustors utilize the air flow available at the inlet, the maximum velocity attained by air flow is limited by combustor operating condition. Hence, final droplet size is limited by entry air flow rate.

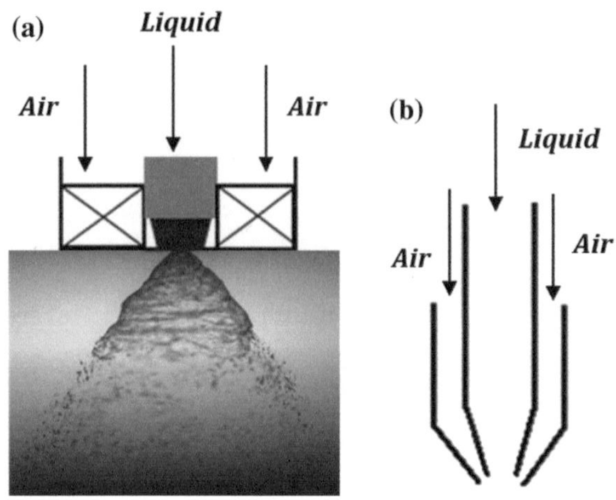

Fig. 8.8 **a** Prefilming airblast atomizer (reproduced with permission from Rajamanickam and Basu 2017); **b** plain jet airblast atomizer

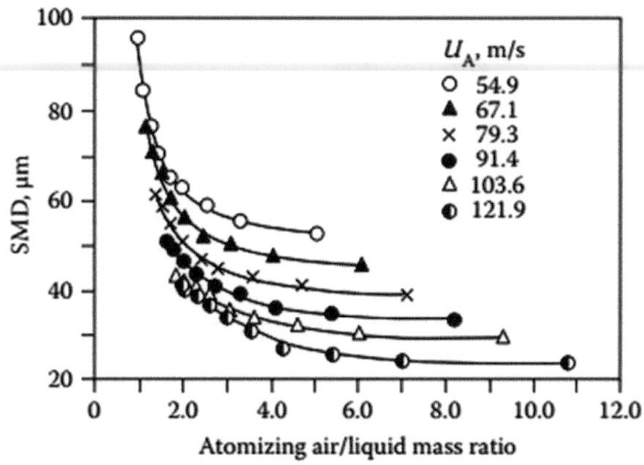

Fig. 8.9 Variation in droplet size as a function of coaxial air velocity (reprinted with permission from Lefebvre 1980)

8.4.2 *Flash-Boiling Atomizer*

This technique relies on superheating the fuel above its boiling point. Subsequently as the liquid exits the nozzle, the sudden depression in pressure causes bubble formation very close to the injector tip (see Fig. 8.10). The bubbles present in the two-phase flow mixture undergo rapid bursting and act as the main energy source for atomizing the neighboring droplets.

Flashing is initiated when the liquid is injected into the medium where the surrounding/ambient pressure (P_∞) is less than the saturation pressure (P_s) of the liquid at that temperature.

$$\Delta P \leq P_s - P_\infty \tag{3}$$

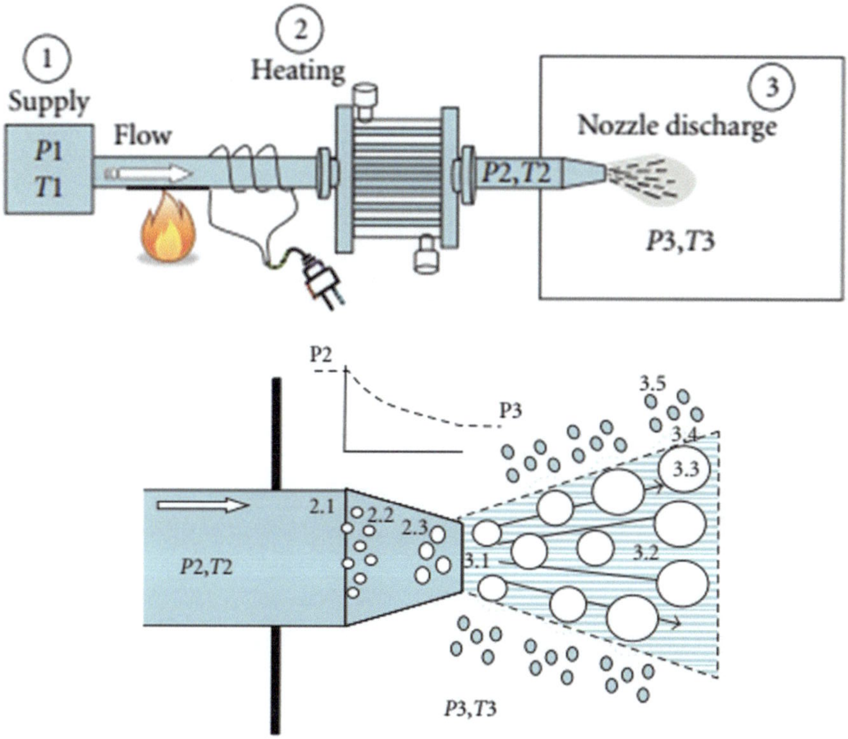

Fig. 8.10 Illustration of flash-boiling atomization process (reprinted with permission from Iyengar et al. 2012)

The major energy input required in this technique involves the degree of superheat required to initiate flash boiling (Fig. 8.11). This will vary as a function of the physical property of the fuel, for instance, high vapor pressure fuels like octane require less amount of superheat. Opposite is true for low vapor pressure fuels (dodecane). This is the reason why flashing phenomenon is often encountered in gasoline injectors. Flashing may be desirable or undesirable with respect to the application. However, in general, flashing yields very fine-sized droplets ensuring rapid evaporation of the fuel. On the other hand, sometimes rapid flashing of the fuel reduces the inertia of the droplets, drastically reducing spray penetration. Hence, this can be implemented in an application which demands wider cone angle and short penetration length.

The resultant bubble size is the major determinant in determining the quality of flash-boiling atomization (Sher et al. 2008). The bubble formation and its growth rate solely depend on the degree of superheat. Once the liquid attains necessary superheat, vapor nucleation occurs. The nucleation may be either heterogeneous (occurs at the interface) or homogeneous (occurs within the liquid). At equilibrium, the bubble size can be related to the pressure balance at vapor–liquid interface. The

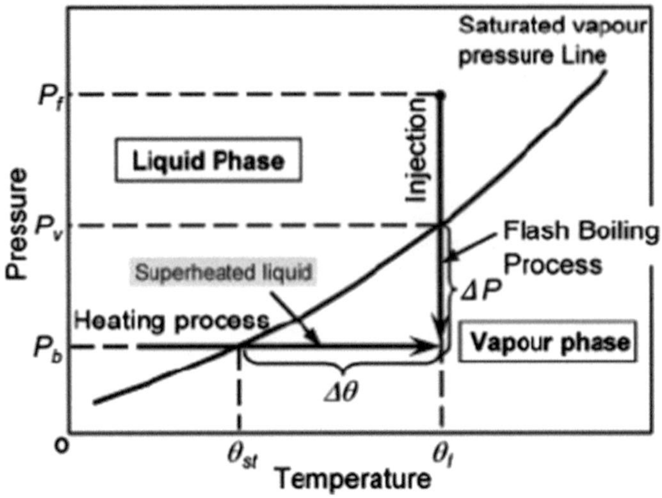

Fig. 8.11 Illustration of flash-boiling phenomenon (reprinted with permission from Senda et al. 2008)

critical bubble radius is calculated as balance between surface tension and pressure difference (Eq. 4)

$$r_c = \frac{2\sigma}{P_s - P_\infty} \tag{4}$$

where σ is surface tension in N/m. The bubble grows if its radius (r_b) is greater than critical radius (r_c).

From Eq. 3, it is evident that flash-boiling atomization is mainly influenced by physical properties of the fuel and pressure of the medium where the fuel has been injected. Zeng et al. (2012) carried out experimental investigation in a spray chamber to delineate the relationship between these properties.

The Mie-scattered spray images acquired for different fuel (n-hexane, methanol, ethanol) at various superheating conditions are depicted in Fig. 8.12. For instance, for the same vapor pressure and fuel superheat temperature, n-hexane and methanol show prominent flash boiling as compared to ethanol (Fig. 8.12iii). This is due to the fact that vapor pressure of n-hexane and methanol is close to the ambient pressure than ethanol. The drastic reduction in droplet size during flash-boiling atomization shows its potential benefits in combustion applications (Oza and Sinnamon 1983; Xu et al. 2013; Park and Lee 1994). This technique has been widely realized in automotive engines, but only very few studies reported the implementation of flash boiling in high-speed combustors like pulse detonation engines (Wen et al. 2012).

Besides the advantage, flash-boiling atomizers suffer from following major barriers; prior flashing (phase change) before the injector orifice often leads to vapor

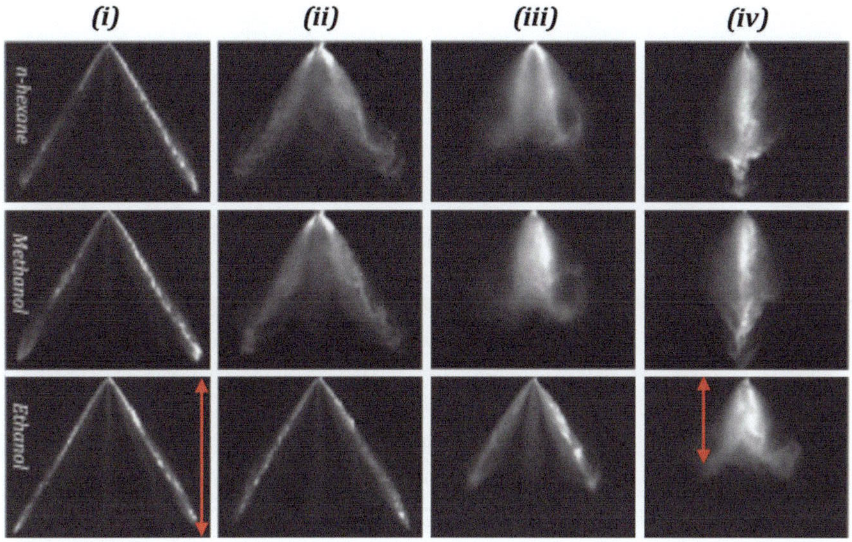

Fig. 8.12 Illustration of spray formation process for different fuels at various superheat temperature and ambient pressure: i. $P_{\infty} = 100$ kPa, $\Delta T = 50$ °C; ii. $P_{\infty} = 40$ kPa, $\Delta T = 50$ °C; iii. $P_{\infty} = 100$ kPa, $\Delta T = 90$ °C; iv. $P_{\infty} = 40$ kPa, $\Delta T = 90$ °C. (reprinted with permission from Zeng et al. 2012)

lock. The major limitations with implementation of flash-boiling atomization in practical system include

- The requirement of high thermal energy in achieving flash boiling with low vapor pressure fuels.
- Chocking of fuel orifice due to vapor lock.
- Safety aspects due to high degree of superheat.

8.4.3 *Effervescent Atomizer*

Another class of injector which adopts bubble as a major source of liquid breakup is effervescent atomizer. The fundamental difference between flash boiling and effervescent injectors lies in the method of initiating the bubbles in the atomizer. Unlike flash-boiling atomizer, this technique involves injection of gas into the liquid phase to initiate bubble formation inside the injector. In general, this atomizer falls under the twin-fluid atomizer family, but unlike other atomizers, here usually gas phase is injected at very low velocity. Here, the role of the gas phase is to initiate the bubble formation without imparting additional kinetic energy.

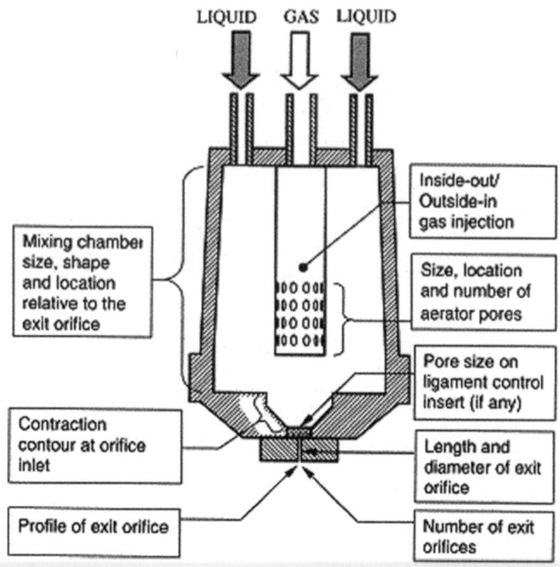

Fig. 8.13 Schematic of effervescent atomizer (reprinted with permission from Sovani et al. 2001)

Schematic of the first effervescent atomizer proposed by Lefebvre et al. (1988) is shown in Fig. 8.13, which utilizes porous tube to introduce gas in the liquid phase. Here, the liquid is injected transversely into the mixing chamber.

8.4.4 Working Principle

The two-phase bubbly flow mixture formed inside the mixing chamber is forced to discharge through the exit orifice (Fig. 8.14). The bubbles injected at the orifice undergoes rapid expansion and ultimately results in bursting. This action leads to shattering of the ligaments into tiny fragments and daughter droplets. These droplets further undergo a secondary atomization either under the influence of aerodynamic effects or neighboring bubbles.

In effervescent mode, usually the liquid exits in the form of an annular jet with bubbles present in the middle (Fig. 8.14). Due to this arrangement, effervescent atomizers can yield primary droplets with sizes even smaller than the exit orifice. It is worthwhile to note that the orifice size in effervescent atomizer is usually kept in the order of ~4–5 mm. The sufficiently big orifice does not demand any pump pressure to push the liquid. Hence, it is evident that effervescent atomizer can yield reasonably better atomization as compared to other twin-fluid atomizers.

The main influencing parameter which decides the final droplet size is gas to liquid ratio (GLR). It can be defined as

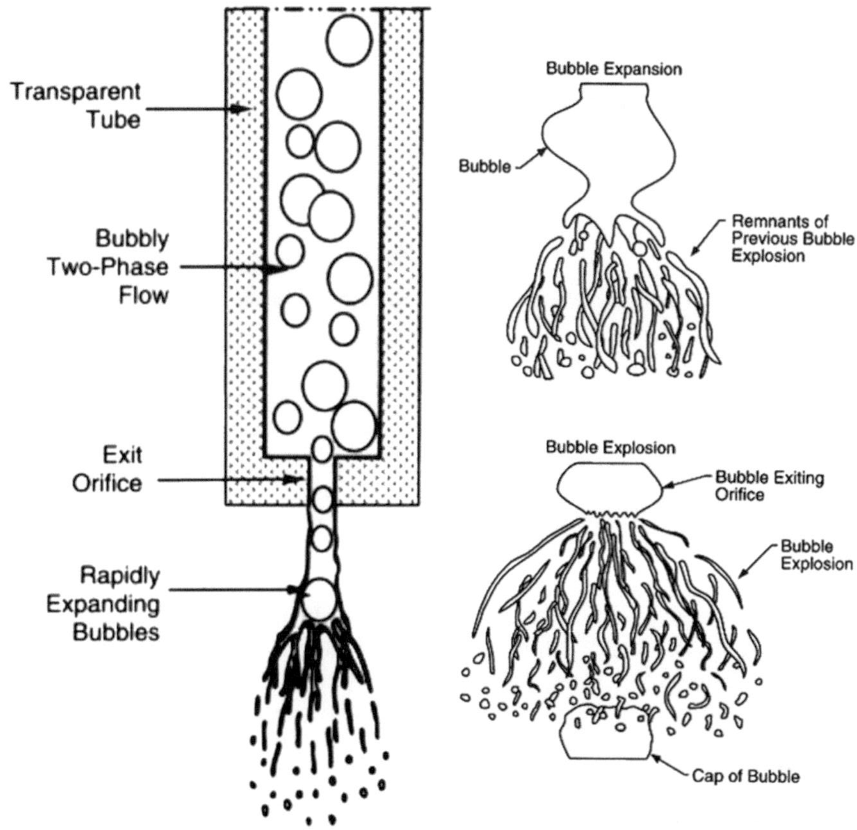

Fig. 8.14 Illustration of atomization process in effervescent atomizer (reprinted with permission from Sovani et al. 2001)

$$GLR = \frac{\dot{m}_g}{\dot{m}_l} \quad (5)$$

Here, $\dot{m}_g$ and $\dot{m}_l$ are gas and liquid phase flow rates (kg/s), respectively. Several studies (Lund et al. 1993; Whitlow and Lefebvre 1993; Buckner and Sojka 1991) have been conducted to investigate the influence of GLR in determining the size of the final droplet. Most of the obtained results show universal trend of decreasing droplet size with increasing GLR (Fig. 8.15). Additionally, other influencing geometrical parameters like orifice size, shape, number and size of aerator holes, mixing chamber length on resulting SMD have been also investigated.

Because of its simple operation and ability to deliver lesser droplet size, effervescent atomizer gained special attention in the combustion community. The multihole effervescent atomizer proposed by Li et al. (1994) shows immense potential toward uniform spray distribution in annual gas turbine combustor.

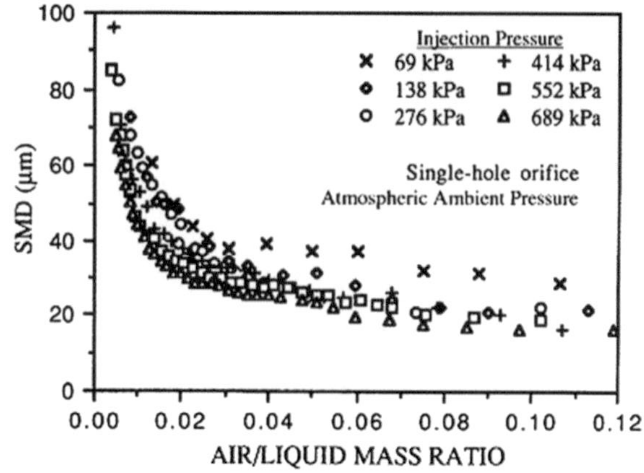

Fig. 8.15 Variation of SMD as a function of GLR (reprinted with permission from Sovani et al. 2001)

Noticeably, an experimental investigation carried out by Sallam et al. (2004) in supersonic crossflow (M ~ 1.94) showed the potential of effervescent atomizer in achieving the very low mean droplet diameter of ~40 μm. Though increasing GLR shows positive results in terms of droplet size, the transition from steady to intermittent bubbly flow is reported (Konstantinov et al. 2010; Kim and Lee 2001). This may be due to the convective transport of bubble instability to the spray (Sen et al. 2014). The detailed review presented by Sovani et al. (2001) elucidated the benefits and possibilities of implementing effervescent atomization in combustion applications.

8.4.5 *Electrospray*

Electrostatic spray utilizes electric charge to disrupt the surface of the liquid jet. The liquid jet breaks up if the pressure induced by applied electric charge is more than the restoring surface tension force. The induced electric pressure over the liquid jet surface can be written as

$$P = \frac{FV^2}{2\pi D^2} \tag{6}$$

The typical electrospray mechanism is illustrated in Fig. 8.16. Here, the fluid inside the nozzle/capillary is supplied with electric charge (mostly positive). The collector plate (negatively charged) is usually kept at a downstream location; this usually acts as pulling source to form the cone-shaped liquid jet at the exit orifice. The cone is famously known as 'Taylor cone.' In the conical portion, like charges repel each other causing the cone surface to expand. This action disturbs the hydrodynamics of the jet and resulting in unstable wave formation over the jet

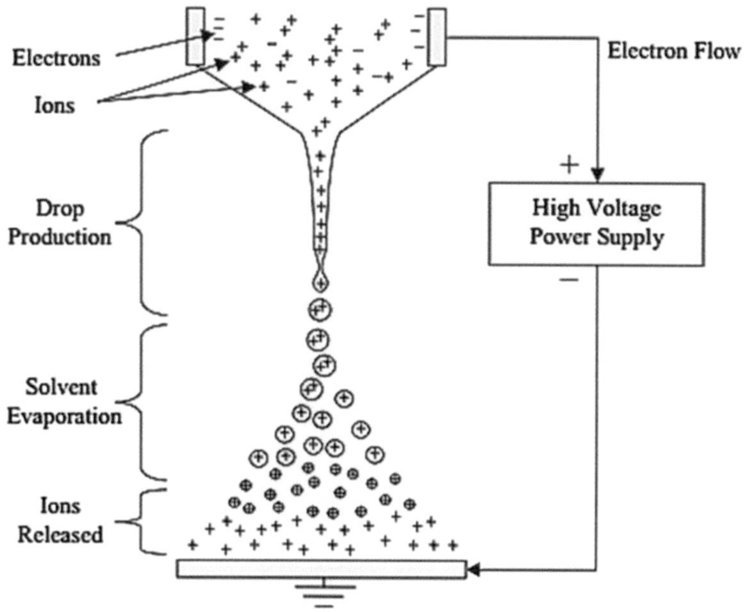

Fig. 8.16 Schematic illustration of process involved in electrostatic atomization (reprinted with permission from Ashgriz 2011)

surface. Hence, this method is also known as electrohydrodynamic atomization (EHD).

In general, charge injection proposed by Shrimpton (2004) is found to be a more apt technique for combustion applications. This involves insertion of charge electrode directly inside the atomizer unit.

Theoretically, the electrostatic force required for atomization can be written as

$$F_{electrostatic} = qE \tag{7}$$

where q is electric charge and E is induced electric field, and the interdependence of the q, E with respect to another parameter is given by

$$q = f(I); E = f(V); I \propto V/Q$$

Hence, for higher electrostatic force, the quantity V and I need to be significant in magnitude. However, for any given applied voltage, any increase in flow rate (Q) results in sudden drop in current (I), which needs to be augmented by increasing the voltage (V). This phenomenon limits the scope of electrostatic atomizers in high flow rate applications. However, electrostatic injectors are used as an assisting technique in existing injectors to enhance atomization. For instance, charge injection is widely implemented in gasoline direct injection system to control the spatial dispersion of the droplets (Anderson et al. 2007; Shrimpton 2003; Agathou and Kyritsis 2012; Hetrick and Parsons 1997).

The additional advantages associated with electrospray nozzle in combustion application include the absence of coalescence due to coulombic repulsion, controlled droplet trajectories to name a few.

8.4.6 *Ultrasonic Atomizer*

Ultrasonic atomizer utilizes high-frequency sound waves over the liquid surface to form the liquid droplets. Since the resultant droplets will be usually scaled with wavelength, the applied high-frequency wave yields very fine-sized droplets (~10 μm).

The widely used ultrasonic atomizer is shown in Fig. 8.17. Here, two piezoelectric disks are placed opposite to each other to generate high-frequency waves (~10^5–10^6 kHz). The waves are in general transferred to the atomizing horn which is positioned in contact with the liquid feed tube. At resonance condition, the maximum sound amplitude leads to the formation of capillary waves over the liquid surface.

The resultant droplet size is usually scaled with length scale of the capillary waves (Eq. 9, Lang 1962)

$$\lambda = \left(\frac{8\pi\sigma}{\rho_f F^2}\right)^{1/3} \tag{8}$$

where σ is surface tension; F is excited sound frequency in Hz. Studies have reported the promising nature of ultrasonic atomizer in achieving droplets size in the range of ≤10 μm. For instance, nebulizers adopted this technique to generate ultrafine droplets. However, like electrostatic technique, this atomizer suffers from the limitation of low flow rate.

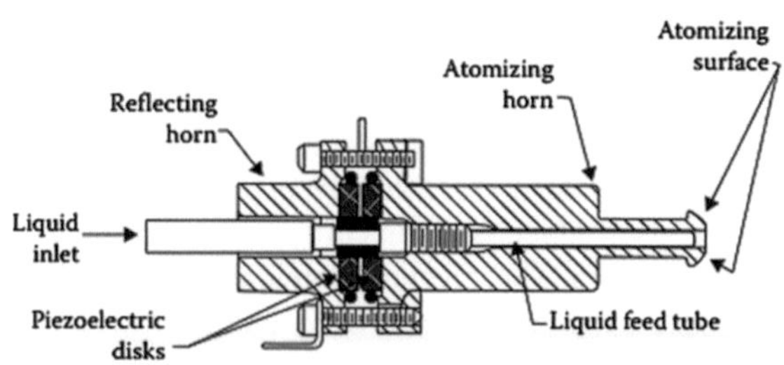

Fig. 8.17 Schematic of ultrasonic atomizer (adapted from Berger 1985)

8.5 Hybrid Atomizers

The comprehensive discussion presented in the previous section clearly shows, besides advantages, the limitation associated with each atomization technique limiting their practical usage in high-speed combustors.

Conventionally used atomizers in combustion systems like airblast and pressure jet offer high momentum, but for the several reasons, the minimum droplet size is limited to ~40–50 μm (Lefebvre 1980, 1990). To meet the size requirements, studies have been carried out to promote atomization in these atomizers by supplementing with other additional techniques. For instance, flash boiling is realized in pressure atomizers by preheating the fuel above its boiling point, before it leaves the orifice. Likewise, airblast atomizer is mated with electrostatic/ultrasonic techniques to reduce the droplet size. Among these, flash-boiling and effervescent atomizer show potential in combustion applications.

8.5.1 Why Hybrid Atomizer?

In standalone airblast and pressure atomizing nozzles, even at very high air velocity/injection pressure, the liquid jet exhibits poor atomization behavior (intact liquid sheet/jet) very close to the injector. This phenomenon leads to bigger droplets in that zone. High shadowgraph-based visualization carried out in coaxial atomizer delineates the above-mentioned phenomena. Though small droplets are seen at very high air velocity (yellow circled in Fig. 8.18iii), at near-nozzle locations, the presence of intact portion of liquid jet normally results in unatomized droplets.

Hence, the main idea behind hybrid atomizer is to disrupt or reduce the intact portion of the liquid jet before it interacts with coaxial air flow. Some researchers have used an electric charge to disrupt the liquid jet, while some have utilized ultrasonic waves to excite the liquid jet/sheet.

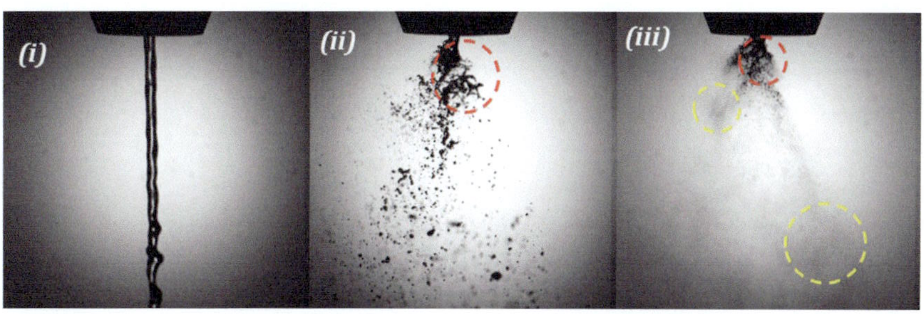

Fig. 8.18 Visualization of jet breakup in coaxial atomizer: i. $U_g \sim 0$ m/s; ii. $U_g \sim 40$ m/s; iii. $U_g \sim 100$ m/s

8.5.2 Effervescent Cum Air-Assist Atomizer

It is seen that effervescent atomizer can yield acceptable droplet size with low injection pressure and low amount of airflow. The hybrid atomizer proposed and tested in this study is schematically shown in Fig. 8.19. This atomizer utilizes combined action of coaxial air and effervescence as primary modes of atomization.

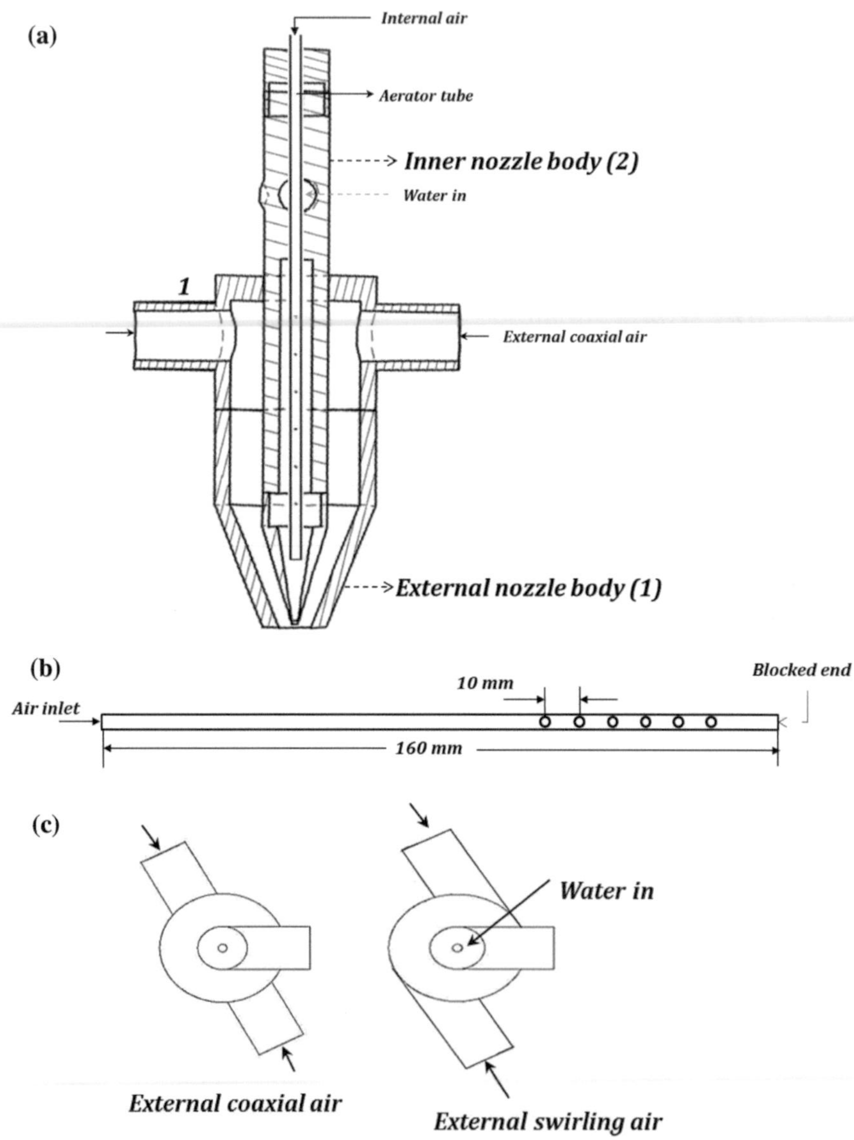

Fig. 8.19 **a** Schematic drawing of the hybrid atomizer; **b** aerator tube; **c** external air flow path

8.5.2.1 Design

The designed atomizer basically contains two major parts, the inner (1) and outer nozzle bodies (2). The essential role of the inner nozzle body is to create two-phase bubbly mixture before the liquid interacts with coaxial air. The aerator tube geometry details are shown in Fig. 8.19b. Here, two types of aerator tubes are used, and the first configuration has 12 holes distanced 180° (opposite to each other) apart. In the second configuration, 24 holes are used with 90° separation between each hole. In both the configurations, the hole size is maintained as 0.75 mm. Further, the external air is admitted in two ways, i.e., coaxial and eccentric/tangential entry with respect to the liquid flow. The air flow field quantified from PIV for the two configurations is shown in Fig. 8.20. The tangential entry essentially leads to swirling flow at the nozzle exit. It is well known that the tangential momentum induced by swirl flows causes radial pressure deficit (Beér and Chigier 1972). This deficit eventually leads to flow reversal/depression in the axial velocity at centerline (see Fig. 8.20a). The combined action of radial and axial velocity in swirling flow results in the wider flow field and spray cone angle. The tangential air entry configuration can be used to effectively disperse the droplets in the radial direction. This configuration can be ported in gas turbine engines, where swirl is mandatory for flame stabilization. On the other hand, coaxial entry yields narrow concentrated axial velocity dominant vector field.

Coaxial entry ensures that the droplets are concentrated in a narrow region of interest.

8.5.2.2 Experimental Conditions and Procedure

The diagnostic tools used to characterize the spray discharged from hybrid atomizer are schematically shown in Fig. 8.21. The diagnostics involve global as well as local measurements. Global measurement is done with the help of high pulse rate

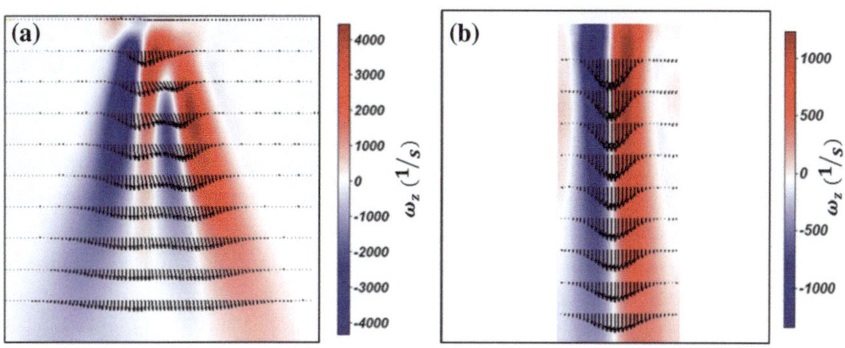

Fig. 8.20 Mean vector field superimposed with vorticity contours: **a** tangential air entry; **b** coaxial air entry

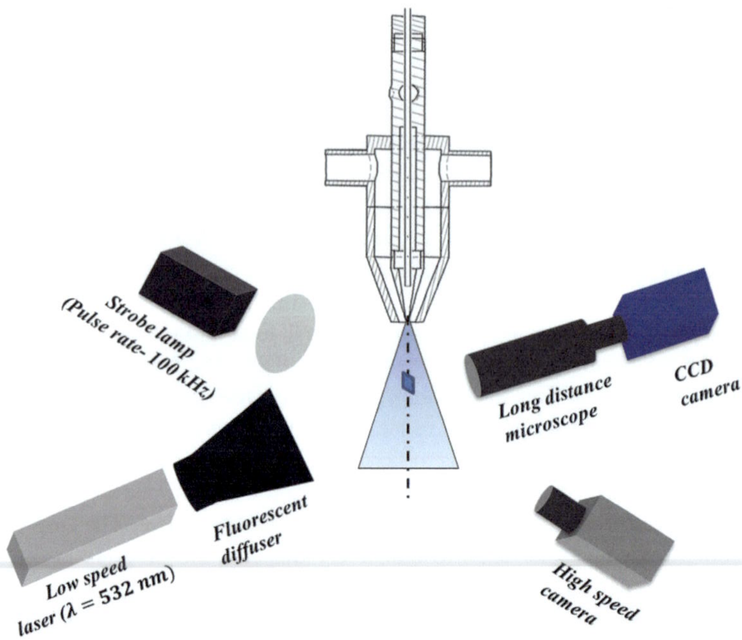

Fig. 8.21 Schematic of diagnostic setup

LED strobe lamp synchronized with camera shutter. The images are acquired at 7500 frames/s at a different field of view (FOV). Large FOV (~80 mm × 80 mm) yields a global picture of the spray, from which breakup length, spray spread, atomization quality of the droplets are quantified. The near-nozzle breakup, particularly bubble formation and breakup, jet oscillations are delineated with small FOV (~40 mm × 40 mm).

The droplets size information at various operating conditions of the spray is acquired with the help of long distance microscope (LDM). To acquire the time-frozen snapshot of the droplet contour, low pulse duration (~5 ns) laser light is used as an illumination source. The LDM is coupled with 2× magnification lens to focus onto a small region of the spray (~1.5 mm × 1.5 mm). Further, the acquired images are post-processed in MATLAB to yield droplet size. Throughout the experiments, water flow rate is maintained constant, while air flow rate across aerator tube and outer nozzle body (coaxial air) is progressively varied in different proportions. The maximum operating conditions at which tests have been carried out are listed in Table 8.1.

The flowchart shown in Fig. 8.22 summarizes the various modes of operation under which studies have been carried out. For instance, to identify the optimum no. of holes in aerator tube, the test has been conducted with 12 and 24 holes. Likewise, to elucidate the mode of air entry, under air-assist configuration, the study is conducted only with tangential and coaxial entry (i.e., absence of effervescent

Table 8.1 Operating conditions

Atomizer	Air			Water
	Internal air pressure (bar)-max	External air pressure (bar)-max	External flow rate (g/s)-max	Flow rate (g/s)
Effervescent	3	–	–	4.33–5.50
External mix	–	5	7.8	4.33–5.50
Hybrid	3	5	7.8	4.33–5.50

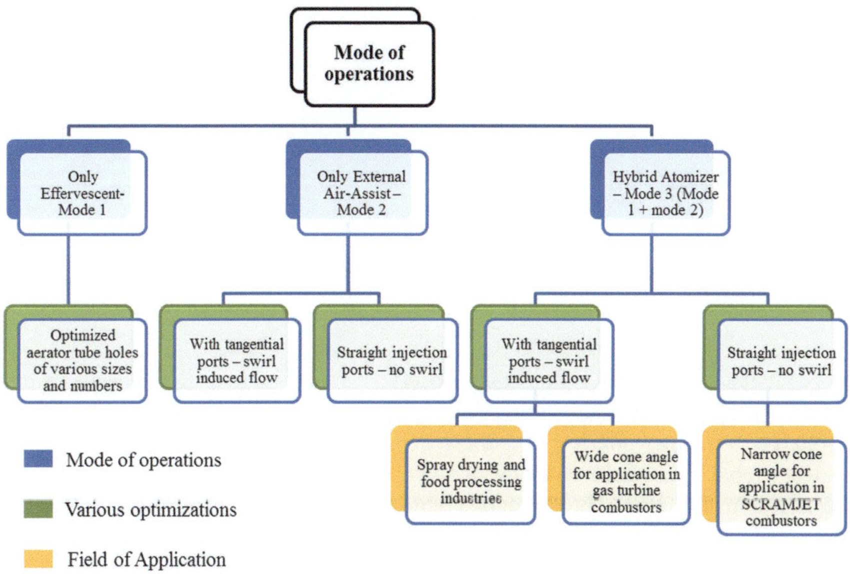

Fig. 8.22 Flowchart illustrates various modes of operation of hybrid atomizer and its possible applications

mode). Later, mode 1 and mode 2 are combined together to yield the hybrid mode of operation (mode 3).

8.5.2.3 Working Principle

In the proposed hybrid atomizer, initially, air is flushed through the aerator tube placed inside the internal nozzle body. The bubbles formed inside the nozzle disrupt the liquid jet while leaving the orifice. In Fig. 8.18, it is seen that an intact portion of liquid jet exists even at high coaxial air velocity. However, a small amount of air injected via effervescence mode results in complete disruption of liquid jet very close to the nozzle exit itself (see Fig. 8.23iv, v). However, in a position little

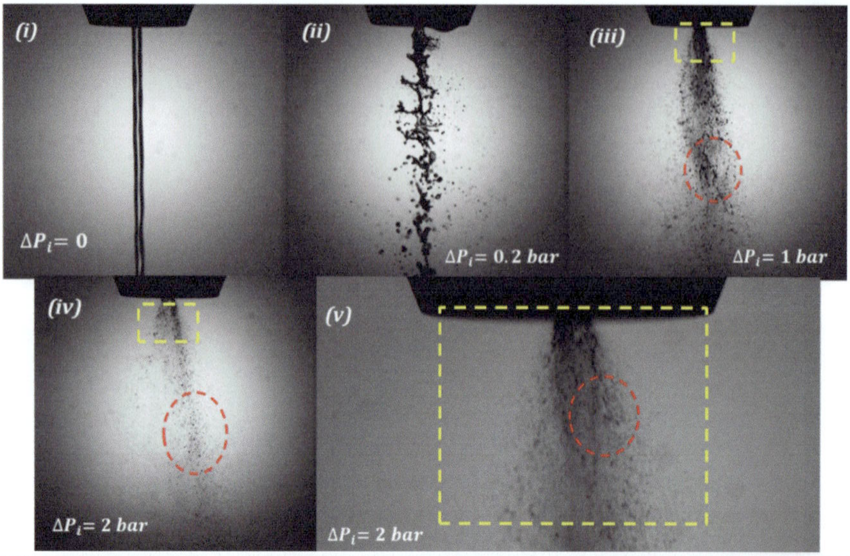

Fig. 8.23 Visualization of jet breakup (effervescent atomization) under the action of various levels of internal air flow rate

downstream, unatomized droplets are observed, mainly due to the absence of any coflowing shear around the droplets. Now, with the incorporation of coaxial air discharged via outer nozzle body, these droplets can be further atomized.

The global spray field observed for hybrid mode over different time instant is illustrated in Fig. 8.24. The combined action of coaxial air and effervescent mode yields very finely atomized spray. The bigger sized droplets observed while operating with only effervescent mode (marked with red circles in Fig. 8.23iii, iv) is completely atomized by high shearing action induced by high-velocity coaxial air. This shows that hybrid combination of effervescence and coaxial airblast modes aids in disrupting the liquid jet close to the nozzle and ensures fine atomization both in the far as well as near fields. In addition, by controlling the degree of tangential

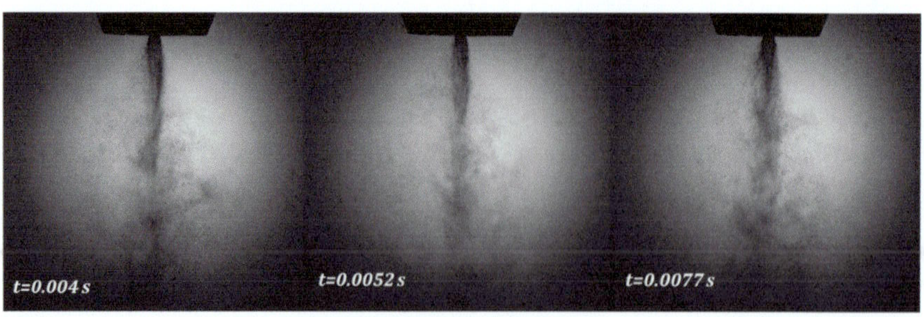

Fig. 8.24 Spray formation in hybrid mode of operation

flow (swirl), the spray cone angle and radial dispersion can be effectively regulated. The swirl also can be used for pilot flame anchoring in high-speed combustors.

8.5.3 *Size Distribution*

The droplet size distribution quantified from the shadowgraphy images at a pre-defined location of the spray is illustrated in Fig. 8.25. While operating the atomizer in standalone mode (i.e., either only effervescent/coaxial air), SMD shows large values ~50 μm, whereas in hybrid mode, a significant reduction in droplet size is observed.

In hybrid mode, majority of droplets are in the range of 5–10 μm. As a result, much lower SMD (~17 μm) is observed. Though only effervescent mode yields lower SMD, the velocity of the droplets is much lower than the coaxial mode. The combined action of coaxial air and effervescent mode not only yields lower droplet size, but also imparts much higher velocity to the droplets (~100 m/s).

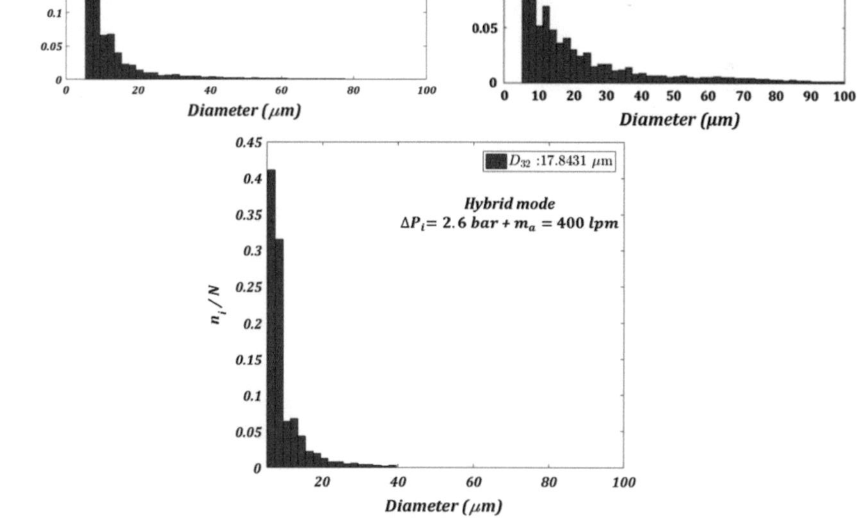

Fig. 8.25 Droplet size distribution observed for different modes of operation; n_i = number of droplets in the given size range; N = total no. of droplets in the experimental set

Table 8.2 Typical scramjet requirements and hybrid atomizer capability

Design parameters	SCRAMJET requirements	Hybrid atomizer
Fuel flow rate	9 g/s (single injector) OR 4.5 g/s each for twin injector configuration	4.67 g/s
Desired supply pressure (Bar)	7	5
Desired droplet velocity (m/s)	>120	>120
Desired droplet size (μm)	≤20	<20

The proposed hybrid atomizer intentionally designed for in-house scramjet combustor (Ma ~ 2.2) closely matches the requirements (see Table 8.2). The values listed in Table 8.2 pertain to the maximum operating condition of the atomizer.

8.5.4 Influence of Design and Operational Parameters

The main design and operational parameters considered in this study include a number of aerator holes (12, 24), air flow entry configuration (coaxial, swirl), external (air-assist) and internal air (effervescent) flow rates. In only effervescent mode, for a given internal air pressure (ΔP_i), increase in number of holes results in reduction of SMD by ~2 μm, whereas increase in internal air pressure for a given hole configuration shows drastic reduction in SMD. This can be understood from graph shown in Fig. 8.26.

8.5.5 Combined Electrostatic and Pressure Jet Atomizer

In pressure jet atomizer, it is well known that even at very high injection pressures intact portion of liquid jet persists in the position very close to the injector. This phenomenon gives rise to the presence of large-sized droplets (equivalent to orifice size) in the primary breakup region. For finer atomization, it is prudent to have a smaller intact length (i.e., breakup length) to disperse the droplets in a homogeneous manner.

Kourmatzis et al. (2012) implemented electric charge injection in the conventional pressure jet atomizer to enhance the turbulence in primary atomization zone (see Fig. 8.27). In this design, the electrode is positioned in such a way that fluid leaving the nozzle get charged by applied electric potential. Upon exiting the orifice, the induced charge in the fluid results in sudden rupture of the liquid jet.

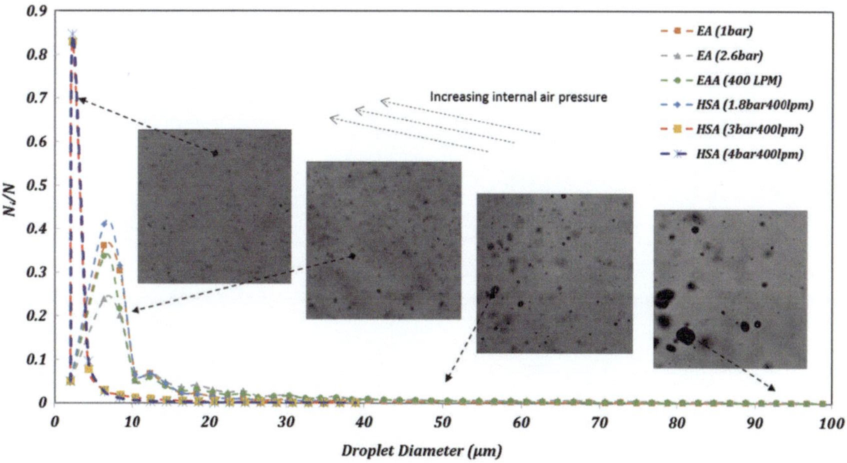

Fig. 8.26 Droplet diameter distribution for various modes of operation. *EA*—only effervescent; *EAA*—only external air-assist; *HSA*—hybrid mode

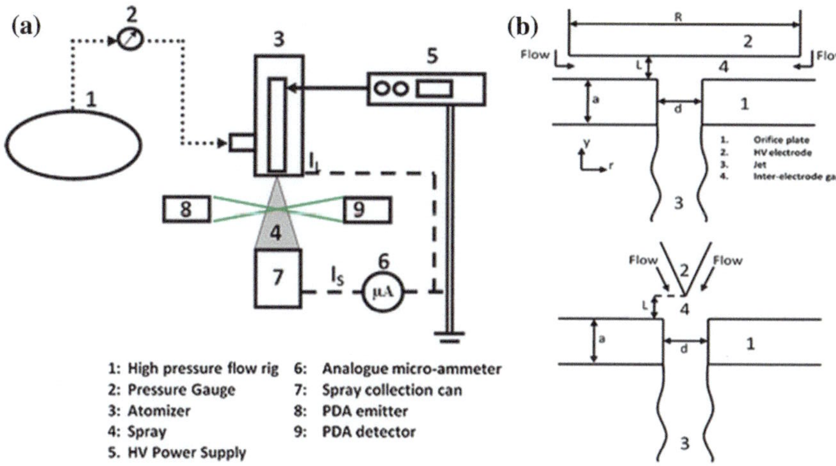

Fig. 8.27 Schematic illustration of hybrid version of pressure jet and electrostatic atomizer (reprinted with permission from Kourmatzis et al. 2012)

For the applied injection pressure (P = 20 bar), in absence of electric field, spray behaves like a laminar jet and exhibits long breakup length (Fig. 8.28i). In this configuration, resultant droplet size will be roughly in the order of orifice size.

However, a significant reduction in breakup length scale and turbulent disruption of liquid jet is observed for nozzle operated in hybrid mode with same injection pressure but varying levels of applied electric potential (Fig. 8.28ii, iii). The

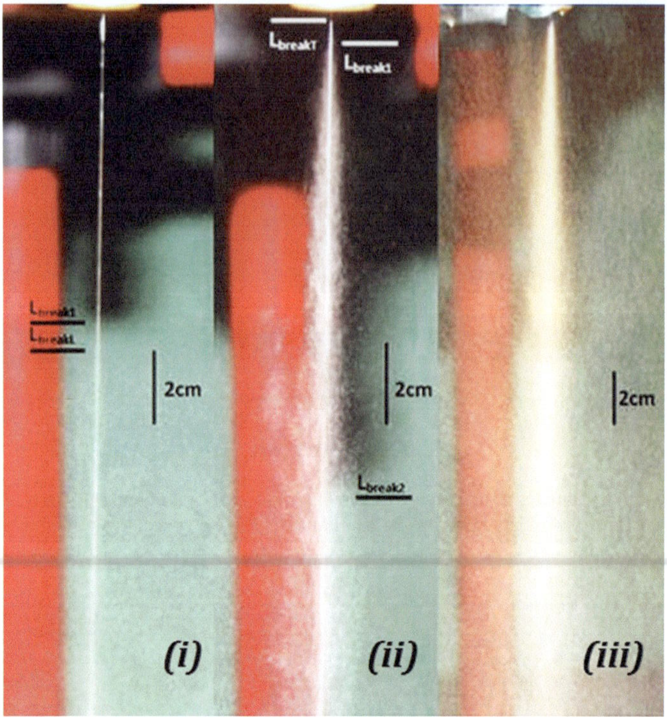

Fig. 8.28 Breakup of liquid under the influence of applied electric field: i. 0 kV; ii. 6 kV; iii. 10 kV (reprinted with permission from Kourmatzis et al. 2012)

combined action of aerodynamic and applied electric force fields results in very fine-sized droplets. The enhanced droplet velocity increases the turbulence level of the spray, thereby promoting the mixing process. Further, studies (Li et al. 2006; Gomez and Chen 1994) also reported acceleration of secondary atomization process in charged droplets.

The action of promoting the atomization in pressure jet nozzle by means of electrostatic force received greater attention in IC engine community (Arkhipov et al. 1998).

8.5.6 *Flash-Boiling Cum Pressure Jet Atomizer*

In previous sections, it is shown that injecting preheated fluid in an environment whose pressure is less than saturation vapor pressure leads to rapid evaporation of the droplets. Studies have been carried out to promote atomization in pressure atomizers by preheating the fuel before injection. Rapid flash boiling in injected

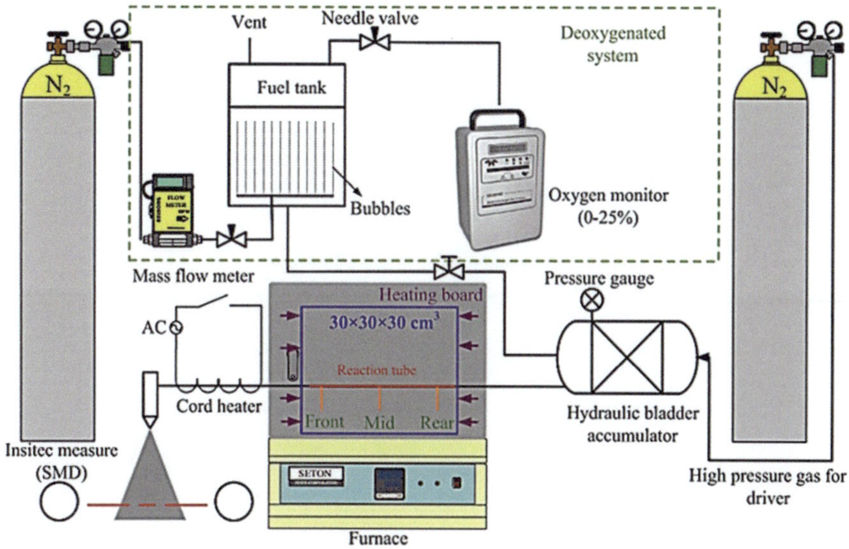

Fig. 8.29 Combined pressure and flash boiling mode of atomization (reprinted with permission from Wen et al. 2012)

spray is reported for preheated fuel in gasoline direct injection (Xu et al. 2013; Zhang et al. 2013).

Wen et al. (2012) implemented the flash-boiling mode in conventional straight jet pressure atomizer (see Fig. 8.29) to enhance the atomization in pulse detonation engines. Here, fuel is preheated in fuel line before being admitted into the fuel nozzle. The injection pressure is maintained at 14 MPa, and fuel temperature is varied from 300 to 500 °C. The results suggest much lower SMD values (~10 μm) for jp-8 fuel while operating the pressure atomizer in hybrid mode.

Senda et al. (2000) proposed gas dissolved concept to achieve flash-boiling atomization in low vapor pressure fuels like dodecane, n-tridecane. Dissolved liquefied CO_2 is mixed with n-tridecane to realize flashing in high-pressure diesel injector. Pressure jet atomizer is operated at 20 bar injection pressure with various mole fraction level of dissolved CO_2. In addition to the injection pressure and CO_2 mole fraction, ambient pressure is also shown to be a key determinant in spray structure. The droplet size distribution clearly depicts the influence of CO_2 in resultant droplet size (Fig. 8.30). Almost 57% reduction in final SMD (~20 μm) is reported with the hybrid mode of operation (pressure jet + dissolved gas). Further, besides this advantage, a simultaneous reduction in NO and soot emission is also realized with the induction of dissolved CO_2 in the main fuel.

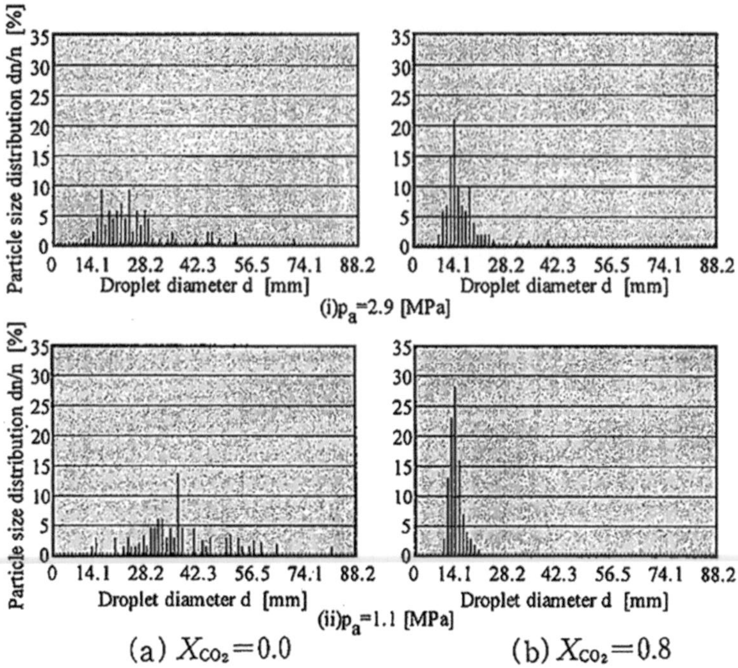

Fig. 8.30 Droplet size distribution for gas dissolved spray in pressure jet atomizer (reprinted with permission from Senda et al. 2000)

8.5.7 *Externally Forced Sprays*

External forcing of jet/spray is common practice in many applications. It is usually done to promote mixing between two streams (fuel and oxidizer) by intensifying the turbulent levels in the jet (Yoshida et al. 2001). Forcing is widely implemented in forms of acoustic excitation of the liquid jet issued from the nozzle. In spray systems, the applied acoustic force is shown to act as aiding parameter in disrupting the liquid jet (Baillot et al. 2009; Rhys 1999). It is also shown that the droplets undergo enhanced vaporization when subjected to acoustic forcing (Sujith 2005).

Baillot et al. (2009) investigated the effect of acoustic excitation of liquid jet in coaxial atomizer. The result shows that transformation of liquid jet into thin sheet occurs under the influence of acoustic waves. Furthermore, promotion of shear instability at liquid–air interface is also reported. However, till date there is no information available on detailed spray characterization (droplet size, velocity) for such acoustic-based hybrid atomizers.

8.6 Summary and Conclusions

The need and challenges of fuel injection systems in typical high-speed combustors like scramjets, pulse detonation engines are briefly presented. It is highlighted that special care is needed in designing fuel injection system since the fluid mechanics and chemical time scales are tightly coupled. The advantages and limitations associated with standalone/conventional atomization technique are reviewed in the context of high-speed combustors. Subsequently, the evolution of hybrid atomizers, recent developments, and its capabilities in high-speed combustors is also presented.

Acknowledgements The financial support received from BRAMHOS aerospace, Swarnajayanti Fellowship and National Centre for Combustion research and Development (NCCRD) is gratefully acknowledged.

References

Agathou MS, Kyritsis DC (2012) Electrostatic atomization of hydrocarbon fuels and bio-alcohols for engine applications. Energy Convers Manag 60:10–17

Anderson EK, Carlucci AP, De Risi A, Kyritsis DC (2007) Synopsis of experimentally determined effects of electrostatic charge on gasoline sprays. Energy Convers Manag 48(11):2762–2768

Arkhipov VI, Emelianova EV, Tak YH, Bässler H (1998) Charge injection into light-emitting diodes: theory and experiment. J Appl Phys 84(2):848–856

Ashgriz N (2011) Handbook of atomization and sprays: theory and applications. Springer Science & Business Media

Baillot F, Blaisot J-B, Boisdron G, Dumouchel C (2009) Behaviour of an air-assisted jet submitted to a transverse high-frequency acoustic field. J Fluid Mech 640:305–342

Bayvel LP (1993) Liquid atomization, vol 1040. CRC Press

Beér JM, Chigier NA (1972) Combustion aerodynamics. New York

Ben-Yakar A, Mungal MG, Hanson RK (2006) Time evolution and mixing characteristics of hydrogen and ethylene transverse jets in supersonic crossflows. Phys Fluids 18(2):026101

Berger HL (1985) Characterization of a class of widely applicable ultrasonic nozzles. In: ILLASS

Billig FS, Schetz JA (1966) Penetration of gaseous jets injected into a supersonic stream. J Spacecr Rockets 3(11):1658–1665

Bogdanoff DW (1994) Advanced injection and mixing techniques for scramjet combustors. J Propul Power 10(2):183–190

Buckner HN, Sojka PE (1991) Effervescent atomization of high-viscosity fluids: Part I. Newtonian liquids. At Sprays 1(3)

Cabell K, Drozda TG, Axdahl EL, Danehy PM (2014) The enhanced injection and mixing project at NASA Langley. In: JANNAF 46th CS/34th APS/34th EPSS/28th PSHS joint subcommittee meeting, Albuquerque, NM, Dec 2014

El-Shanawany M, Lefebvre AH (1978) Airblast atomization: the effect of linear scale on the mean drop size. Cranfield Institute of Technology

Engelbert C, Hardalupas Y, Whitelaw JH (1995) Breakup phenomena in coaxial airblast atomizers. Proc R Soc Lond A: Math Phys Eng Sci 451:189–229

Ferri A (1973) Mixing-controlled supersonic combustion. Annu Rev Fluid Mech 5(1):301–338

Funk J, Orth R (1967) An experimental and comparative study of jet penetration in supersonic flow. J Spacecr Rockets 4(9):1236–1242

Gomez A, Chen G (1994) Charge-induced secondary atomization in diffusion flames of electrostatic sprays. Combust Sci Technol 96(1–3):47–59

Guerrassi N, Dupraz P (1998) A common rail injection system for high speed direct injection diesel engines. SAE technical paper

Gutmark EJ, Schadow KC, Yu KH (1995) Mixing enhancement in supersonic free shear flows. Annu Rev Fluid Mech 27(1):375–417

Hede PD, Bach P, Jensen AD (2008) Two-fluid spray atomisation and pneumatic nozzles for fluid bed coating/agglomeration purposes: a review. Chem Eng Sci 63(14):3821–3842

Hetrick RE, Parsons MH (1997) Electrospray for fuel injection. SAE technical paper

Iyengar V, Simmons H, Ransom D (2012) Flash atomization: a new concept to control combustion instability in water-injected gas turbines. J Combust 2012

Jasuja AK (1979) Atomization of crude and residual fuel oils. ASME J Eng Power 101(2):250–258

Kim JY, Lee SY (2001) Dependence of spraying performance on the internal flow pattern in effervescent atomizers. At Sprays 11(6)

Konstantinov D, Marsh R, Bowen PJ, Crayford A (2010) Effervescent atomization for industrial energy-technology review. At Sprays 20(6)

Kourmatzis A, Ergene EL, Shrimpton JS, Kyritsis DC, Mashayek F, Huo M (2012) Combined aerodynamic and electrostatic atomization of dielectric liquid jets. Exp Fluids 53(1):221–235

Lang RJ (1962) Ultrasonic atomization of liquids. J Acoust Soc Am 34(1):6–8

Lasheras JC, Villermaux E, Hopfinger EJ (1998) Break-up and atomization of a round water jet by a high-speed annular air jet. J Fluid Mech 357:351–379

Lee MP, McMillin BK, Palmer JL, Hanson RK (1992) Planar fluorescence imaging of a transverse jet in a supersonic crossflow. J Propul Power 8(4):729–735

Lefebvre AH (1980) Airblast atomization. Prog Energy Combust Sci 6(3):233–261

Lefebvre A (1988) Atomization and sprays, vol. 1040. CRC Press

Lefebvre AH (1990) Energy considerations in twin-fluid atomization. In: ASME 1990 international gas turbine and aeroengine congress and exposition, pp V003T06A002–V003T06A002

Lefebvre AH (1992) Twin-fluid atomization: factors influencing mean drop size. At Sprays 2(2)

Lefebvre AH (1998) Gas turbine combustion. CRC Press

Lefebvre AH, Wang XF, Martin CA (1988) Spray characteristics of aerated-liquid pressure atomizers. J Propul Power 4(4):293–298

Leibovich S (1984) Vortex stability and breakdown-survey and extension. AIAA journal 22 (9):1192–1206

Li J, Lefebvre AH, Rollbuhler JR (1994) Effervescent atomizers for small gas turbines. In: ASME 1994 international gas turbine and aeroengine congress and exposition, pp V003T06A048–V003T06A048

Li L, Yu S, Hu Z (2006) Theoretical and experimental studies of electrospray for IC engine. SAE technical paper

Lund MT, Sojka PE, Lefebvre AH, Gosselin PG (1993) Effervescent atomization at low mass flow rates. Part I: The influence of surface tension. At Sprays 3(1)

Masuya G et al (1995) Ignition and combustion performance of scramjet combustors with fuel injection struts. J Propul Power 11(2):301–307

Mayer WOH (1994) Coaxial atomization of a round liquid jet in a high speed gas stream: a phenomenological study. Exp Fluids 16(6):401–410

Menon S (1989) Shock-wave-induced mixing enhancement in scramjet combustors. In: 27th AIAA aerospace sciences meeting

Northam G, Greenberg I, Byington C (1989) Evaluation of parallel injector configurations for supersonic combustion. In: 25th joint propulsion conference, p 2525

Oza RD, Sinnamon JF (1983) An experimental and analytical study of flash-boiling fuel injection. SAE technical paper

Park BS, Lee SY (1994) An experimental investigation of the flash atomization mechanism. At Sprays 4(2)

Rajamanickam K, Basu S (2017) Insights into the dynamics of spray–swirl interactions. J Fluid Mech 810:82–126

Rhys NO (1999) Acoustic excitation and destruction of liquid sheets

Riggins DW, Vitt PH (1995) Vortex generation and mixing in three-dimensional supersonic combustors. J Propul Power 11(3):419–426

Rogers R, Capriotti D, Guy R (1998) Experimental supersonic combustion research at NASA Langley. In: 20th AIAA advanced measurement and ground testing technology conference, p 2506

Sallam K, Aalburg C, Faeth G, Lin K, Carter C, Jackson T (2004) Breakup of aerated-liquid jets in supersonic crossflows. In: 42nd AIAA aerospace sciences meeting and exhibit, p 970

Segal C (2009) The scramjet engine: processes and characteristics, vol 25. Cambridge University Press

Seiner JM, Dash SM, Kenzakowski DC (2001) Historical survey on enhanced mixing in scramjet engines. J Propul Power 17(6):1273–1286

Sen D, Balzan MA, Nobes DS, Fleck BA (2014) Bubble formation and flow instability in an effervescent atomizer. J Vis 17(2):113–122

Senda J, Asai T, Kawaguchi B, Fujimoto H (2000) Characteristics of gas-dessolved diesel fuel spray: spray characteristics and simulating flash boiling process. JSME Int J Ser B 43(3):503–510

Senda J, Wada Y, Kawano D, Fujimoto H (2008) Improvement of combustion and emissions in diesel engines by means of enhanced mixture formation based on flash boiling of mixed fuel. Int J Engine Res 9(1):15–27

Sher E, Bar-Kohany T, Rashkovan A (2008) Flash-boiling atomization. Prog Energy Combust Sci 34(4):417–439

Shrimpton JS (2003) Pulsed charged sprays: application to DISI engines during early injection. Int J Numer Meth Eng 58(3):513–536

Shrimpton JS (2004) Design issues concerning charge injection atomizers. At Sprays 14(2)

Sovani SD, Sojka PE, Lefebvre AH (2001) Effervescent atomization. Prog Energy Combust Sci 27 (4):483–521

Stumpp G, Ricco M (1996) Common rail-an attractive fuel injection system for passenger car DI diesel engines. SAE technical paper

Sujith RI (2005) An experimental investigation of interaction of sprays with acoustic fields. Exp Fluids 38(5):576–587

Sunami T, Scheel F (2002) Analysis of mixing enhancement using streamwise vortices in a supersonic combustor by application of laser diagnostics. In: AIAA/AAAF 11th international space planes and hypersonic systems and technologies conference, p 5203

Sunami T, Magre P, Bresson A, Grisch F, Orain M, Kodera M (2005) Experimental study of strut injectors in a supersonic combustor using OH-PLIF. In: AIAA/CIRA 13th international space planes and hypersonics systems and technologies conference, p 3304

Swithenbank J (1967) Hypersonic air-breathing propulsion. Prog Aerosp Sci 8:229–294

Wang H, Li P, Sun M, Wei J (2017) Entrainment characteristics of cavity shear layers in supersonic flows. Acta Astronaut 137:214–221

Wen C-S, Chung K-M, Lu FK, Lai W-H (2012) Assessment of flash-boiling for pulse detonation engines. Int J Heat Mass Transf 55(11):2751–2760

Whitlow JD, Lefebvre AH (1993) Effervescent atomizer operation and spray characteristics. At Sprays 3(2)

Xu M, Zhang Y, Zeng W, Zhang G, Zhang M (2013) Flash boiling: easy and better way to generate ideal sprays than the high injection pressure. SAE Int J Fuels Lubr 6 (2013-01-1614):137–148,

Yoshida H, Koda M, Ooishi Y, Kobayashi KP, Saito M (2001) Super-mixing combustion enhanced by resonance between micro-shear layer and acoustic excitation. Int J Heat Fluid Flow 22(3):372–379

Yuan LL, Street RL, Ferziger JH (1999) Large-eddy simulations of a round jet in crossflow. J Fluid Mech 379:71–104

Zeng W, Xu M, Zhang G, Zhang Y, Cleary DJ (2012) Atomization and vaporization for flash-boiling multi-hole sprays with alcohol fuels. Fuel 95:287–297

Zhang G, Xu M, Zhang Y, Hung DL (2013) Characteristics of flash boiling fuel sprays from three types of injector for spark ignition direct injection (SIDI) engines. In: Proceedings of the FISITA 2012 world automotive congress, pp 443–454

Chapter 9
Experimental Investigation of Spray Characteristics of Kerosene, Ethanol, and Ethanol-Blended Kerosene Using a Gas Turbine Hybrid Atomizer

Amlan Garai, Shinjan Ghosh, Swarnendu Sen and Achintya Mukhopadhyay

Abstract Gas turbines have wide applications as prime movers in transportation and power-generating sectors, most of which are currently driven by fossil fuels. The problem of air pollution can be associated with the use of conventional fuels, and their prolonged use has caused the fuel reserves to get depleted gradually. The addition of ethanol in conventional fossil fuel leads to better spraying characteristics and decreases air pollution as well. The present work is done for knowing the spray characteristics of pure kerosene, pure ethanol, and ethanol-blended kerosene (10 and 20% ethanol-blended kerosene by volume) by using a hybrid atomizer. The novelty of the hybrid atomizer lies in the fact that the fuel stream is sandwiched between two annular air streams. Tangential inlets are used for both fuel and air stream; however, the inner air stream can be used in axial configuration. A high swirling effect is produced outside the nozzle due to the tangential inlet of the flow direction. The direction of the fuel flow and both the air streams in the atomizer may be configured in the same direction or in opposite directions, respectively. The inner and outer air flow rates are varied continuously. Here, backlight imaging technique is used for capturing the spray images. Various spray breakup regimes like distorted pencil, onion, tulip, and fully developed spray regimes have been observed. The breakup length, cone angle, and sheet width of the fuel stream are analyzed from the images for different fuels and air flow rates. It is observed that

A. Garai (✉) · S. Sen · A. Mukhopadhyay
Department of Mechanical Engineering, Jadavpur University, Kolkata 700032, India
e-mail: amlangarai@gmail.com

S. Sen
e-mail: sen.swarnendu@gmail.com

A. Mukhopadhyay
e-mail: achintya.mukho@gmail.com

S. Ghosh
Department of Mechanical and Aerospace Engineering, University of Central Florida, Orlando, FL 32826, USA
e-mail: shinjandgp025@gmail.com

S. Basu et al. (eds.), *Droplets and Sprays*, Energy, Environment, and Sustainability,
https://doi.org/10.1007/978-981-10-7449-3_9

breakup length decreases for ethanol-blended kerosene due to low viscosity of ethanol. It is also observed that at higher air flow rate, breakup length decreases due to turbulent nature of the fuel stream.

9.1 Introduction

The rapid depletion in the fossil fuel reserves and the emission of greenhouse gases from vehicle exhausts are pressing issues for the global environment. There has been an upsurge in the environmental pollution since the last century which in turn has led to global warming and erratic climatic changes. To prevent further damage to the environment, researchers have been trying to develop alternative fuel technologies with low exhaust emission. In this regard, ethanol may be treated as an appropriate alternative, due to its lower viscosity and low combustion emission. The use of ethanol blended with diesel was a subject of research in the 1980s, and it was shown that ethanol–diesel blends were technically acceptable for existing diesel engines (Hansen et al. 2005). Ethanol can be produced from crops with high sugar or starch contents. Some of these crops are sugarcane, sorghum, corn, barley, cassava, sugar beets, etc. Ethanol can be identified as one of the feasible alternative fuels (Yahya and Goering 1977). This type of research work was started from 1970s. Gasohol (a mixture of 10% alcohol with 90% gasoline) is now a commercial

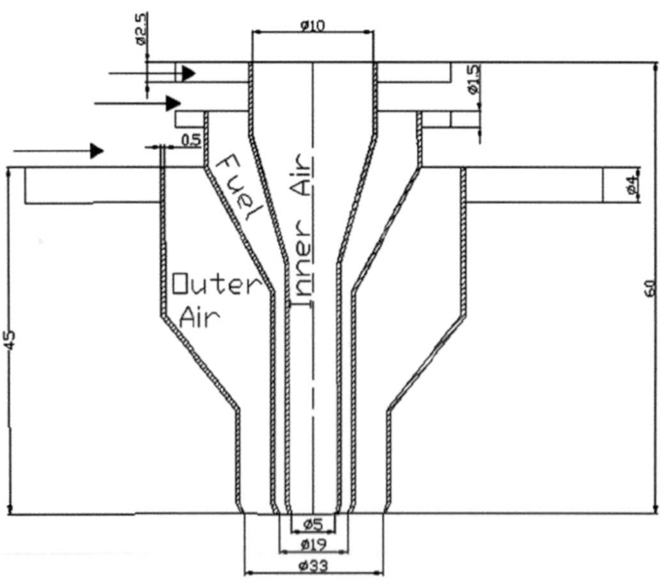

Fig. 9.1 Cross-sectional view of the atomizer all dimension in mm (Chatterjee et al. 2015; Garai et al. 2017)

fuel in over 35 countries of the world including the USA, Canada, and France (Ajav and Akingbehin 2002). In Brazil, car engines have been modified for running on neat alcohol (Reeser et al. 1995) .

A few research projects are going on for comparative study on emission characteristics and ethanol-based fuel in diesel engine. The impacts of ethanol on emissions vary with engine operating conditions and ethanol content. At high load operating conditions, ethanol-blended diesel fuels have stronger effects on smoke, NOx, acetaldehyde emissions, and unburned ethanol emissions, but at low loads,

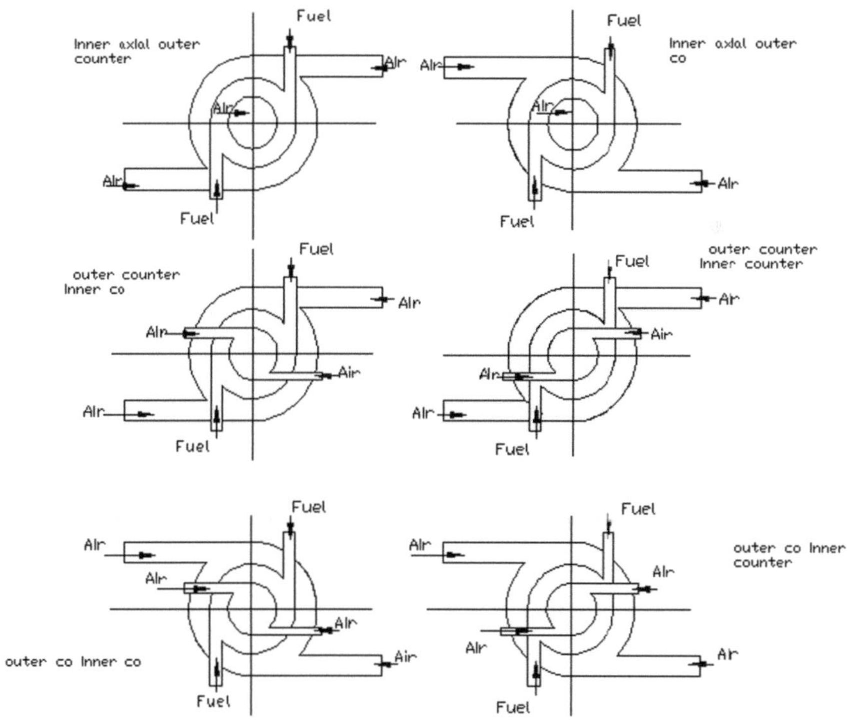

Fig. 9.2 Inlet configuration of the atomizer

Table 9.1 Property of the blended fuel and pure fuel

Particulate	Viscosity (cSt)	Surface tension (dynes/cm)	Calorific value (MJ/Kg)
Kerosene (KE0)	1.38	25.6	46.7
10% ethanol-blended kerosene (KE10)	1.34	25.3	44.2
20% ethanol-blended kerosene (KE20)	1.26	25.1	43.01
Ethanol (KE100)	1.24	24.6	29.5

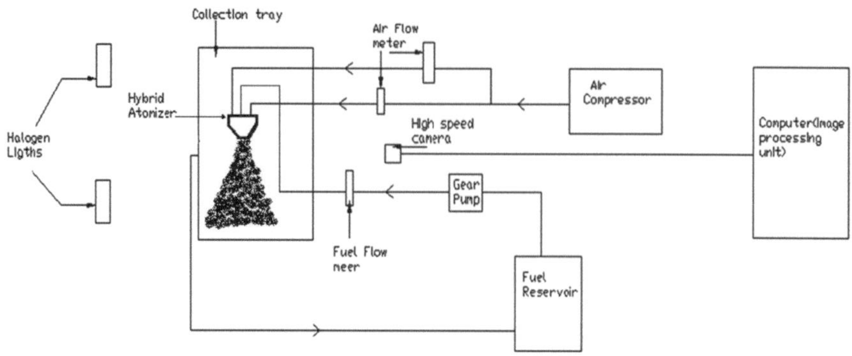

Fig. 9.3 Block diagram of experimental setup

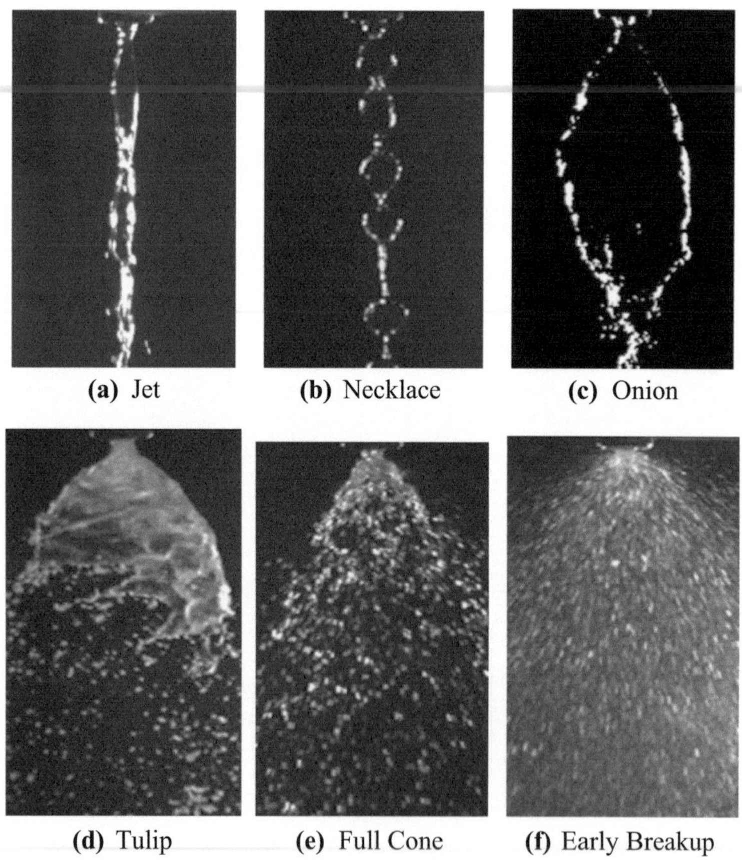

Fig. 9.4 Different breakup stages during spray formation

the blends have slight effects on smoke reduction on a four-cylinder direct injection diesel engine (He et al. 2003). Increasing of ethanol percentage in diesel ethanol blended fuels, DS (dry soot) in the PM (particular matter) and smoke decrease significantly, but the SOF (soluble organic fraction) mass percentage increases above 20% ethanol blended diesel at all lodes (Chen et al. 2007). Xing-Cai et al. (2004) studied about the cetane number improvement additive and also studied the emission and heat release rate on ethanol-blended diesel fuel. Ethanol-blended diesel fuel can be used as a gas turbine fuel.

Khan et al. (2013) studied the thermal efficiency and fuel consumption rates of cooking stove using 5, 10, 15, and 20% ethanol-blended kerosene. Dioha et al. (2012) have also studied the performance of the cooking stove by calculating the boiling time of specified volume of water using different blending of ethanol–kerosene blends. Patra et al. (2015) studied the flame characteristics and combustor performance of the kerosene and kerosene–ethanol blends. They also observed that the soot formation in flame decreases with increasing blend percentage of ethanol. Khalil and Gupta (2013) studied the combustion characteristics of gaseous and liquid fuels for finding an alternative fuel. They used methane, diluted methane, hydrogen-enriched methane, and propane as gaseous fuels and kerosene and

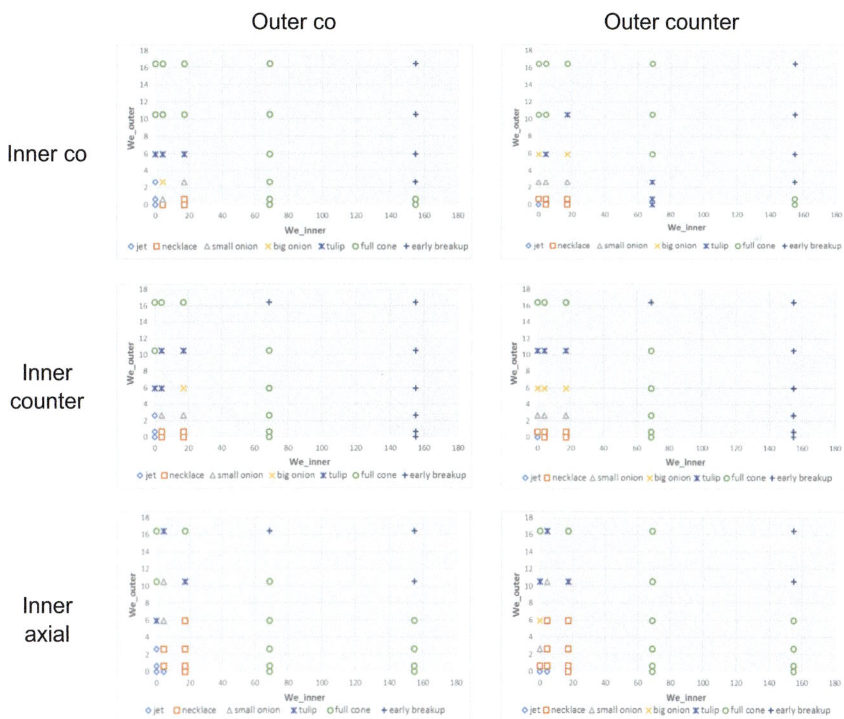

Fig. 9.5 Breakup regimes for developing spray on 1 LPM kerosene flow

ethanol as liquid fuels. They reported that NO and CO emissions were reduced for using the alternative fuels. Without any modifications of the injectors, ethanol-blended kerosene can be used. Pure jet A fuel was blended with ethanol and butanol at varying volume fractions, and performance characteristics along with emission characteristics were studied by Mendez et al. (2012, 2014). They reported that CO and NO_x emissions were reduced for using both types of blended fuels. Lower emissions were also detected in case of fossil fuel blends with four pure vegetable oils in an experiment performed on a 30 KWe commercial micro-gas turbine by Chiariello et al. (2014).

The present study has been concentrated on ethanol-blended diesel and ethanol-blended kerosene fuel. Also, the spray character of blended fuel is compared with the pure fuel. Spray characteristics study has also played a significant role for combustion study. For this study, a hybrid atomizer is used for knowing the inner and outer air flow effects on spray field.

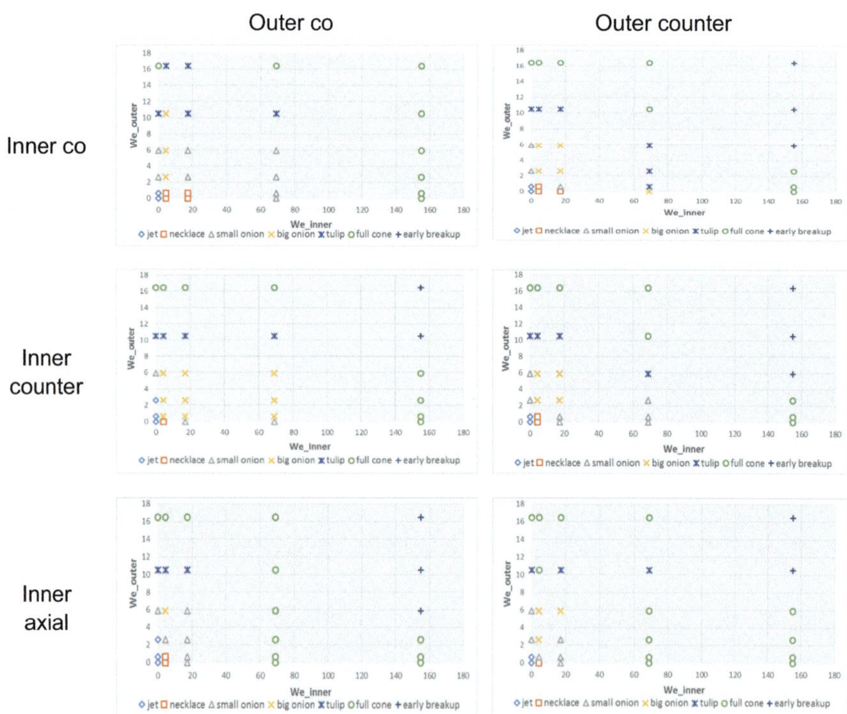

Fig. 9.6 Breakup regimes for developing spray on 2 LPM kerosene flow

9.2 Equipment and Method

9.2.1 Atomizer

The atomizer is a device which is used for discharging a high-velocity liquid into a relatively slow-moving air or gas stream or a low-velocity liquid stream discharge into relatively high-velocity air stream. Major adverbial atomizers are pressure atomizer, simplex atomizer, rotary atomizer, air-assist atomizer, and air-blast atomizer. Different types of nozzles produce three canonical geometrical shapes of liquid: jet, planar sheet, and annular sheet (Lin 2003). Our study is concentrated on a simple gas turbine atomizer spray at low air flow regime. Pressure-swirl atomizers and the air-blast atomizers are commonly used in gas turbine engines. In a pressure-swirl atomizer, a high-pressure liquid discharges into a slow-moving environment. This generally leads to a lot of soot formation. In an air-blast nozzle, liquid is discharged at low velocity into a fast-moving gas (air) flow resulting in a fine spray with good air–fuel mixture. For solving such problems, hybrid atomizers (Chin et al. 1999; Rizk et al. 1996) have been designed that combine the features of

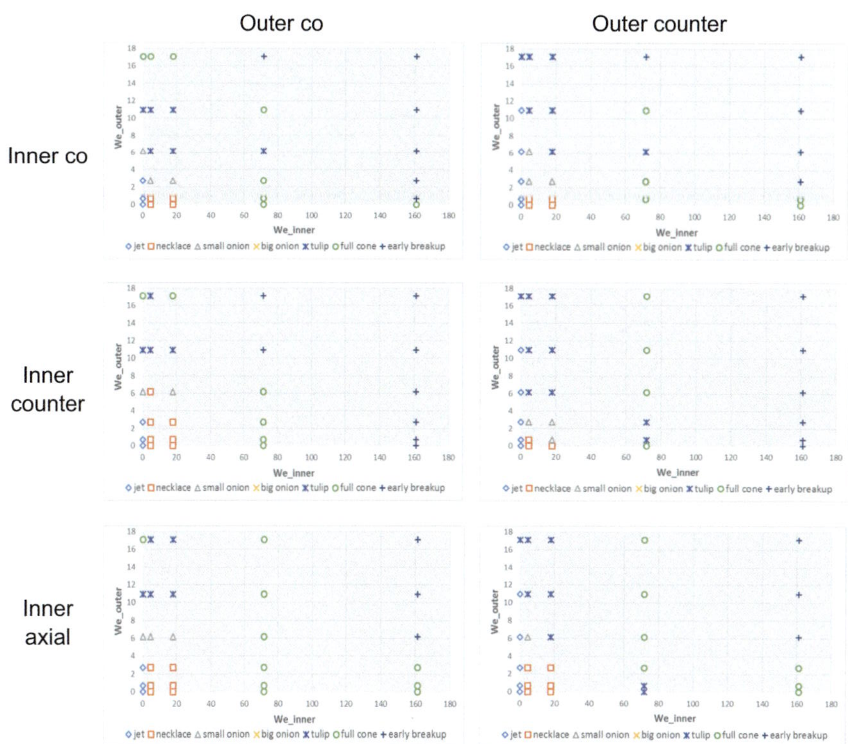

Fig. 9.7 Breakup regimes for developing spray on 1 LPM ethanol flow

both pressure-swirl and air-blast atomizers using the combined effects of air momentum and fuel pressure. Recently, air-blast atomizer is an area of immense interest in industrial and aircraft field due to its reduced emission characteristics and reduced injection pressures. In the present study, a modified hybrid atomizer is used for knowing the effect of inner and outer air at liquid sheet breakup. The different breakup regimes in the spray zone have also been studied. Rizk et al. (1996) and Chin et al. (1999) studied the spray characteristics of a hybrid atomizer, where liquid was flowing into center part of the nozzle and air is flowing into outer periphery of the nozzle. Leboucher et al. (2010) studied the different breakup processes on an annular liquid sheet assisted by coaxial gaseous flow.

In this present study, a hybrid atomizer is used. The hybrid is a combination of pressure-swirl and air-blast atomizers. In this atomizer, annular liquid sheet is sandwiched between two air streams. Tangential inlet ports are used in liquid inlet and outer air and inner air inlets. The tangential inlet is used for producing swirling effect in the spray field. Ma (2002) did an experimental work for studying the effect of tangential inlet on the spray regime. Ibrahim and Jog (2007) studied the nonlinear

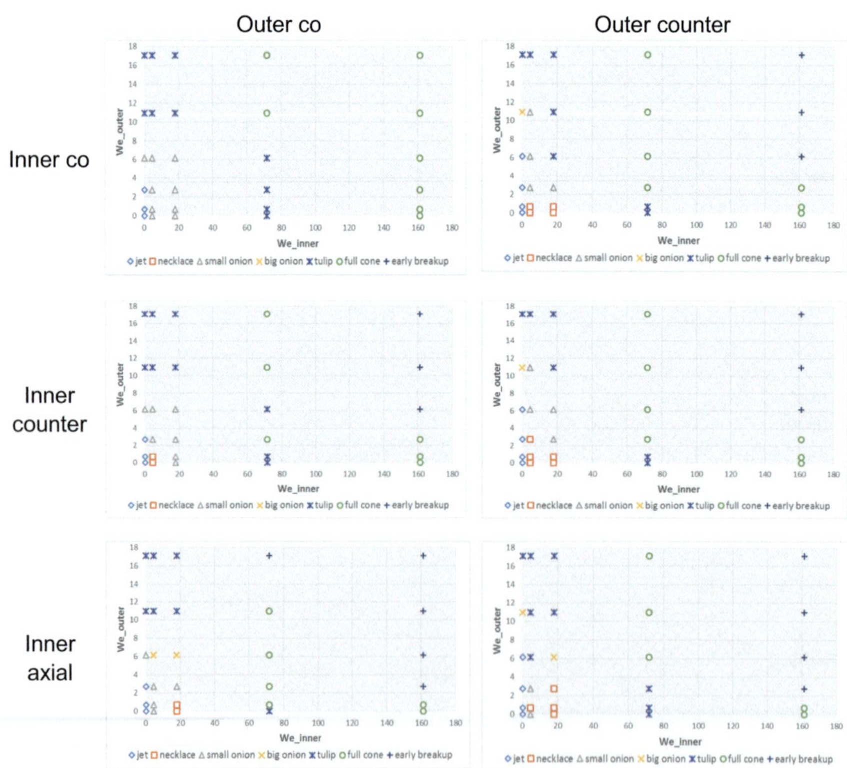

Fig. 9.8 Breakup regimes for developing spray on 2 LPM ethanol flow

breakup model for a pressure-swirl atomizer. They also used tangential inlet for producing the swirling effect in the spray regime.

Figure 9.1 Cross–sectional view of the atomizer (all dimension in mm). The dimension of the present nozzle is very much similar with the nozzle used by Chatterjee et al. (2013, 2014, 2015) and Carvalho and Heitor (1998). In the present study, six different configurations are studied by combining the above inlet arrangements (Fig. 9.2). The swirl direction of fuel is kept constant while that of inner and outer air streams are changed. The swirl directions of air streams are either the same as those of the fuel stream (defined as 'co') or opposite to the direction of the fuel sheet (defined as 'counter'). The inner air is also fed in axial direction for one configuration (defined as 'inner axial'). The flow rate of inner air has been varied between 0 and 30 LPM in steps of 10, and the flow rate of fuel stream is 1 and 2 LPM. The variation of spray characteristics with outer air stream flow rate has been studied by varying the outer air flow rate from 0 to 50 LPM in steps of 10. The air flow rate is less than actual gas turbine engine air flow rate. This study has been done for knowing the effect of the spray at low air flow rate. This study was made to understand the atomization effect on the liquid sheet. The flow rates are not sufficient for gas turbine combustion. Figure 9.2 shows the nozzle inlet configuration.

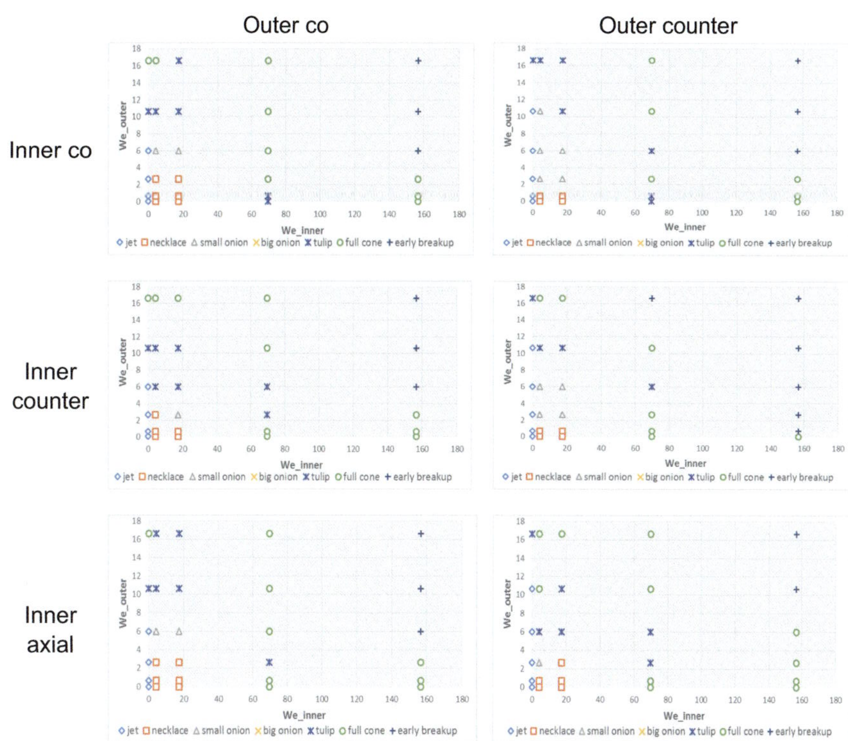

Fig. 9.9 Breakup regimes for developing spray on 1 LPM 10% ethanol-blended kerosene flow

9.2.2 *Blend*

In the present experiment, the spray atomizer was used to ascertain the characteristics of kerosene, pure ethanol, and two types of ethanol–kerosene blends. For the preparation of the blend, 99.9% concentrated ethanol (manufactured by MARK Germany) was used. During this experimental process, ethanol has been mixed with kerosene at 10 and 20% by volume ratio. The blend has been made at 30 °C by using 2% cosolvent tetrahydrofuran (C_4H_8O) for the 20% blend. When the blend is produced, it is observed that initially, the ethanol is properly mixed up with the kerosene. However, a week later, the blend separates into two layers for the 20% ethanol-blended kerosene fuel. At low temperature, water separation occurs from the blended fuel. The water content of the blended fuel also plays an effective role in phase separation of the blended fuel (Lapuerta et al. 2007; Li et al. 2005).

In order to ascertain the degree of deviation of the blended fuel from the conventional fuel (kerosene in this case), certain properties need to be determined by experimentation such as viscosity, surface tension, and blend stability from the spray and atomization point of view. Ajav and Akingbehin (2002) studied the fuel

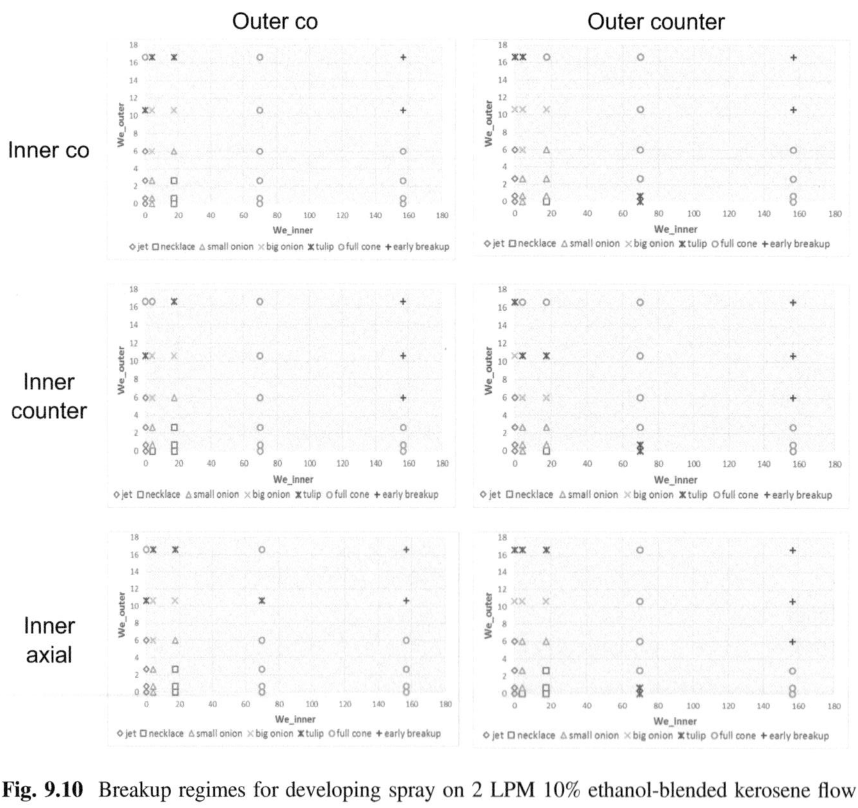

Fig. 9.10 Breakup regimes for developing spray on 2 LPM 10% ethanol-blended kerosene flow

properties like relative density, cloud and pour point, viscosity, and calorific value for 5% to 30% ethanol-blended diesel fuel. They reported that the relative density and viscosity of the blends decreased upon increasing the ethanol percentage in blended fuel. He et al. (2003) also studied the carbon content of the blended fuel and C/H ratio of the ethanol-blended diesel fuel. Brake specific fuel consummation, brake thermal efficiency, and the emission characteristics of the ethanol-blended diesel fuel were studied in a water-cooled single-cylinder direct injection (DI) diesel engine by Li et al. (2005).

In the present study, viscosity, surface tension, and calorific value of the blended fuel and the pure fuel are measured. The Ostwald viscometer (also known as U-tube viscometer or capillary viscometer) (Vesztergom 2014), the du Noüy ring tensiometer, and the bomb calorimeter (Parr, model-6100) are used for measuring the viscosity, surface tension, and calorific values, respectively, for all types of fuel at room temperature. Table 9.1 shows the measured values of the aforesaid fuel properties for pure and blended fuels. It can be observed that all three properties for kerosene are slightly higher than those of the blended fuels and ethanol. It can be seen that calorific value of ethanol is very low compared to that of the other fuels

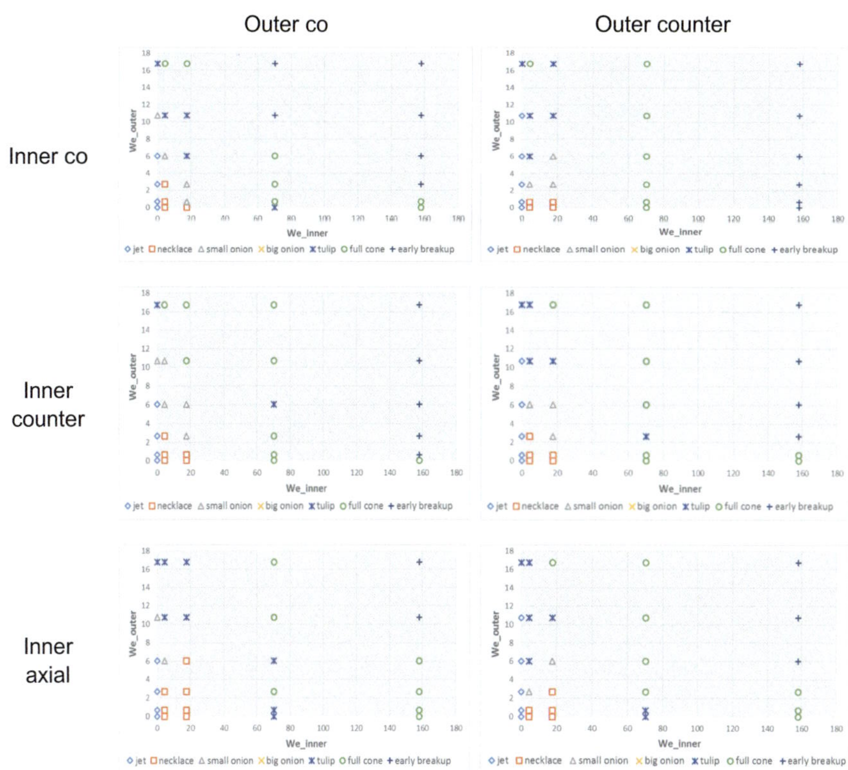

Fig. 9.11 Breakup regimes for developing spray on 1 LPM 20% ethanol-blended kerosene flow

which make it an unsuitable candidate. Sayin (2010) reported that for the lower viscosity, alcohol can be atomized easily in air, producing less emission for its high stoichiometric air–fuel ratio, high oxygen content, and low sulfur content. Drop size decreases with increasing air/liquid mass ratio and air velocity and increases with the increasing liquid viscosity and surface tension (Carvalho and Heitor 1998).

9.2.3 Experimental Setup and Image Capturing Technique

A variety of image capturing techniques can be used for capturing the spray image. For example, Wahono et al. (2008) used two types of image capturing techniques. One was volume lighting mode, and another was backlighting mode. For volume lighting technique, four sets of halogen lights (total power of 4 kW) were used to provide volume illumination of the spray. And for backlighting technique, a halogen light (2 kW) was placed directly behind the spray. Duke et al. (2010) used a photographic measurement technique by using a backlit arrangement illuminating the focal plane of the camera optics in which the liquid sheet is placed.

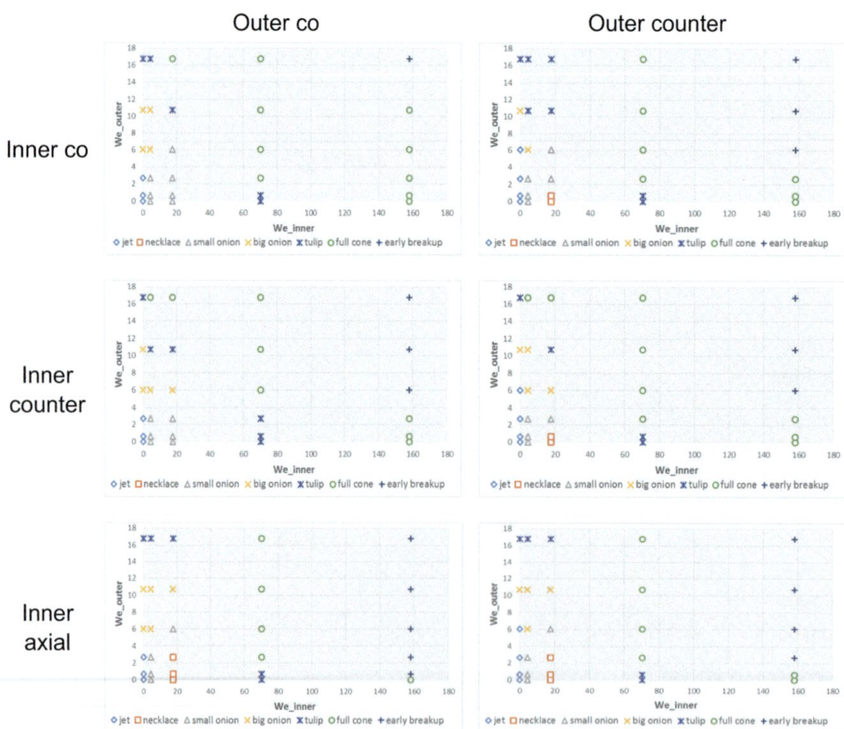

Fig. 9.12 Breakup regimes for developing spray on 2 LPM 20% ethanol-blended kerosene flow

For this present study, the backlit imaging technique is used with two halogen lights (1 KW each) for illumining the spray field. The position and angle of the light are optimized for best image contrast. A high-speed camera (Prosilica high-speed camera, 230 fps, model—CV1280, digital machine vision camera, 1280 × 1024 monochrome, 1394 DCAM) is used for capturing the spray images (Fig. 9.3).

The fuel is fed from a gear pump (manufactured by Rotodyne pump and system Ahmedabad (India), 1440 rpm, MAX. DISP. Vol. = 20 LPM and driven by single phase 50 Hz 370 W 4.4 A ½ HP motor), and both air streams are fed from a compressor (manufactured by ELGI Equipment Ltd., model—SA OF 01 080 OF, DISP. Vol. = 92 LP and driven by a single phase, 1 HP, 1450 rpm AC motor.). Rotameters (0–4 LPM kerosene rotameter is used after calibrating with used fuel for measuring the different fuels' flow rates and air rotameter 0–50 LPM used for measuring the air flow rate) are used in all fluid inlet lines for flow rate measurement.

9.3 Results and Discussion

The present study is done with the aid of a hybrid atomizer. This atomizer is a combination of air-blast atomizer and a pressure-swirl atomizer. In a simplex-type pressure-swirl atomizer, an air-cored vortex is formed within the swirling liquid. This air core plays a major role in producing a hollow cone spray (Lefebvre 1989).

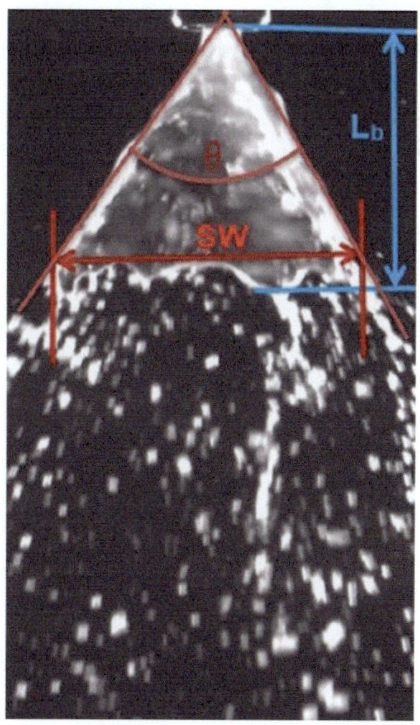

Fig. 9.13 Macroscopic spray characteristics (SW = sheet width, L_b = breakup length, θ = cone angle)

In the present study, the annular liquid sheet is sandwiched between two swirling and inner axial air streams. The annular liquid sheet also has a swirling velocity. The relative magnitude of the tangential and axial components of the swirling velocity at the nozzle outlet plays a significant role in the development of the spray cone angle. For the present experiment, liquid flow is maintained at 1 and 2 LPM flow rate. Outer air flow rate is maintained on 0, 5, 10, 20, 30 LPM, and inner air flow rate is maintained on 0–50 LPM with 10 LPM interval.

For understanding the effect of surface tension on the air–liquid interface at different breakup stages, we calculate the Weber number for inner air (**We_inner**) and outer air (**We_outer**) as follows:

$$\textbf{We\_inner} = \frac{\rho v^2 l_i}{\sigma} \tag{9.1}$$

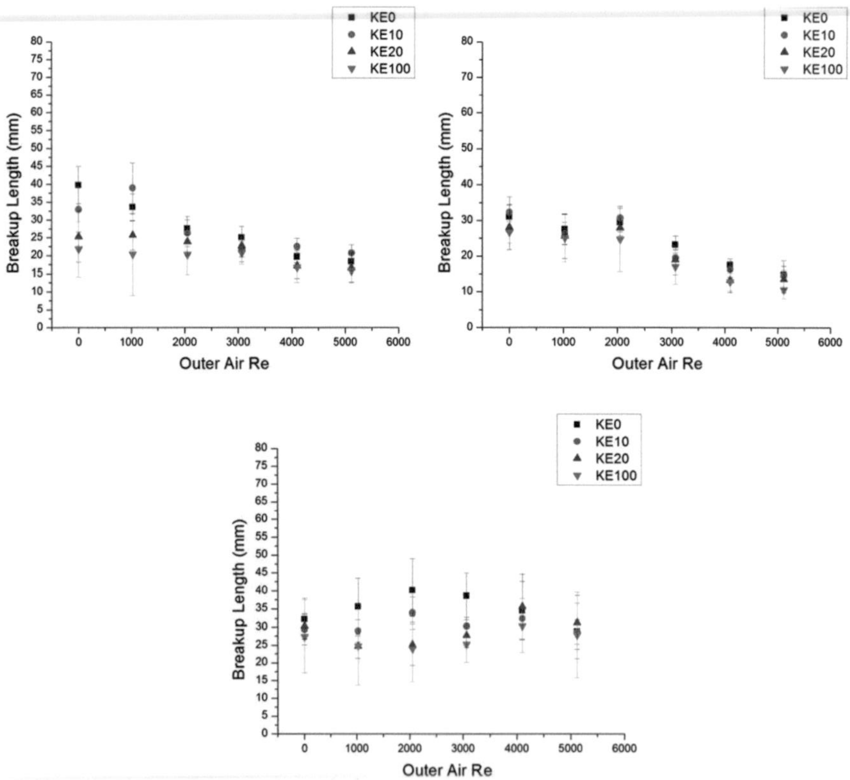

Fig. 9.14 Breakup length variation for different blended fuels with increasing inner air for outer co-swirl configuration and inner co-swirl (upper left), inner counter-swirl (upper right), and inner axial (below) configurations at 1 LPM fuel flow rate

$$\textbf{We\_outer} = \frac{\rho v^2 l_o}{\sigma} \tag{9.2}$$

In order to measure the macroscopic spray characteristics like breakup length, cone angle, and sheet width, we calculate the Reynolds number (**Re_outer**) for outer air flow as follows:

$$\textbf{Re\_outer} = \frac{\rho v l_o}{\mu} \tag{9.3}$$

Here, ρ is the density of air, v is the velocity of the air, σ is the surface tension of the fuel–air interface, μ is the viscosity of the air, l_i is the characteristic length of the inner part of the nozzle (taken as inner diameter), and l_o is the characteristic length of the outer part of the nozzle (taken as outer diameter of the nozzle).

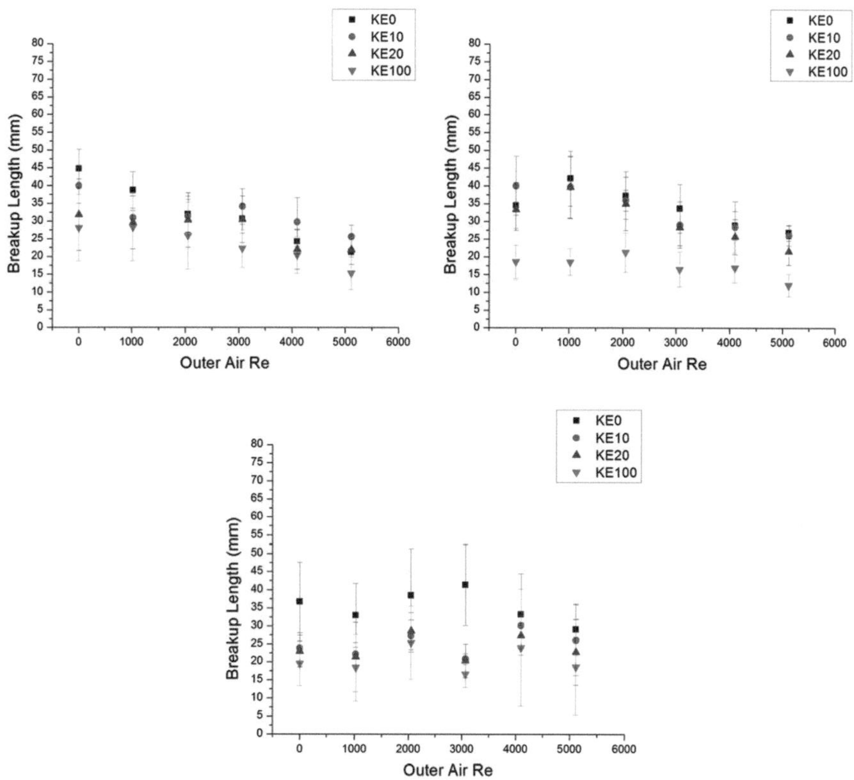

Fig. 9.15 Breakup length variation for different blended fuels with increasing inner air for outer counter-swirl configuration and inner co-swirl (upper left), inner counter-swirl (upper right), and inner axial (below) configurations at 1 LPM fuel flow rate

9.3.1 Breakup Phases

Different breakup regimes are found during the spray development. The relative magnitudes of various velocity components play a vital role in producing the different breakup regimes. These breakup regimes are mainly jet, necklace, onion, tulip, full cone, and early breakup regime. Most preferable breakup regimes are fuel cone and early breakup regimes. Lefebvre reported that these spray stages are formed due to the injection pressure of the liquid (Lefebvre 1989). Figure 9.4 shows the different breakup stages during spray formation.

The liquid sheet which emanates without any major perturbations on the liquid outer surface is called the jet-type liquid sheet as shown in Fig. 9.4a. The nozzle outer air region is stationary, and liquid flow rate is low without any inner air flow. This type of liquid sheet travels a long distance, and the breakup length is very high. The liquid inlet pressure is low, and the nozzle inlet–outlet pressure difference is

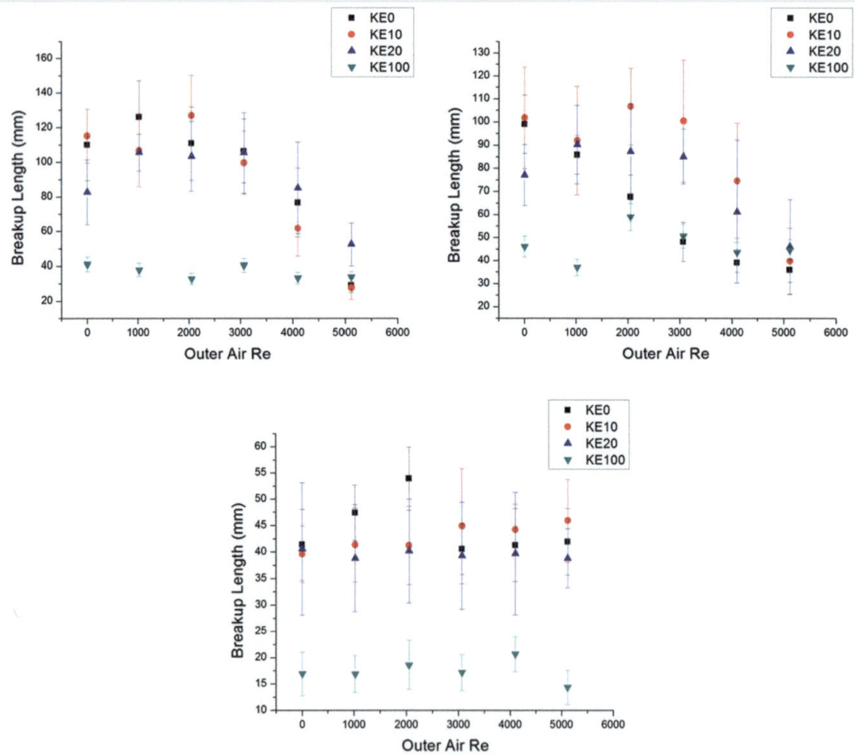

Fig. 9.16 Breakup length variation for different blended fuels with increasing inner air for outer co-swirl configuration and inner co-swirl (upper left), inner counter-swirl (upper right), and inner axial (below) configurations at 2 LPM fuel flow rate

low. The liquid sheet instability growth rate is also very low. The instability is primarily caused by the outer air.

The necklace-type spray shape as shown in Fig. 9.4b is normally formed in situations of low pressure difference between nozzle inlet and nozzle outlet. At very low inner air flow rate, the inner air gets entrapped between the annular liquid sheets resulting in the formation of this shape. Also at very low outer air flow, the outer air creates a low-pressure regime outside the annular liquid sheets creating instability in the outer surface of the annular liquid sheet. For the counter-swirling inner flow, the opposite directions of swirls on the two sides of the liquid sheet lead to modulation of the sheet along the axial direction. This gives rise to a necklace-like structure, particularly at low inner air flow rates and moderate outer air flow rates. Similar structure has also been observed by Zhao et al. (2015). However, in their configuration, there was no outer air flow. Necklace formation is favorable for better spray characteristics.

The onion-type shape as shown in Fig. 9.4c is also formed due to low pressure difference between nozzle inlet and nozzle outlet but the pressure difference in this

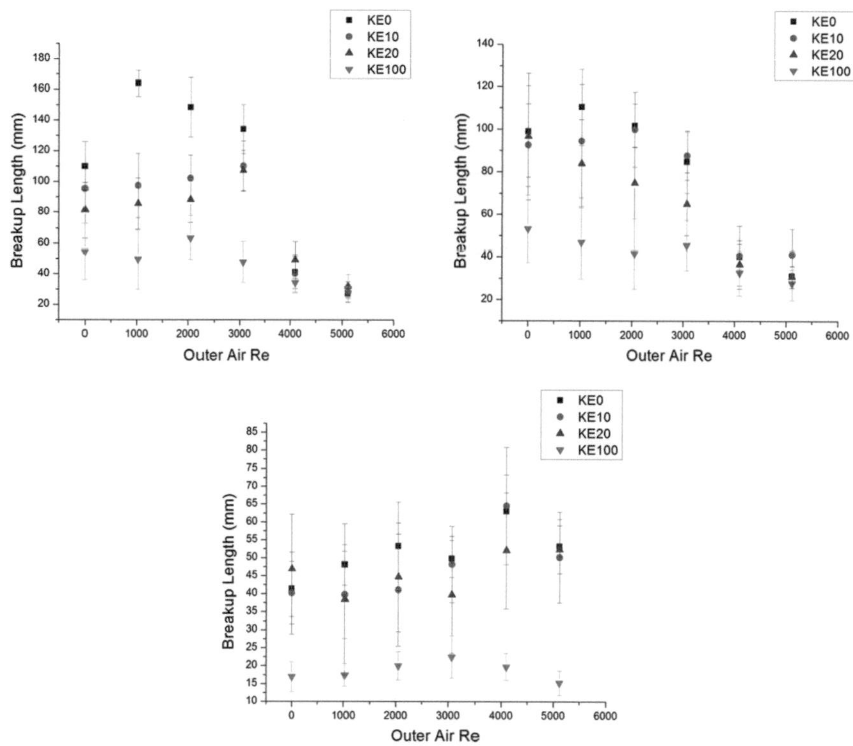

Fig. 9.17 Breakup length variation for different blended fuels with increasing inner air for outer counter-swirl configuration and inner co-swirl (upper left), inner counter-swirl (upper right), and inner axial (below) configurations at 2 LPM fuel flow rate

case is higher than that in case of the necklace-like shape. In the present study, the inner and outer air flow rates are higher than the necklace-type inner and outer air flow rates. The high outer flow creates a low-pressure regime outside the spray regime, which causes the annular liquid sheet to form a bulge of the jet. The swirl direction plays an important role in the formation of big and small bulges. The outer air flow rate also plays a crucial role here. From the spray snapshots, we observe that the inner co-swirl and counter-swirl configurations have higher breakup lengths at lower flow rates due to the formation of a stable onion structure. A slight bulge of the jet is observed at low inner and outer air flow rates. A bigger bulge is created as the outer flow rate increases, and the higher strength of the outer swirling flow induces greater tangential momentum to the annular liquid sheet. This can be observed at the onset of the onion structure formation. Onion shape is usually observed in pressure-swirl atomizers, where the swirling liquid sheet surrounds an air core whose velocity is not so high.

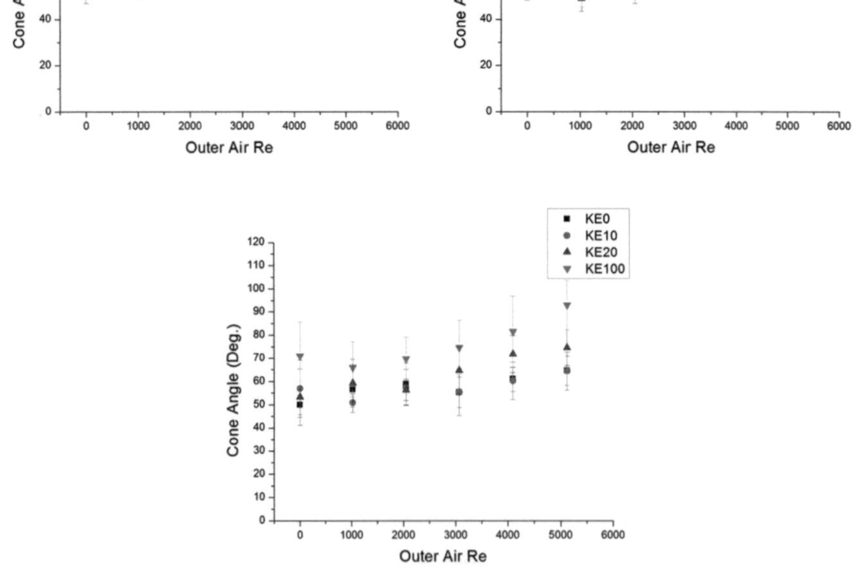

Fig. 9.18 Cone angle variation for different blended fuels with increasing inner air for outer co-swirl configuration and inner co-swirl (upper left), inner counter-swirl (upper right), and inner axial (below) configurations at 1 LPM fuel flow rate

The tulip shape is formed when the inner and outer flow rates are higher those that in case of onion shape as shown in Fig. 9.4d. The collapse of this bulge can be observed at some distance (along the axis) downstream. This occurs when no further expansion of the liquid sheet is allowed by the outer air. This distance is mostly found to decrease with increase in outer air flow rates. It can be observed from Fig. 9.4d that the outer periphery of the tulip-shaped spray is curved in nature unlike the conical shape observed for full cone spray as shown in Fig. 9.4e.

Figure 9.4e shows the spray regime known as the full cone-type spray. As shown in the figure, this regime has a conical shape at the outer periphery. An increase in the inner and outer air flow rates from that of the tulip spray regime results in a shorter breakup length. The breakup happens at a point closer to the nozzle outlet, and very short liquid sheet is present at the outlet of the nozzle leading to the formation of the full cone spray.

The early breakup regime as shown in Fig. 9.4f is formed at even high inner and outer air flow rates than those required for the formation of the full cone spray regime. The high swirling causes a high tangential velocity and creates a

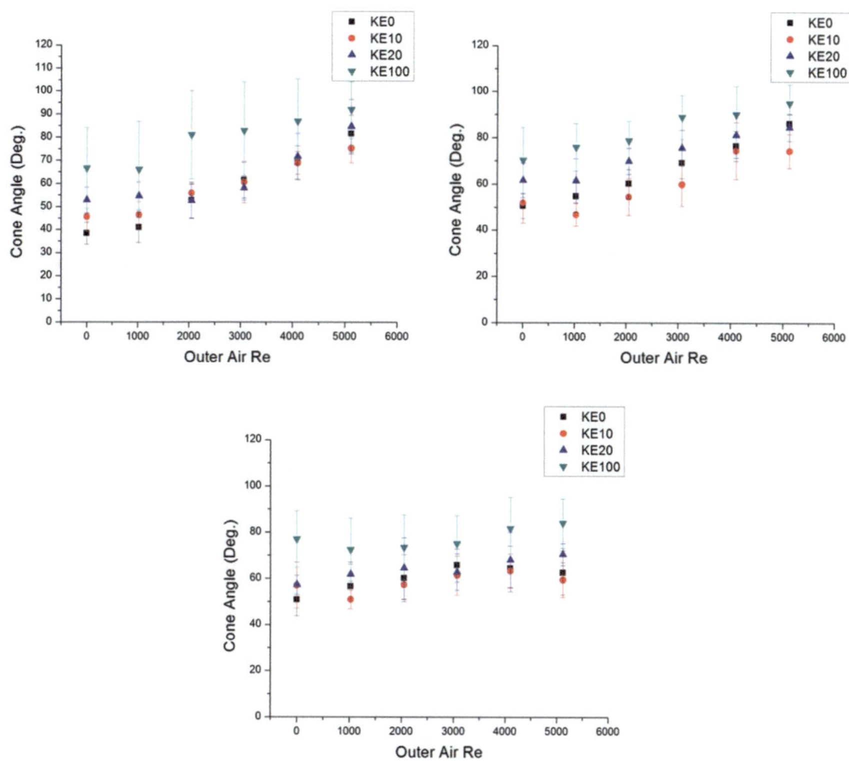

Fig. 9.19 Cone angle variation for different blended fuels with increasing inner air for outer counter-swirl configuration and inner co-swirl (upper left), inner counter-swirl (upper right), and inner axial (below) configurations at 1 LPM fuel flow rate

low-pressure regime outside the nozzle. The liquid sheet breaks at a point closer to the nozzle outlet creating a mist-type spray.

The early breakup regime and full cone-type spray are almost similar to each other. They only differ by the fact that the full cone-type spray shape has a small liquid sheet at the nozzle exit, whereas for the early breakup stage, this sheet breaks up very close to the nozzle exit.

Figure 9.5 shows the different breakup regime plots for 1 LPM kerosene flow for different inner and outer air flow rates and different flow configurations. Increase in the Weber number denotes the increase in air flow rate. At inner axial configuration, the early breakup stage shows up at high outer and inner air flow. However, for the inner co-swirl and counter-swirl configurations, early breakup stage occurs at high inner air flow and low outer air flow rate. Inner air flow and outer air flow play an important role in the formation of different spray breakup stages. Big onion-type shape is not present at inner axial flow condition. Onion-type, tulip-type, and full cone-type stages start to occur with increase in outer air flow at zero inner

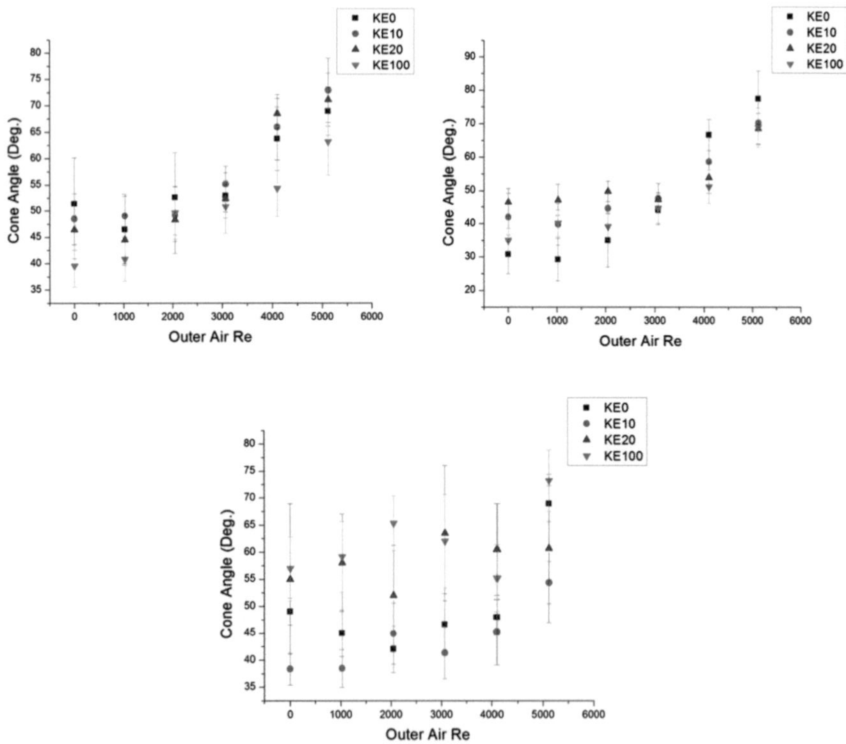

Fig. 9.20 Cone angle variation for different blended fuels with increasing inner air for outer co-swirl configuration and inner co-swirl (upper left), inner counter-swirl (upper right), and inner axial (below) configurations at 2 LPM fuel flow rate

air Weber number. Necklace-type shape is formed at lower inner and outer air Weber number.

Figure 9.6 shows the different breakup regime plots for 2 LPM kerosene flow for different inner and outer air flow rates and different flow configurations. It can be observed that the early breakup regime is not present in the inner and outer air co-swirl configurations; however, for other inner and outer air swirl configurations, early breakup regime is present. Full cone-type spray shape happens quickly in inner axial configuration than that in inner air swirl configuration at low outer air Weber number and 70 inner air Weber number. Big onion-type spray regime is present in more air flow configurations on inner counter-swirl and outer co-swirl configurations.

Figure 9.7 shows the different breakup regime plots for 1 LPM ethanol flow for different inner and outer air flow rates and different air flow configurations. Big onion-type breakup regime is not found for the ethanol 1 LPM flow. Early breakup regime arises quickly in inner air swirling configuration. At high inner and outer air flow conditions, early breakup regimes are present in all types of air flow configurations. Onion-type spray regime and full cone spray regime are not present in outer counter-swirl configuration at zero inner air Weber number.

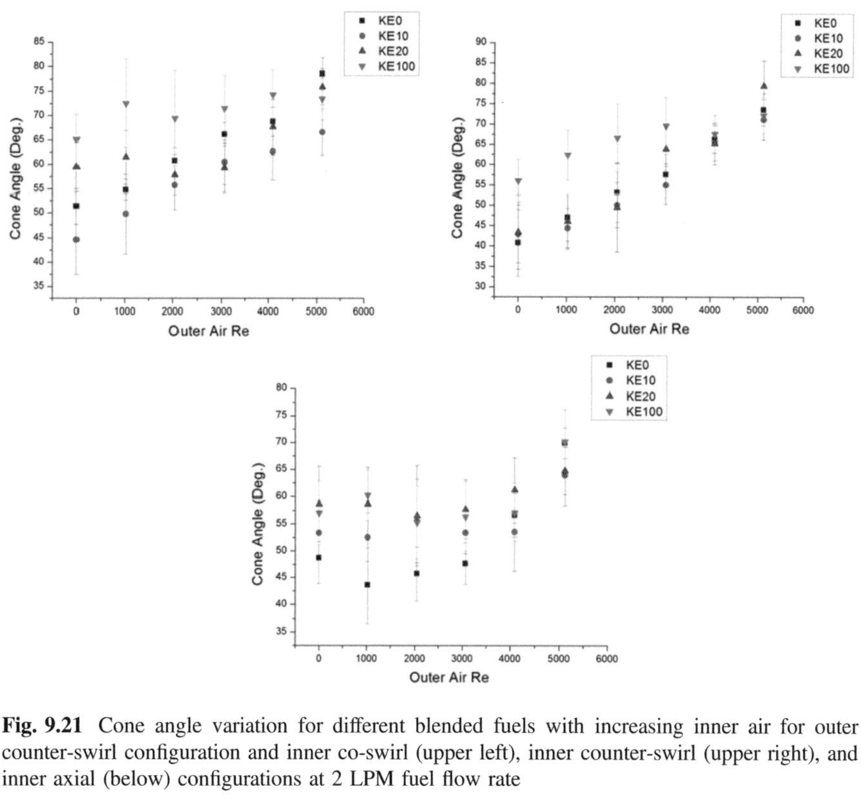

Fig. 9.21 Cone angle variation for different blended fuels with increasing inner air for outer counter-swirl configuration and inner co-swirl (upper left), inner counter-swirl (upper right), and inner axial (below) configurations at 2 LPM fuel flow rate

Figure 9.8 shows the different breakup regime plots for 2 LPM ethanol flow for different inner and outer air flow rates and different air flow configurations. Big onion-type regime is present in inner axial configuration and zero Weber number for outer counter-swirl configuration. The difference between tulip and full cone regime is almost negligible. Similarly, a trivial difference is present between the full cone and early breakup regimes. Early breakup regime is present at high inner and outer air flow conditions for all air flow configurations without inner co-swirl and outer co-swirl configurations. Necklace-type breakup regime is not present in inner co-swirl and outer co-swirl configurations.

Figure 9.9 shows the different breakup regime plots for 1 LPM 10% ethanol-blended kerosene flow for different inner and outer air flow rates and different air flow configurations. Big onion-type breakup regime formation is not found in 10% ethanol-blended kerosene fuel for 1 LPM fuel flow rate. At zero inner air Weber number with increasing outer air Weber number, the onion-type breakup phase is not present in any air flow configuration. The development of the spray phase is very rapid from the jet-type breakup regime to tulip-type breakup regime at

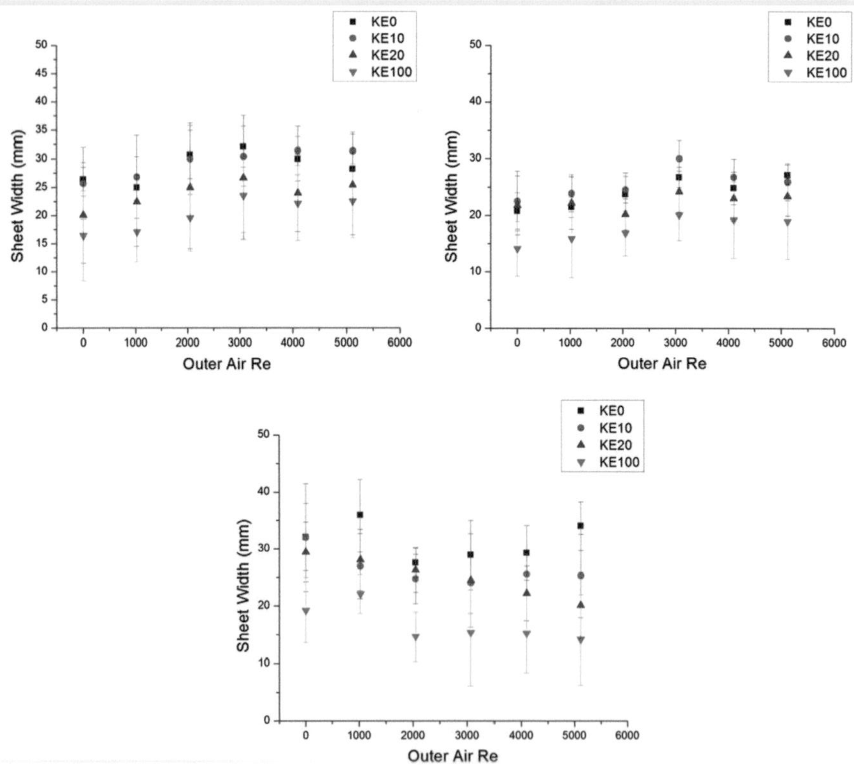

Fig. 9.22 Sheet width variation for different blended fuels with increasing inner air for outer co-swirl configuration and inner co-swirl (upper left), inner counter-swirl (upper right), and inner axial (below) configurations at 1 LPM fuel flow rate

zero inner air Weber number with increasing outer air Weber number. Early breakup regime is found at high inner air flow and outer air flow.

Figure 9.10 shows the different breakup regime plots for 2 LPM 10% ethanol-blended kerosene flow for different inner and outer air flow rates and different air flow configurations. In this case, the nature of the breakup regime is almost similar to that of other fuel regime plots.

Figures 9.11 and 9.12 show the different breakup regime plots for 1 LPM and 2 LPM 20% ethanol-blended kerosene flow, respectively, for different inner and outer air flow rates and different air flow configurations. The nature of the breakup regimes is similar to that of regimes observed earlier. The effect of variation in fuel properties does not seem to have much effect on the formation of the spray breakup regimes. The reason for this unusual behavior may be attributed to the fact that the properties like viscosity and surface tension differ very slightly for kerosene, ethanol, and ethanol-blended fuels.

The spray regimes depend on the swirling configuration and the fuel property only up to a certain degree. However, they are mainly dependent on the flow

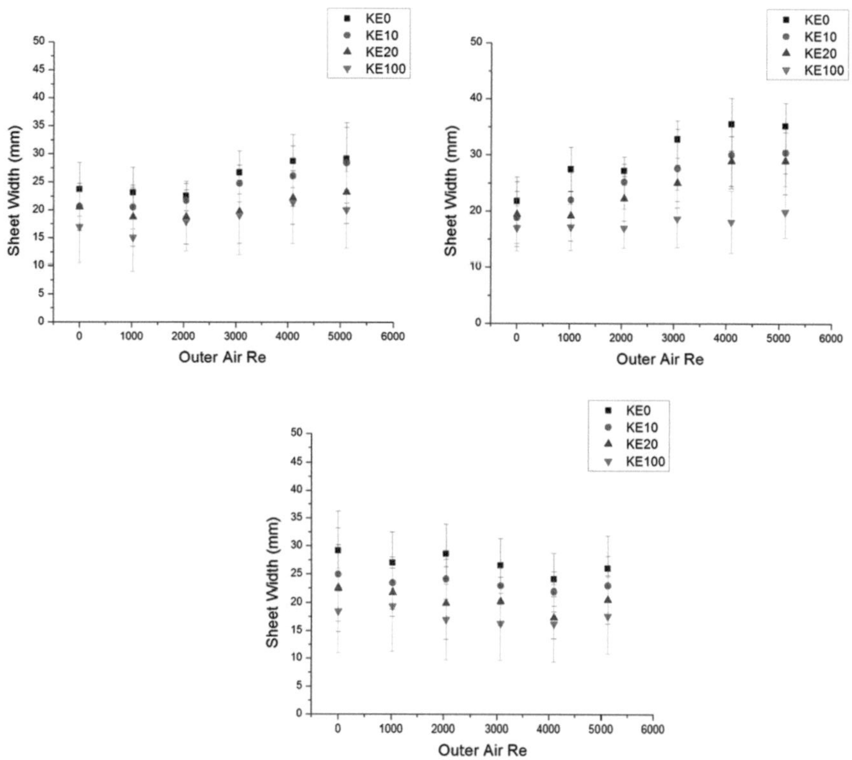

Fig. 9.23 Sheet width variation for different blended fuels with increasing inner air for outer counter-swirl configuration and inner co-swirl (upper left), inner counter-swirl (upper right), and inner axial (below) configurations at 1 LPM fuel flow rate

velocity. At low inner and outer air flow rates, initially, the jet flow regime appears. Upon increasing both the air flow rates, the necklace-type and onion-type regimes follow. Upon further increasing the air flow rates, tulip-type, full cone-type, and early breakup-type spray regimes are formed.

9.3.2 *Macroscopic Spray Characteristics*

Macroscopic spray characteristics of the spray are mainly breakup length, sheet width, and cone angle. Breakup of liquid sheets occurs when a discontinuity is observed. A continuous sheet gives rise to droplets and ligaments at the point of breakup. The breakup length is defined as the axial distance between the outlet of the nozzle and the sheet breakup point. Cone angle is measured by calculating the angle between the two lines made by the spray cone in the images obtained from high-speed camera. Sheet width is the maximum width of the liquid fuel sheet before breakup occurs. For this macroscopic measurement, inner air 20 LPM flow

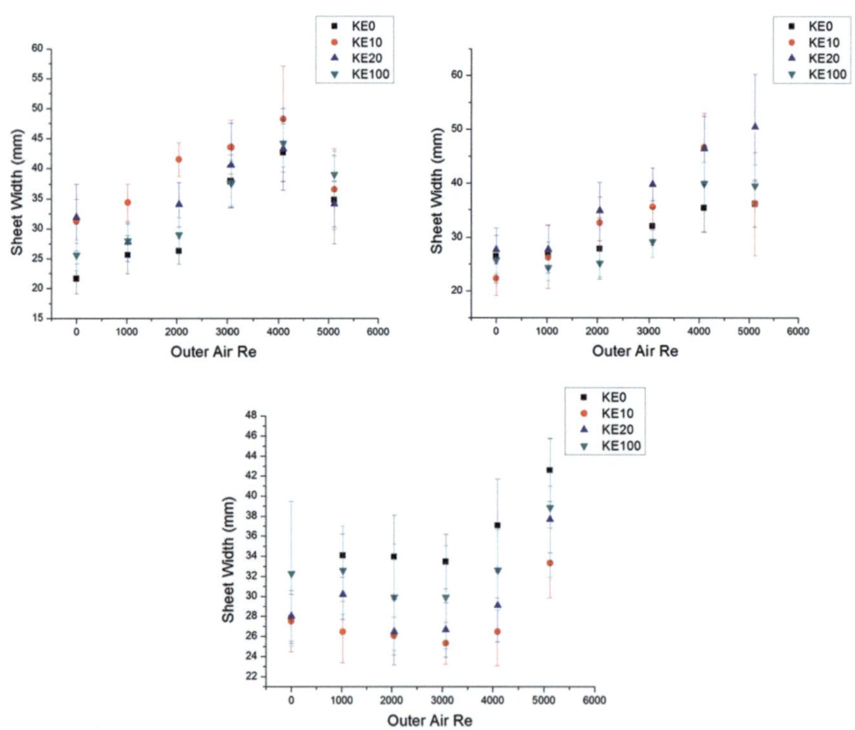

Fig. 9.24 Sheet width variation for different blended fuels with increasing inner air for outer co-swirl configuration and inner co-swirl (upper left), inner counter-swirl (upper right), and inner axial (below) configurations at 2 LPM fuel flow rate

rate is considered. At 30 LPM inner air flow, maximum spray belongs to early break-type spray phase. Primary breakup length measurement in early breakup stage is not so easy. Lower than 20 LPM inner air flow, the spray structure is not developed. For comparing the macroscopic spray characteristics for different fuels, the tests are done on 20 LPM inner air flow condition with 0–50 LPM outer air flow condition. Figure 9.13 shows the macroscopic spray characteristics measurement technique.

Figure 9.14 shows the breakup length variation for different blended fuels with increasing outer air flow at 1 LPM liquid flow for outer co-swirl configuration with inner axial, co-swirl, and counter-swirl configurations. At low air flow, breakup length variation is more in inner co-swirl and outer co-swirling configurations for all used fuels. But for inner axial configuration, the breakup length variation is more at intermediate outer air flow. At high air flow rate, the breakup length is almost similar for kerosene, kerosene-blended fuel, and ethanol. In outer co-swirl and inner counter-swirl configurations, the breakup length variation is very small for different used fuels.

Figure 9.15 shows the breakup length variation for different blended fuels with increasing outer air flow at 1 LPM liquid flow for outer counter-swirl configuration

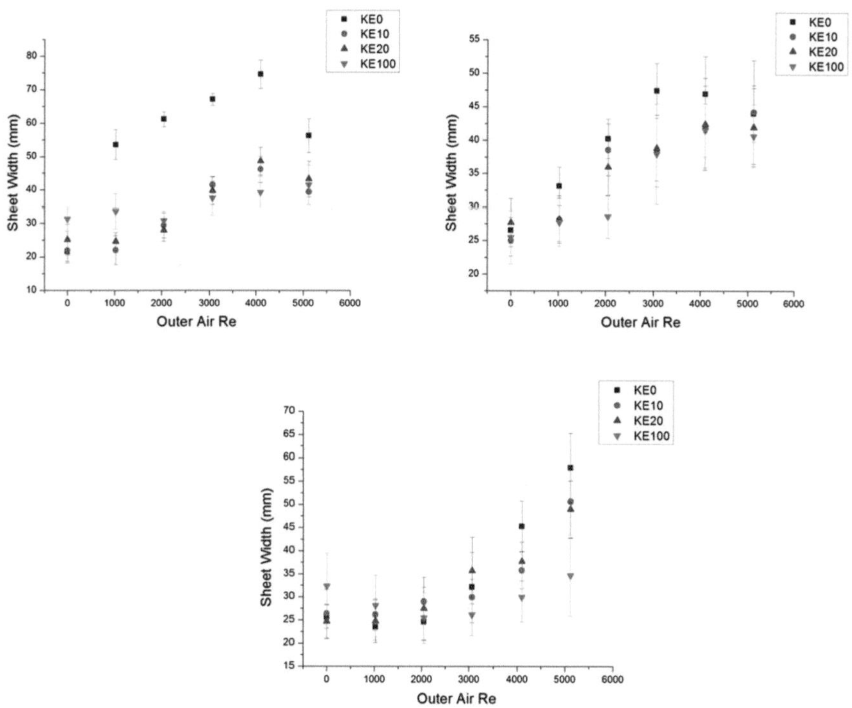

Fig. 9.25 Sheet width variation for different blended fuels with increasing inner air for outer counter-swirl configuration and inner co-swirl (upper left), inner counter-swirl (upper right), and inner axial (below) configurations at 2 LPM fuel flow rate

with inner axial, co-swirl, and counter-swirl configurations. Inner counter outer counter swirling configuration ethanol breakup length is minimum when compared to other used fuels, but in inner axial configuration, kerosene breakup length is maximum among all cases. Ethanol breakup is minimum in every configuration due to lower viscosity and surface tension. Kerosene, ethanol and ethanol blended kerosene fuel surface tension and viscosity difference is quite low, due to this reason the breakup length variation is small. Ethanol (KE100) breakup is small for almost every configuration than other used fuels.

Figure 9.16 shows the breakup length variation for different blended fuels with increasing outer air flow at 2 LPM liquid flow for outer co-swirl configuration with inner axial, co-swirl, and counter-swirl configurations. At 2 LPM liquid flow ethanol, breakup length is small, compared to that of other fuel breakup lengths. The breakup length pattern is almost similar to diesel and ethanol-blended diesel 2 LPM pattern. Figure 9.17 shows the breakup variation for outer counter-swirl configuration for 2 LPM fuel flow rate. With increasing blend percentages, the breakup length decreases in almost every condition. Ethanol (KE100) breakup length is minimum than other blended fuels. Kerosene and ethanol-blended kerosene breakup lengths are almost same but only in inner co and outer counter-swirl configuration kerosene breakup length is more than ethanol-blended kerosene at low outer air flow rate but at high outer flow condition breakup length variation is less for kerosene and ethanol blended kerosene fuel.

Figure 9.18 shows the outer air co-swirling effect at 1LPM liquid flow on kerosene, ethanol, and ethanol-blended kerosene. Here, the ethanol cone angle is maximum compared to kerosene and ethanol-blended kerosene fuel. The cone angle increases with increasing outer air flow rate. Cone angle increasing nature with outer air flow represents that the spray divergence increases with increasing outer air flow rate.

The cone angle trend is the same for outer air counter-swirl configuration (Fig. 9.19). The viscosity effect is prominent. For this reason, ethanol cone angle is more than corresponding cases of kerosene and ethanol-blended kerosene. But, Kerosene and ethanol blended kerosene have very close values of viscosity and surface tension. Hence, the cone angle variation is less. The variation of cone angle on inner air axial configuration for both outer air configurations is less for used fuel with increasing outer air flow than inner air co-swirl and counter-swirl configurations due to the inner air axial velocity.

Figures 9.20 and 9.21 show the outer air co-swirl and counter-swirl effect and inner air co-swirl and counter-swirl and axial direction effect on cone angle at 2 LPM air flow rate. Cone angle increases with increasing outer air flow. Viscosity difference in kerosene, ethanol, and ethanol-blended kerosene is very low. This may be the reason for differences in cone angle pattern with other cases. But the ethanol cone angle is more than kerosene and ethanol-blended kerosene fuel on outer counter-swirl configuration. The viscosity effect is not so prominent in this cone angle with outer air plot.

Figures 9.22 and 9.23 show the effect of outer air co-swirl and counter-swirl for three inner air flow configurations on sheet width at 1 LPM fuel flow. Here, the kerosene sheet width is the highest and ethanol sheet width is the lowest for every flow condition. The blends lie between the kerosene and ethanol sheet widths. Here, with increasing blend percentage, the sheet width is decreasing. Kerosene and 10% ethanol- blended kerosene sheet width is almost same for outer air co and inner air co and counter swirl configurations as concluded from the experimental data.

Figures 9.24 and 9.25 show the effect of outer air co-swirl and counter-swirl for three inner air flow configurations on sheet width at 2 LPM fuel flow. For this condition, the plot does not follow any discernible pattern. Ethanol and kerosene and ethanol-blended kerosene do not have a high difference in viscosity and surface tension. May be for this reason the nature of the plot is very random. In inner co-swirl for both outer co-swirl and counter-swirl configuration, the nature of the plot is almost the same for all used fuels; sheet width increases with increasing outer air flow at low outer air flow rates, and then suddenly, sheet width decreases with increasing outer air flow.

9.4 Conclusions

The motivation behind the present experimental study is to ascertain the spray characteristics of kerosene, ethanol, and ethanol-blended kerosene. Different breakup phases are studied during this study. The breakup phases are very much dependent on the inlet flow velocity. The phases are less dependent on the viscosity and surface tension of the fuels used. With increasing inner and outer air flow rates, the breakup regime develops accordingly, from jet-type breakup regime to fully developed spray regime.

The present experiments have also been done for understanding the macroscopic behavior of spray for kerosene, ethanol, and ethanol-blended kerosene. A comparative study is done for different blended fuels with the conventional fuel. For this experimental investigation, we used a hybrid atomizer for determining the inner and outer air effect on the swirling annular liquid sheet. From the results of the current study, we broadly conclude that with increasing percentage of ethanol in blended fuels, the breakup length decreases and the cone angle increases. Ethanol-blended diesel and kerosene fuel spray characteristics are almost similar to conventional fuel. But on increasing blend percentage, the spray characteristics were found to be better than those of conventional fuel. Inner air axial configuration for both outer air swirl conditions, the spray characteristics is different compared to the inner air swirl configuration.

References

Ajav EA, Akingbehin OA (2002) A study of some fuel properties of local ethanol blended with diesel fuel. Agric Eng Int CIGR J Sci Res Dev IV:1–9

Chatterjee S et al (2013) Effect of Co and counter-swirling inner air on the spray dynamics of an annular fuel sheet in a hybrid atomizer. In: Ninth Asia-Pacific conference on combustion, The Combustion Institute (Korean section), Paper 2013-129, Pittsburgh, PA

Carvalho IS, Heitor MV (1998a) Liquid film break-up in a model of a prefilming airblast nozzle. Exp Fluids 1998(24):408–415

Carvalho IS, Heitor MV (1998b) Liquid film break-up in a model of a prefilming airblast nozzle. Exp Fluids 24(5):408–415

Chatterjee S et al (2014) Effect of geometric variations on the spray dynamics of an annular fuel sheet in a hybrid atomizer. Atomization Sprays 24(8):673–694

Chatterjee S et al (2015) Experimental investigation of breakup of annular liquid sheet in a hybrid atomizer. J Propul Power 2015(31):1232–1241

Chen H, Shi-Jin S, Jian-Xin W (2007) Study on combustion characteristics and PM emission of diesel engines using ester–ethanol–diesel blended fuels. Proc Combust Inst 31(2):2981–2989

Chiariello F et al (2014) Gaseous and particulate emissions of a micro gas turbine fueled by straight vegetable oil–kerosene blends. Exp Thermal Fluid Sci 56(july):16–22

Chin JS, Rizk NK, Razdan MK (1999) Study on hybrid airblast atomization. J Propul Power 15 (2):241–247

Chin JS, Rizk NK, Razdan MK (1999b) Study on hybrid airblast atomization. J Propul Power 15 (2):241–247

Dioha IJ et al (2012) Comparative studies of ethanol and kerosene fuels and cook stoves performance. J Nat Sci Res 2(6). http://www.iiste.org

Duke D, Honnery D, Soria J (2010) A cross-correlation velocimetry technique for breakup of an annular liquid sheet. Exp Fluids 49(2):435–445

Garai A et al (2017) Experimental investigation of spray characteristics of kerosene and ethanol-blended kerosene using a gas turbine hybrid atomizer. Sadhana 42(7):543–555

Hansen AC, Zhang Q, Lyne PWL (2005) Ethanol–diesel fuel blends—a review. Biores Technol 96(3):277–285

He B-Q et al (2003) The effect of ethanol blended diesel fuels on emissions from a diesel engine. Atmos Environ 37(35):4965–4971

Ibrahim AA, Jog MA (2007) Nonlinear breakup model for a liquid sheet emanating from a pressure-swirl atomizer. J Eng Gas Turbines Power 129(4):945–953

Khalil AEE, Gupta AK (2013) Fuel flexible distributed combustion for efficient and clean gas turbine engines. Appl Energy 109:267–274

Khan MY, Khan FA, Beg MS (2013) Ethanol-kerosene blends: fuel option for kerosene wick stove. Int J Eng Res Appl (IJERA) 3(1):464–466

Lapuerta M, Armas O, Garcıa-Contreras R (2007) Stability of diesel–bioethanol blends for use in diesel engines. Fuel 2007(86):1351–1357

Leboucher N, Roger F, Carreau J-L (2010) Disintegration process of an annular liquid sheet assisted by coaxial gaseous coflow(s) 20(10):847–862

Lefebvre AH (1989) Atomization and sprays. Hemisphere Publishing Corporation

Li D-G et al (2005) Physico-chemical properties of ethanol–diesel blend fuel and its effect on performance and emissions of diesel engines. Renew Energy 30(6):967–976

Lin SP (2003) Breakup of liquid sheets and jets. Cambridge University Press

Ma Z (2002) Investigation on the internal flow characteristics of pressure swirl atomizers. University of Cincinnati

Mendez CJ, Parthasarathy RN, Gollahalli SR (2012) Performance and emission characteristics of a small scale gas turbine engine fueled with ethanol/jet A blends. In: 50th aerospace sciences meeting, AIAA, Nashville, Tennessee 2012. AIAAA, p 0522

Mendez CJ, Parthasarathy RN, Gollahalli SR (2014) Performance and emission characteristics of butanol/Jet A blends in a gas turbine engine. Appl Energy 118:135–140

Patra J et al (2015) Studies of combustion characteristics of kerosene ethanol blends in an axi-symmetric combustor. Fuel 144:205–213

Reeser LG, Acra APL, Lee T (1995) Covering solar energy into liquid fuels. Resource engineering & technology for a sustainable world, vol 2(1). ASAE, pp 8–11

Rizk NK, Chin JS, Razdan MK (1996) Influence of design configuration on hybrid atomizer performance. AIAA 1996. Paper 96-2628

Rizk NK, Chin JS, Razdan MK (1996) Influence of design configuration on hybrid atomizer performance. In: 32nd joint propulsion conference and exhibit. American Institute of Aeronautics and Astronautics

Sayin C (2010) Engine performance and exhaust gas emissions of methanol and ethanol–diesel blends. Fuel 2010(89):3410–3415

Vesztergom (2014) M.Z.U.T.V.F.p.w.w.b.S. Determination of viscosity with Ostwald viscometer. http://foundation01.chem.elte.hu/Chemistry_BSc_English_Group/BLOCK_08/Viscosity_2014BSc1_eng.pdf

Wahono S et al (2008) High-speed visualisation of primary break-up of an annular liquid sheet. Exp Fluids 2008(44):451–459

Xing-Cai L et al (2004) Effect of cetane number improver on heat release rate and emissions of high speed diesel engine fueled with ethanol–diesel blend fuel. Fuel 83(14–15):2013–2020

Yahya RK, Goering CE (1977) Some trends in fifty-five years of Nebraska tractor test data. ASAE paper no MC-77-503, 1977. ASAE, St. Joseph, Ml 49085

Zhao H et al (2015) Breakup morphology of annular liquid sheet with an inner round air stream. Chem Eng Sci 137:412–422

Chapter 10
Two-Phase Characterization for Turbulent Dispersion of Sprays: A Review of Optical Techniques

Srikrishna Sahu, M. Manish and Yannis Hardalupas

Abstract In liquid-fueled combustion, the interaction of spray droplets with surrounding turbulent air flow is crucial since it influences the evaporation rate of the fuel droplets and the process of air and fuel vapor mixture preparation. For detail understanding on the droplet-turbulence interaction mechanisms as well as to validate numerical simulations of sprays, simultaneous measurement of both dispersed and carrier phases of the spray is essential. This way the vortical interaction of droplets can be fully characterized such that important issues such as local spray unsteadiness and spatiotemporal inhomogeneity of droplet concentration due to droplet clustering and group evaporation of droplets can be addressed. This review focuses on the advances in the optical diagnostics (especially the planar techniques) in the last few decades to meet these requirements. Due to broad size and velocity distributions of spray droplets, the application of the two-phase measurement techniques to sprays encounters challenges especially in phase discrimination. Additionally, for sprays, it is not sufficient to simultaneously measure two-phase velocities but the droplet size is also important since the dynamic drag on droplets is according to their size and velocity relative to the surrounding gas flow. The techniques with such capability are explored. The sources of uncertainty and advantages and limitations of different two-phase measurement approaches are analyzed according to their application to dense and dilute sprays.

S. Sahu (✉) · M. Manish
Department of Mechanical Engineering, Indian Institute of Technology Madras, Chennai, India
e-mail: ssahu@iitm.ac.in

M. Manish
e-mail: manishxpress@gmail.com

Y. Hardalupas
Department of Mechanical Engineering, Imperial College, London, UK
e-mail: y.hardalupas@imperial.ac.uk

S. Basu et al. (eds.), *Droplets and Sprays*, Energy, Environment, and Sustainability,
https://doi.org/10.1007/978-981-10-7449-3_10

10.1 Introduction

Background

Research on dispersion of spray droplets is motivated by broad range of engineering, environmental and medical applications. The interaction of droplets with turbulent flows represents a class of combined turbulence and multi-phase flows, which is known to be one of the most challenging areas of research in fluid mechanics (Balachandar and Eaton 2010). The complexity of the problem escalates for applications involving combustion of the liquid fuel sprays due to additional effects owing to droplet evaporation, ignition, and burning of the air-fuel mixture. In addition, droplet collision, coalescence, and breakup can play vital roles in the dispersion of spray droplets. Nevertheless, detail investigation on turbulent spray dispersion is extremely important for combustion applications including internal combustion engines, gas turbines, liquid-fueled rocket engines, and industrial burners. For instance, in gasoline direct injection (GDI) or direct injection spark ignition (DISI) engines, the process of air-fuel mixture preparation is intimately related to the in-cylinder gas flow, spray characteristics, and the distribution of droplets within the combustion chamber. This process must ensure that the fuel-air mixture in the vicinity of the spark plug remains in the flammable range at the optimum ignition timing. Thus, for accurate prediction of the combustion process, a detailed knowledge on the spray-air interaction process is essential (Zhao et al. 1997). Similarly, in the design and development of gas turbine combustion chambers, achieving low pattern factor, which means consistent uniform distribution of temperature in the efflux gases discharging into the turbine, is one of the most important and difficult problems (Lefebvre 1984). Since the fuel spray characteristics and spray penetration strongly influence the temperature in the primary zone, which contributes to the pattern factor, so understanding spray dispersion within the primary zone is of utmost importance. Figure 10.1 presents two examples of instantaneous spray structures from a multi-hole GDI injector and a model prefilming airblast atomizer for gas turbines. The photographs highlight the unique droplet dispersion characteristics depending on the injector. For both cases, strong spatial inhomogeneity in droplet concentration is evident. However, in spite of its practical importance, the mechanisms of interaction of spray droplets with surrounding turbulent air flow are partially known even for sprays operating under atmospheric conditions and when the influences of droplet evaporation and chemical reactions on spray characteristics are weak/absent.

The origin behind the complexity of the two-phase interaction mechanisms is attributed to two aspects inherent to such systems: polydispersity of the spray (presence of wide range of droplet sizes typically 1–100 μm) and carrier phase turbulence kinetic energy spectrum (presence of turbulent eddies characterized by wide range of length and time scales). The way a droplet reacts to various turbulent eddy sizes depends on the initial size and momentum of the droplet. The response of a droplet to the turbulent flow is usually judged according to its Stokes number (*St*), which is defined as the ratio of aerodynamic time constant to some time scale of the flow. Usually, the *St* based on eddy turnover time is considered to estimate the response

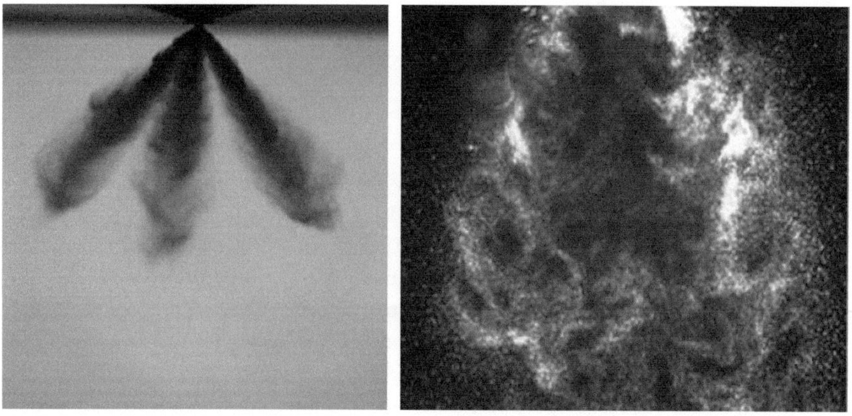

(a) Max Paschke. (2015) (b) Batarshe, Feras Zaki Musa. (2009)

Fig. 10.1 Instantaneous spray structure for **a** pulsed sprays from a multi-hole GDI injector (Paschke 2015) **b** a spray from a model prefilming airblast atomizer relevant to gas turbines (Batarseh 2009)

of droplets to large-scale turbulent eddies, which provides an effective means for dispersing the droplets. However, the role of small-scale turbulent structures cannot be neglected. In high Reynolds number flows, small turbulence eddies may significantly influence the dispersion of droplets of size comparable to those eddies (Bachalo 1994). Thus, yet there is no general consensus on the 'most relevant' turbulent length scale for droplet dispersion. Also, as a consequence of interaction with the air flow turbulent eddies, the droplets are centrifuged out of the vortex core due to their larger inertia and thus tend to accumulate in the peripheral regions of the neighboring eddies leading to 'clustering.' This results in inhomogeneous distribution of spray droplets (regions where droplet concentration is very high above the mean value as well as regions devoid of droplets) leading to spatial and temporal fluctuations in local droplet number density within sprays. Droplet clustering has been identified in several experiments on non-reacting and reacting sprays (Allen and Hanson 1988; McDonell et al. 1992; Nakabe et al. 1991; Akamatsu et al. 1996; Aliseda et al. 2002; Sahu et al. 2016). However, the role of droplet clustering on local spray unsteadiness and droplet group evaporation/combustion are not yet well explored. For deeper understanding on turbulent droplet dispersion in sprays, the characteristics of both dispersed and carrier phases are required to be known. However, this is often hampered at higher liquid mass loading due to limitations in the current experimental diagnostics to probe into dense spray region as well as the difficulty in predictions of the flow due to two-way coupling between the two phases. The droplets modify the turbulent kinetic energy (TKE) of the carrier flow and alter the energy distribution across the eddies of different sizes and this back influence cannot be neglected (Elghobashi 1991, 1994; Kyle 1991; Kulick 1994; Boivin et al. 1998; Sundaram and Lance 1999). In addition, the wake behind large droplets may affect the gas turbulence as well, which has not been quantified yet. However, the above situations are typical to sprays in practical combustors.

Motivation

A comprehensive knowledge on the turbulent spray dispersion process either through experiments or via modeling and simulations has been and remains as a challenge. In spite of significant advances in optical diagnostic tools as well as computational power in the last few decades, the required knowledge relevant to practical scenarios and operating conditions is still insufficient. Since the atomization and spray processes in practical combustion engines refers to a three-dimensional, time-dependent system involving multi-phase turbulent flow and chemical reactions, and the models for the interphase coupling and the turbulence-chemistry interaction are only conditionally successful, the computational fluid dynamic (CFD) study is more difficult than CFD analysis in many other areas (Xi Jiang and Siamas 2010). Even after ignoring the influences of liquid jet/sheet breakup, secondary atomization, droplet-wall interaction, etc., the prediction of turbulent dispersion of droplets is not easy due to the difficulty stemming out, especially, from the necessity of modeling the source terms representing interphase coupling due to mass, momentum, and energy exchange between the droplet and gas phases. Nevertheless, CFD is playing an increasingly important role as a design tool in industry due to its cost-effectiveness and also impracticability of conducting detail experiments under realistic operating conditions. At this juncture, the role of experimentalists is crucial in two respects:

1. First, to generate high quality and reliable experimental data, considering flow configurations and operating conditions with well-defined boundary conditions appropriate for CFD computations in order to validate the modeling approaches (e.g., dilute spray laden jet flow experiments by Chen (2006); Gounder (2012)).
2. Secondly, to be able to measure for operating conditions for which conducting simulations are heavily model dependent or computationally expensive and to provide higher order statistical measurements apart from basic statistics. For example, in addition to mean droplet size/velocity, the measurement of size conditioned droplet-gas slip velocity and joint fluid-droplet statistics (e.g., covariance of droplet and gas velocity, droplet concentration, and droplet and/or gas velocity) are essential for development of new models for the unclosed terms appearing in the governing equations of the carrier phase.

Focusing on both category of experiments as outlined above, the current review intends to discuss the development and the application of some of the optical diagnostic tools for two-phase measurements in sprays. Though, measurement of the dispersed phase (mostly droplet size and velocity) is common in spray community, measurement of the gaseous phase is equally important to study the two-way interaction between the droplets and the turbulent gaseous phase. Although detail characterization of the carrier phase is desirable including measurements of velocity and temperature of the air flow entrained into the spray as well as concentration of fuel vapor or combustion products, the measurement of the gas velocity is more important for understanding droplet dispersion. The droplet-gas slip velocity not only influences droplet concentration distribution but also influences mass, momentum, and energy exchange between the two phases. Hence, gas velocity measurement around spray droplets has received special attention in past. In addition, the droplet dispersion in sprays being a highly unsteady process demands instantaneous and simultaneous measurement of both phases, which possess tough challenges to experimentalists when the spray droplet size distribution is wide (as it is usually) and/or the droplet-gas slip velocity is high (usually close to injector exit).

Scope of the review paper

The present article reviews the optical diagnostics developed in past for two-phase measurements especially for sprays relevant to combustion applications. The focus of the paper is on the optical techniques capable of simultaneously measuring the velocity of the droplets and the gas phase. Depending on the location of measurement within the spray, for instance, either dense spray close to the injector or dilute spray away from the injector exit, the gas velocity is measured either around the spray or individual droplets. Accordingly, the measurement challenges are different. This review considers especially the planar optical techniques for two-phase spray measurements, which are advantageous for instantaneous visualization of the flow field, though a brief discussion on single-point techniques is also included. As compared to mono-sized particle-laden jets, the two-phase measurements for sprays possess additional challenges especially for phase discrimination, which are emphasized. Apart from velocities of the two phases, simultaneous measurement of droplet size is also crucial for sprays. Hence, optical techniques with such capability are also discussed. Finally, the present review is intended to compliment the excellent reviews on spray measurements appeared in past (Bachalo (2000); Fansler (2014)) and provide current status on the capability of the optical techniques to provide simultaneous two-phase measurements in sprays. Throughout the paper, 'particle' refers to solid particles while 'droplet' refers to liquid particles.

The paper is organized as follows. Section 10.2 outlines the challenges encountered for simultaneous optical measurement of droplet and gas phases while studying turbulent dispersion of droplets in sprays. The review of planar two-phase techniques appears in Sect. 10.3. The techniques are discussed separately for dense sprays applications (pulsed sprays as in automotive applications) and dilute sprays (sparse region in sprays or sprays with low liquid mass loading ratios). Section 10.4 presents a summary and outlook on future prospects. It is noted here that the primary atomization of liquid jets/sheets is not considered in the present review, and the focus is on droplets downstream of the injector exit assuming no further breakup of droplets (secondary atomization) occur.

10.2 Challenges for Two-Phase Measurements in Sprays

10.2.1 Droplet Measurement Techniques

Due to their inherent non-intrusive feature, the laser-based techniques have garnered significant attention in multi-phase flow research, and last few decades have seen tremendous development in the diagnostic tools. A detail account of optical techniques for particle characterization for measuring size, velocity, temperature, and composition of particles can be found in a recent review by Tropea (2011). Fansler and Parrish Fansler (2014) have reviewed wide range of optical techniques especially for spray measurements. By far the most popular technique for measurement of spray droplets (which are particularly of spherical shape) is the phase Doppler anemometry (PDA) technique. The operational principle of the PDA technique has been described by several researchers in past (Bachalo and Houser (1984); Saffman et al. (1984); Hardalupas et al. (1992)). PDA is a single-particle counter technique, which provides simultaneous measurement of size and velocity of a droplet passing through the probe volume located at the intersection of two laser beams. The distinct advantage of PDA is that the instrument is almost calibration free. However, several other optical techniques have been developed in past for spray droplet measurements. While measurements of droplet temperature and composition (for multi-component droplets) principally based on laser-induced fluorescence (LIF) techniques have also gained attention (for instance, see Maqua et al. (2006); Hishida and Sakakibara (2000); Lavieille et al. (2001); Maqua et al. (2007); Deprédurand et al. (2008)), the experimental techniques mostly focused on the measurements of droplet size and/or velocity. These techniques include droplet sizing by direct imaging (e.g., shadowgraphy and glare-point technique König et al. (1986); Hess and L'Esperance (2009), line-of-sight-averaged droplet sizing by laser diffraction technique (e.g., Malvern size analyzer, Dodge et al. (1987)), planar droplet size and velocity measurement based on light interference principle, for example, Interferometric Laser Imaging for Droplet Sizing or ILIDS (Glover et al. 1995; Maeda et al. 2000, 2002; Matsuura et al. 2006; Hardalupas, Taylor and Zarogoulidis 2004; Sugimoto et al. 2006), global phase Doppler or GPD (Damaschke et al. 2005), digital holographic techniques (Müller et al. 2004; Palero et al. 2005, 2007; Garcia-Sucerquia et al. 2006), planar LIF-Mie droplet sizing technique (Jermy and Greenhalgh 2000; Domann and Hardalupas 2001; Roland and Yannis 2001) and planar droplet velocity measurement by Particle Image Velocimetry (PIV) or Particle Tracking Velocimetry (PTV) techniques.

10.2.2 Two-Phase Measurement Techniques

As far as measurements of both droplet and gas phases (especially gas phase velocity) are concerned, the ultimate objectives are

1. To measure droplet and gas phases at the 'same' time. Since droplets always occupy some finite space, this requirement is satisfied when both phases are measured simultaneously at spatially different locations.
2. To measure droplet size simultaneously with velocity of both phases. Since each droplet experiences different dynamic drag and also evaporation rate according to its size, hence information on droplet size is absolutely essential for spray dispersion studies, and measurement of two-phase velocities alone is not sufficient.

While both objectives hold true also for any two-phase flows, it is the second requirement that makes two-phase measurements more demanding for sprays as compared to the frequently reported turbulent flows laden with nearly mono-sized solid particles as the dispersing medium.

We note that the above requirements cannot be met by utilizing any one of the droplet measurement technique alone. PDA being a well-established technique for measurement of droplet size and velocity, obviously its application for two-phase measurements is attractive. For this purpose, usually very small ($<5\,\mu$m) seeding particles are introduced in the air flow, which is assumed to follow the gas phase faithfully (though this assumption is strictly true when the corresponding Stokes number based on large-scale eddies is much smaller than unity, $St \approx 0.01$). However, the scattering intensity from droplets is proportional to square of droplet sizes. This demands wider measurable droplet size limit while ensuring reliable signal processing of the tracer particles with relatively poor signal-to-noise ratio Gillandt et al. (2001). In addition, dense seeding can lead to multiple scattering, which disturbs the signals emitted by the drops and introduces errors in the measurements of sizes. Application of PDA for droplet and gas measurements in sprays have been a standard approach reported by several researchers in past (McDonell et al. (1992); Gillandt et al. (2001); McDonell et al. (1993); McDonell and Samuelsen (1993); Gillandt et al. (1998); Prevost et al. (1996); Ferrand (2003); Borée et al. (2001); Horender and Hardalupas (2003); Sanchez et al. (2000) to name few). However, the two-phase measurements are not simultaneous since a single particle or droplet is detected within the probe volume. Thus, the gas velocity is not measured at the 'same' time as a droplet. This is essential to calculate the slip velocity as well as the droplet-gas velocity correlation at zero-time delay in order to estimate turbulence modulation by droplets. To obtain gas velocity 'seen' by droplets, some researchers Prevost et al. (1996); Ferrand (2003); Borée et al. (2001) considered reconstruction of the temporal velocity of the gas phase of a dilute two-phase jet by interpolating the signal of the fluid tracers. However, in this approach uncertainties remain with accurate interpolation of the gas phase velocity especially when data rates for tracer particles are low Horender and Hardalupas (2003). To calculate the two-phase velocity correlation at zero-time delay, (Wang et al. 2010) extrapolated the cross-correlation at nonzero-time delay to zero-delay, while (Horender and Hardalupas 2003) relied on a stochastic eddy interaction model in addition to the measurements. Sanchez et al. (2000) measured both droplet and gas phases for an unsteady diesel spray. The PDA measurements were acquired for temporal window of 50 μs (at definite time from start of injection) during which the spray flow was assumed to be frozen. The mean gas flow field was obtained by considering spray droplets below 5 μm as tracers. However, the statistical uncertainty is heavily dependent on data validation rate.

While PDA is undoubtedly popular for two-phase measurements in sprays, however, as mentioned earlier, instantaneous and simultaneous measurement of both phases is not easy. In addition, visualization of the droplet distribution within a spray is not possible by single-particle counters like PDA. Use of particle image velocimetry (PIV) is advantageous here. The PIV is a widely accepted and established optical technique for planar flow velocity measurement for single-phase fluid flows and has been described in detail in some of the classic articles by Raffel et al. (2013); Adrian and Westerweel (2011); Willert and Gharib (1991); Adrian (2005). Because of its ability for whole field characterization, PIV has received wide acceptance among multi-phase researchers as well. The application of PIV for measurement of spray droplets' velocity is principally same as that of the single-phase flows. However, while PIV can be used to measure droplet velocity when droplet concentration is high, for sparse sprays a similar approach known as particle tracking velocimetry (PTV) is used to track individual droplets (instead of group of droplets) to obtain droplet velocity. The planar droplet measurement by PIV/PTV facilitates direct comparison with spray computations. However, unfortunately, the two requirements for two-phase measurements in sprays, as outlined earlier in this section, are still not fully satisfied. For simultaneous gas phase measurements, the air flow must be seeded with tracer particles. Then separating the signals from the droplets and tracers possess tougher challenges as compared to phase separation by PDA technique in which a particle is identified as a droplet or a tracer depending on its size. In addition, obtaining droplet size by PIV is not accurate. Below we review the approaches taken in the past for two-phase measurements based on the PIV techniques focusing on the advantages and the limitations of the experimental methods developed in past for phase discrimination. Also, the experimental approaches which combine PIV with other techniques for droplet sizing will be discussed in detail.

10.3 Planar Techniques for Two-Phase Measurements in Sprays

Mostly the planar two-phase velocity measurements in sprays are based on the PIV technique or its derivatives. For application of the PIV technique to measure the gas velocity in sprays, the gas phases are seeded with particles or droplets much smaller than the average droplet size though some overlap between size distributions of droplets and particles is always possible. However, in such cases straight forward application of the technique would result in bias since the Mie scattering images may contain both droplets and seeding particles, and most of the time the velocity field calculated results from the displacement of large drops. This means that the velocity field in regions where the two phases are present is usually more representative of the velocity field of the spray than that of the gas flow (Kosiwczuk et al. 2005).

As far as application of PIV for measuring the velocity of spray droplets is concerned, it is noted such measurements can provide only a representative droplet velocity rather than the actual velocity of droplets. This is because droplets of different sizes may have significantly different velocity such that the velocity vectors determined by PIV does not represent trajectory of a particular droplet.

Although PIV is popularly used for single-phase flow velocity measurement, its application to two-phase flow brings up several challenges the most important one being the phase separation to distinguish the signals from the dispersed and the continuous phases. Because PIV is based on Mie scattering from droplets/particles and the scattering intensity is proportional to surface area of particles ($\propto d^2$), it is difficult to image both largest size droplets and small seeding particles while avoiding camera saturation and simultaneously obtain reasonable signal strength from small particles. Thus, size- and/or intensity-based discrimination between the two phases in the spray PIV images is not easy. An alternative approach for phase discrimination, which gained popularity for two-phase measurements, relies on tagging either the droplets or the tracers with fluorescence dyes such that when illuminated with appropriate laser sources the corresponding laser-induced fluorescence lights are distinct from the Mie scattering, and this way individual phases can be imaged separately using suitable optical filters. Such methods are also termed as 'LIF-PIV' techniques. The choice of introducing fluorescent dyes to the dispersed and/or the carrier phase brings about different routes for phase discrimination depending on if it is achieved during image acquisition or post-processing of the images (each method has its own strengths and limitations). However, due to limitations in the transmission efficiency of the optical filters, some cross-talk between the two phases always exists. In addition, since the fluorescence emission depends on volume of droplets ($\propto d^3$), the difficulty in imaging simultaneously large and small droplets still persists. Apart from the difficulty in the phase separation, there are several other issues which also need attention for simultaneous two-phase measurements as outlined below,

1. presence of significant droplet-gas slip velocity.
2. degradation of the continuous phase signal in the vicinity of droplets as a result of intense scattering/fluorescence close to droplet surfaces.
3. multiple scattering due to tracer particles.
4. collision and coalescence of tracer particles/droplets with the spray droplets.
5. uncertainty in processing either Mie scattering images containing both droplets and tracers (for droplet velocity measurement) or droplet-filtered PIV images containing tracers (for gas velocity measurement).

Below we discuss various approaches taken in past to obviate one or more of the above-mentioned problems. The experimental techniques can be classified in various ways according to applications involving large or small droplet-gas slip velocity, ability to measure in dense or dilute regions of a spray, phase separation approach, optical setup (one or two cameras, camera orientation with respect to laser sheet, laser sources), adopted image processing methods, ability to measure droplet size etc. In the following section, the previous works on simultaneous two-phase

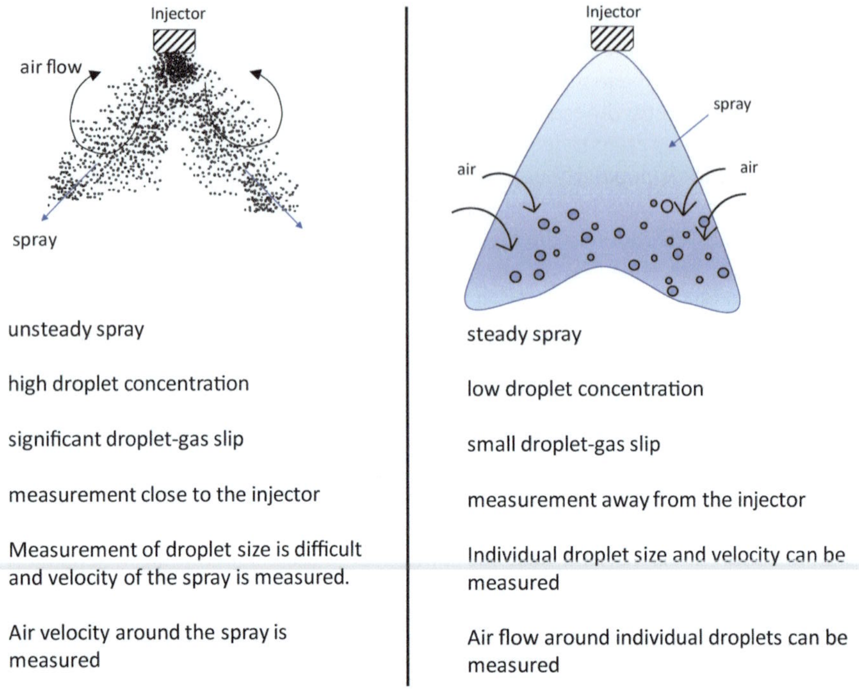

Fig. 10.2 Different scenarios for planar and simultaneous two-phase measurements in sprays (left) dense spray near the injector of a Diesel or GDI engine (right) dilute region of a steady spray

measurements in sprays are discussed according to the classification based on the application of the experimental techniques to either dense sprays (typically pulsed spray of a high pressure injector for a diesel or DISI engine; the region of interest is close to the injector exit) or dilute regions of sprays (operating under steady-state conditions; the region of interest is locations where droplet concentration is not too high). The two cases are shown in Fig. 10.2. This classification is justifiable since even if the basic approach for phase discrimination remains same for both cases, the significant difference in the droplet number density and slip velocity for the two cases leads to different challenges in the application of the techniques and the image processing methods as well as the measured statistical quantities.

10.3.1 Dense Spray Measurements

Previous research on two-phase measurements for dense sprays mostly referred to characterization of the time varying gas flow field around automotive fuel sprays for studying mixing process in internal combustion engines (Diesel and DISI engines). Since pulsed injection is used, the measurements must be obtained during spray development stage at different times after the start of injection. An inherent challenge for simultaneous two-phase velocity measurement in this case is due to significantly large velocity difference between the injected droplets and the entrained air flow (for the spray, the velocity is of the order of 100 ms^{-1} in comparison to the gas velocity which is about 1–10 ms^{-1}). Accordingly, the laser separation time for PIV application is smaller than 10 μs for the spray, while that for the gas flow lies in the range from tens of microseconds to hundreds of microseconds (Hillamo et al. 2008; Driscoll et al. 2003; Rottenkolber et al. 2002; Zhang et al. 2014). The choice of same separation time for both phases can lead to inaccuracy in velocity measurements for both phases. Also, due to high droplet concentration within the spray, reliable measurement of gas phase is possible away from the spray boundary, while measuring gas velocity around individual droplets is not possible. Thus, the application of conventional PIV results in strong biases in the measured gas velocity especially close to the spray boundary (Rhim and Farrell 2000). Hence, the phase discrimination is achieved based on LIF-PIV techniques as mentioned earlier. While some of the earlier researchers relied on separating the two phases during post-processing, it can be achieved during image acquisition itself.

Phase separation in post-processing

In this case, usually a single laser source (typically 532 nm) is used to illuminate the spray within a high pressure chamber. The air flow within the chamber is seeded with very small tracer particles/droplets for gas velocity measurements. Two charge coupled device (CCD) PIV cross-correlation cameras are used one for each phase. The phase separation is achieved by tagging either the spray droplets or the tracers with appropriate fluorescent dye (see Fig. 10.3).

In the first approach (for example, see Boedec and Simoens (2001); Le Moyne et al. (2007)), the fluorescent light from the spray droplets (the wavelength depends on the dye used to dope the liquid) alone is imaged in a CCD camera operating in conjunction with a suitable long pass optical filter, while, Mie scattering light from both droplets and particles appeared on a second camera (Fig. 10.3a). The tracers can be either micron-sized solid particles or droplets. The contribution due to droplets on the second image is separated by comparison of the fluorescence and Mie scattering images of the same droplets in both images. The reliable identification of the corresponding droplets (small as well as large) on both Mie and LIF images is key to successful application of the image processing method. Thus, the images containing only droplets and only tracers can be processed to obtain the respective velocity fields. However, poor fluorescent signal from small droplets and the lack of tracer particles (where droplet number density is high) can lead to bias in gas velocity measurement. Since the liquid phase is being separately imaged, the geometrical size of the liquid droplets (which were mostly non-spherical) can be characterized from the binarized droplet LIF images (Boedec and Simoens 2001; Le Moyne et al. 2007).

In the second approach (Fig. 10.3b), only the tracers (preferably small aerosol droplets) are tagged with a fluorescent dye (Rottenkolber et al. 2002; Lee and Nishida 2003). Hence, the camera for the carrier phase velocity measurement captures the fluorescent light from the tracers alone. The other camera images Mie scattering

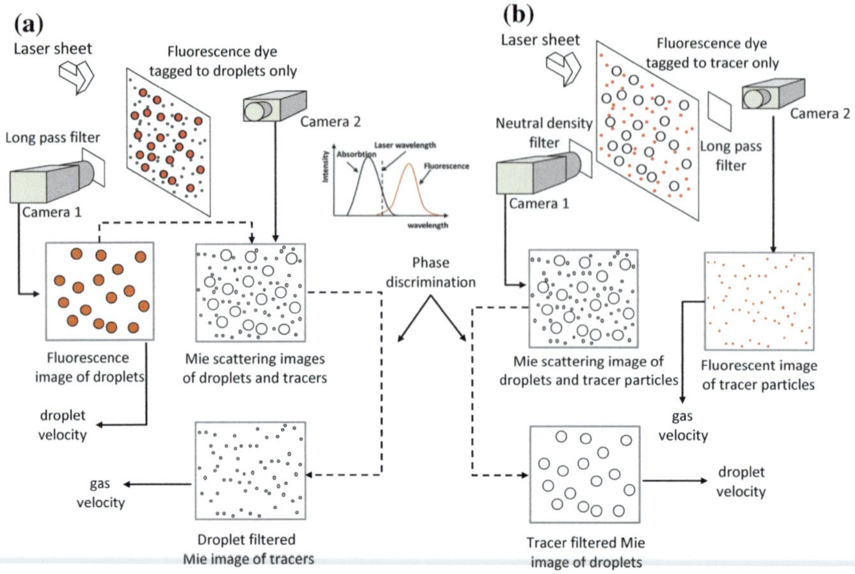

Fig. 10.3 Principle of phase discrimination in LIF-PIV technique: fluorescent dye tagged to **a** dispersed phase or droplets **b** tracers representing the carrier phase

light from both droplets and tracers. Rottenkolber et al. (2002) applied such two-phase PIV technique to the pulsed hollow-cone spray of a high pressure swirl injector for a DISI engine. However, in their case use of an adequate optical high-pass filter allowed transmission of the fluorescent light from the particles, while the Mie scattering from droplets was attenuated. Thus, both phases were imaged using a single camera and separated by post-processing. However, this way small droplets may not be imaged due to low scattering intensity. The authors described two different phase discrimination methods: mask and peak separation techniques, each having some limitations. The authors reported that the phase discriminatory measurement is much more robust for single phases (when the gas or droplet phase alone are measured) in comparison to simultaneous two-phase measurements. While the benefit of the above approach is in the use of a single camera to record both phases, further developments in the image processing are essential for realization of better accuracy.

Phase separation during image acquisition
Achieving successful phase separation by the methods described earlier is heavily dependent on the image processing algorithms employed. This can be avoided by separately capturing the images of the droplets and the tracers during image acquisition itself (Towers et al. 1999; Driscoll et al. 2003; Kosiwczuk et al. 2005). However, such approaches demands for additional instrumentation (thus offers expensive options) or special selection of fluorescent dyes (which may be specific to the selection of the liquid for the spray droplets or tracers). Below we discuss three different

approaches reported in the literature and based on the principle of the LIF-PIV technique for phase separation during image acquisition.

For obtaining simultaneous two-phase velocity measurements in a gasoline direct injection fuel spray, Driscoll et al. (2003) developed a new diagnostic, which utilizes a two-laser (double-pulse) two-camera (double-frame) setup (Fig. 10.4a). A 532 nm laser source is used to illuminate the spray droplets, and the Mie scattering signals from the droplets are collected in a CCD camera equipped with an interference filter. The gas phase is seeded with small seeding particles of ethylene glycol doped with Lambdachrome Stilbene 3 (428-nm peak fluorescence, 415–439 nm range, 15% fluorescence efficiency) prior to the experiment. Using a separate laser source at 355 nm, the fluorescent light from the seeding particles is imaged in a second CCD camera equipped with an optical band-pass filter to suppress the Mie scattering from seeding particles and droplets from both lasers. However, this way scattering by seeding particles (from 532 nm laser) also appears on the droplet image, though this is not a limitation. Since the droplet and gas phases are independently imaged, the delay time between the pulses of each laser can be set separately (8 μ.s and 80 μs, respectively, were used in their experiments). However, often financial constraints do not allow use of two pulsed laser sources. Moreover, usually double pulsed 355 nm laser is not readily available, and aligning two laser sheets for illuminating the same area within the spray is not easy.

Kosiwczuk et al. (2005) described a rather simpler yet unique approach (Fig. 10.4b) for phase separation using a single laser source (double pulsed 355 nm) by tagging different fluorescent dyes Rhodamine 610 and Stilbene 420, respectively, to spray droplets and small tracer droplets (the liquid was Methanol for both cases). This way the peak of the fluorescence spectrum for the spray droplets is around 580 nm, which is distinct from that of the seeding droplets for which the peak is about 430 nm. Hence, with suitable band-pass filters the two phases can be separately and simultaneously imaged in two different PIV cameras. The technique was applied to spray droplets from an ultrasonic atomizer. The gas phase velocity was determined using standard PIV algorithm. Since the droplet distribution was non-uniform, the spray droplet velocity mesh is not regular. Hence, to avoid difficulty in comparing the velocity fields of the two phases the velocities of the seeding droplets were determined on the spray droplet locations by interpolating the regular velocity field. An issue with this method is that the dyes for the spraying liquid as well as tracer droplets must be appropriately selected through spectroscopic analysis as carried out by the authors.

Recently, Zhang et al. (2014) developed a unique method for simultaneous velocity measurement of an eight-hole DISI fuel spray and its ambient gas based on a single laser source and camera. This was possible by utilizing a high-speed PIV laser (527 nm) in conjunction with a high-speed camera equipped with an image doubler to divide the camera chip into two halves focusing on the same field of view through careful adjustment. The method is depicted in Fig. 10.4c. Fluorescent seeding particles were used as flow tracers. The program for timing the acquisition of droplet and particle images is central to successful application of this approach. While one double exposure (with suitable separation time) was used to capture Mie scattering from

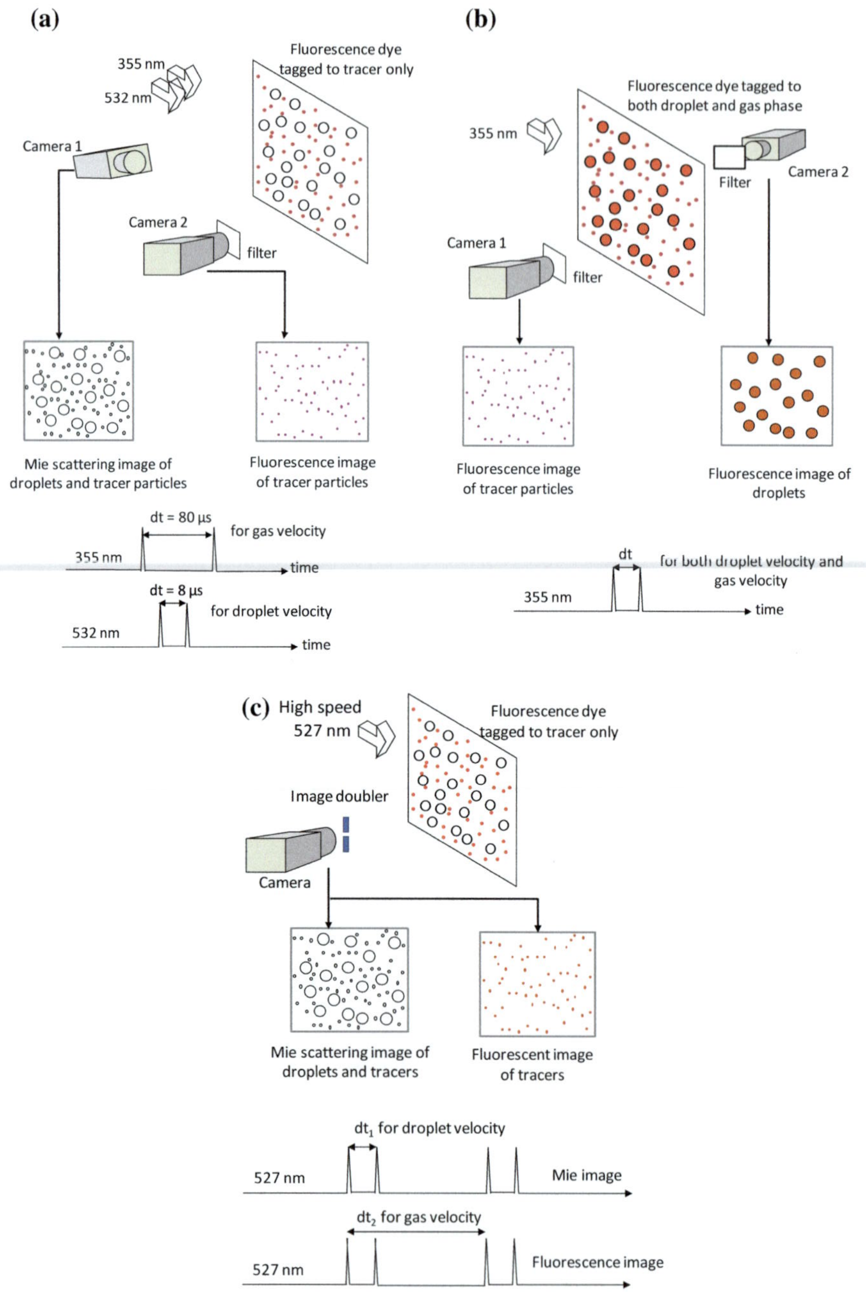

Fig. 10.4 Phase discrimination during image acquisition by different approaches based on **a** two lasers and two cameras by Driscoll et al. (2003), **b** single laser and two cameras by Kosiwczuk et al. (2005), **c** a high-speed laser and camera (with image doubler) by Zhang et al. (2014)

droplets (as well as tracer particles) for droplet velocity, the first pulses of consecutive double exposure were used to correlate the LIF images of the tracer particles to obtain the gas velocity. This way simultaneous and separate imaging of both phases could be obtained in spite of large difference between velocities of the two phases.

It is noted that the application of the optical techniques based on the PIV technique as described in this section is suitable for measurement of velocities of both droplet and gas phases, while reports on simultaneous measurement of droplet size is rare, in spite of being important. Although, PDA measurement of droplet sizes in diesel and DISI sprays have been reported by several authors earlier. The droplet size is measured for different times after the start of the spray injection. However, high droplet number density leading to multiple scattering often restricts applicability of the conventional droplet sizing techniques close to the injector especially during early stage of injection. Additionally, the droplets may not be spherical and secondary breakup is not over. Hence, reliability of droplet size measurement is ensured only away from the injector exit and for late injection stages.

10.3.2 Dilute Spray Measurements

The two-phase measurement techniques as described earlier for dense sprays are certainly applicable to dilute regions of the spray or sprays with low liquid mass loading ratios. However, since the droplet number density of the dispersed phase is much lower, the interrogation window size for PIV cross-correlation calculation for obtaining droplet velocity vectors must be large enough to contain sufficient number of droplets such that the correlation peak is high enough to allow reliable measurement. This leads to poor spatial resolution of the droplet velocity vectors. Instead, individual particles can be tracked between a pair of images, to obtain droplet velocity and this approach is known as the particle tracking velocimetry (PTV). Since the droplet concentration limit demarcating dense and dilute spray is not unique, one needs to exercise caution while using PTV since higher droplet number density may lead to erroneous pairing up of droplets and so incorrect estimation of the droplet velocity.

Earlier several works have been reported on two-phase measurements in turbulent flows carrying nearly mono-sized solid particles, as the dispersed medium (e.g., Gui and Merzkirch 1996; Gui et al. 1997; Kiger and Pan 2000; Khalitov and Longmire 2002). The solid particle concentration was dilute in these studies: The mass loading ratio was low. The size of these dispersed particles were usually much larger than that of the tracers introduced for PIV application. The Mie scattering images from both solid and tracer particles are imaged in one camera, and phase discrimination was achieved through image processing. In these studies, the size of the dispersed particles are usually much larger than that of the tracers introduced for PIV application. Consequently, substantial difference in the light scattering intensity between the dispersed particles and tracers is realized, which exceeds the dynamic range of the imaging sensor. Although the imaging parameters (i.e., lens aperture, exposure time, laser power) are adjusted to be able to image both solid particles and tracers, the resulting image quality is often poor. In addition, the large particles may partially block the illuminating laser light influencing the image quality. Finally, the above results in poor accuracy of the phase separation algorithm and uncertainty in the measurement of velocity of both phases (Poelma 2004). One way to overcome these limitations is to image the dispersed and the carrier phases separately by using an additional camera in conjunction with fluorescent tracer particles and color filtering as described earlier in the context of dense spray measurements (Fig. 10.3b). This has been earlier demonstrated for bubbly two-phase flows (Philip et al. 1994; Gerbrand Deen et al. 2000) or particle-laden grid-generated turbulent flow within a water tunnel (Poelma et al. 2006). In particular, we mention Poelma et al. (2006) who have identified few important issues related to the two-phase PIV techniques.

1. The dispersed-phase-filtered PIV images of the carrier phase flow are 'non-ideal.' As the dispersed particles occupy finite space, the phase discrimination (either at the imaging stage or due to post-processing) results in the corresponding region devoid of images of tracer particles (see Fig. 10.3). This non-uniform distribution of tracers can cause only fraction of PIV interrogation cells to be occupied by tracers, which lead to erroneous estimation of fluid phase velocity vectors. This effect is expected not to be significant for calculation of basic statistics such as the average flow field if the mass loading of the dispersed phase is low (though this limit has not been quantified yet), however, locally the fluid phase velocity vectors is still wrongly estimated if the data loss within a PIV cell is significant. Hence, application of the conventional PIV cross-correlation algorithm is not suitable (still several works reported earlier have used this approach). This is demonstrated in Fig. 10.5 using the numerically simulated PIV images by Kl Okamoto and Nishio (2000). It is assumed that particles of different sizes are randomly located within the PIV images, (to simulate two-phase PIV image of a spray) and the intensity values at the corresponding locations are assigned 'zero' values. A sample image pair (Fig. 10.5a) and the same pair after removal of particles (Fig. 10.5b) are processed using the conventional PIV algorithm. As an example, the results are presented for the interrogation and search windows shown in Fig. 10.5a, b. It can be clearly observed that the peak of the cross-correlation function, which is evident in case of the original image pair (Fig. 10.5a), is suppressed for the particle-filtered image pair, and for the latter case the peak appears somewhere else (Fig. 10.5b) indicating erroneous estimation of fluid velocity. In order to accurately measure the fluid phase velocity in the vicinity of the dispersed phase particles, Gui and co. Gui and Merzkirch (1996); Gui et al. (1997); Gui (1998); Gui et al. (2003) adopted a digital masking technique in conjunction with PIV evaluation algorithm. The digital mask is a binarized two-dimensional array such that the values are 'zero' for pixels corresponding to the dispersed phase and 'one' otherwise. The image processing technique is essentially based on considering the contribution from the region corresponding to the digital value of 'one' while evaluating the correlation spectrum. Figure 10.5c highlights the advantage of the digital masking technique by Gui et al. (2003). The masks correspond-

ing to Fig. 10.5b are shown in Fig. 10.5c, which also shows the corresponding correlation function evaluated based on the masking approach considering the particle-filtered images in conjunction with the respective masks. It is interesting to observe that the suppressed peak in Fig. 10.5b reappears in Fig. 10.5c ensuring correct estimation of the fluid velocity vector. However, the accuracy of any algorithm for processing the particle-filtered PIV images depends the size of the interrogation cells relative to the size of the masked area. Smaller the interrogation cell size higher is the uncertainty in velocity estimation. Gui et al. (2003) demonstrated the potential of the masking technique for different kinds of flows, and the error due to conventional algorithm (without the use of mask) was highlighted. When the positions of the dispersed particles vary from frame to frame as in the case of sprays, the digital mask must be dynamically created (Sahu 2011), for example, in Fig. 10.5 the locations of the same particles are different in the first and second images of the image pair, hence the corresponding masks are different. Sahu (2011) developed an algorithm by combining the digital masking technique with an advanced PIV evaluation algorithm by Ronneberger et al. (1998) for faster processing of the two-phase PIV images for application to sprays.

2. The second reason behind the reduced PIV image quality is the presence of dispersed particles in the line of sight between the field of view and the CCD sensor. As mentioned earlier partial blocking due to the dispersed particles as well as because of multiple scattering, the signal-to-noise ratio of the tracer images drops significantly. The loss of signal from the tracer particles due to this effect is unavoidable and hence it sets a limit up to which the two-phase techniques can be successfully employed. Also, increasing the tracer density (for better accuracy in gas velocity measurement) has negative effect on signal-to-noise ratio of droplets. On the other hand, reducing the tracer density may result in oversampling, which may introduce a systematic error in the turbulence statistics (Poelma et al. 2006).
3. The application of the two-phase PIV techniques is associated with mostly qualitative flow analysis rather than quantitative analysis, for example, to obtain joint fluid-particle turbulence statistics. In the literature, only few studies have reported measurements pertaining to two-point gas-particle and particle-particle velocity correlations (Sakakibara et al. 1996; Kiger and Pan 2002) and particle-gas slip velocity (Khalitov and Longmire 2002) or longitudinal power spectra for fluid and particle velocity (Poelma et al. 2006). However, application of planar two-phase techniques to sprays for measurements of such higher order droplet-gas statistics has been rarely reported (for example, Sahu et al. 2014, 2016).

Earlier research on simultaneous planar measurement of droplet and gas phase velocity and also droplet size for dilute sprays has been rather sparse in comparison to particle-laden flows. However, such measurements are very important for validation of spray models (Laurent and Massot 2001; Reveillon and Vervisch 2005; Xi Jiang and Siamas 2010). Though, intuitively, application of the planar phase separation techniques to dilute or relatively denser sprays seems to be straightforward, this is not always the case as explained below:

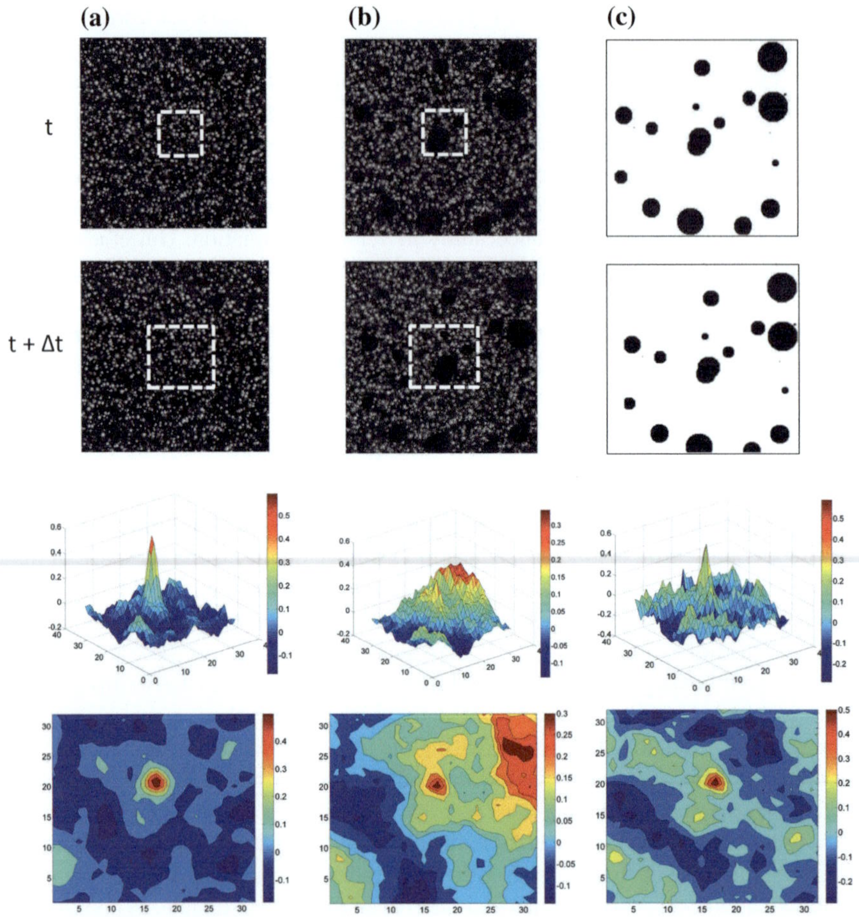

Fig. 10.5 **a** A sample PIV image pair (the dashed boxes denote the interrogation and the search windows), **b** Same image pair with particles removed from random locations, **c** Masks corresponding to the image pair in (**b**). The surface and the contour plots of the cross-correlation function for each case are also shown. The cross-correlation calculation for (**a**) and (**b**) is based on the conventional PIV algorithm, while that for (**c**) is based on the masking technique proposed by Gui et al. (1997)

1. For dense sprays, especially, pulsed high pressure sprays generated from DISI or diesel injectors, mostly the air flow surrounding the spray is measured. Thus, apart from the boundary region of the spray, the coexistence of the two phases within the viewing area is not of concern elsewhere: Away from the spray core region droplets concentration is much lower, while within the spray core the droplet concentration is too high to allow accurate measurement of gas velocity. However, for either pulsed or steady sprays and away from the injector exit, droplets and tracers coexist throughout. Also, since secondary breakup of droplets

is completed the droplet size distribution is populated by small droplets such that overlapping of the size distributions of droplet and seeding particles is almost inevitable. Hence, the possibility of cross-talk between the two phases is significant. While use of fluorescent seeding particles offers the advantage of phase separation during image acquisition itself (Fig. 10.4), some cross-talk due to Mie scattering is always possible. In addition, it is expensive to use. At the same time, phase separation on Mie images of droplets and tracers based on size, shape, and intensity is not easy. Hence, some compromise must be made either way.
2. Away from the injector exit the droplet-gas slip velocity, though not negligible, is not as substantially higher as at the near nozzle locations. Hence, the velocity difference between the two phases cannot be used to discriminate between the two phases. However, for dilute sprays the stringent requirement of significantly different laser pulse separation time for droplet and gas phases (as for pulsed high pressure sprays) is no more important.
3. Unlike studies based on the particle-laden turbulent flows (where mostly mono-sized particles are used as dispersed medium), in the present case the two-phase statistics is meaningful when measured conditional on the droplet size. This is also important for comparison with numerical simulation and developing models for spray computation.

The above discussion implies that achieving phase separation for measurement of two-phase velocity is not sufficient and simultaneous measurement of the droplet size is essential. This means PIV alone is not sufficient for this task and the application of the optical techniques for whole field or planar measurement of droplet size is necessary. One way to achieve this is by employing the multi-dimensional droplet sizing techniques such as the interferometric laser imaging for droplet sizing (ILIDS) or global phase Doppler (GPD) techniques (Damaschke et al. 2005), which are based on out-of-focus imaging of droplets illuminated by laser light sheets of finite width. The ILIDS technique, also known as interferometric particle imaging (IPI), can provide instantaneous measurements of the spatial distribution of individual droplet size and velocity in polydisperse sprays (Horie 2007; Zarogoulidis 2012; Hishida 2013; Hardalupas 2014). The principle of the technique is based on the interference of the reflected and the first-order refracted light from individual droplets when collected under defocused optical setting at an optimum forward scattering angle. This results in a fringe pattern corresponding to each droplet such that the droplet size is proportional to the number of fringes within the fringe pattern (Glover et al. 1995). Though overlapping of the droplet fringe patterns is a major issue with this technique, the incorporation of a pair of cylindrical lenses in the imaging path reduces the fringe overlapping substantially in relatively denser sprays and avoids the complexity of the evaluation of fringe spacing (Maeda et al. 2002). However, as the droplet concentration is higher, validation ratio reduces due to overlapping of droplet fringe patterns (Tropea 2011). In contrast to the ILIDS technique, the GPD technique relies on the interference fringes from the like scattering orders, and also, two intersecting light sheets are used. Though the GPD technique has certain advantages over the ILIDS technique (for instance, there is no restriction on viewing angle), its application to two-phase measurements is yet to be explored.

It should be noted that though the droplet concentration limit up to which the two techniques can be applied for the droplet measurement is dependent on the design of the receiving optics and features of the image processing software, currently this limit is far below than that for the phase Doppler technique (Damaschke et al. 2002). Nevertheless, as explained below, the planar measurement of droplets has some distinct advantages over the single-point technique such as PDA. By assuming the measured small droplets (< 5–10 μm) to follow the surrounding gas flow, simultaneous droplet and gas flow velocity along with droplet size can in principle be obtained with the ILIDS technique alone. However, the droplet validation rate in the technique must be good and also large number of image samples must be acquired in order to obtain droplet statistics with low uncertainty. This feature is important to study preferential accumulation of droplets of different sizes (Lacour et al. 2011). Also, ILIDS is capable of measuring the inter-droplet-distance conditioned on droplet size, which is a quantity of significance for measurement of the two-point droplet velocity correlation as well as the radial distribution function (RDF) for the study clustering of droplets (Sahu et al. 2014). In addition, the ILIDS technique is advantageous for its ability to be coupled with other planar techniques such as PIV and LIF for simultaneous measurement of carrier phase velocity as well as concentration of droplet vapor. Below, we describe two different approaches (see Fig. 10.6), reported in the literature for simultaneous two-phase measurement in dilute sprays based on the combination of ILIDS and PIV techniques.

Hardalupas et al. (2010) developed a novel optical instrument by combining the 'out-of-focus imaging' technique ILIDS for planar droplet size and velocity measurements with PIV for gas phase velocity measurements in far downstream locations of a confined spray. The coflowing air around the spray was seeded with tracer particles for PIV measurements. The optical arrangement is shown in Fig. 10.6. While on the ILIDS image the defocused images from tracer particles appear on the ILIDS camera, but without fringe patterns, the PIV image contains both droplet glare points and seeding particles. The principle of the combined ILIDS and PIV technique is based on identifying the same droplet on the instantaneous ILIDS and PIV images (i.e., the fringe pattern on the ILIDS image and the corresponding glare points on the PIV image is identified). This way it is possible to associate the droplet size/velocity measured by ILIDS to the glare points on the PIV image. The droplet glare points on the PIV image are separately identified and filtered out so that the image contains the tracer particles only. Instead of relying on intensity-based thresholding, which may lead to ambiguity since both droplets and particles may saturate the camera, the phase discrimination in the PIV image is achieved by application of the continuous wavelet transform (CWT). The wavelet scale is appropriately selected based on the size of the droplet images such that the peaks produced by the tracer images in the CWT spectrum is suppressed. Thereafter, by selecting a suitable threshold on the spectrum, only glare points can be identified and the signals from the seeding particles are eliminated. However, (Hardalupas and Sahu 2010) identified that the application of the ILIDS technique results in the 'center discrepancy' in the location

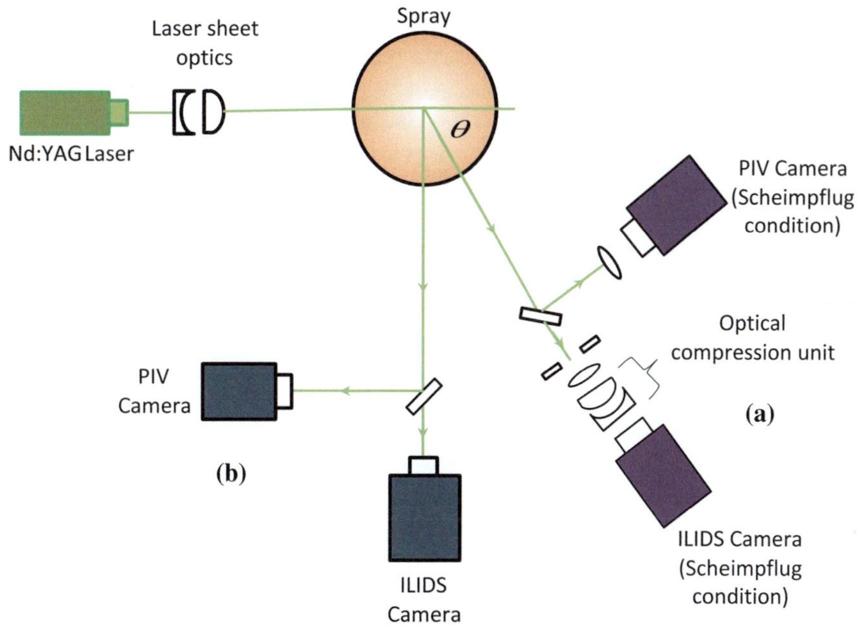

Fig. 10.6 Simultaneous planar measurement of droplet and gas velocity and droplet size in dilute sprays by combining ILIDS (or IPI) with PIV **a** non-90 viewing at forward scattering angle (Hardalupas et al. (2010)) **b** 90 viewing angle by Lacour et al. (2011)

of the geometric center of a droplet due to defocusing. This may lead to erroneous pairing up of droplet glare points and fringe patterns during application of the combined technique. The authors found that, the center discrepancy varies approximately linearly across the image along the direction of propagation of the laser sheet for a given defocus setting of ILIDS. However, the center discrepancy error can be calibrated based on ILIDS+PIV measurements of droplets in a very dilute region of the spray (without seeding particles), which can be subtracted from the droplet center identified in ILIDS images from the spray. This approach was shown to result in order of magnitude reduction of the discrepancy between PIV and ILIDS droplet centers. Later, a similar technique was developed by combining ILIDS with LIF for simultaneous droplet and vapor mass fraction measurements in an evaporating spray (Sahu et al. 2014).

The simultaneous two-phase measurements approach by Hardalupas and Sahu (2010) facilitates measurement of several statistical quantities, importantly, conditional on droplet size, which are useful for further understanding on two-way interaction between the dispersed and the carrier phases, i.e., turbulent dispersion of the droplets by the entrained air flow and air turbulence modulation due to droplets (Sahu et al. 2014, 2016). These quantities include droplet-gas slip velocity (by considering gas velocity measured close to droplets), two-point spatial correlations of droplet-gas velocity fluctuations (signifies strength of coupling between the two phases), droplet velocity fluctuations (represents droplet kinetic stress) and correlations of droplet number density and droplet/gas velocity fluctuations (signifies consequence of preferential accumulation of droplets on gas phase turbulence) etc. Since the gas velocity was measured independently, Sahu et al. (2014, 2016) characterized the gas phase turbulence in presence of the droplets. The gas phase turbulence (in presence of droplets) could be characterized since the gas velocity could be independently measured. This facilitated detail investigation on the interaction of droplets specifically with large-scale eddies, which is of utmost importance in particle-laden flows.

Lacour et al. (2011) used combined IPI and PIV technique to study droplet-vortex interaction mechanisms for a cloud of evaporating droplets injected into a laminar, axisymmetric, and pulsed jet flow. While for both this study and (Hardalupas and Sahu 2010), the optical configuration includes two cameras imaging the same scattered light using a beam splitter and the laser is perpendicularly polarized, the scattering angle is 90° for the former (Fig. 10.6a) in contrast to 69° for the latter (Fig. 10.6b). Though a non-90° viewing for ILIDS or IPI leads to the inconvenience of varying degree of focus/defocus across the imaging plane (which, though, is not of concern if Scheimpflug condition is implemented, (Sugimoto et al. 2006)), the signal-to-noise ratio of the droplet fringe patterns are higher due to maximum interference between refracted and reflected rays. Lacour et al. (2011) obtained detailed measurements of dynamics of decane and n-heptane spray (weakly and strongly evaporating droplets, respectively) conditioned by droplet size, which can be used for validation of numerical models.

10.4 Summary and Outlook

Simultaneous measurement of droplet and gas phases in sprays is vital for our understanding on turbulent dispersion of droplets, which has direct consequences on the air-fuel mixture preparation process for spray combustion applications. Though in the last few decades various optical experimental methods have been developed for particle-laden turbulent flows, measurements in sprays offer additional challenges due to broad droplet size and velocity distributions. Thus, detail investigation on turbulent spray dispersion demands simultaneous measurement of droplet size in addition to the velocities of the droplet and the gas phases. This review article focused on the planar optical diagnostic techniques (based on the PIV technique) capable of providing such information. As far as two-phase velocity measurements are concerned, successful discrimination between signals from the droplets and the tracers is very important. This is achieved considering either size-/intensity- based discrimination on the Mie scattering images of both phases or by tagging one of the phases with a fluorescent marker (LIF-PIV). In the latter case, the phase separation can be achieved during the image acquisition itself. However, several factors critically influence the success of phase discrimination approaches and two-phase velocity measurements, which are highlighted in the paper. The techniques are reviewed according to their application to dense or dilute sprays. The dense spray measurements are relevant to diesel or gasoline direct injection fuel sprays where substantial difference in droplet and gas velocity exists. In such case, experiments are performed close to the injector exit to measure droplet and gas velocity to study the time varying spray dispersion. However, simultaneous measurement of droplet size has been rarely reported. On the other hand, it is possible to measure droplet size/velocity as well as gas velocity around individual droplets. However, extension of such techniques to even relatively dense sprays is inhibited due to measurement uncertainty.

In spite of the advances in the optical diagnostic tools for spray measurements, the capability of the techniques to simultaneously characterize both droplet and gas phases to study droplet dispersion is still limited, and more so under practically relevant conditions. In addition to the qualitative information on visualization of droplet and gas flow fields, it is very important to achieve quantitative measurements of the joint droplet-gas statistics. For understanding local spray unsteadiness such as clustering of droplets, the instantaneous droplet number density must also be measured apart from the droplet size. Also, it is highly desirable to extend the two-phase techniques to study laboratory scale sprays under evaporating and burning conditions. There are several issues which needs attention, for example, it is not yet clear how preferential accumulation of droplets influences spatial and temporal fluctuations of air-fuel mixture ratio within the spray, and if clustering of droplets leads to group evaporation and group burning of droplets and in such case its influence on the emission characteristics. However, this demands, apart from droplet size and velocity of the two phases, simultaneous measurement of concentration of droplets and vapor phases as well as droplet and gas temperature. Application of the two-phase techniques for studying turbulent breakup of droplets (secondary atomization) is also vital. In such cases, non-spherical liquid droplets are often encountered and the role of air turbulence in the disintegration of those to fine spherical droplets is not yet well understood.

References

Adrian RJ (2005) Twenty years of particle image velocimetry. Exp Fluids 39(2):159–169

Adrian RJ, Westerweel J (2011) Particle image velocimetry, no 30. Cambridge University Press

AH Lefebvre (1984) Fuel effects on gas turbine combustion-liner temperature, pattern factor, and pollutant emissions. J Aircr 21(11):887–898

Akamatsu F, Miutani Y, Katsuki M, Tsushima S, Cho YD (1996) Measurement of the local group combustion number of droplet clusters in a premixed spray stream. In: Symposium (international) on combustion, vol 26. Elsevier, pp 1723–1729

Aliseda A, Cartellier A, Hainaux F, Lasheras JC (2002) Effect of preferential concentration on the settling velocity of heavy particles in homogeneous isotropic turbulence. J Fluid Mech 468:77–105

Allen MG, Hanson RK (1988) Digital imaging of species concentration fields in spray flames. In: Symposium (international) on combustion, vol 21. Elsevier, pp 1755–1762

Bachalo W (2000) Spray diagnostics for the twenty-first century. Atomization and Sprays 10(3–5)

Bachalo WD, Houser MJ (1984) Development of the phase doppler spray analyzer for liquid drop size and velocity characterizations

Bachalo WD (1994) Experimental methods in multiphase flows. Int J Multiph Flow 20:261–295

Balachandar S, Eaton JK (2010) Turbulent dispersed multiphase flow. Ann Rev Fluid Mech 42:111–133,

Batarseh FZM (2009) Spray generated by an airblast atomizer: atomization, propagation and aerodynamic instability. PhD thesis, Technische Universität

Boedec T, Simoens S (2001) Instantaneous and simultaneous planar velocity field measurements of two phases for turbulent mixing of high pressure sprays. Exp Fluids 31(5):506–518

Boivin M, Simonin O, Squires KD (1998) Direct numerical simulation of turbulence modulation by particles in isotropic turbulence. J Fluid Mech 375:235–263

Borée J, Ishima T, Flour I (2001) The effect of mass loading and inter-particle collisions on the development of the polydispersed two-phase flow downstream of a confined bluff body. J Fluid Mech 443:129–165

Chen Y-C, Stårner SH, Masri AR (2006) A detailed experimental investigation of well-defined, turbulent evaporating spray jets of acetone. Int J Multiph Flow 32(4):389–412

Damaschke N, Nobach H, Tropea C (2002) Optical limits of particle concentration for multi-dimensional particle sizing techniques in fluid mechanics. Exp Fluids 32(2):143–152

Damaschke N, Nobach H, Nonn TI, Semidetnov N, Tropea C (2005) Multi-dimensional particle sizing techniques. Experiments in fluids 39(2):336–350

Deprédurand V, Miron P, Labergue A, Wolff M, Castanet G, Lemoine F (2008) A temperature-sensitive tracer suitable for two-colour laser-induced fluorescence thermometry applied to evaporating fuel droplets. Meas Sci Technol 19(10):105403

Dodge LG, Rhodes DJ, Reitz RD (1987) Drop-size measurement techniques for sprays: comparison of malvern laser-diffraction and aerometrics phase/doppler. Appl Opt 26(11):2144–2154

Domann R, Hardalupas Y (2001) Spatial distribution of fluorescence intensity within large droplets and its dependence on dye concentration. Appl Opt 40(21):3586–3597

Driscoll KD, Sick V, Gray C (2003) Simultaneous air/fuel-phase piv measurements in a dense fuel spray. Exp Fluids 35(1):112–115

Elghobashi S (1991) Particle-laden turbulent flows: direct simulation and closure models. Appl Sci Res 48(3):301–314

Elghobashi S (1994) On predicting particle-laden turbulent flows. Appl Sci Res 52(4):309–329

Fansler TD, Parrish SE (2014) Spray measurement technology: a review. Meas Sci Technol 26(1):012002

Ferrand V, Bazile R, Borée J, Charnay G (2003) Gas-droplet turbulent velocity correlations and two-phase interaction in an axisymmetric jet laden with partly responsive droplets. Int J Multiph Flow 29(2):195–217

Garcia-Sucerquia J, Xu W, Jericho SK, Klages P, Jericho MH, Kreuzer HJ (2006) Digital in-line holographic microscopy. Appl Opt 45(5):836–850

Gerbrand Deen N, Solberg T, Helge Hjertager B (2000) Numerical simulation of the gas-liquid flow in a square cross-sectioned bubble column. In: Proceedings of 14th International congress of chemical and process engineering: CHISA (Praha, Czech Republic, 2000)

Gillandt I, Schulze T, Fritsching U, Bauckhage K (1998) Simultaneous measurement of continuous and dispersed phase in a two phase jet flow. Flow Meas Instrum 8(3):199–206

Gillandt I, Fritsching U, Bauckhage K (2001) Measurement of phase interaction in dispersed gas/particle two-phase flow. Int J Multiph Flow 27(8):1313–1332

Glover AR, Skippon SM, Boyle RD (1995) Interferometric laser imaging for droplet sizing: a method for droplet-size measurement in sparse spray systems. Appl Opt 34(36):8409–8421

Gounder JD, Kourmatzis A, Masri AR (2012) turbulent piloted dilute spray flames: flow fields and droplet dynamics. Combust Flame 159(11):3372–3397

Gui L (1998) Methodische Untersuchungen zur Auswertung von Aufnahmen der digitalen Particle image velocimetry, Shaker

Gui L, Lindken R, Merzkirch W (1997) Phase-separated piv measurements of the flow around systems of bubbles rising in water. ASME FEDSM 97:22–26

Gui L, Merzkirch W (1996) Phase-separation of piv measurements in two-phase flow by applying a digital mask technique. ERCOFTAC Bull 30:45–48

Gui L, Wereley ST, Kim YH (2003) Advances and applications of the digital mask technique in particle image velocimetry experiments. Meas Sci Technol 14(10):1820

Hardalupas Y, Horender S (2003) Fluctuations of particle concentration in a turbulent two-phase shear layer. Int J Multiph Flow 29(11):1645–1667

Hardalupas Y, Taylor AMKP, Zarogoulidis K (2004) Sizing spheroidal droplets by ilids (paper 66)

Hardalupas Y, Srikrishna S, Taylor AMKP, Zarogoulidis K (2010) Simultaneous planar measurement of droplet velocity and size with gas phase velocities in a spray by combined ilids and piv techniques. Exp Fluids 49(2):417–434

Hardalupas Y, Taylor AMKP, Zarogoulidis K (2014) Detection and evaluation of droplet and bubble fringe patterns in images of planar interferometric measurement techniques using the wavelet transform. In: International Conference on Optical Particle Characterization (OPC 2014). International Society for Optics and Photonics, pp 92320E–92320E

Hardalupas Y, Taylor AMKP, Whitelaw JH (1992) Particle dispersion in a vertical round sudden-expansion flow. Philos Trans R Soc Lon A Math Phys Eng Sci 341(1662):411–442

Hess CF and L'Esperance D (2009) Droplet imaging velocimeter and sizer: a two-dimensional technique to measure droplet size. Exp Fluids 47(1):171–182

Hillamo H, Kaario O, Larmi M (2008) Particle image velocimetry measurements of a diesel spray. Technical report, SAE Technical Paper

Hishida K, Sakakibara J (2000) Combined planar laser-induced fluorescence-particle image velocimetry technique for velocity and temperature fields. Exp Fluids 29:S129–S140

Hishida K, Hardalupas Y, Zarogoulidis K, Hayashi K, Taylor AMKP (2013) Characterization of a hollow-cone spray impinging on an obstacle using a single-camera double plane interferometric imaging technique. In 8th Intern. Conference of Multiphase Flows ICMF-2013

Horie S, Sugimoto D, Hishida K (2007) Measurements of droplet size and three components velocity by interferometric laser imaging techinique (a flow structure of a spray in a swirling flow). In: National Heat Transfer Symposium of Japan, 44:111. Japan Heat Transfer Society; 1999

Jeekuen L, Keiya N (2003) Simultaneous flow field measurement of di gasoline spray and entrained ambient air by lif-piv technique. Technical report, SAE technical paper

Jermy MC, Greenhalgh DA (2000) Planar dropsizing by elastic and fluorescence scattering in sprays too dense for phase doppler measurement. Appl Phys B Lasers and Opt 71(5):703–710

Jiang X, Siamas GA, Jagus K, Karayianni TG (2010). Physical modelling and advanced simulations of gas-liquid two-phase jet flows in atomization and sprays. Prog Energy Combust Sci 36(2):131–167

Khalitov DA, Longmire EK (2002) Simultaneous two-phase piv by two-parameter phase discrimination. Exp Fluids 32(2):252–268

Kiger KT, Pan C (2000) Piv technique for the simultaneous measurement of dilute two-phase flows. J Fluids Eng 122(4):811–818

Kiger KT, Pan C (2002) Suspension and turbulence modification effects of solid particulates on a horizontal turbulent channel flow. J Turbulence 3(19):1–17

König G, Anders K, Frohn A (1986) A new light-scattering technique to measure the diameter of periodically generated moving droplets. J Aerosol Sci 17(2):157–167

Kosiwczuk W, Cessou A, Trinite M, Lecordier B (2005) Simultaneous velocity field measurements in two-phase flows for turbulent mixing of sprays by means of two-phase piv. Exp Fluids 39(5):895–908

Kulick JD, Fessler JR, Eaton JK (1994) Particle response and turbulence modification in fully developed channel flow. J Fluid Mech 277:109–134

Kyle D (1991) Squires and John K Eaton. Preferential concentration of particles by turbulence. Phys Fluids A Fluid Dyn 3(5):1169–1178

Lacour C, Durox D, Ducruix S, Massot M (2011) Interaction of a polydisperse spray with vortices. Exp Fluids 51(2):295–311

Laurent F, Massot M (2001) Multi-fluid modelling of laminar polydisperse spray flames: origin, assumptions and comparison of sectional and sampling methods. Combust Theory Mod 5(4):537–572

Lavieille P, Lemoine F, Lavergne G, Lebouché M (2001) Evaporating and combusting droplet temperature measurements using two-color laser-induced fluorescence. Exp Fluids 31(1):45–55

Le Moyne L, Guibert P, Roy R, Jeanne B (2007) Fluorescent-piv spray/air interaction analysis of high-pressure gasoline injector. Technical report, SAE technical paper

Maeda M, Akasaka Y, Kawaguchi T (2002) Improvements of the interferometric technique for simultaneous measurement of droplet size and velocity vector field and its application to a transient spray. Exp Fluids 33(1):125–134

Maqua C, Depredurand V, Castanet G, Wolff M, Lemoine F (2007) Composition measurement of bicomponent droplets using laser-induced fluorescence of acetone. Exp Fluids 43(6):979–992

Maqua C, Castanet G, Lemoine F, Doué N, Lavergne G (2006) Temperature measurements of binary droplets using three-color laser-induced fluorescence. Exp Fluids 40(5):786

Masanobu M, Tatsuya K, Koichi H (2000) Novel interferometric measurement of size and velocity distributions of spherical particles in fluid flows. Meas Sci Technol 11(12):L13

Matsuura K, Zarogoulidis K, Hardalupas Y, Taylor AMKP, Kawaguchi T, Sugimoto D, Hishida K (2006) Simultaneous planar measurement of size and three-component velocity of droplets in an aero-engine airblast fuel spray by stereoscopic interferometric laser imaging technique. In: 10th International Conference on Liquid Atomization and Spray Systems, ICLASS 2006

McDonell VG, Adachi M, Samuelsen GS (1992) Structure of reacting and non-reacting swirling air-assisted sprays. Combust Sci Technol 82(1–6):225–248

McDonell V, Adachi M, Scott Samuelsen G (1993) Structure of reacting and nonreacting, non-swirling, air-assisted sprays, part ii: drop behavior. Atomization Sprays, 3(4)

McDonell V, Scott Samuelsen G (1993) Structure of vaporizing pressure atomized sprays. Atomization Sprays, 3(3)

Müller J, Kebbel V, Jüptner W (2004) Characterization of spatial particle distributions in a spray-forming process using digital holography. Meas Sci Technol 15(4):706

Nakabe K, Yukio M, Tomoyuki H, Hiroyuki F (1991) An experimental study on detailed flame structure of liquid fuel sprays with and without gaseous fuel. Combust Flame 84(1):3–14

Okamoto K, Nishio S, Saga T, Kobayashi T (2000) Standard images for particle-image velocimetry. Meas Sci Technol 11(6):685

Palero V, Lobera J, Arroyo MP (2005) Digital image plane holography (diph) for two-phase flow diagnostics in multiple planes. Exp Fluids 39(2):397–406

Palero V, Arroyo MP, Soria J (2007) Digital holography for micro-droplet diagnostics. Exp Fluids 43(2–3):185–195

Paschke M (2015) Recording & characterization of fuel spray of injectors for gasoline direct injection. Bachelor's thesis, Institut für Photogrammetrie - Universität Stuttgart, June 2015

Philip OG, Schmidl WD, Hassan YA (1994) Development of a high speed particle image velocimetry technique using fluorescent tracers to study steam bubble collapse. Nucl Eng Des 149(1–3):375–385

Poelma C (2004) Experiments in particle-laden turbulence. PhD thesis, Delft University of Technology

Poelma C, Westerweel J, Ooms G (2006) Turbulence statistics from optical whole-field measurements in particle-laden turbulence. Exp Fluids 40(3):347–363

Prevost F, Boree J, Nuglisch HJ, Georges C (1996) Measurements of fluid/particle correlated motion in the far field of an axisymmetric jet. Int J Multiph Flow 22(4):685–701

Raffel M, Willert CE, Wereley S, Kompenhans J (2013) Particle image velocimetry: a practical guide. Springer

Reveillon J, Vervisch L (2005) Analysis of weakly turbulent dilute-spray flames and spray combustion regimes. J Fluid Mech 537:317–347

Rhim D-R, Farrell PV (2000)Characteristics of air flow surrounding non-evaporating transient diesel sprays. Technical report, SAE Technical paper

Roland D, Yannis H (2001) A study of parameters that influence the accuracy of the planar droplet sizing (pds) technique. Part Part Syst ms Charact 18(1):3–11

Ronneberger O, Raffel M, Kompenhans J (1998) Advanced evaluation algorithms for standard and dual plane particle image velocimetry. In: Proceedings of the 9th international symposium on applied laser techniques to fluid mechanics, Lisbon, Portugal, paper, vol 10, pp 13–16

Rottenkolber G, Gindele J, Raposo J, Dullenkopf K, Hentschel W, Wittig S, Spicher U, Merzkirch W (2002) Spray analysis of a gasoline direct injector by means of two-phase piv. Exp Fluids 32(6):710–721

Saffman M, Buchhave P, Tanger H (1984) Simultaneous measurement of size, concentration and velocity of spherical particles by a laser doppler method. In: Proceeding of the 2nd international symposium on applications of laser anemometry to fluid mechanics, LADOAN Lisbon

Sahu S (2011) Experimental study of isothermal and evaporative sprays. PhD thesis, Imperial College London

Sahu S, Hardalupas Y, Taylor AMKP (2014) Droplet-turbulence interaction in a confined polydispersed spray: effect of droplet size and flow length scales on spatial droplet-gas velocity correlations. J Fluid Mech 741:98–138

Sahu S, Hardalupas Y, Taylor AMKP (2016) Droplet-turbulence interaction in a confined polydispersed spray: effect of turbulence on droplet dispersion. J Fluid Mech 794:267–309

Sahu S, Hardalupas Y, Taylor AMKP (2014) Simultaneous droplet and vapour-phase measurements in an evaporative spray by combined ilids and plif techniques. Exp Fluids 55(2):1673

Sakakibara J, Wicker RB, Eaton JK (1996) Measurements of the particle-fluid velocity correlation and the extra dissipation in a round jet. Int J Multiph Flow 22(5):863–881

Sanchez ML, Castro F, Tinaut FV, Melgar A (2000) Considerations on the gas-phase velocity field in a nonevaporating diesel spray. Atomization Sprays, 10(6)

Sugimoto D, Zarogoulidis K, Kawaguchi T, Matsuura K, Hardalupas Y, Taylor AMKP, Hishida K (2006) Extension of the compressed interferometric particle sizing technique for three component velocity measurements. In: 13th international symposium on applications of laser techniques to fluid mechanics Lisbon, Portugal, pp 26–29

Sundaram S, Collins LR (1999) A numerical study of the modulation of isotropic turbulence by suspended particles. Journal of Fluid Mechanics 379:105–143

Towers DP, Towers CE, Buckberry CH, Reeves M (1999) A colour piv system employing fluorescent particles for two-phase flow measurements. Meas Sci Technol 10(9):824

Tropea C (2011) Optical particle characterization in flows. Ann Rev Fluid Mech 43:399–426

Wang X, Huang Z, Kuti OA, Zhang W, Nishida K (2010) Experimental and analytical study on biodiesel and diesel spray characteristics under ultra-high injection pressure. Int J Heat Fluid Flow 31(4):659–666

Willert CE, Gharib M (1991) Digital particle image velocimetry. Exp Fluids 10(4):181–193

Zarogoulidis K, Hardalupas Y, Taylor AMKP (2012) Theoretical evaluation of droplet concentration limits for interferometric droplet sizing measurements, paper 378. Proceedings of the 16th Int Symp on Applications of Laser Techniques to Fluid Mechanics, Lisbon, Portugal, pp 9–12

Zhang M, Xu M, Hung DLS (2014) Simultaneous two-phase flow measurement of spray mixing process by means of high-speed two-color piv. Meas Sci Technol 25(9):095204

Zhao F-Q, Lai M-C, Harrington DL (1997) A review of mixture preparation and combustion control strategies for spark-ignited direct-injection gasoline engines. Technical report, SAE Technical Paper

Part IV
Turbulent Spray Combustion

Chapter 11
Turbulent Spray Combustion

Seong-Young Lee, Ahmed Abdul Moiz and Khanh D. Cung

Abstract Understanding turbulence is one of the most difficult topics in science and engineering. This is because turbulent spray combustion involves many areas of physics and chemistry which accompany a variety of mathematical challenges. Defining the various length and timescales existing in turbulent flow provides a better way to understand and characterize this chaotic phenomenon. However, the degree of complexity increases when there is a strong interaction between turbulence flow and chemistry. Here, characteristic times of chemical reaction in a molecular level (chemical) and fluid-mechanic level (physical) determine which of these are more dominant. This interaction remains as one of the most important and challenging aspects of turbulent reacting spray. In the present chapter, we begin with a general discussion on turbulence. The following section covers description of key features involved in a spray combustion scenario. Concepts involving higher fidelity in description of turbulent combustion are covered by discussion of interaction of turbulence and combustion. In most actual spray combustion applications, the combustion is dominantly non-premixed. There is a minor aspect of premixed combustion too which are discussed in this chapter. New advanced combustion modes such as partially premixed combustion (PPC) and multiple injections, topics with growing interests, are introduced and discussed later. Finally, numerically simulating these aspects is a key area of combustion research. It is of utmost important to optimize the combustion system using computer-based simulations to avoid higher cost for experimentally parametric study. Reynolds-averaged Navier–Stokes (RANS) models are mostly used in commercial sector for computationally tractable simulation time. Large-eddy simulation (LES) offers a higher fidelity approach. With the advent of higher computational resources, LES approaches are becoming more popular for obtaining solutions of turbulent combustion. Aspects of both RANS and LES relevant to spray combustion scenarios are discussed. Although usually requiring very high computational power, direct numerical simulation (DNS) can provide an actual representative of many chemical and physical

S.-Y. Lee (✉) · A. A. Moiz · K. D. Cung
Mechanical Engineering-Engineering Mechanics, Michigan Technological University, Houghton, MI, USA
e-mail: sylee@mtu.edu

S. Basu et al. (eds.), *Droplets and Sprays*, Energy, Environment, and Sustainability,
https://doi.org/10.1007/978-981-10-7449-3_11

aspects of spray combustion such as evaporation and auto-ignition, which are discussed at the end of this chapter.

11.1 Introduction

Turbulent combustion associated with liquid fuels occurs in many energy generation sectors including industrial power systems, ground, and aviation transportation sectors. Examples are liquid fuel boilers, internal combustion engines, both aviation and land-based turbine engines, and liquid-propelled rockets. In those energy devices, liquid fuels are typically delivered into the system by means of a highly turbulent spray formation, which would undergo complex processes of the heat and mass transfers, phase changes and eventually mix with the surrounding high-temperature oxidizer and subsequently yield chemical reactions. Moreover, those processes are strongly coupled each other, not easily isolated from the individuals. This chapter, turbulent spray combustion, will focus the attention on high-injection pressure liquid fuel spray emanating from submicron nozzles into a pressurized quasi-state and/or pre-turbulence chamber. The important processes include the liquid injection core, atomization through various instabilities and generation of smaller droplets detached from the intact liquid core via primary and secondary breakups, vaporization and mixing with the surrounding gas. The primary breakup results from the shear force between the gas and liquid phases, called Kelvin–Helmholtz (KH) instability and subsequently, secondary breakup process separates the light fluid from the dense, heavy fluid through Rayleigh–Taylor (RT) instability. These processes occur very quickly with turbulence and are defined as "spray formation region" (primary breakup region near the nozzle exit) and further downstream "dilute spray region" where a collection of drops in air is formed. Atomization process is significantly enhanced by the high-injection pressure through interaction with turbulence and the homogeneous air–fuel mixing further yields conditions for chemical heat releases. Therefore, essentials of turbulence spray combustion are the coupling of a collection of droplet evaporation where their flow fields are dominant in turbulence, leading to chemical energy release when optimized with proper air–fuel ratio for combustion.

A background of single-injection diesel sprays is provided here in understanding of some of the spray and spray combustion concepts. Spray injection has a tremendous influence on the vaporization of injected fuel, which impacts the thermochemical interaction with the surrounding oxidizer, and eventually affects combustion process. Knowledge of non-reacting sprays is necessary to investigate the near-nozzle flow (which are mostly not reacting) and also the mixture formation until the ignition event. For example, the quality of fuel delivery is affected by the nozzle hesitation mainly caused by a delay in beginning transverse motion of the needle. The other factor affecting spray characteristics is the injector type and more specifically the injector geometry. Another factor closely related to the injector geometry is the cavitation. Cavitation occurs by a different mechanism in non-sac and sac type injector nozzles; with the former observing a "conventional" cavitation and the latter a "vortex"-type cavitation (Afzal et al. 1999; Bae and Kang 2006). In addition, there are effects from the fuel pressure, fuel temperature, fuel properties, ambient pressure and density, ambient temperature which greatly have effects on the spray progression. With higher injection pressures, smaller droplets are created, liquid momentum is increased and this enables the liquid droplets to have an intense interaction with the ambient air. This interaction occurs throughout the lateral and front periphery of the spray plume which increases the air entrainment. Fuel temperature affects the spray breakup behavior by dominantly causing cavitation (Zeng et al. 2012). Increasing fuel temperatures causes more cavitation, which helps in spray breakup, faster liquid atomization and increased air entrainment.

In typical high-pressure spray combustion, i.e., auto-ignited flame, combustion takes place with the combination of partially premixed combustion and mixing controlled diffusion-like combustion in which the degree of these individual combustion modes is dependent on the degree of the air–fuel mixing rate. Better mixing leads to higher premixed combustion mode. The typical pattern of heat release per unit time is shown in Fig. 11.1 (left) when the injected spray burns in a diesel engine combustion scenario, while the well-known corresponding turbulent spray combustion structure (Fig. 11.1 (right)) with emissions formation is also seen (Dec 1997). After the fuel gets injected, it starts vaporizing in the hot ambient. After a duration of an "ignition time delay," auto-ignition of the vaporized fuel takes place. Depending on the ambient conditions, this ignition might be two-staged: first event is a low-temperature heat release event (cool flame) called first-stage ignition and the second event is the high-temperature heat release event called second-stage ignition. During the high-temperature ignition event, there is a premixed combustion peak where bulk combustion of the vaporized fuel–air mixture takes place. The late the ignition event, the more the time for the fuel vapor to mix with the ambient oxidizer which results in a higher premix combustion peak. After the premix combustion phase, as the fuel is being injected, there is mixing or injection-rate controlled combustion phase. This phase has a positive slope as long as the fuel injection is active. This phase is strictly controlled by (spray) mixing which is induced by the injection event. As soon as the injection ends, fuel stops entering the chamber and the heat release drops rapidly while it continues burning the rest of the fuel in the chamber.

A specific validation of the CFD model is needed by considering accurate predictability of important outcomes such as ignition time of fuel–air mixture, flame stabilization location, heat release from the combustion event, intermediate/stable species level, and spread across the spray flame domain. Reactive spray simulations in IC engines are of high Reynolds numbers, i.e., turbulent intensities. These simulations are performed using RANS or LES methods, whereas the turbulence equations are solved for average or filtered quantities, respectively, where in most cases, the timescales of the fluid flow are comparable to the timescales of the chemical rate of reactions (Peters 2000). Thus, chemistry and flow (turbulence) cannot be separated since they actually occur simultaneously on similar timescales. This is a concept of a *turbulence–chemistry interaction* in reacting flow simulations.

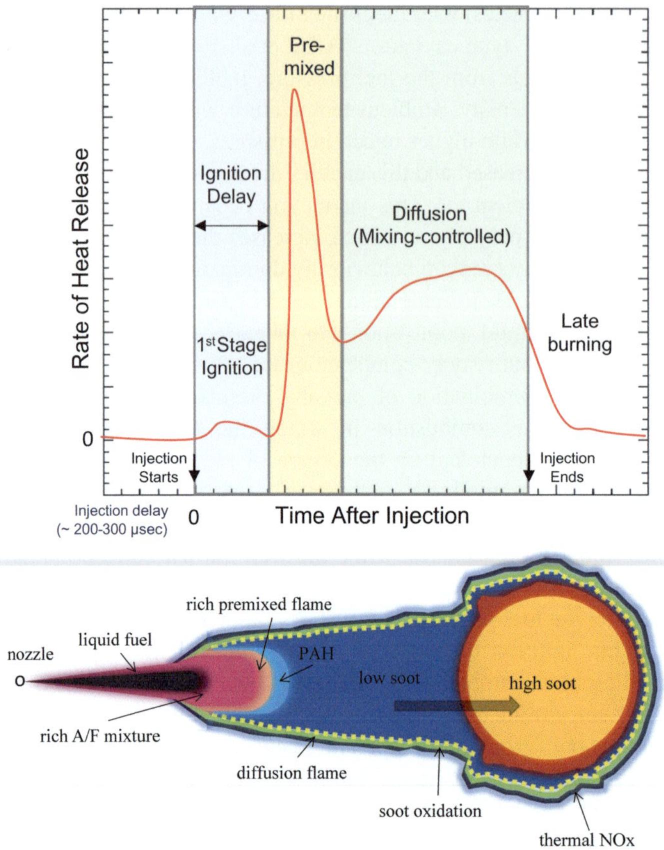

Fig. 11.1 Illustration of heat release rate (left) and diesel spray flame structure (right) (figure on right adapted from Dec 1997)

Typically, a closure for the chemical source term is one of the major challenges in the interaction forms due to the nonlinearity of chemistry (obeying Arrhenius-type exponential expression). Overall, a TCI closure necessity has not been established yet, but utilization of non-TCI models has given inferior results in context of RANS. Using high-fidelity LES models has been found to bridge the gap since the sub-grid fluctuations have been well treated in LES models (Pei et al. 2015b, c). In a non-TCI model, the mean chemical source term is calculated using the mean temperature and composition with no consideration of turbulent fluctuations. In spite of the physically inadequate assumption in a non-TCI model, various studies have successfully applied them for diesel engine conditions (Han et al. 1996; Moiz et al. 2015; Senecal et al. 2014; Singh et al. 2006; Som et al. 2012; Vishwanathan and Reitz 2009; Wang et al. 2013).

In contrast to a simplistic non-TCI model, the TCI family of combustion models has been applied to diesel engine conditions from the early 1990s, especially after the introduction of flamelet-based models. Some of the TCI models in present use include flamelet models (Pitsch et al. 1996), transported probability density function model (Pope 1985), conditional moment closure model (Klimenko and Bilger 1999), progress variable model (Pierce and Moin 2004), flamelet-generated manifold model (Oijen and Goey 2000), flame surface density-based models (including ECFM3Z) (Tap and Veynante 2005; Colin et al. 2003), and partially stirred reactor model (Golovitchev et al. 2000). Two of the popular and fairly successful TCI models are flamelet model and transported PDF model, which have been briefly explained further. The most used TCI model was the one of RANS-based laminar flamelet model developed by Peters (Peters 1984) as steady laminar flamelet model. This was extended to use a LES turbulence modeling approach (Pitsch and Steiner 2000). In spite of this, the inherent deficiency in the flamelet models has been that the cross-scalar dissipation rate does not have an accurate closure with the simplified expressions being employed. The transported PDF method (Pope 1985) has been proved to be a pretty accurate solution methodology. A good review of this has been presented in references (Haworth 2010; Pope 2001). This approach solves the transport equation for the one-time one-point joint PDF of velocity and composition (in some cases composition only). In this approach, no closure problems arise. This is due to nonlinear terms resulting from averaging of one-point quantities pertaining to chemistry. But this is also the reason why application of this method is very time-consuming. Nonetheless, this model has been successfully applied for diesel spray combustion cases (Pei et al. 2013, 2015a; Bhattacharjee and Haworth 2013).

This chapter will firstly discuss optical visualization of turbulent spray flame taken in the constant volume combustion vessel. The characteristics of combustion vessel will be present along with key optical diagnostics for flame visualization including high-speed shadow/Schlieren method, Mie scattering, and planar laser-induced fluorescence of the key intermediate species. Secondly, non-premixed combustion mode in the reacting flow will be discussed with various ambient conditions including injection pressure, fuel, ambient density, and temperatures. Results of the key spray properties from both single injection and multiple injections will be discussed along with the newly developed imaging analysis where main focus is the discussion of low-temperature combustion influenced by turbulence. Thirdly, partially premixed combustion will be discussed, particularly focusing on the flame ignition where high turbulence enhances air–fuel mixing to promote auto-ignition. Fourthly, various numerical simulations conducted in the spray flame will be introduced for feasibility including the RANS, LES, and DNS models.

The sequence of events from fuel injection from a nozzle to its combustion involves various sub-events which are interrelated in some way or the other. The conceptual image in Fig. 11.2 shows fuel being injected from a nozzle after undergoing effects of cavitation. In diesel sprays, the injection velocities can be as high as 150 m/s (corresponding to a nozzle with diameter of 100 μm at 1500 bar injection pressure). Such high velocities when injected into a relatively quieter bulk flow are the main source of turbulence.

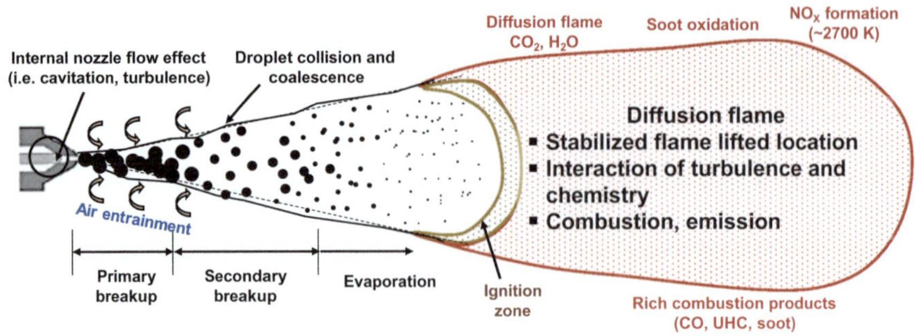

Fig. 11.2 A schematic of spray combustion formation through multiple steps

This is termed "spray-induced turbulence," and spray is one of the major sources of turbulence in conventional diesel injections affecting the near-nozzle thermos-fluidic events. Spray-induced turbulence also causes entrainment of the surrounding oxidizer for the fuel to use and combust. Atomization of the liquid fuel into tiny droplets with events like primary and secondary breakup is also a result of high-velocity fuel injection and its interaction with the relatively quieter ambient bulk flow. The tinier the droplets, the more the net surface area to volume ratio of the fuel injected which in-turn increases its vaporization characteristics. Thus, spray-induced turbulence has an impact on vaporization characteristics. Together with the entrained air, the vaporized fuel mixes and starts to undergo low-temperature chemistry at slightly richer fuel–air equivalence ratios. This paves the way for high-temperature ignition event which is usually termed as an "auto-ignition event." As the fuel is being injected steadily from the nozzle, a flame gets formed and in conventional diesel spray combustion scenarios are often stabilized at a location termed as "liftoff length." Upstream of this liftoff length is the high-temperature flame which burns to produce, heat energy to be harnessed in power generation devices together with emission which needs to tackled and lowered. The region around flame stabilization and around high-temperature flame itself is turbulent and is chemically very active. Often, thus turbulence and chemistry effects are mutually linked to be termed as "turbulence–chemistry interaction" which was explained earlier in this section.

11.2 Turbulent Combustion Characteristics

11.2.1 Optical Diagnostics and Analysis

Visualization of spray and combustion can be obtained from optical engines (Zhao and Ladommatos 2001), rapid compression machine (Donovan et al. 2004), but the application for the technique is usually complicated due to uncertainty and limited in pressure and temperature. Therefore, combustion chambers (Siebers 1998; Nesbitt et al. 2011; Kim and Ghandhi 2001; Baert et al. 2009) have been designed to better understand the fundamental of combustion at relevant engine conditions. These chambers, usually with constant volume, provide a well-controlled environment of temperature, pressure, and mixture composition. For example, as shown in Fig. 11.3, the combustion vessel at Michigan Tech (Nesbitt et al. 2011) can withstand maximum temperature of 1400 K and maximum pressure 350 bar. The vessel was equipped with multiple highly configurable windows for optical accessibility. Typically, pre-burn mixture is used to achieve high-temperature and high-pressure environment prior to injection. As the products from pre-burn combustion cool down, fuel is injected at defined pressure (measured by pressure transducer) and reacts with the remaining oxygen in the pre-burn products. If there is no oxygen, liquid fuel is only vaporized without any chemical reaction, hence is applicable for experiment of characterizing liquid and spray profile. In an oxidizer available condition, fuel is vaporized and reacted with oxygen until auto-ignition. Pressure rise in combustion chamber is recorded and sometimes used to determine ignition delay and overall combustion process. Several references (Cung et al. 2016a, c; Moiz et al. 2015, 2016a, b; Cung 2015; Zhang et al. 2013; Bhagat et al. 2013; Moiz 2016) are recommended for more related work using combustion vessel system.

Spray (liquid/vapor) visualization

Fuel liquid and vapor are shown in example images captured from a high-speed camera in Fig. 11.4. Liquid fuel can be seen from Mie scattering measurement

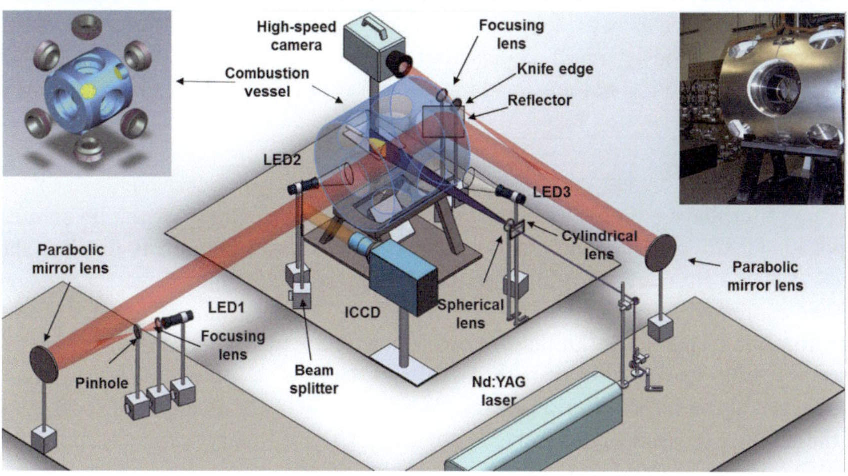

Fig. 11.3 Combustion vessel and example of some optical arrangements (Moiz et al. 2016b) to study spray and combustion phenomenon including shadow/Schlieren, Mie scattering, chemiluminescence, and laser application. High-speed camera can also be used to capture flame luminosity directly

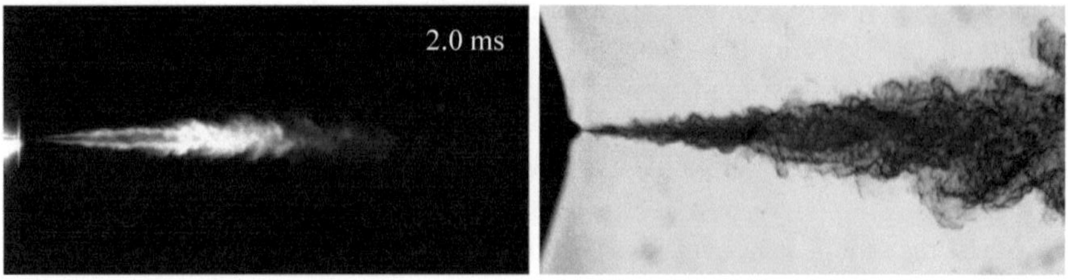

Fig. 11.4 Near-simultaneous visualization of liquid (left) and vapor (right) using corresponding Mie/Schlieren (hybrid) measurement (Cung 2015)

based on Mie theory (Mie 1908) of scattering phenomenon of the interaction between light and transparent particles such as fuel droplets. The light scattering from the subject is then captured by high-speed camera. Using this method, the entire liquid profile (spatially and temporally) can be examined and further, imaging process can extract information of spray such as liquid penetration and spray cone angle. The principle of Schlieren is about capturing the diffraction of light through an inhomogeneous medium. The technique is able to distinguish phases of fuel spray: liquid and vapor due to its density gradient (Settles 2012). The density gradient is usually referred to as refractive index due to their directly linear relationship. Schlieren technique has been used widely to detect the gradients in refractive index of transparent medium (Lillo et al. 2012; Pickett et al. 2009, 2011). The modified Z-shape Schlieren setup was previously demonstrated in Fig. 11.3. Moreover, the setup was configured further to capture both liquid and vapor for the same single spray event. This hybrid imaging system was performed as a combination of both Mie scattering and Schlieren along the same line of sight of the Schlieren setup. This was achieved through specific operation of the two LED lights. A continuously on/off sequence of the two LEDs was set to provide light source for Mie and Schlieren imaging that are captured by the camera frame by frame.

Natural luminosity

Effect of turbulence on combustion can also be reflected partially from the emitted soot luminosity of the reacting spray. This is because soot luminosity displays the overall flame structure, therefore can provide spatial and temporal characterization of the entire spray and combustion processes from mixing to auto-ignition, to flame stabilization. The flame luminosity signal can be captured through high-speed imaging with good quality of both spatial and temporal resolution.

To retain the information from images such as intensity as a function of spray axial distance and time, an approach known as IXT plot was developed (Cung et al. 2015a). IXT contains temporal and spatial details and can be summarized as the following equation

$$IXT(x,t) = \int_{-R}^{R} I(x,r,t)dr$$

where “x” is defined as the spray axial direction, “r” is radial direction, and “R” is the intensity boundary width. An example of IXT plot is shown in Fig. 11.5.

Due to the original line of sight, the normal direction to the radial direction “r” in image is often neglected. Therefore, IXT(x, t) is usually referred to illustrate the evolution of the combustion. A pre-defined threshold was selected to clean up background noise while not affecting the interpretation of flame luminosity signal. The usefulness of IXT plot is its capability in capturing many different spray and combustion characteristics from start of injection to the end of combustion recession into a single 2-D plot. The spray combustion progresses including start of injection, ignition delay detection, flame liftoff length, flame height, and high-intensity region are shown in Fig. 11.5.

Liquid fuel was detected after approximately 1.5 ms after start of injection (ASOI) hydraulic delay (physical delay from the electronically trigger of the injector). About 0.75 ms later, flame signal becomes visible and is shown in the IXT plot with increasing flame height. Flame upstream region remained relatively stable at a certain length which is also known as liftoff length. Note that liftoff length is actually defined from OH* chemiluminescence which is a different optical technique that uses intensified camera to capture excited radical OH*. OH* imaging reflects better the flame structure than combustion luminosity because many studies have confirmed that OH* formation is an evidence of high-temperature reaction.

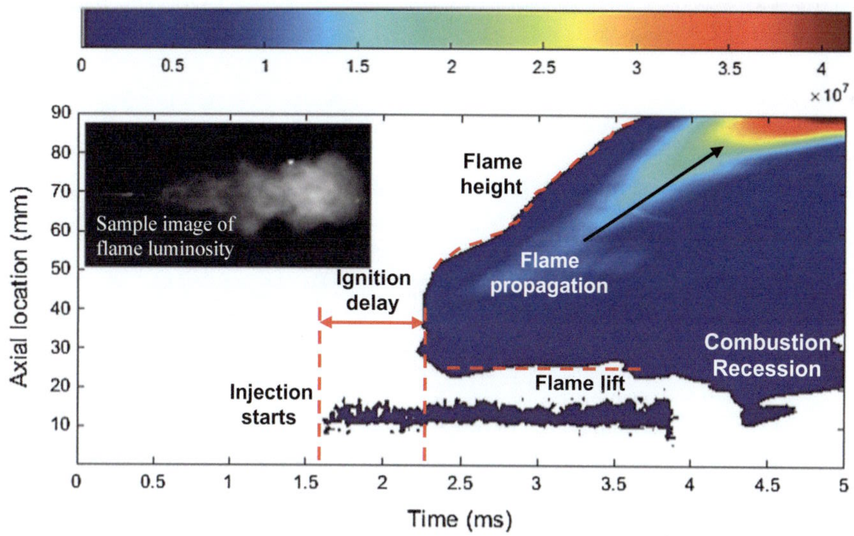

Fig. 11.5 Flame intensity plot as function of axial distance and time using IXT analysis (Cung et al. 2015a; Moiz et al. 2016a)

Unfortunately, OH* signal is usually weak and can only be captured in UV length range, particularly 310 nm cutoff (Pickett et al. 2005; Higgins and Siebers 2001). More discussion of OH* imaging is described at later section.

One of the important observations from IXT is the gap difference (called "mixing gap") between liquid signal and flame liftoff length. This region of about 8 mm as shown in Fig. 11.6 is where vaporized fuel and air are mixed with each other to eventually reach ignitable conditions of temperature and composition. The boundary of flame region from IXT plot was shown in different dash lines for different injection pressure conditions which are shown in Fig. 11.6. As injection pressure increases, liquid length remains almost constant, while mixing gap expands slightly. At this case, higher injection pressure also shortened the injection delay as shown in the vertical line for start of injection. This is basically related to the hydraulic operation of the injection system that required less time for liquid fuel ejected the injector nozzle at higher fuel injection pressure.

Flame liftoff length

Using band-pass filter (310 nm centered, 10 nm FWHW) and UV lenses f/4.5, high-speed imaging for OH* chemiluminescence can be captured by an intensified CCD camera with appropriate exposure time for visualization of stabilized lifted flame. The exposure time and the time of record are important to obtain an "averaged" image during the period of stable injection, hence a steady flame liftoff length.

The image post-analysis basically includes filtering intensity to remove background noise, followed by finding the distances between injector tip and first axial location of upper and lower maximum intensity profile. The averaged intensity profile was analyzed with certain threshold value to obtain LOL. The method is similar to ECN standard methodology for OH* chemiluminescence image

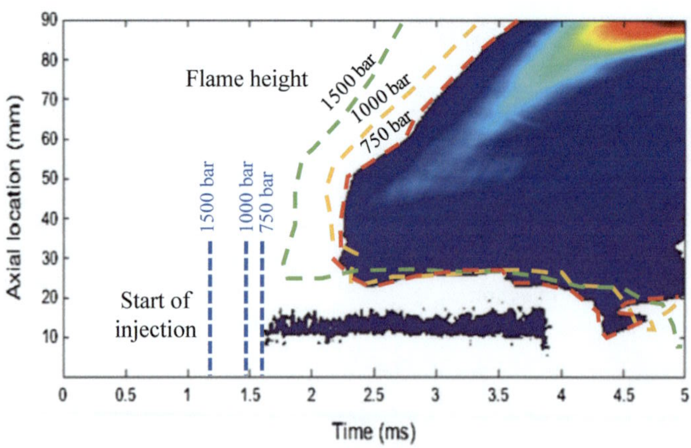

Fig. 11.6 Effect of injection pressure on flame structure as shown in spatial and temporal flame luminosity signal of IXT plot (Cung 2015)

processing. It was found that with threshold of 10% peak intensity, LOL can be specified reasonably for different conditions of ambient temperature, pressure, and ambient density. Similar selection of 8% peak intensity was also used for grade #2 diesel fuel. The sensitivity of intensity threshold has minimum effect on LOL determination with threshold of 20% peak intensity resulting in less than 2.5% increase in LOL. A sample imaging process is shown in Fig. 11.7 with LOL located at the "knee" in intensity profile.

Among different parameters that can affect liftoff length, injection pressure is perhaps the most direct factor with almost linear relationship with flame liftoff length as described by experimental-based empirical equation in reference (Siebers et al. 2002; Pickett et al. 2005). Higher injection pressure is responsible for high spray momentum which pushes the fuel further downstream with enhanced fuel–air mixing. The latter outcome from high-injection pressure can lead to either earlier ignition with improved mixing or retarded ignition if overly mixed. Depending on the competition between physical and chemical timescales, ignition event can occur at a specific time and axial location. In this case, spray-induced turbulence dominated by higher injection pressure resulted in longer liftoff length and faster increase in flame height as shown in Fig. 11.8.

Laser Application*: *PLIF–CH_2O

While further downstream of liftoff length is characterized by a significant heat release, high-temperature product, and soot growth (Dec 1997; Higgins and Siebers 2001), spray upstream prior to LOL region is known to responsible for cool flame reaction. Several studies have shown that formaldehyde (CH_2O) is also formed

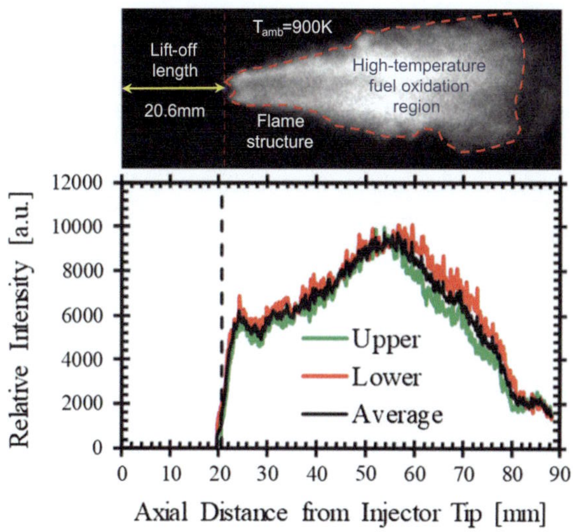

Fig. 11.7 Chemiluminescence OH* (310 ± 10 nm) image and sample image processing analysis for flame liftoff length determination (Cung et al. 2015b)

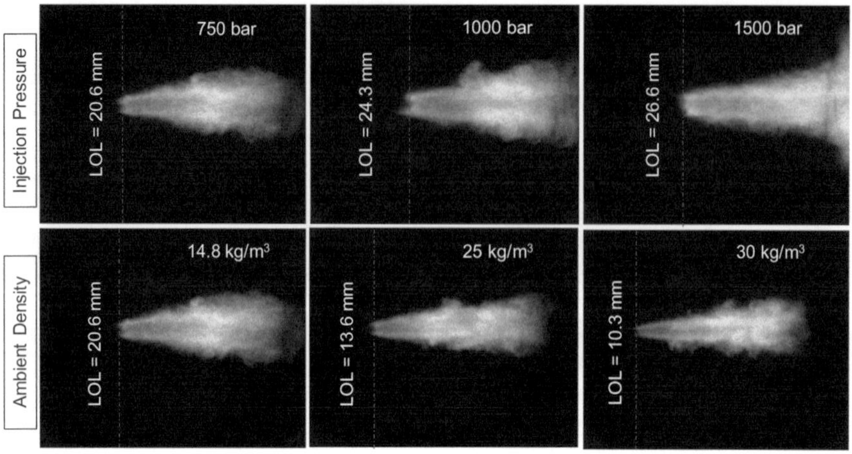

Fig. 11.8 Flame structure and liftoff length from OH* chemiluminescence for different injection pressures (top) and ambient densities (bottom). Images were taken with 1.0 ms exposure time (Cung 2015)

upstream during the initial premixed combustion period (Dec and Espey 1998; Bruneaux 2008). Visualization of formaldehyde has been assisted by laser application of planar laser-induced fluorescence at 355 nm excitation captured by ICCD camera with the detection wavelength range (optical filter) of 380–450 nm (Brackmann et al. 2003; Särner et al. 2005; Idicheria and Pickett 2006; Haessler et al. 2012; Skeen et al. 2015b).

The development of formation and consumption of CH_2O is shown in Fig. 11.9 for sequential after start of injection (ASOI) ASOI times. The first image (260 μs) shows the initial formation of CH_2O near-nozzle extending to X/D = 150. CH_2O continue to penetrate further into hot ambient up to approximately X/D = 330 according to image at 985 μs ASOI. High intensity of CH_2O occurs between times 665–865 μs ASOI. The averaged ignition delay from repeated runs (baseline condition) calculated based on pressure trace was 1.04 ms (with increase pressure threshold of 0.025 bar used in ECN network in reference (Engine Combustion Network). It can be seen here that after 985 μs ASOI, CH_2O region starts to decrease and eventually comes to near-full depletion after 2985 μs.

11.2.2 *Reacting Spray: Fundamental and Characteristics*

There are many aspects of turbulence which effect spray combustion processes. Some of them are discussed in this section. The discussion uses experimental and CFD techniques to examine aspects of (single injection) spray combustion in conventional and low-temperature combustion environments and moving ahead to

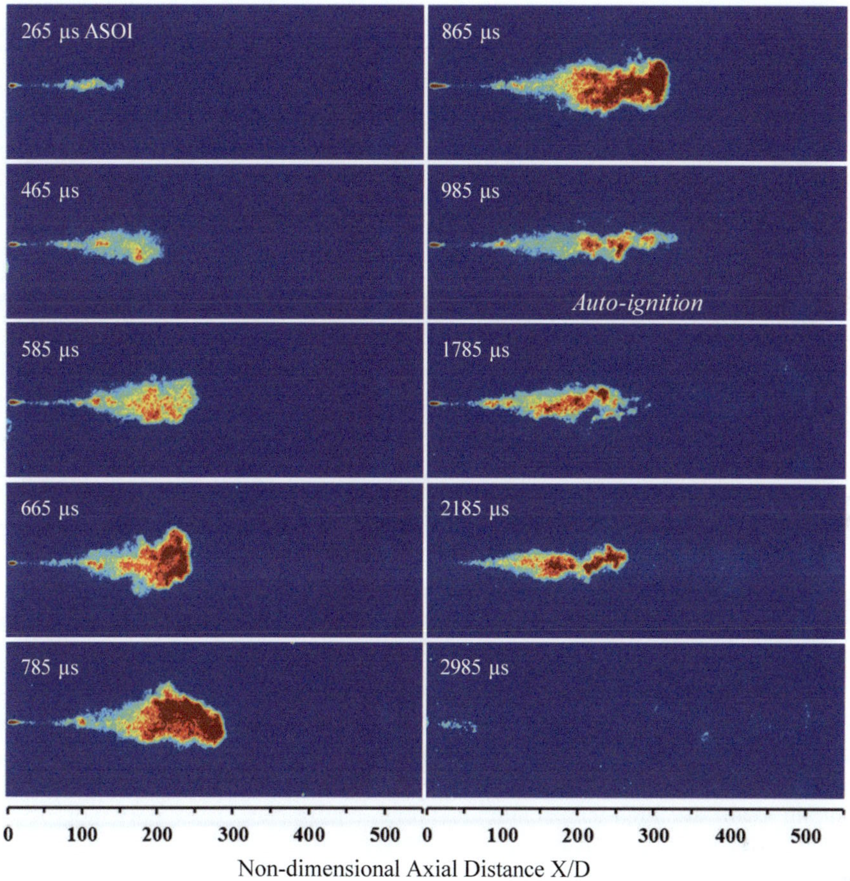

Fig. 11.9 Evolution of CH_2O formation and consumption from false-color PLIF imaging. The red color indicates high intensity of CH_2O (Cung et al. 2016c)

discuss how turbulence effects multiple injection strategies and partially premixed combustion scenarios.

Single injection

With higher injection pressures, smaller droplets are created, liquid momentum is increased and this enables the liquid droplets to have an intense interaction with the ambient air. This interaction occurs throughout the lateral and front periphery of the spray plume which increases the air entrainment. With an increase in injection pressure as well as the chamber pressures, a reduction of breakup time and breakup length is observed (Karimi 2007). In addition to that, an increase in spray-induced turbulence via increasing injection pressure decreases the soot (luminosity) which corresponds to reduced soot emissions. Reducing the nozzle diameter also has a similar effect since it results in smaller droplets too (Wang et al. 2011).

The effect of ambient pressure (/density) and temperatures is correlated and often oppose each other. With an increase in ambient density, an increase in plume angle and reduction in liquid length are observed (Zeng et al. 2012; Hiroyasu and Arai 1990; Zhu et al. 2013). On other hand, observations from Hiroyasu and Arai (1990) claim that hot ambient air can cause higher fuel vaporization at the spray peripheries which results in lower spray angles. Siebers (1999) studied the dependence of liquid length on the ambient conditions using injectors with orifice sizes from 100 to 500 μm and with various fuels.

Figure 11.10 shows the impact of the ambient temperature and density on the liquid length parameter which effects piston wall wetting. The lower gray region represents the range of liquid lengths expected for warm running engine conditions. The upper gray region represents the range of liquid lengths expected for cold start engine conditions. The cross-hatched horizontal band across the middle of the Fig. 11.10 represents typical maximum distances from the injector to the piston bowl wall. The light gray curves are liquid lengths for lines of constant density which are included in reference (Siebers 1999).

Most of the above concepts are related to conventional combustion of single injected diesel spray. There are a few deviations when the spray combustion occurs in a low-temperature combustion environment. Figure 11.11 shows the combustion event in an equivalence ratio and temperature space, outlining the differences in a conventional event and other technologies like LTC, PPCI, and HCCI.

In the equivalence ratio,temperature plot of Fig. 11.11, soot and NOx are plotted. There is little room for a combustion regime to place itself in a spot which is free from both soot and NOx. In reality, some of the space in the plot cannot be actually inhabited by any spray combustion event, e.g., the top right portion containing both the high-temperature and high-equivalence ratio does not occur simultaneously. By increasing the injection pressure or spray-induced turbulence, a lowering of the local equivalence ratio is observed which reduces the soot output from spray combustion. But in the case of rich combustion of the formed fuel–air

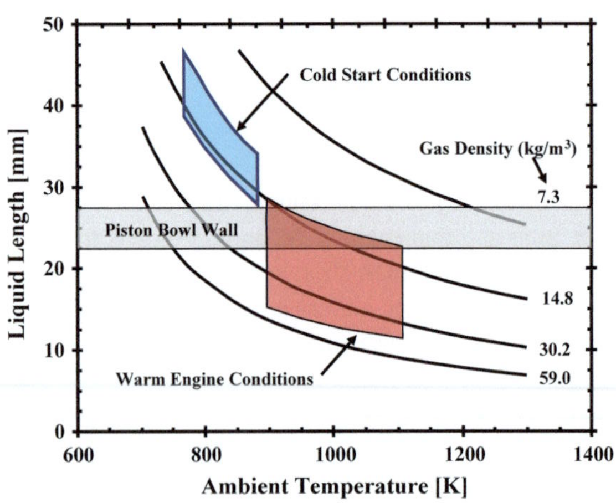

Fig. 11.10 Liquid lengths expected in the heavy-duty diesel engine (figure adapted from Siebers 1999)

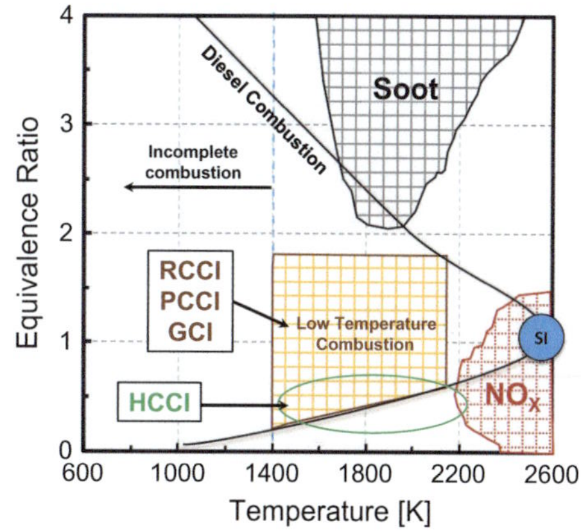

Fig. 11.11 New strategies for diesel engine emission control (figure concept adapted from Dempsey et al. 2016; Kitmura et al. 2002)

mixture (due to the ambient conditions), there is little room to escape from the soot peninsula in conventional diesel combustion. This is not the case with low-temperature combustion where the ambient in which the combustion occurs is different than conventional combustion. It can be seen from Fig. 11.11 that the combustion zone has similar shape as in conventional diesel combustion but moves down toward fuel-lean mixtures and moves left toward lower temperatures.

Low-temperature combustion

The conventional diesel spray combustion model shown previously in Fig. 11.1 has been modified in Fig. 11.12 by Musculus et al. (2013a) to account for the cool flame and accommodate low-temperature combustion concept. The original conceptual model was only for the quasi-steady spray combustion period and so did not describe the combustion process after end of injection.

The diesel spray takes similar characteristics in the initial few time instants for both the conventional and LTC conditions. Since LTC conditions involve lower ambient pressure and temperature, the liquid length is usually ~10 mm longer for LTC than conventional diesel case. However, this may not be true for late-injection LTC where fuel injection takes place near TDC in an engine. Soon after vaporization of the liquid fuel takes place, first-stage ignition happens in the conventional diesel case. Due to the nature of faster ignition near the liquid zone, there is high amount of soot and soot precursors forming during the ignition times and continue to form in the quasi-steady spray combustion periods. The degree of air entrained from beneath the liftoff is critical in conventional diesel spray combustion.

This causes enough premixing to cause the fuel-air mixture to burn lean downstream. Unlike conventional diesel combustion, in low-temperature combustion, high-temperature ignition happens after a very long time due to the nature of the ambient conditions employed in LTC. The first-stage ignition starts late in LTC case and prevails for time until after injection. The temperature arising from the low-temperature heat release reduces the liquid length of the LTC spray which was ~10 mm longer the conventional until the ignition event. A gradual blending

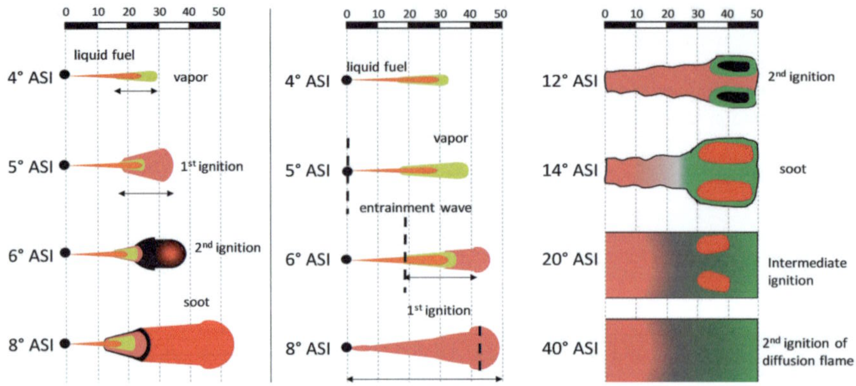

Fig. 11.12 Conceptual combustion model for conventional diesel (left) and partially premixed low-temperature diesel (right) (figure adapted from Musculus et al. 2013b)

of the air–fuel mixing field is observed with wave-like patterns. Second-stage ignition occurs later in the combustion cycle and causes low amount of soot and soot precursors further downstream (in some fuel-rich pockets) which get oxidized later by mixing with the ambient. Later in the combustion cycles, there might be dribble of liquid droplets from the nozzles or sac volumes. This causes locally fuel-rich spots and results in UHC and CO emissions.

Multiple injection

A background of single-injection diesel sprays is already provided earlier for ease in understanding of some of the spray and spray combustion concepts. Going toward split injections, there would be added complexity layered over that of single-injection concepts since there will be consequent injection of fuel after the first injection event within some short interval of time. Limiting ourselves to double-injection case, the second injection has different behaviors than the first injection. Discussion coming further injections will help understanding the differences.

Effect of increased spray turbulence via increased injection pressure: A higher injection pressure increases the local turbulence in the combustion zone of the first injection and that increases its local temperature as it burns. When the second injection enters this burning zone, the overall combustion event has a higher efficiency. The effect of increased turbulence via mechanism of increased pressure is shown in this section. Figure 11.13 shows the set simultaneous images at two injection pressures (1200 and 1500 bar) at 750 K ambient injection. The test fuel is n-dodecane at ambient density of 22.8 kg/m^3 with equal times of first, second injection duration and dwell (0.5/0.5/0.5 ms) (ref, Abdul thesis or publication).

For the 1200 bar injection pressure shown in Fig. 11.13 (left), no cool flame is observed in the PLIF images until approximately 1.8 ms ASOI. At the last image in the sequence, there is significant formaldehyde formation as well as flame

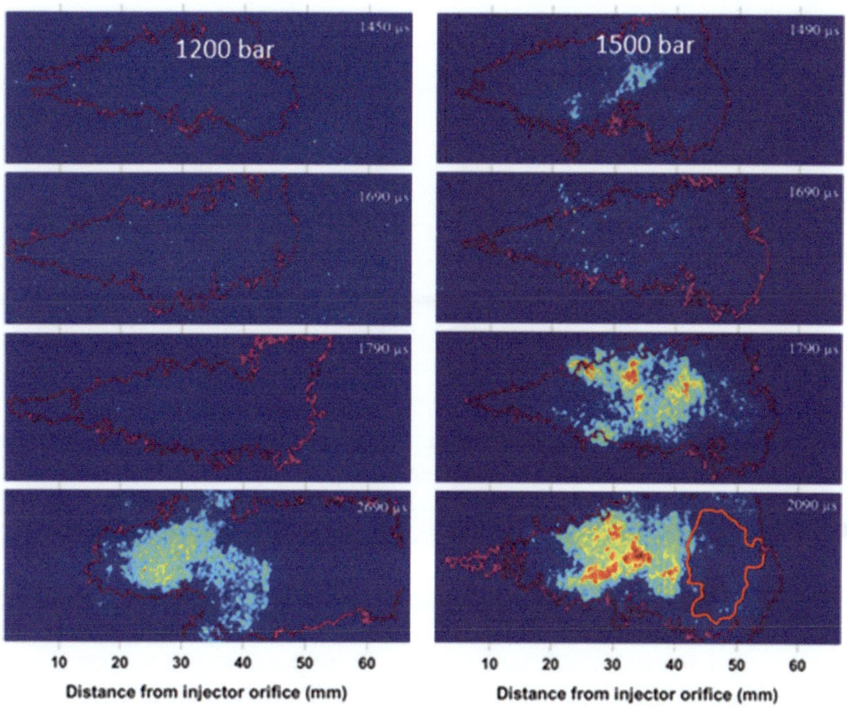

Fig. 11.13 Simultaneous Schlieren-PLIF imaging of the 0.5/0.5/0.5 ms injection spray combustion event at 1200 bar (left) and 1500 bar (right) injection pressure for a 750 K ambient. The RGB intensity denotes formaldehyde presence and the dotted spray boundaries are the schlieren boundaries denoting spray–vapor extremities (Moiz et al. 2016b)

propagation, as denoted by the Schlieren boundaries at further downstream locations. Ignition has long occurred at 2690 μs, and the fuel injected is left mostly unreacted upstream which will result in the presence of unburned hydrocarbons and hence low combustion efficiencies. For the same condition of 750 K ambient, at a higher injection pressure of 1500 bar, formaldehyde signal was observed at a relatively lesser ASOI time as evident from Fig. 11.13 (right). At 1790 μs, there is a strong formaldehyde signal, which was not observed in the case with 1200 bar pressure, Fig. 11.13 (left) at the same thermodynamic ambient conditions. Also, a high-intensity emission was observed at around ~2090 μs for the 1500 bar injection case, whereas in the 1200 bar injection case, it was not observed all throughout the combustion process implying the increased turbulence of the high-injection pressure helps in better combustion in LTC conditions for a split injection event.

A comparison of the mean heat release rate profiles of the 0.5/0.5/0.5 ms injection sequence at 1200 and 1500 bar injection pressures reveals interesting trends about the characteristics of combustion and its dependence upon mixing-chemistry relations at relatively low-temperature condition of 750 K depending on ambient turbulent conditions. It can be seen in Fig. 11.14 that for the

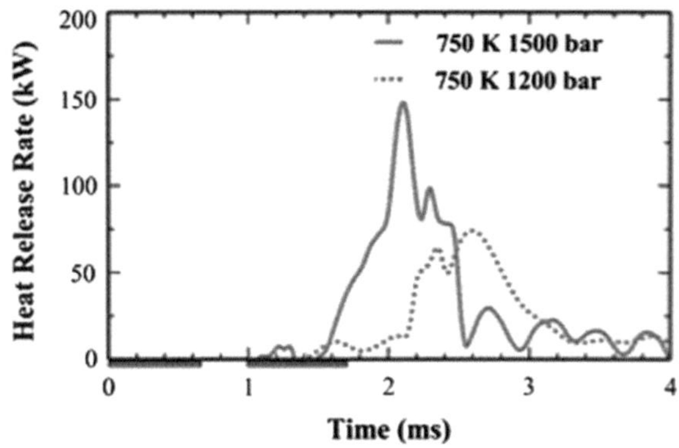

Fig. 11.14 Heat release rate comparison of the 1200 and 1500 bar injection pressure cases of the 0.5/0.5/0.5 ms injection event at the 750 K ambient temperature case. The actual injection durations highlighted in the time axis (Moiz et al. 2016b)

750 K case, the ignition delay is ~0.8 ms later in the case of 1200 bar than the 1500 bar injection case. There is a delayed rapid rise in heat release for the 1200 bar case and an early gradually increasing rise in heat release for the 1500 bar case. Similar heat release rate profile was observed for the 1500 bar case by Skeen et al. (2015a). Thus, there is a significant increase in ignition delay (~0.8 ms) in the 750 K case with a decrease in the injection pressure by 300 bar (from 1500 to 1200 bar). This can be attributed to enhanced spray–gas mixing due to the high-injection pressure, which aids in ignition of the main spray. Since 750 K is on the lower end of a low-temperature combustion (LTC) regime, the extra turbulence induced by a higher injection pressure might be critical to affect the relatively slow chemistry involved in low-temperature reactions. Reasoning for the early ignition of main injection can be obtained from studies of Felsch et al. (2009), where the authors concluded that the main ignition occurs by a premixed flame controlled mechanism. Unlike pilot ignition, the ignition of main is influenced by the higher spay-induced turbulence due to its premixed flame controlled nature. Additionally, an increase in mixing, by increasing the injection pressure, will further reduce the ignition times following this theory.

For a qualitative comparison of the performance of sprays at different spray-induced turbulence conditions, a combustion efficiency term based on normalizing the cumulative heat release with the product of mass injection and lower heating value of the fuel was calculated. This calculation also assumes no heat loss from the combustion event to the chamber walls. For the 1200 bar and 1500 bar injection 750 K ambient cases, the normalized combustion efficiency came to 59% and 68%, respectively. This indicates better combustion with usage of higher injection pressure for the lowest (750 K) case. Thus, the energy content of the pilot injected fuel, which is sufficiently heated until low-temperature reaction intermediates form, affects the ignition and combustion of the main spray. As seen in the PLIF images of Fig. 11.13 (left), there is no formaldehyde signal until late ~1.8 ms, indicating no low-temperature chemistry in effect until then for the 1200 bar injection case. Whereas in Fig. 11.13 (right), formaldehyde signal appears at earlier times of ~1.5 ms giving indication of early low-temperature chemistry initiation and hence more chances of early ignition for the 1500 bar injection case. These trends are consistent with the initial rise in heat release times for the 750 K case in Fig. 11.14.

Effect of increased flame–flame turbulence via mutual interactions: The influence of increased turbulence on split injection via CFD analysis is also characterized. Here we study how the turbulence from the first injection helps in cleaner combustion of the second injection. Since the ignition and flame lift-off have an impact on the soot formation processes of the second injection (and possibly on the first injection due to mutual flame interactions), a brief analysis using mixture fraction (Z) and temperature (T) plots is provided below to help understand the level of soot in relation to oxygen presence and turbulent kinetic energy (TKE) availability within the spray flame.

It is possible to correlate a Z-T plot, to actual location in a flame up to some degree of accuracy. An illustrative diagram is provided in Fig. 11.15, with Z-T plot in the bottom and actual locations of the flame in the top. It can be seen that this plot has a delta-like (Δ) shape.

Figure 11.16 shows a set of oxygen distribution Z-T plots in the top and turbulent kinetic energy (TKE) Z-T plots in the bottom to gauge the effect of turbulence on oxygen distribution which had impacted higher soot formation in main and reduced the soot production in pilot by main-induced pilot soot oxidation. Volume percent of oxygen between 0 and 0.5% is considered for oxygen Z-T plot, whereas a 0–0.5 of TKE fraction (fraction of TKE/TKEmax at any time instant in the computational domain) is plotted for the TKE Z-T plot.

It can be observed from Fig. 11.16 that at 0.6 ms, when the pilot injection is about to end and when there is a quasi-steady liftoff developed, the region just upstream of the liftoff has indications of oxygen presence (from oxygen Z-T plot)

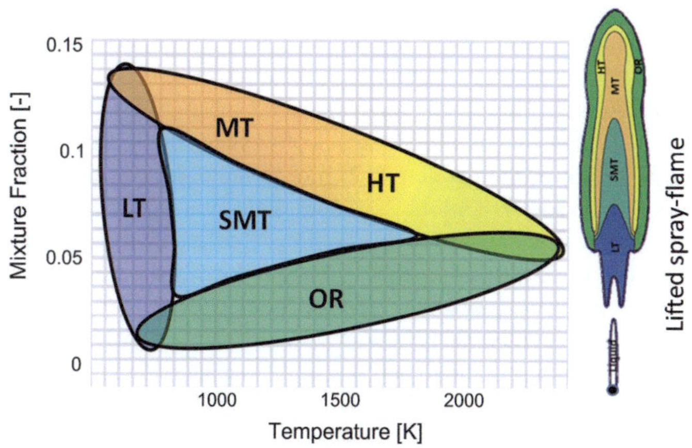

Fig. 11.15 A Z-T plot with legends given as follows—LT: Low Temperature (700–1000 K); SMT: Slightly Medium Temperature (1000–1400), MT: Medium Temperature (1400–1800), HT: High Temperature (1800–2200), OR: Outer Region (Moiz 2016)

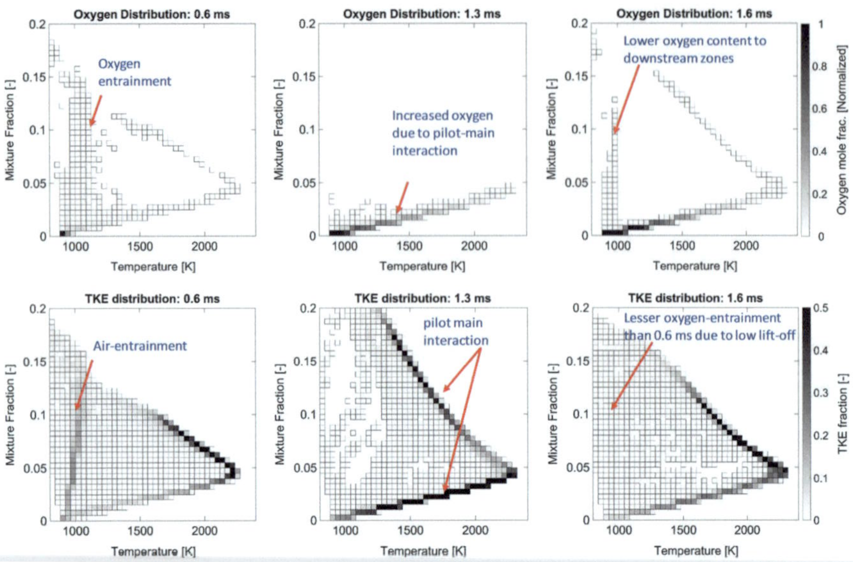

Fig. 11.16 Z-T plot with oxygen and TKE fraction distribution (Moiz 2016)

and higher TKE present (from TKE Z-T plot). The dark TKE-binned region to the left in 0.6-ms is the region upstream of the lifted flame where mixing between injection fuel and air takes place. It can be observed that the TKE and oxygen level in this region are higher at 0.6 ms than at 1.6 ms indicating a higher level of air entrainment at 0.6 ms. In other words, pilot flame has higher level of air entrainment than the main, perhaps due to its higher low liftoff than the main flame. To notice in all the plots is the level of TKE in the bins at the bottom region which constitute the far outer region of the flame and the bins in the upper-right side which indicate the high-temperature flame region. At 1.6 ms, these bins show both higher magnitude and spread of TKE fraction than the ones in 0.6 ms. This is due to the effect of induced turbulence of the main injection on the pilot after start of main injection times. This results in a higher level of oxygen supply as evident from the bins in bottom region of the oxygen TKE plot in the corresponding time instants. As said before in the soot Z-T plot discussion, the interaction effect of the main on the pilot results in higher oxygen availability (increased mixing), perhaps by breaking into the diffusion flame sheet of the pilot flame and entraining more oxygen from the ambient to oxidize the pilot soot.

This effect is more evident at 1.3 ms in the TKE Z-T plot when the main starts interacting with the pilot soot cloud and where there is increased magnitude of TKE presence in the far-upper-right-side and bottom-side bins indicating the central region and the far outer region of the flame, respectively. Lesser oxygen and TKE availability of the main flame (as in 1.6 ms) than the pilot flame (as in 0.6 ms) are observed due to the lower main flame liftoff than the pilot.

11.2.3 Partially Premixed Combustion

Basically, limitation in meeting restrictions of both NO_x and soot from conventional diesel and SI engines has led to the development of some advanced engine concepts including homogeneous charge compression ignition (HCCI) (Christensen et al. 1999), partial premixed combustion (PPC), also known as gasoline compression ignition (GCI) (Noehre et al. 2006; Kalghatgi et al. 2006, 2007), and reactivity controlled combustion ignition (RCCI) (Kokjohn et al. 2011). Researchers (Cung et al. 2016b; Ciatti 2015; Kolodziej et al. 2015, 2016; Dec 2009, 2015; Reitz and Duraisamy 2015; Kavuri et al. 2016; Tanov et al. 2015; Sellnau et al. 2016) investigated these concepts through experiment and simulation in order to gain better understanding of the physical and chemical fuel–air interaction mechanism, and ultimately design more fuel-efficient engines. Interaction of fuel–air charge and cylinder geometry is also highly important because thermodynamic and heat transfer loss can result from poor fuel–air distribution resulting emissions characteristics of high unburned hydrocarbon (HC) or high carbon monoxide (CO). Simply speaking, these advanced combustion concepts center with low-temperature combustion approach which includes several pathways of using high level of dilution or EGR and/or partially premixed combustion. Depending on the level of mixing, combustion mode can be categorized as fully premixed (HCCI) or partially premixed (GCI, RCCI). Only GCI work is discussed mainly in this section due to the growing interests in this field. HCCI and RCCI have some disadvantages in terms of maximum achievable engine load without knock and an expensive and complicated duel-fuel system. Nonetheless, many experiments (Dec et al. 2015; Hanson et al. 2016; Kokjohn et al. 2011) demonstrated the benefits of these systems to achieve low emission and high-efficiency engine performance.

Diesel has a CN of 40–55 (Law 2006) while CN of gasoline is less than 30 (Kalghatgi 2014; Manente et al. 2011). Gasoline, a much lower reactivity fuel as compared to diesel, allows longer mixing time prior to ignition, therefore can reduce soot formation from locally rich region in the combustion chamber. By using a single fuel with low reactivity, GCI relies mostly tailoring in-cylinder fuel stratification level to control combustion phasing as well as heat release rate profile. Dempsey et al. (2016), through computational fluid dynamics (CFD) simulations, showed the importance of fuel stratification on GCI ignition. The author described different levels of stratification by different approaches including partially fuel stratification (PFS), moderate fuel stratification (MFS), and heavy fuel stratification (HFS). The different levels of fuel stratification are basically characterized by the SOIs and split ratio in a multiple-injection scheme. This can be demonstrated in start of injection (SOI) sweep as shown in Fig. 11.17. With early injection prior to −60° aTDC, time of ignition and combustion varies insignificantly. This is because fuel and air are given a very long mixing time, and the combustion occurs mostly through the auto-ignition of a "quasi" well-mixed mixture. As fuel is injected at later time in the SOI sweep, time of ignition and combustion changed with earliest ignition around −30° aTDC while the ignition follows almost linearly with

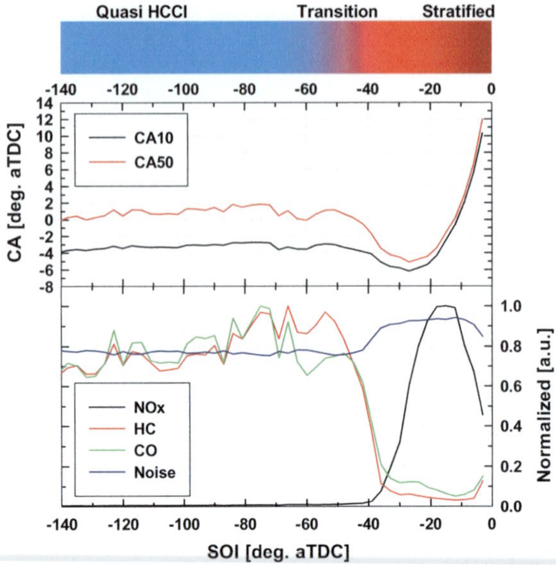

Fig. 11.17 Characteristics of combustion modes by start of injection sweep using single injection from −141° aTDC to TDC (Cung et al. 2016b)

injection timing as SOI approaches TDC. Clearly, the effect of fuel stratification became more dominant as fuel is injected later in the compression event. Higher fuel stratification promoted ignition as happened at −30° aTDC. More retarded SOI cases can be considered to be similar to diffusion flame of conventional diesel combustion with ignition follows closely after start of injection. GCI application often involved multiple-injection strategy because it allows the best combination of low soot (from early injection) and combustion phasing for noise control (from late injection) to achieve a wide range of engine loads. This leads to the scenario when part of the fuel is premixed and the remaining fuel is introduced later with more effect from spray-induced turbulence. The combustion can be said to be taking place at a stratified medium (Peters 2000).

GCI concept has been studied extensively in all-metal engine testing; hence, very limited numbers of work have been done regarding optical diagnostics of the cylinder combustion process of GCI. Tanov et al. (2014) showed that flame signal was more homogeneous or wide spread under triple injection as compared to single- or double-injection condition. The combustion luminosity of diesel was also reported in reference (Lu et al. 2015) to be much stronger than gasoline under the same condition. They also reported a faster evaporation in gasoline as compared to diesel which leads to its wider in-cylinder spray distribution. While these studies were done in optical engine, endoscope is one of the other optical techniques that can be used to visualize the spray and combustion processes in the engine chamber. Typically, endoscope measurement offers insights on soot formation and oxidation process via flame naturally luminosity images.

The endoscope system consists of a charge coupled device (CCD) camera (PCO Pixelfly), an air cooling system for the endoscope, and a rigid industrial endoscope with desirable angle of view (i.e., 30° from endoscope axis). The camera was connected to an AVL VisioscopeTM optical system with frame rate of 10 Hz and minimum exposure time of 5 µs. Images with example specification of 640 × 480 pixel and 12-bit resolution were processed by AVL's Thermo VisionTM software to give post-processed soot and temperature distributions using two-color optical pyrometry method. More details of the endoscope system and measurement technique can be found in other references (Product Guide 2004, Miers et al. 2005). The endoscope schematic is shown in Fig. 11.18.

The endoscope system allowed to record images by 0.5 CAD interval from certain CAD range (i.e., −5° to 30° aTDC), usually covering the all combustion process as also seen by the rise and fall of HRR profile. Note that, the entire visualized combustion event was based on a collection of images from different cycles. This really depends on the camera's capability. However, because good combustions stability was maintained, high repeatability was acquired for all cycles. Image distortion, also known as "fish-eye view" or "spherical distortion" (Dierksheide et al. 2002), could be present through the lens system. Image correction can be performed by applying light source to capture initial background with grid lines as done in reference (Tanov et al. 2014).

Flame luminosity images in Fig. 11.19 indicate that combustion began shortly after TDC. In this certain testing condition, this indicates a very low to no signals from the premixed combustion of early injection (~ −50° aTDC). Soon after the later injection (right before TDC), combustion signals became visible. Flame luminosity signals occurred mostly between 4° and 12° aTDC. Some spray

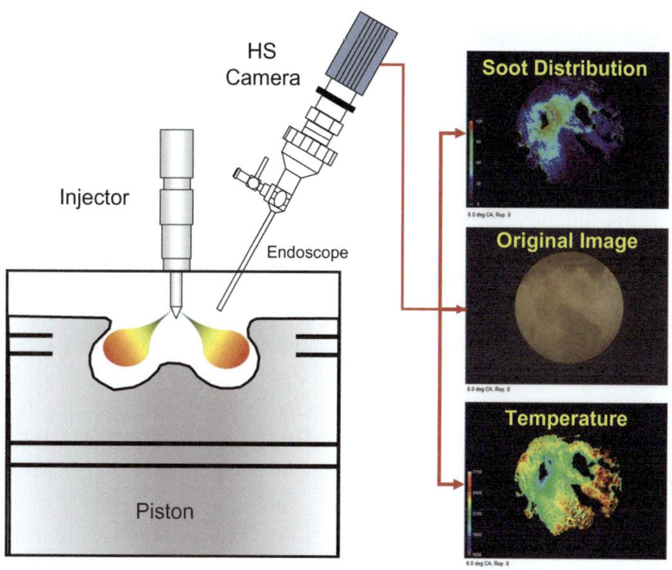

Fig. 11.18 Endoscope images and post-processed temperature and soot distribution (Miers et al. 2005)

Fig. 11.19 Endoscope images of combustion luminosity over different CAD

scattering near the injector nozzle was also observed as seen at 3° aTDC which was possible because of emitted light from the combustion.

Endoscope imaging has been demonstrated as a robust optical technique that typically requires simple implementation on the engine body. Temporal and spatial soot and temperature distribution post-processed natural luminosity images can provide valuable information for validation of CFD simulation which is quite unique for partially premixed combustion like GCI. However, challenge lies on limited to no visualization of spray and pre-ignition intermediates species due to the lack of optical diagnostics such as Schlieren and laser application.

11.3 Numerical Approaches

To gain further insights into the combustion process, it is common practice to perform a numerical simulation of the combustion process at hand and analyze the entire suite of outputs. Some of the common modeling approaches based on their treatment of turbulence are mentioned here briefly along with their applications specific to spray-induced turbulent combustion problems. The common approaches in treatment of turbulence are direct numerical simulations (DNS), large-eddy simulation (LES) and Reynolds-averaged Navier-Stokes (RANS).

Direct numerical simulations basically solve all scales of instantaneous quantities and thus are the most time-consuming to simulate. With the recent development of spectral codes together with high-performance computing, these are starting to become computationally tractable, but their use will still be limited to small-scale problems even in the next few decades. Large-eddy simulations use a filtering technique to filter out the smallest length and timescales, effectively causing some spatial and time averaging of instantaneous quantities.

Comparing to RANS this technique considers smaller time and length scales and thus need finer grid resolution to be effectively implemented and thus still time-consuming. However, with the ease in availability of high-performance computing resources, LES is becoming a popular option. From a spray-induced spray combustion standpoint, having this better treatment of turbulence via LES causes better mixing of the liquid core with the ambient and better ignition of that fuel–air mixture since in true sense ignition begins in the smaller scales of turbulence and modeling them is the ideal way to treat them. Since the instantaneous

Navier–Stokes equations (which describe the motion of fluid) are impossible to solve, specific operators (commonly called RANS operators) are used to decompose into them into simpler quantities. An important aspect of RANS equations is that the instantaneous quantities (e.g., velocities) are decomposed into mean and fluctuating quantities and are eventually give ensemble-averaged solution of the Navier–Stokes equations. Ensemble averaging in terms can be viewed as averaging of a common time frame when the test is repeated over several cycles.

A simpler way to look at spray combustion process is by 1-D modeling using codes like Two-stage Lagrangian (TSL) model. The TSL model was developed originally by Broadwell and Lutz (Broadwell and Lutz 1998) to study chemical process in diesel spray combustion for a very time efficient manner even with the use of a very large chemical mechanism. TSL modeling is unique in permitting the inclusion of mixing processes with detailed kinetics in a computationally efficient method, thereby providing continuous reactions while entrainment and other essential flow aspects are also considered (Han et al. 1999). The structure of the TSL program consists of a flame sheet reactor to represent the diffusion-like flame and homogeneous reactor to represent the jet core. Figure 11.20 shows the simple diagram of the TSL model.

To simply explain the species evolution along the spray axis, TSL model output is shown in Fig. 11.21. Fuel is mixed with air initially without ignition source in mixing model. Hence, a fuel/air composition prior to lifted length is known and then used as input in TSL lifted flame model with reactor is "on" at LOL location.

As the result in lifted flame (solid lines in Fig. 11.21), a spontaneous temperature rise is followed by increasing in formation of CH_2O, and HCO between X/D of 40–70 (high-temperature reaction zone). This is different in mixing model in which low-temperature and flame sheet residence time were used providing a visualization of multi-stage ignition process as seen in temperature profile (red dash line). The formation of CH_2O is shown clearly to have strong dependence with premixed ignition region or cool flame. It is consumed eventually downstream of LOL

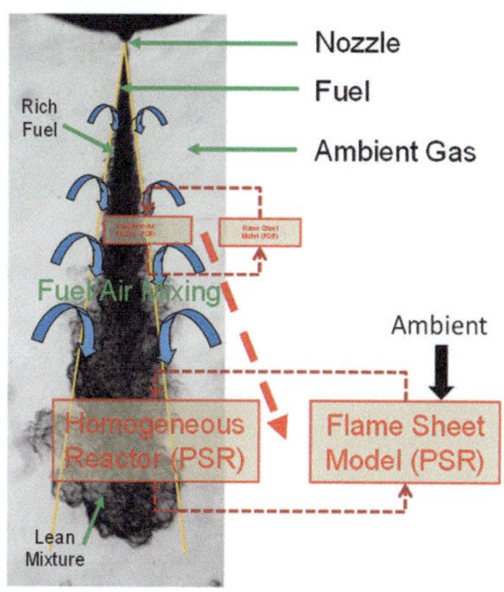

Fig. 11.20 Schematic of TSL conceptualization model for spray simulation with two reactors representing entrainment of ambient gas to fuel injection (Broadwell and Lutz 1998; Cung et al. 2013)

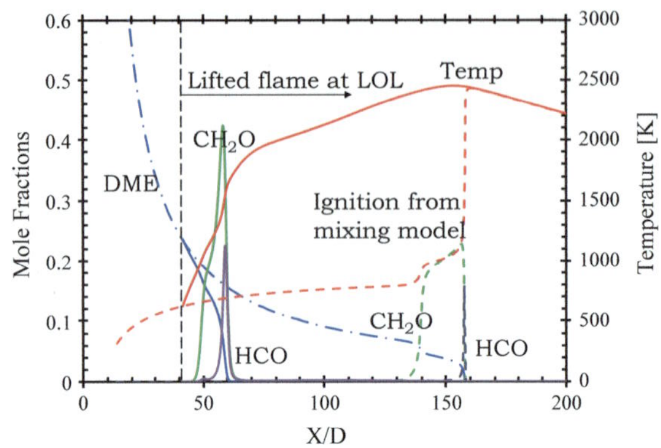

Fig. 11.21 Evolution of species from lifted and mixing flame models (TSL) (Cung et al. 2016c)

location. In both models, the peak value of HCO is achieved when fuel depletion gradient and derivative of second-stage ignition temperature are maximum. The mixing model captured very well the two-stage ignition phenomena: (1) initial ignition reactions for slow-rate and low-temperature combustion, (2) cool flame region where NTC (negative temperature coefficient) effect is present, and finally (3) ignition reaction accelerates rapidly as the reaction of vaporized mixture of fuel and ambient gas become more exothermic with heat release exceeding the vaporization of fuel, hence usually marked as the end of ignition delay where combustion is started.

Since RANS and LES codes are the widely used simulation choices, the upcoming separate subsection explains their application briefly followed by some insights of DNS spray combustion aspects.

11.3.1 RANS and LES Modeling

Unlike RANS, LES is capable of predicting key transient features occurring in spray combustion events. In the following section, some results are presented based on DME spray and combustion experimentation performed in a constant volume combustion vessel. CFD work was done for non-reacting sprays first to ensure proper spray mixing, later analyzing the key DME ignition processes using LES turbulence model with a 62.5 μm fine grid and 25 million peak cell counts shown in Fig. 11.22.

Validation with experimental data pertaining to cool flame species formation (formaldehyde) and spray flame generation (ignition and stabilization) was performed. The injection duration of liquid DME was 2.3 ms at 750 bar rail pressure, and ambient temperature/density was 900 K/14.8 kg/m^3 in an 18% oxygen ambient. For both experiments and CFD, ignition delay (ID) is calculated based on apparent heat release rate (AHRR) (pressure derived), where 10% of rise in AHRR corresponds to high-temperature ignition. For CFD, liftoff length (LOL) follows the

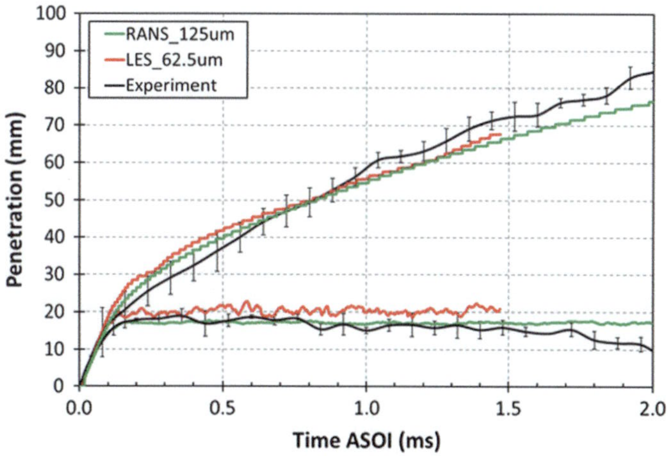

Fig. 11.22 Penetration with different modeling approaches compared with experiment

traditional definition of diesel spray flames corresponding to an axial extent of 2% of mass fraction of hydroxyl species in the upstream region. LES calculations were performed using three realizations. The ID and LOL variations along with the averaged value are tabulated in Table 5.1. The LOLs predicted are ~4 mm higher than experiments considering the deviations, which can be due to the LOL definition employed by CFD and considering the fact that OH was used as measurement species and not OH* as in experiments.

Stabilization of the flame is related to the formation of cool flame since upstream zone prior to LOL is where the fuel–air is mixed and heat is released. It is important to consider cool flame generation since it gives key information of ignition process and flame stabilization. Formaldehyde species are the commonly considered species signifying cool flame. Figure 11.23 shows experimental formaldehyde profiles compared with LES. It can be seen that the overall extent and temporal formation of formaldehyde are captured well by the simulations. Experiments show some formaldehyde signal in the far-upstream regions near the injector, but that is mostly due to liquid scattering of the incident laser. The upstream formation of formaldehyde near the end of the liquid length (~10 mm) is an indicator of low-temperature chemistry initiation in those zones. This paves the way for

Table 11.1 Ignition delay and liftoff data of LES simulations and experiments

	Simulation				Experiment
	RANS	LES_1	LES_2	LES_3	
ID (ms)	1.01	0.915	0.95	0.93	1.03 ± 0.015
LOL (mm)	30.56	25.96	30.59	27.92	20.87 ± 2.20
		Averaged ID = 0.93 ± 0.017 ms Averaged LOL = 28.16 ± 2.32 mm			

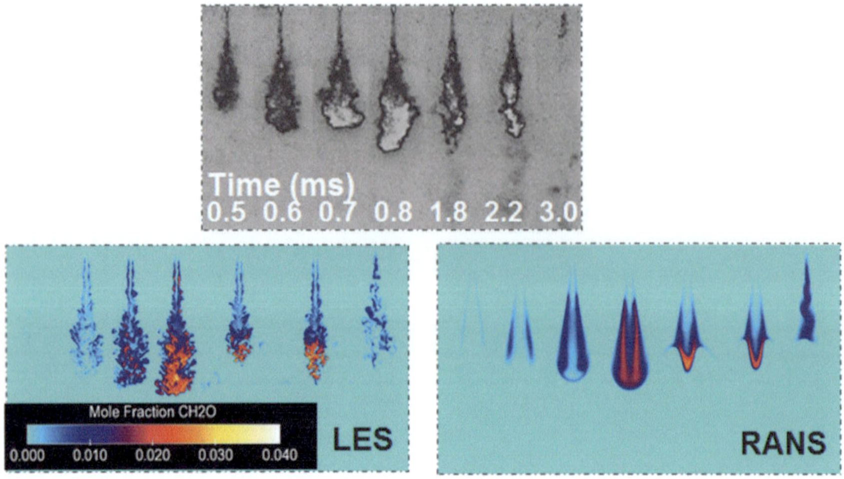

Fig. 11.23 Formaldehyde structures comparison between experiment, LES and RANS

generation of high-temperature ignition kernels upstream, which would in-turn stabilize the flame.

Difference between LES and RANS formaldehyde upstream presence might be due to the fact that RANS probably has larger turbulent diffusivities at the upstream locations leading to higher dissipation of formaldehyde or the precursors of formaldehyde. Also, LES has much finer cells, i.e., more cells with very rich mixture (e.g., the maximum Z in RANS was recorded as ~0.3, whereas maximum Z in LES was ~0.5), which may lead to larger formaldehyde formation at upstream locations.

Some of the key features of DME spray combustion are shown in Fig. 11.24. As seen in the experimental image accompanying the CFD image in the figure for the top and middle CFD snapshots, we see high-temperature ignition kernels generated out of an auto-ignition event in the boundary of low-temperature species (formaldehyde here) and high-temperature species (here OH). This auto-ignition kernels are what keep the flame stabilized causing a quasi-steady liftoff length and may be quenched and re-ignited depending on the surrounding spray-induced turbulence field. In the last CFD snapshot, an event termed as combustion recession event is seen to take place (as also found in the accompanying experimental image). This recession event is generated when all the low-temperature species (here formaldehyde) is consumed by the high-temperature species like OH after the fuel injection event has ended.

Figure 11.25 shows Z-T scatter plots of some important time instants during the DME spray combustion event. The black dashed line passing through the scatter is the mean temperature conditional on mixture fraction (T|Z), and the vertical gray line is the stoichiometric mixture fraction. At 0.4 ms, for both RANS and LES, the slanted line is the mixing line where fuel vapor mixes with air and exchanges

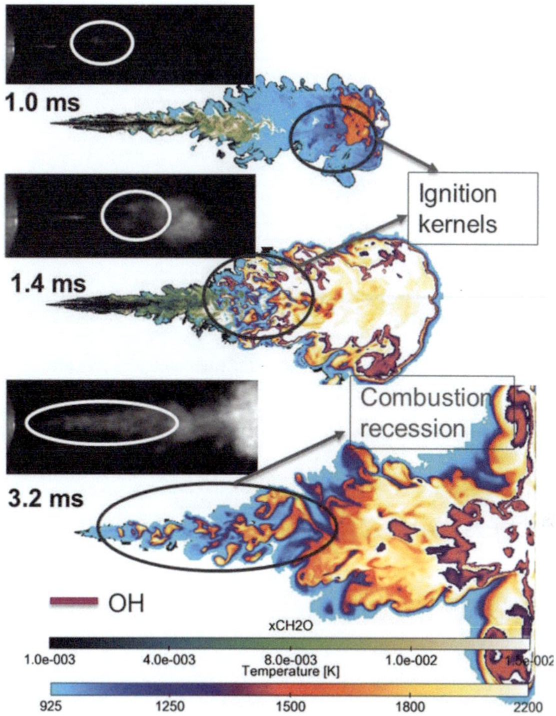

Fig. 11.24 Key transient DME flame development features, OH, CH_2O, T

energy and momentum. Around time instants of 0.8–1.0 ms, there is low-temperature heat release occurring in Z's greater than Z_{st} signifying that combustion initiates in the rich fuel–air mixtures. For RANS, at 1 ms, where most of the heat is released around a narrow band of Z, LES has a much broader brush signifying multiple ignition locations controlling the high-temperature heat release, which was also seen in experiments (not shown here). High-temperature ignition shortly follows after 1 ms, and flame is stabilized at around 1.4 ms for the LES case. Unlike LES, where the T|Z line attains peak temperature around Z_{st} location, RANS T|Z line is still colder and attains a peak temperature during EOI times of 2 ms. This signifies that for RANS flame stabilization occurs in later instants possibly near EOI.

11.3.2 Direct Numerical Simulation (DNS)

DNS simulations are usually not simulated using spray in computational approaches. The traditional approach is to use a gas jet approximation and simulate a gaseous injection of the fuel. Most of the DNS simulations are limited to smaller domains in the order of 2 × 2 cm and are two-dimensional in nature. This is due to

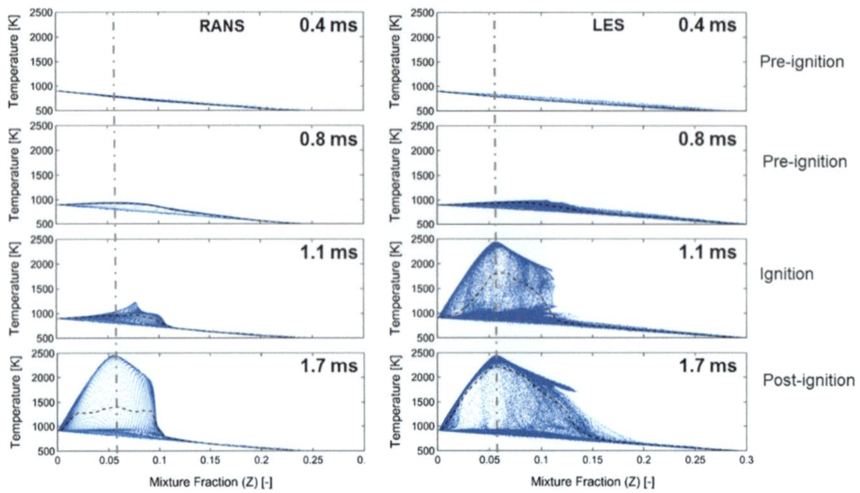

Fig. 11.25 Temperature (T) and mixture fraction (Z) scatters for RANS and LES

the fact that resolving all the scales for a bigger domain is too computationally intensive and is a tough problem even for the supercomputers of this age. But there has been recent progress using spectral codes which make DNS computationally tractable; although they are still two-dimensional, but they are simulated for a larger domain.

11.4 Further Discussion

While there is tremendous progress in our understanding of turbulent spray combustion science to date some of the most challenging questions are still difficult to answer for lack of experimental and numerical methods. For example, the flow inside the injector is difficult to visualize due to high-density steel-type structure blocking the diagnostics, but very high-power X-ray measurement is one way out which enables user to study needle wobble and cavitation like features (Powell et al. 2004a, b). The other common challenge is understanding of combustion chemistry which is usually not done while considering turbulence; most of the mechanisms used in CFD codes are developed in shock-tube experiment where there is minor involvement of spray-induced turbulence. That being said, chemistry model used in CFD codes is still with skeletal/reduced species; although there is some recent progress with detailed tabulated chemistry mechanism usage in CFD (Ameen et al. 2016; Kundu et al. 2017). With the advent of supercomputers, CFD calculations are now, however, much more computationally tractable and give us a view of micro-phenomenon in turbulent combustion environments.

11.5 Summary

Turbulent spray combustion is a complex process involving the heat and mass transfers, two physical phases, fluidic-mixing, and reaction chemistry, which is experimentally challenging for simultaneous quantity measurement and numerically challenging. Further research on non-reacting and reacting flow is needed. Because of the tremendous computing capabilities nowadays, the modeling of turbulent, spray atomization, and detailed chemical reaction mechanisms is partially feasible in identifying the averaged and instantaneous spray flame properties. However, a number of open issues such as the nozzle internal flow which is the boundary condition for spray formation, the dense spray near the injector nozzle exit, consideration of advanced chemical reaction schemes for use in reacting spray flows should be researched. Moreover, because of limitation of current computational capabilities, chemical reaction species ($> \sim 2000$) and reaction steps ($> \sim 10{,}000$) are the biggest challenge for the CFD integration. New development of efficient use in chemistry is needed along with inclusion of more fidelity in capturing turbulence phenomena in a higher detail.

11.6 Disclaimer and Funding Acknowledgement

The research was supported by the NSF/DOE Advanced Engine Combustion Program, NSF-CBET 1258720. Authors wish to thank Ruey-Hung Chen at NSF and Gurpreet Singh at DOE, for their support. Some work in the chapter was also supported by US DOE Office of Vehicle Technologies, Office of Energy Efficiency and Renewable Energy under Contract No. DE-AC02-06CH11357. Authors wish to thank Gurpreet Singh and Leo Breton, program managers at DOE, for their support. This research used resources of the Argonne Leadership Computing Facility (ALCF), which is a DOE Office of Science User Facility supported under contract DE-AC02-06CH11357. We would like to thank Dr. Stephen Ciatti of Argonne National Laboratory for allowing us to use the PPCI content as part of this book section. We would also like to acknowledge computational resources provided by the Argonne National Laboratory Computing Resource Center.

References

Afzal H, Arcoumanis C, Gavaises M, Kampanis N (1999) Internal flow in diesel injector nozzles: modelling and experiments. IMechE Pap S 492:25–44

Ameen MM, Kundu P, Som S (2016) Novel tabulated combustion model approach for lifted spray flames with Large Eddy Simulations. SAE Int J Engines 9(2016-01-2194):2056–2065

Bae C, Kang J (2006) The structure of a break-up zone in the transient diesel spray of a valve-covered orifice nozzle. Int J Engine Res 7(4):319–334

Baert RSG, Frijters PJM, Somers B, Luijten CCM, de Boer W (2009) Design and operation of a high pressure, high temperature cell for HD diesel spray diagnostics: guidelines and results

Bhagat M, Cung K, Johnson J, Lee S-Y, Naber J, Barros S (2013) Experimental and numerical study of water spray injection at engine-relevant conditions. SAE Technical Paper

Bhattacharjee S, Haworth DC (2013) Simulations of transient n-heptane and n-dodecane spray flames under engine-relevant conditions using a transported PDF method. Combust Flame 160 (10):2083–2102

Brackmann C, Nygren J, Bai X, Li Z, Bladh H, Axelsson B, Denbratt I, Koopmans L, Bengtsson P-E, Aldén M (2003) Laser-induced fluorescence of formaldehyde in combustion using third harmonic Nd: YAG laser excitation. Spectrochim Acta Part A Mol Biomol Spectrosc 59 (14):3347–3356

Broadwell JE, Lutz AE (1998) A turbulent jet chemical reaction model: NOx production in jet flames. Combust Flame 114(3):319–335

Bruneaux G (2008) Combustion structure of free and wall-impinging diesel jets by simultaneous laser-induced fluorescence of formaldehyde, poly-aromatic hydrocarbons, and hydroxides. Int J Engine Res 9(3):249–265

Christensen M, Hultqvist A, Johansson B (1999) Demonstrating the multi fuel capability of a homogeneous charge compression ignition engine with variable compression ratio. SAE Technical Papers. https://doi.org/10.4271/1999-01-3679

Ciatti S (2015) Compression ignition engines–revolutionary technology that has civilized frontiers all over the globe from the industrial revolution into the twenty-first century. Front Mech Eng 1 (5)

Colin O, Benkenida A, Angelberger C (2003) 3D modeling of mixing, ignition and combustion phenomena in highly stratified gasoline engines. Oil Gas Sci Technol 58(1):47–62

Cung K, Bhagat M, Zhang A, Lee S.-Y. (2013) Numerical study on emission characteristics of high-pressure dimethyl ether (DME) under different engine ambient conditions. SAE Technical Papers 2. https://doi.org/10.4271/2013-01-0319

Cung K, Moiz A, Johnson J, Lee S-Y, Kweon CB, Montanaro A (2015a) Spray-combustion interaction mechanism of multiple-injection under diesel engine conditions. Proc Combust Inst 35(3):3061–3068. https://doi.org/10.1016/j.proci.2014.07.054

Cung K, Moiz AA, Zhu X, Lee S-Y (2016a) Ignition and formaldehyde formation in dimethyl ether (DME) reacting spray under various EGR levels. In: Proceedings of the combustion institute, vol 36 https://doi.org/10.1016/j.proci.2016.07.054

Cung K, Rockstroh T, Ciatti S, Cannella W, Goldsborough SS (2016b) Parametric study of ignition and combustion characteristics from a gasoline compression ignition engine using two different reactivity fuels. In: ASME 2016 internal combustion engine fall technical conference. American Society of Mechanical Engineers, pp V001T003A011–V001T003A011

Cung K, Zhang A, Lee S-Y (2015b) Ignition and formaldehyde formation in dimethyl ether spray combustion: experiment and chemical modeling. In: 9th U.S. national combustion meeting, Cincinnati, Ohio, USA

Cung K, Zhu X, Moiz AA, Lee S-Y, De Ojeda W (2016b) Characteristics of formaldehyde (CH2O) formation in dimethyl ether (DME) spray combustion using PLIF imaging. SAE Int J Fuels Lubr 9(1):138–148. https://doi.org/10.4271/2016-01-0864

Cung K (2015) Spray and combustion characteristics of dimethyl ether under various ambient conditions: an experimental and modeling study. PhD thesis, Michigan Technological University

Dec JE (1997) A conceptual model of DI diesel combustion based on laser-sheet imaging*. SAE Technical Paper

Dec JE (2009) Advanced compression-ignition engines—understanding the in-cylinder processes. Proc Combust Inst 32(2):2727–2742. https://doi.org/10.1016/j.proci.2008.08.008

Dec JE, Espey C (1998) Chemiluminescence imaging of autoignition in a DI diesel engine

Dec JE, Yang Y, Dernotte J, Ji C (2015) Effects of gasoline reactivity and ethanol content on boosted, premixed and partially stratified low-temperature gasoline combustion (LTGC). SAE Int J Engines 8(3):935–955. https://doi.org/10.4271/2015-01-0813

Dempsey AB, Curran SJ, Wagner RM (2016) A perspective on the range of gasoline compression ignition combustion strategies for high engine efficiency and low NOx and soot emissions: effects of in-cylinder fuel stratification. Int J Engine Res 1468087415621805

Dierksheide U, Meyer P, Hovestadt T, Hentschel W (2002) Endoscopic 2D particle image velocimetry (PIV) flow field measurements in IC engines. Experiments in Fluids 33(6):794–800. https://doi.org/10.1007/s00348-002-0499-3

Donovan MT, He X, Zigler BT, Palmer TR, Wooldridge MS, Atreya A (2004) Demonstration of a free-piston rapid compression facility for the study of high temperature combustion phenomena. Combust Flame 137(3):351–365. https://doi.org/10.1016/j.combustflame.2004.02.006

Engine_Combustion_Network. http://www.sandia.gov/ECN/

Felsch C, Gauding M, Hasse C, Vogel S, Peters N (2009) An extended flamelet model for multiple injections in DI Diesel engines. Proc Combust Inst 32(2):2775–2783

Golovitchev VI, Nordin N, Jarnicki R, Chomiak J (2000) 3-D diesel spray simulations using a new detailed chemistry turbulent combustion model. SAE Technical Paper

Haessler H, Bockhorn H, Pfeifer C, Kuhn D (2012) Formaldehyde-LIF of dimethyl ether during auto-ignition at elevated pressures. Flow Turbul Combust 89(2):249–259. https://doi.org/10.1007/s10494-011-9374-8

Han D, Mungal M, Zamansky V, Tyson T (1999) Prediction of NOx control by basic and advanced gas reburning using the Two-Stage Lagrangian model. Combust Flame 119(4):483–493

Han Z, Uludogan A, Hampson GJ, Reitz RD (1996) Mechanism of soot and NOx emission reduction using multiple-injection in a diesel engine. SAE Technical Paper

Hanson R, Ickes A, Wallner T (2016) Use of adaptive injection strategies to increase the full load limit of RCCI operation. J Eng Gas Turbines Power 138(10):102802

Haworth D (2010) Progress in probability density function methods for turbulent reacting flows. Prog Energy Combust Sci 36(2):168–259

Higgins B, Siebers DL (2001) Measurement of the flame lift-off location on DI diesel sprays using OH chemiluminescence

Hiroyasu H, Arai M (1990) Structures of fuel sprays in diesel engines. SAE Technical Paper

Idicheria CA, Pickett LM (2006) Formaldehyde visualization near lift-off location in a diesel jet. SAE Technical Paper

Kalghatgi GT (2014) Fuel/engine interactions

Kalghatgi GT, Risberg P, Ångström H-E (2006) Advantages of fuels with high resistance to auto-ignition in late-injection, low-temperature, compression ignition combustion

Kalghatgi GT, Risberg P, Ångström H-E (2007) Partially pre-mixed auto-ignition of gasoline to attain low smoke and low NOx at high load in a compression ignition engine and comparison with a diesel fuel

Karimi K (2007) Characterisation of multiple-injection diesel sprays at elevated pressures and temperatures. University of Brighton

Kavuri C, Paz J, Kokjohn SL (2016) A comparison of reactivity controlled compression ignition (RCCI) and gasoline compression ignition (GCI) strategies at high load, low speed conditions. Energy Convers Manag 127:324–341. https://doi.org/10.1016/j.enconman.2016.09.026

Kim T, Ghandhi JB (2001) Quantitative 2-D fuel vapor concentration measurements in an evaporating diesel spray using the exciplex fluorescence method

Kitamura T, Ito T, Senda J, Fujimoto H (2002) Mechanism of smokeless diesel combustion with oxygenated fuels based on the dependence of the equivalence ration and temperature on soot particle formation. Int J Engine Res 3(4):223–248

Klimenko AY, Bilger RW (1999) Conditional moment closure for turbulent combustion. Prog Energy Combust Sci 25(6):595–687

Kokjohn SL, Hanson RM, Splitter DA, Reitz RD (2011) Fuel reactivity controlled compression ignition (RCCI): a pathway to controlled high-efficiency clean combustion. Int J Engine Res 12(3):209–226. https://doi.org/10.1177/1468087411401548

Kolodziej C, Kodavasal J, Ciatti S, Som S, Shidore N, Delhom J (2015) Achieving stable engine operation of gasoline compression ignition using 87 AKI gasoline down to idle

Kolodziej CP, Sellnau M, Cho K, Cleary D (2016) Operation of a gasoline direct injection compression ignition engine on naphtha and E10 gasoline fuels. SAE Int J Engines 9(2). https://doi.org/10.4271/2016-01-0759

Kundu P, Ameen M, Unnikrishnan U, Som S (2017) Implementation of a tabulated flamelet model for compression ignition engine applications. SAE Technical Paper

Law CK (2006) Combustion physics. Cambridge University Press, ISBN

Lillo PM, Pickett LM, Persson H, Andersson O, Kook S (2012) Diesel spray ignition detection and spatial/temporal correction. SAE Int J Engines 5(3):1330–1346. https://doi.org/10.4271/2012-01-1239

Lu P, Zhao H, Herfatmanesh MR (2015) In-cylinder studies of high injection pressure gasoline partially premixed combustion in a single cylinder optical engine

Manente V, Johansson B, Cannella W (2011) Gasoline partially premixed combustion, the future of internal combustion engines? Int J Engine Res 12(3):194–208

Mie G (1908) Beiträge zur Optik trüber Medien, speziell kolloidaler Metallösungen. Annalen der physik 330(3):377–445

Miers SA, Ng H, Ciatti SA, Stork K (2005) Emissions, performance, and in-cylinder combustion analysis in a light-duty diesel engine operating on a fischer-tropsch, biomass-to-liquid fuel

Moiz AA (2016) Low temperature split injection spray combustion: ignition, flame stabilization and soot formation characteristics in diesel engine conditions. PhD thesis, Michigan Technological University

Moiz AA, Ameen MM, Lee S-Y, Som S (2016a) Study of soot production for double injections of n-dodecane in CI engine-like conditions. Combust Flame 173:123–131. https://doi.org/10.1016/j.combustflame.2016.08.005

Moiz AA, Cung KD, Lee S-Y (2016b) Simultaneous Schlieren-PLIF studies for ignition and soot luminosity visualization with close-coupled high pressure double injections of n-dodecane. J Energy Res Technol. https://doi.org/10.1115/1.4035071

Moiz AA, Som S, Bravo L, Lee S-Y (2015) Experimental and numerical studies on combustion model selection for split injection spray combustion. SAE Technical Papers 2015-April. https://doi.org/10.4271/2015-01-0374

Musculus MPB, Miles PC, Pickett LM (2013a) Conceptual models for partially premixed low-temperature diesel combustion. Prog Energy Combust Sci 39(2–3):246–283. https://doi.org/10.1016/j.pecs.2012.09.001

Musculus MPB, Miles PC, Pickett LM (2013b) Conceptual models for partially premixed low-temperature diesel combustion. Prog Energy Combust Sci 39(2–3):246–283. https://doi.org/10.1016/j.pecs.2012.09.001

Nesbitt JE, Johnson SE, Pickett LM, Siebers DL, Lee S-Y, Naber JD (2011) Minor species production from lean premixed combustion and their impact on autoignition of diesel surrogates. Energy Fuels 25(3):926–936. https://doi.org/10.1021/ef101411f

Noehre C, Andersson M, Johansson B, Hultqvist A (2006) Characterization of partially premixed combustion

Oijen Jv, Goey LD (2000) Modelling of premixed laminar flames using flamelet-generated manifolds. Combust Sci Technol 161(1):113–137

Pei Y, Hawkes ER, Kook S (2013) Transported probability density function modelling of the vapour phase of an n-heptane jet at diesel engine conditions. Proc Combust Inst 34(2):3039–3047

Pei Y, Hawkes ER, Kook S, Goldin GM, Lu T (2015a) Modelling n-dodecane spray and combustion with the transported probability density function method. Combust Flame 162(5):2006–2019. https://doi.org/10.1016/j.combustflame.2014.12.019

Pei Y, Som S, Kundu P, Goldin GM (2015b) Large eddy simulation of a reacting spray flame under diesel engine conditions. SAE Technical Paper

Pei Y, Som S, Pomraning E, Senecal PK, Skeen SA, Manin J, Pickett LM (2015c) Large eddy simulation of a reacting spray flame with multiple realizations under compression ignition engine conditions. Combust Flame 162(12):4442–4455. https://doi.org/10.1016/j.combustflame.2015.08.010

Peters N (1984) Laminar diffusion flamelet models in non-premixed turbulent combustion. Prog Energy Combust Sci 10(3):319–339

Peters N (2000) Turbulent combustion. Cambridge university press

Pickett LM, Kook S, Williams TC (2009) Visualization of diesel spray penetration, cool-flame, ignition, high-temperature combustion, and soot formation using high-speed imaging. SAE Int J Engines 2(1):439–459. https://doi.org/10.4271/2009-01-0658

Pickett LM, Manin J, Genzale CL, Siebers DL, Musculus MPB, Idicheria CA (2011) Relationship between diesel fuel spray vapor penetration/dispersion and local fuel mixture fraction. SAE Int J Engines 4(1):764–799. https://doi.org/10.4271/2011-01-0686

Pickett LM, Siebers DL, Idicheria CA (2005) Relationship between ignition processes and the lift-off length of diesel fuel jets. SAE Technical Paper

Pierce CD, Moin P (2004) Progress-variable approach for large-eddy simulation of non-premixed turbulent combustion. J Fluid Mech 504:73–97

Pitsch H, Barths H, Peters N (1996) Three-dimensional modeling of NOx and soot formation in DI-diesel engines using detailed chemistry based on the interactive flamelet approach. SAE Technical Paper

Pitsch H, Steiner H (2000) Scalar mixing and dissipation rate in large-eddy simulations of non-premixed turbulent combustion. Proc Combust Inst 28(1):41–49

Pope SB (1985) PDF methods for turbulent reactive flows. Prog Energy Combust Sci 11(2):119–192

Pope SB (2001) Turbulent flows. IOP Publishing

Powell C, Ciatti S, Cheong S, Liu J, Wang J (2004a) X-ray characterization of diesel sprays and the effects of nozzle geometry. In: Proceedings of the diesel engine emission reduction conference

Powell CF, Ciatti SA, Cheong S-K, Liu J, Wang J (2004b) X-ray absorption measurements of diesel sprays and the effects of nozzle geometry

Product Guide AVL Thermovision advanced, (2004)

Reitz RD, Duraisamy G (2015) Review of high efficiency and clean reactivity controlled compression ignition (RCCI) combustion in internal combustion engines. Prog Energy Combust Sci 46:12–71. https://doi.org/10.1016/j.pecs.2014.05.003

Särner G, Richter M, Aldén M, Hildingsson L, Hultqvist A, Johansson B (2005) Simultaneous PLIF measurements for visualization of formaldehyde- and fuel- distributions in a DI HCCI engine

Sellnau M, Foster M, Moore W, Sinnamon J, Hoyer K, Klemm W (2016) Second generation GDCI multi-cylinder engine for high fuel efficiency and US Tier 3 emissions. SAE Int J Engines 9(2):1002–1020. https://doi.org/10.4271/2016-01-0760

Senecal P, Mitra S, Pomraning E, Xue Q, Som S, Banerjee S, Hu B, Liu K, Rajamohan D, Deur J (2014) Modeling fuel spray vapor distribution with large eddy simulation of multiple realizations. In: ASME 2014 internal combustion engine division fall technical conference. American Society of Mechanical Engineers, pp V002T006A002–V002T006A002

Settles GS (2012) Schlieren and shadowgraph techniques: visualizing phenomena in transparent media. Springer Science & Business Media

Siebers DL (1998) Liquid-phase fuel penetration in diesel sprays

Siebers DL (1999) Scaling liquid-phase fuel penetration in diesel sprays based on mixing-limited vaporization. SAE Technical Paper

Siebers DL, Higgins B, Pickett L (2002) Flame lift-off on direct-injection diesel fuel jets: oxygen concentration effects

Singh S, Reitz RD, Musculus MPB (2006) Comparison of the characteristic time (CTC), representative interactive flamelet (RIF), and direct integration with detailed chemistry combustion models against optical diagnostic data for multi-mode combustion in a heavy-duty DI diesel engine

Skeen S, Manin J, Pickett LM (2015a) Visualization of ignition processes in high-pressure sprays with multiple injections of n-dodecane. SAE Int J Engines 8(2):696–715. https://doi.org/10.4271/2015-01-0799

Skeen SA, Manin J, Pickett LM (2015b) Simultaneous formaldehyde PLIF and high-speed schlieren imaging for ignition visualization in high-pressure spray flames. Proc Combust Inst 35(3):3167–3174. https://doi.org/10.1016/j.proci.2014.06.040

Som S, Longman DE, Luo Z, Plomer M, Lu T, Senecal PK, Pomraning E (2012) Simulating flame lift-off characteristics of diesel and biodiesel fuels using detailed chemical-kinetic mechanisms and large eddy simulation turbulence model. J Energy Res Technol 134(3):032204

Tanov S, Collin R, Johansson B, Tuner M (2014) Combustion stratification with partially premixed combustion, PPC, using NVO and split injection in a LD—diesel engine. https://doi.org/10.4271/2014-01-2677

Tanov S, Wang Z, Wang H, Richter M, Johansson B (2015) Effects of injection strategies on fluid flow and turbulence in partially premixed combustion (PPC) in a light duty engine

Tap F, Veynante D (2005) Simulation of flame lift-off on a diesel jet using a generalized flame surface density modeling approach. Proc Combust Inst 30(1):919–926

Vishwanathan G, Reitz RD (2009) Modeling soot formation using reduced polycyclic aromatic hydrocarbon chemistry in n-heptane lifted flames with application to low temperature combustion. J Eng Gas Turbines Power 131(3):032801

Wang H, Jiao Q, Yao M, Yang B, Qiu L, Reitz RD (2013) Development of an n-heptane/toluene/polyaromatic hydrocarbon mechanism and its application for combustion and soot prediction. Int J Engine Res 14(5):434–451. https://doi.org/10.1177/1468087412471056

Wang X, Huang Z, Zhang W, Kuti OA, Nishida K (2011) Effects of ultra-high injection pressure and micro-hole nozzle on flame structure and soot formation of impinging diesel spray. Appl Energy 88(5):1620–1628

Zeng W, Xu M, Zhang G, Zhang Y, Cleary DJ (2012) Atomization and vaporization for flash-boiling multi-hole sprays with alcohol fuels. Fuel 95:287–297

Zhang A, Cung K, Lee S-Y, Naber J, Huberts G, Czekala M, Qu Q (2013) The impact of spark discharge pattern on flame initiation in a turbulent lean and dilute mixture in a pressurized. https://doi.org/10.4271/2013-01-1627

Zhao H, Ladommatos N (2001) Engine combustion instrumentation and diagnostics, vol 842. Society of Automotive Engineers, Warrendale, PA

Zhu J, Kuti OA, Nishida K (2013) An investigation of the effects of fuel injection pressure, ambient gas density and nozzle hole diameter on surrounding gas flow of a single diesel spray by the laser-induced fluorescence–particle image velocimetry technique. Int J Engine Res 14(6):630–645

Chapter 12
Modelling of Variance and Co-variance in Turbulent Flame–Droplet Interaction: A Direct Numerical Simulation Analysis

Sean P. Malkeson, Daniel H. Wacks and Nilanjan Chakraborty

Abstract The advancement of high-performance computing has made computational fluid dynamics (CFD) simulations a viable alternative to expensive experimentation. Most industrial flows are innately turbulent, and the modelling of turbulent flow remains a challenging task. This complexity is augmented in turbulent droplet combustion simulations by the complex interaction of heat and mass transfer associated with evaporation, fluid dynamics, combustion and heat release. As a result of this, the fidelity of CFD simulations of turbulent reacting flows remains sensitive to the accuracy of the combustion modelling, which principally focuses on the prediction of mean chemical reaction and heat release rates. The closure of the mean reaction rate in the context of Reynolds-averaged Navier–Stokes (RANS) simulations in the combustion of turbulent droplet-laden mixtures often requires the knowledge of the variances of the fuel mass fraction Y_F fluctuations $\left(\widetilde{Y_F''^2}\right)$, the mixture fraction ξ fluctuations $\left(\widetilde{\xi''^2}\right)$ and co-variance $\left(\widetilde{Y_F''\xi''}\right)$, where $\bar{q}$, $\tilde{q}=\overline{\rho q}/\bar{\rho}$ and $q''=q-\tilde{q}$ are Reynolds average, Favre mean and Favre fluctuation of a general quantity q and ρ is the gas density. Algebraic and transport equation-based closures of $\widetilde{Y_F''^2}$, $\widetilde{\xi''^2}$ and $\widetilde{Y_F''\xi''}$ have previously been considered in the context of purely gaseous phase combustion where variations in equivalence ratio exist. Whilst limited effort has been directed to the modelling of the fuel mass fraction variance $\widetilde{Y_F''^2}$ and mixture fraction variance $\widetilde{\xi''^2}$ for turbulent combustion in

S. P. Malkeson
MMI Engineering Ltd., The Brew House, Wilderspool Park, Greenall's Avenue, Warrington WA4 6HL, UK
e-mail: smalkeson@mmiengineering.com

D. H. Wacks
School of Engineering and Computing Sciences, Durham University, Durham DH1 3LE, UK
e-mail: daniel.h.wacks@durham.ac.uk

N. Chakraborty (✉)
School of Mechanical and Systems Engineering, University of Newcastle, Newcastle Upon Tyne NE1 7RU, UK
e-mail: nilanjan.chakraborty@ncl.ac.uk

S. Basu et al. (eds.), *Droplets and Sprays*, Energy, Environment, and Sustainability,
https://doi.org/10.1007/978-981-10-7449-3_12

droplet-laden mixtures, the statistical behaviour of $\widetilde{Y_F''\xi''}$ and its transport in turbulent spray flames are yet to be considered in detail. Furthermore, the validity of existing closures for $\widetilde{Y_F''\xi''}$ and the unclosed terms of its transport equation, which were originally proposed for purely gaseous phase combustion, are yet to be assessed for turbulent spray flames. These gaps in the existing literature are addressed by analysing the statistical behaviours of $\widetilde{Y_F''^2}$, $\widetilde{\xi''^2}$ and $\widetilde{Y_F''\xi''}$ as well as the terms of their transport equations using a three-dimensional compressible Direct Numerical Simulation (DNS) database of statistically planar turbulent flames propagating into droplet-laden mixtures where the fuel is supplied in the form of monodisperse droplets ahead of the flame. This chapter focuses on the effects of droplet diameter a_d and droplet equivalence ratio ϕ_d (i.e. fuel in liquid droplets to air ratio by mass, normalized by fuel-to-air ratio by mass under stoichiometric conditions) on the statistical behaviours of $\widetilde{Y_F''^2}$, $\widetilde{\xi''^2}$ and $\widetilde{Y_F''\xi''}$ and their transport in detail. Furthermore, the validity of the existing models for the unclosed terms of $\widetilde{Y_F''^2}$, $\widetilde{\xi''^2}$ and $\widetilde{Y_F''\xi''}$ transport equations, which were originally proposed for gaseous stratified mixture combustion, has been assessed for turbulent combustion in droplet-laden mixtures. Based on this exercise, either the modification of existing models has been suggested or new models are proposed, wherever necessary, based on the physical insights extracted from DNS data.

Nomenclature

Arabic

a_d	Droplet diameter
A	Coefficient which determines fuel mass fraction distribution on Burke–Schumann diagram
B_d	Spalding mass transfer number
c	Reaction progress variable
C_P	Specific heat at constant pressure
C_u	Correction for drag coefficient
C_v	Specific heat at constant volume
$C_{LR}, C_Y, C_\xi, C_{Y\xi}, C_{T_4}$	Model parameters
D	Mass diffusivity
D_0	Diffusivity in unburned gas
D_{Y1}	Molecular diffusion term in the fuel mass fraction variance transport equation
D_{Y2}	Dissipation term in the fuel mass fraction variance transport equation
$D_{\xi 1}$	Molecular diffusion term in the mixture fraction variance transport equation
$D_{\xi 2}$	Dissipation term in the mixture fraction variance transport equation

$D_{Y\xi 1}$	Molecular diffusion term in the co-variance transport equation
$D_{Y\xi 2}$	Dissipation term in the co-variance transport equation
Da	Damköhler number
$\tilde{k}$	Turbulent kinetic energy
L_{11}	Integral length scale for turbulent velocity fluctuation
L_V	Latent heat of droplet evaporation
m	Model parameter
Nu_c	Corrected Nusselt number for droplets
p	Pressure
p_F^S	Partial pressure at the droplet surface
$P(Y_F)$	PDF of fuel mass fraction Y_F
$P(\xi\|Y_F)$	PDF of mixture fraction ξ conditional on fuel mass fraction Y_F
$P(Y_F, \xi)$	Joint PDF between fuel mass fraction Y_F and mixture fraction ξ
$\tilde{P}(Y_F, \xi)$	Favre joint PDF between fuel mass fraction Y_F and mixture fraction ξ
Pr	Prandtl number
$\bar{q}$	Reynolds-averaged value of a general quantity
$\tilde{q}$	Favre-averaged value of a general quantity
q''	Favre fluctuation of a general quantity
Re_d	Droplet Reynolds number
s	Ratio of oxidizer to fuel by mass under stoichiometric condition
S	Segregation factor
S_{mod}	Modified segregation factor
Sc	Schmidt number
Sh_c	Corrected Sherwood number
$S_{b(\phi_g)}$	Unstrained laminar burning velocity at equivalence ratio ϕ_g
t	Time
t_{chem}	Chemical time scale
t_e	Initial turbulent eddy turnover time
T	Non-dimensional temperature
$\hat{T}$	Dimensional temperature
$T_{ad(\phi_g)}$	Adiabatic flame temperature
T_d	Dimensional droplet temperature
T_0	Unburned gas temperature
T_{Y1}	Turbulent transport term in the fuel mass fraction variance transport equation
T_{Y2}	Generation/destruction term in the fuel mass fraction variance transport equation due to scalar flux

T_{Y3}	Reaction rate contribution to the fuel mass fraction variance transport equation
T_{Y4}	Droplet evaporation contribution to the fuel mass fraction variance transport equation
$T_{\xi 1}$	Turbulent transport term in the mixture fraction variance transport equation
$T_{\xi 2}$	Generation/destruction term in the mixture fraction variance transport equation due to scalar flux
$T_{\xi 3}$	Droplet evaporation contribution to the mixture fraction variance transport equation
$T_{\xi 4}$	Droplet evaporation contribution to the mixture fraction variance transport equation
$T_{Y\xi 1}$	Turbulent transport term in the co-variance transport equation
$T_{Y\xi 2}$	Generation/destruction term in the co-variance transport equation due to scalar flux
$T_{Y\xi 3}$	Generation/destruction term in the co-variance transport equation due to scalar flux
$T_{Y\xi 4}$	Droplet evaporation contribution to the co-variance transport equation
$T_{Y\xi 5}$	Droplet evaporation contribution to the co-variance transport equation
u_i	Ith component of non-dimensional fluid velocity
u'	Root-mean-square fluctuation velocity
$\vec{u}_d$	Droplet velocity vector
W_F, W_O	Molecular weight of fuel and oxidizer
$\vec{x}_d$	Droplet position vector
x_i	Ith Cartesian co-ordinate
Y_F	Fuel mass fraction
$Y_{F\infty}$	Fuel mass fraction in pure fuel stream
Y_{Fst}	Fuel mass fraction under stoichiometric condition
Y_{max} and Y_{min}	Maximum and minimum values of fuel mass fraction according to the Burke–Schumann relation
Y_O	Oxidizer mass fraction
$Y_{O\infty}$	Oxidizer mass fraction in pure airstream

Greek

α	Heat release parameter
α_T	Thermal diffusivity
α_W	Parameter in the presumed joint PDF $\tilde{P}(Y_F, \xi)$
$\alpha_1, \alpha_2, \alpha_4$	Model parameters
$\beta_1, \beta_2, \beta_4, \beta_\epsilon$	Model parameters

γ	Ratio of specific heats of constant pressure to constant volume in gaseous phase
γ_4	Model parameter
δ_{th}	Thermal laminar premixed flame thickness for the stoichiometric mixture
$\tilde{\varepsilon}$	Dissipation rate of turbulent kinetic energy
$\tilde{\varepsilon}_Y$	Dissipation rate of fuel mass fraction variance
$\tilde{\varepsilon}_\xi$	Dissipation rate of mixture fraction variance
$\tilde{\varepsilon}_{Y\xi}$	Dissipation rate of fuel mass fraction and mixture fraction co-variance
η	Kolmogorov length scale
λ	Thermal conductivity of the gaseous phase
λ_W	Parameter in the presumed joint PDF $\tilde{P}(Y_F, \xi)$
μ	Dynamic viscosity
μ_t	Eddy viscosity
ξ	Mixture fraction
ξ^{max} and ξ^{min}	Maximum and minimum values of mixture fraction within the domain of definition
ξ_{st}	Mixture fraction under stoichiometric condition
ψ, ψ_1	General primitive variable
ρ	Gas density
ρ_d	Droplet density
ρ_0	Unburned gas density
σ	Turbulent Schmidt number
τ	Heat release parameter
τ_d^p, τ_d'' and τ_d^T	Relaxation/decay timescales for droplet velocity, diameter and temperature
ϕ_d	Droplet equivalence ratio
ϕ_g	Equivalence ratio in gaseous phase
$\dot{\omega}_F$	Reaction rate of fuel
$\dot{\omega}_A$ and $\dot{\omega}_B$ ($\dot{\omega}_C$ and $\dot{\omega}_D$)	Fuel reaction rates when the fuel mass fraction values are given by Y_{F11} and Y_{F12} (Y_{F21} and Y_{F22}), respectively, at a mixture fraction ξ_{41} (ξ_{42})
Ω_Y	The term given by $\left[\overline{\dot{\omega}_F Y_F} - \overline{\dot{\omega}_F}\tilde{Y}_F\right]$

Subscript

d	Droplet (i.e. in liquid phase)
g	Gaseous phase
l	Liquid phase
ref	Reference value

Superscript

g Gaseous phase
s Saturated state

Acronyms

CFD Computational fluid dynamics
DNS Direct numerical simulation
PDF Probability density function
RANS Reynolds-averaged Navier–stokes

12.1 Introduction

Flame propagation into turbulent droplet-laden mixtures is of pivotal importance to several engineering applications. These range from internal combustion (IC) engines to aero-gas turbines and the prediction and control of hazards (Aggarwal 1998; Heywood 1998; Lefebvre 1998). A number of experimental (Hayashi et al. 1976; Burgoyne and Cohen 1954; Ballal and Lefebvre 1981; Szekely and Faeth 1983; Aggarwal and Sirignano 1985; Faeth 1987; Nomura et al. 2000; Lawes and Saat 2011), analytical (Silverman et al. 1993; Greenberg et al. 1998) and computational (Watanbe et al. 2007; Watanbe et al. 2008; Fujita et al. 2013; Wacks et al. 2016a) analyses have concentrated on flame propagation in turbulent droplet-laden mixtures and have indicated that the complex physical interactions of evaporative heat and mass transfer, fluid dynamics and combustion thermochemistry are simultaneously at play in turbulent spray combustion. The experimental evidence suggested that there could be significant differences between the overall equivalence ratio (considering fuel in both liquid and gaseous phases) and gaseous equivalence ratio due to incomplete evaporation and that this discrepancy can lead to significant differences in flame behaviour. Furthermore, droplet inertia along with the difference between overall and gaseous equivalence ratios could give rise either to augmentation or reduction of burning rate in quiescent and low turbulence conditions, but that these effects disappear for sufficiently high turbulence intensity (Aggarwal and Sirignano 1985; Nomura et al. 2000; Lawes and Saat 2011). Advancements in high-performance computing have made Direct Numerical Simulation (DNS) a viable alternative to expensive experimentation. However, the modelling of turbulent fluid flow, in general, remains a challenging task. This complexity is augmented in the case of the simulation of turbulent spray flames due to the aforementioned complex physical interaction of heat and mass transfer associated with evaporation, fluid dynamics, combustion and heat release. Notwithstanding this complexity, DNS-based investigations have made significant contributions both in terms of the physical understanding and modelling of the combustion of turbulent droplet-laden mixtures (Miller and Bellan 1999; Reveillon and Vervisch 2000; Wang and Rutland 2005; Reveillon and Demoulin 2007; Sreedhara and Huh 2007; Wandel et al. 2009; Luo et al. 2011; Wandel 2013, 2014; Neophytou et al. 2011, 2012). For example, Neophytou and Mastorakos (2009) analysed the effects of volatility, droplet diameter and droplet equivalence ratio on burning velocity in one-dimensional laminar flames where fuel is supplied in the form of monodisperse droplets. Wacks et al. (2016a) and Wacks and Chakraborty (2016a) recently extended the analysis of Neophytou and Mastorakos (2009) for turbulent flames by carrying out three-dimensional compressible DNS of freely propagating turbulent flame propagation into droplet-laden mixtures.

Most industrial flows involving spray combustion are turbulent in nature and are therefore subject to the full complexity of the flame–droplet interaction described above. Thus, the fidelity of industrial computational fluid dynamics (CFD) simulations of turbulent reacting flows remains sensitive to the accuracy of combustion modelling, which principally focuses on the prediction of mean chemical reaction and heat release rates. The closure of the mean reaction rate in the context of Reynolds-averaged Navier–Stokes (RANS) simulations in turbulent combustion often requires knowledge of the variance of the fuel mass fraction (Y_F) fluctuations $\widetilde{Y_F''^2} = \overline{\rho Y_F''^2}/\bar{\rho}$ (Topaldi et al. 2000; Ashayek and Jacobs 2001; Li and Zhou 2012; Ribert et al. 2005; Robin et al. 2006), the variance of the mixture fraction (ξ) fluctuations $\widetilde{\xi''^2} = \overline{\rho \xi''^2}/\bar{\rho}$ (Robin et al. 2006) and the co-variance of the fuel mass fraction (Y_F) and mixture fraction (ξ) fluctuations $\widetilde{Y_F''\xi''} = \overline{\rho Y_F''\xi''}/\bar{\rho}$ (Robin et al. 2006; Malkeson and Chakraborty 2013), where $\bar{q}$, $\tilde{q} = \overline{\rho q}/\bar{\rho}$ and $q'' = q - \tilde{q}$ are Reynolds average, Favre mean and Favre fluctuation of a general quantity q and ρ is the gas density. Algebraic and transport equation-based closures of $\widetilde{Y_F''^2}$ (Ribert et al. 2005; Robin et al. 2006; Mura et al. 2007; Malkeson and Chakraborty 2010a, b), $\widetilde{\xi''^2}$ (Robin et al. 2006; Mura et al. 2007; Malkeson and Chakraborty 2010b) and $\widetilde{Y_F''\xi''}$ (Mura et al. 2007; Malkeson and Chakraborty 2013) have previously been considered in the context of purely gaseous phase combustion where variations in equivalence ratio exist.

Previous studies on droplet combustion analysed the modelling of the mixture fraction variance $\widetilde{\xi''^2}$ (Reveillon and Vervisch 2000; Sreedhara and Huh 2007). Whilst the statistical behaviour of $\widetilde{Y_F''^2}$ and its transport in the context of the combustion of droplet-laden mixtures have recently been assessed (Wacks et al. 2016b), the statistical behaviour of $\widetilde{Y_F''\xi''}$ and its transport is yet to be examined in detail in the context of the combustion of droplet-laden mixtures. Furthermore, the validities of existing closures of $\widetilde{Y_F''^2}$ and the unclosed terms of its transport equation, which were originally proposed for purely gaseous phase combustion, have been assessed for turbulent spray flames (Wacks et al. 2016b), but this is yet to be done for $\widetilde{Y_F''\xi''}$.

These gaps in the existing literature have been addressed in this chapter by analysing the statistical behaviours of $\widetilde{Y_F''^2}$, $\widetilde{\xi''^2}$ and $\widetilde{Y_F''\xi''}$ and the terms of their transport equation using a three-dimensional compressible Direct Numerical Simulation (DNS) database (Wacks et al. 2016a; Wacks and Chakraborty 2016a, b) of statistically planar turbulent flames propagating into droplet-laden mixtures where the fuel is supplied in the form of monodisperse droplets ahead of the flame. In this respect, it is worth noting that the description given by "DNS" applies only for the carrier gaseous phase in this chapter, whereas the droplets are considered as monodisperse spherical particles which are tracked in Lagrangian sense and treated as point sources instead of resolving them. This chapter considers selected cases from a large database (Wacks et al. 2016a, Wacks and Chakraborty 2016a, b) so that the effects of droplet diameter a_d and droplet equivalence ratio ϕ_d (i.e. fuel in liquid droplets to air ratio by mass, normalized by fuel-to-air ratio by mass in gaseous phase under stoichiometric condition) on the statistical behaviours of $\widetilde{Y_F''^2}$, $\widetilde{\xi''^2}$ and $\widetilde{Y_F''\xi''}$ and their transport can be analysed in detail. The main objectives of this chapter are:

(a) To analyse the statistical behaviours of $\widetilde{Y_F''^2}$, $\widetilde{\xi''^2}$ and $\widetilde{Y_F''\xi''}$ and the various unclosed terms of their transport equation for turbulent spray flames in the context of RANS.

(b) To assess the validity of the existing models for the unclosed terms of $\widetilde{Y_F''^2}$, $\widetilde{\xi''^2}$ and $\widetilde{Y_F''\xi''}$ transport equations for turbulent combustion of droplet-laden mixtures.

12.2 Mathematical Formulation of Flame–Droplet Interaction

The mathematical formulation is similar to the one previously followed and discussed in Wacks et al. (2016a) and Wacks and Chakraborty (2016a, b), but is presented here for completeness. The extensive parametric analysis presented in Wacks et al. (2016a) and Wacks and Chakraborty (2016a, b) has been performed using a single-step irreversible Arrhenius-type chemical mechanism in order to mitigate the exorbitant computational cost which such analysis typically engenders:

$$\text{Fuel} + \text{s} \cdot \text{Oxidiser} \rightarrow (1 + \text{s}) \cdot \text{Products} \quad (12.1)$$

where s is the oxidizer-to-fuel ratio by mass (i.e. the mass of oxygen consumed per unit mass of fuel). The fuel reaction rate $\dot{\omega}_F$ is expressed as:

$$\dot{\omega}_F = -\rho B^* Y_F Y_O \exp\left(-\frac{\beta(1-T)}{1-\alpha(1-T)}\right) \tag{12.2}$$

where Y_F and Y_O are the fuel and oxidizer mass fractions, respectively, and ρ is the gas density. In Eq. 12.2, T, the non-dimensional temperature, β, the Zeldovich number, α, a heat release parameter, and B^*, the normalized pre-exponential factor, are given by the following expressions:

$$T = \left(\frac{\hat{T} - T_0}{T_{ad(\phi_g=1)} - T_0}\right) \tag{12.3a}$$

$$\beta = \frac{E_{ac}(T_{ad(\phi_g=1)} - T_0)}{R_0 T^2_{ad(\phi_g=1)}} \tag{12.3b}$$

$$\alpha = \frac{\tau}{1+\tau} = \frac{(T_{ad(\phi_g=1)} - T_0)}{T_{ad(\phi_g=1)}} \tag{12.3c}$$

$$B^* = B\exp\left(-\frac{\beta}{\alpha}\right) \tag{12.3d}$$

where $\hat{T}$ is the instantaneous dimensional temperature, T_0 is the unburned gas temperature, $T_{ad(\phi_g=1)}$ is the adiabatic flame temperature for the stoichiometric mixture, E_{ac} is the activation energy, R_0 is the universal gas constant, B is the pre-exponential factor, and $\tau = (T_{ad(\phi_g=1)} - T_0)/T_0$ is a heat release parameter. A modified single-step chemical mechanism proposed by Tarrazo et al. (2006) has been utilized by Wacks et al. (2016a) and Wacks and Chakraborty (2016a, b), and in this framework the activation energy, E_{ac}, and the heat of combustion are taken to be functions of the gaseous equivalence ratio, ϕ_g, which results in more accurate predictions of the equivalence ratio ϕ_g dependence of the unstrained laminar burning velocity $S_{b(\phi_g)}$ in hydrocarbon–air flames, especially for fuel-rich mixtures. According to Tarrazo et al. (2006), the Zel'dovich number, β, is expressed as $\beta = 6f\left(\phi_g\right)$, where:

$$f(\phi_g) = \begin{cases} 1.0 + 8.250(\phi_g - 1.00)^2 & ,\phi_g \le 0.64 \\ 1.0 & ,0.64 < \phi_g < 1.07 \\ 1.0 + 1.443(\phi_g - 1.07)^2 & ,\phi_g \ge 1.07 \end{cases} \tag{12.4}$$

Also according to Tarrazo et al. (2006), the heat release per unit mass of fuel $H_{\phi_g} = \left[\left(T_{ad(\phi_g)} - T_0\right)C_P\right] / \left[Y_{F0(\phi_g)} - Y_{Fb(\phi_g)}\right]$ is given by $H_{\phi_g}/H_{\phi_g=1} = 1$ for $\phi_g \le 1$ and $H_{\phi_g}/H_{\phi_g=1} = 1 - \alpha_H(\phi_g - 1)$ for $\phi_g > 1$, where $\alpha_H = 0.18$ and $Y_{F0(\phi_g)}$

and $Y_{Fb(\phi_g)}$ are the fuel mass fraction in the unburned gas and fully burned gas, respectively, for a premixed flame of equivalence ratio ϕ_g. For the investigation of Wacks et al. (2016a) and Wacks and Chakraborty (2016a, b), the Lewis numbers of all species were taken to be equal to unity and all species in gaseous phase were considered to be perfect gases. Standard values have been taken for the ratio of specific heats ($\gamma = C_P^g/C_V^g = 1.4$, where C_P^g and C_V^g are the specific heats at constant pressure and volume for the gaseous mixture, respectively) and Prandtl number ($Pr = \mu C_P^g/\lambda = 0.7$, where μ and λ are the dynamic viscosity and thermal conductivity, respectively).

The droplet transport equations used in several previous analyses (Reveillon and Vervisch 2000; Wang and Rutland 2005; Reveillon and Demoulin 2007; Sreedhara and Huh 2007; Wandel et al. 2009; Luo et al. 2011; Wandel 2013, 2014; Neophytou et al. 2011, 2012) have been adopted by Wacks et al. (2016a) and Wacks and Chakraborty (2016a, b). The position, $\vec{x}_d$, velocity, $\vec{u}_d$, diameter, a_d, and temperature, T_d, of individual droplets are tracked in the Lagrangian manner, and the relevant transport equations for these quantities are given by:

$$\frac{d\vec{x}_d}{dt} = \vec{u}_d \tag{12.5}$$

$$\frac{d\vec{u}_d}{dt} = \frac{\vec{u}(\vec{x}_d, t) - \vec{u}_d}{\tau_d^P} \tag{12.6}$$

$$\frac{da_d^2}{dt} = -\frac{a_d^2}{\tau_d^u} \tag{12.7}$$

$$\frac{dT_d}{dt} = -\frac{\hat{T}(\vec{x}_d, t) - T_d - B_d L_v/C_P^g}{\tau_d^T} \tag{12.8}$$

where L_v is the latent heat of vaporization, and τ_d^P, τ_d^u and τ_d^T are relaxation timescales associated with droplet velocity, diameter and temperature, respectively, which are given by:

$$\tau_d^P = \frac{\rho_d a_d^2}{18 C_u \mu} \tag{12.9a}$$

$$\tau_d^u = \frac{\rho_d a_d^2}{4\mu} \frac{Sc}{Sh_c} \frac{1}{\ln(1 + B_d)} \tag{12.9b}$$

$$\tau_d^T = \frac{\rho_d a_d^2}{6\mu} \frac{Pr}{Nu_c} \frac{B_d}{\ln(1 + B_d)} \frac{C_P^L}{C_P^g} \tag{12.9c}$$

where ρ_d is the droplet density, C_p^L is the specific heat for the fuel in liquid phase, and C_u is the correction for drag coefficient and is expressed as (Clift et al. 1978):

$$C_u = 1 + \frac{1}{6}\mathrm{Re}_d^{2/3} \tag{12.10}$$

In Eqs. 12.9a-12.10, Re_d is the droplet Reynolds number, Sc is the Schmidt number, B_d is the Spalding mass transfer number, Sh_c is the corrected Sherwood number, and Nu_c is the corrected Nusselt number, which are expressed as (Clift et al. 1978; Reveillon and Vervisch 2000; Wandel et al. 2009; Luo et al. 2011; Wandel 2013, 2014; Neophytou et al. 2011, 2012):

$$\mathrm{Re}_d = \frac{\rho\left|\vec{u}(\vec{x}_d, t) - \vec{u}_d\right| a_d}{\mu} \tag{12.11}$$

$$B_d = \frac{Y_F^s - Y_F(\vec{x}_d, t)}{1 - Y_F^s} \tag{12.12}$$

$$\mathrm{Sh}_c = \mathrm{Nu}_c = 2 + \frac{0.555\mathrm{Re}_d\mathrm{Sc}}{(1.232 + \mathrm{Re}_d\mathrm{Sc}^{4/3})^{1/2}} \tag{12.13}$$

where Y_F^s is the value of fuel mass fraction Y_F on the droplet surface. It should be noted that the unity Lewis number assumption has been implicitly invoked in Eqs. 12.11–12.13. Furthermore, the Clausius–Clapeyron relation for the partial pressure of the fuel vapour at the droplet surface, p_F^s, is utilized to calculate the Spalding number B_d:

$$p_F^s = p_{ref}\exp\left(\frac{L_v}{R_0}\left[\frac{1}{T_{ref}^s} - \frac{1}{T_d^s}\right]\right); Y_F^s = \left(1 + \frac{W_O}{W_F}\left[\frac{p(\vec{x}_d, t)}{p_F^s} - 1\right]\right)^{-1} \tag{12.14}$$

where T_{ref}^s is the boiling point of the fuel at reference pressure p_{ref}, and the droplet surface temperature T_d^s is taken to be T_d, and W_O and W_F are the molecular weights of oxidizer and fuel, respectively.

The conservation equations in the gaseous phase can be generically written as (Reveillon and Vervisch 2000; Wandel et al. 2009; Wandel 2013, 2014; Neophytou et al. 2011, 2012):

$$\frac{\partial\rho\psi}{\partial t} + \frac{\partial\rho u_j\psi}{\partial x_j} = \frac{\partial}{\partial x_j}\left(\Gamma_\psi\frac{\partial\psi_1}{\partial x_j}\right) + \dot{\omega}_\psi + \dot{S}_g + \dot{S}_\psi \tag{12.15a}$$

where $\psi = \{1, u_j, e, Y_F, Y_O\}$ for the transport equations of mass, momentum, energy and mass fractions, respectively, $\psi_1 = \{1, u_j, \hat{T}, Y_F, Y_O\}$ for $\psi = \{1, u_j, e, Y_F, Y_O\}$

and $\Gamma_\psi = \mu/\sigma_\psi$ and λ for $\psi = \{u_j, Y_F, Y_O\}$ and $\psi = e$ respectively, with u_j and $e = \int_{T_0}^{\hat{T}} C_v^g d\hat{T} + u_k u_k/2$ being the velocity in the jth direction and the specific stagnation internal energy, respectively. The quantity σ_ψ is an appropriate Schmidt number corresponding to ψ. The $\dot{\omega}_\psi$ term in Eq. 12.15a arises due to the chemical reaction rate, and $\dot{S}_g$ and $\dot{S}_\psi$ are the appropriate source terms in the gaseous phase and due to droplet evaporation, respectively. The droplet source term $\dot{S}_\psi$ is trilinearly interpolated from the droplet's sub-grid position, $\vec{x}_d$, to the eight surrounding nodes. The droplet source term for any variable ψ may be expressed as (Wandel et al. 2009; Wandel 2013, 2014; Neophytou et al. 2011, 2012):

$$\dot{S}_\psi = -\frac{1}{V}\sum_d \frac{dm_d \psi_d}{dt} \tag{12.15b}$$

where V is the cell volume, $m_d = \rho_d \frac{1}{6}\pi a_d^3$ is the droplet mass, and the summation is carried out over all droplets in the vicinity of each node. As indicated in Eq. 12.15a, the variable ψ is identified as $\psi = \{1, u_j, e, Y_F, Y_O\}$; however, since within the droplets $Y_F = 1.0$, the source term for both the continuity and the fuel mass fraction equations is identical.

Droplet evaporation leads to mixture inhomogeneities, which are characterized by the mixture fraction in the following manner (Bilger 1988):

$$\xi = \frac{(Y_F - Y_O/s + Y_{O\infty}/s)}{(Y_{F\infty} + Y_{O\infty}/s)} \tag{12.16}$$

where $Y_{F\infty} = 1.0$ is the fuel mass fraction in the pure fuel stream and $Y_{O\infty} = 0.233$ is the oxidizer mass fraction in the pure airstream. The fuel used by Wacks et al. (2016a) and Wacks and Chakraborty (2016a, b) is n-heptane, C_7H_{16}, for which $s = 3.52$ and the stoichiometric fuel mass/mixture fraction is given as $Y_{Fst} = \xi_{st} = 0.0621$. Using the definition of the mixture fraction ξ, a reaction progress variable c can be defined in the following manner (Wacks et al. 2016a; Wacks and Chakraborty (2016a, b); Wandel 2013, 2014; Wandel et al. 2009):

$$c = \frac{(1-\xi)Y_{O\infty} - Y_O}{(1-\xi)Y_{O\infty} - \max(0, [\xi_{st} - \xi]/\xi_{st})Y_{O\infty}} \tag{12.17}$$

It should be noted that the reaction progress variable *c* rises monotonically from 0 in the unburned gases to 1 in the fully burned products.

12.2.1 Fuel Mass Fraction Variance $\widetilde{Y_F''^2}$ Transport Equation

In the context of RANS simulations, the transport equation of $\widetilde{Y_F''^2}$ for droplet-laden mixture combustion takes the following form (Wacks et al. 2016b):

$$\frac{\partial\left(\bar{\rho}\widetilde{Y_F''^2}\right)}{\partial t}+\frac{\partial\left(\bar{\rho}\tilde{u}_j\widetilde{Y_F''^2}\right)}{\partial x_j}=\underbrace{\frac{\partial}{\partial x_j}\left(\bar{\rho}\tilde{D}\frac{\partial\widetilde{Y_F''^2}}{\partial x_j}\right)}_{D_{Y1}}-\underbrace{\frac{\partial\left(\overline{\rho u_j''Y_F''^2}\right)}{\partial x_j}}_{T_{Y1}}-\underbrace{2\overline{\rho u_j''Y_F''}\left(\frac{\partial\widetilde{Y}_F}{\partial x_j}\right)}_{T_{Y2}}+\underbrace{2\left(\overline{\dot{\omega}_F Y_F}-\overline{\dot{\omega}_F}\widetilde{Y}_F\right)}_{T_{Y3}}$$
$$+\underbrace{\overline{(2Y_F-Y_F^2)\Gamma}-(2\widetilde{Y}_F-\widetilde{Y_F^2})\overline{\Gamma}}_{T_{Y4}}-\underbrace{2\bar{\rho}\tilde{\varepsilon}_Y}_{D_{Y2}} \quad (12.18)$$

where D is the mass diffusivity, $\dot{\omega}_F$ is the reaction rate of the fuel, Γ is the source term in the mass conservation equation due to evaporation, and $\tilde{\varepsilon}_Y=\left[\overline{(\rho D(\partial Y_F''/\partial x_j)(\partial Y_F''/\partial x_j))}\right]/\bar{\rho}$ is the scalar dissipation rate of fuel mass fraction. The first term in the left-hand side of Eq. 12.18 is the transient term, whilst the second term is the mean advection term. The first term in the right-hand side of Eq. 12.18 is the molecular diffusion term D_{Y1}, whilst the second term is the turbulent transport term T_{Y1}. The term denoted as T_{Y2} represents the generation/destruction of $\widetilde{Y_F''^2}$ by the mean scalar gradient $\partial\widetilde{Y}_F/\partial x_j$. The term denoted as T_{Y3} is the reaction rate contribution term, the term T_{Y4} arises due to the evaporation of the droplets, and the term D_{Y2} accounts for the molecular dissipation of $\widetilde{Y_F''^2}$. In the context of second-moment closure, the terms T_{Y1}, T_{Y3}, T_{Y4} and D_{Y2} are unclosed and it is necessary for these terms to be modelled. The term T_{Y2} can be considered to be not unclosed in the context of second-moment closure, but the accurate estimation of this term depends upon the accurate modelling of the turbulent flux $\overline{\rho u_j''Y_F''}$. The modelling of the terms T_{Y1}, T_{Y3}, T_{Y4} and D_{Y2} will be addressed in Sect. 12.4.2 of this chapter.

12.2.2 Mixture Fraction Variance $\widetilde{\xi''^2}$ Transport Equation

Using the conservation equations of momentum components and mixture fraction, the transport equation of $\widetilde{\xi''^2}$ for droplet-laden mixture combustion can be obtained in the following form:

$$\frac{\partial\left(\bar{\rho}\widetilde{\xi''^2}\right)}{\partial t}+\frac{\partial\left(\bar{\rho}\tilde{u}_j\widetilde{\xi''^2}\right)}{\partial x_j}=\underbrace{\frac{\partial}{\partial x_j}\left(\bar{\rho}\tilde{D}\frac{\partial\widetilde{\xi''^2}}{\partial x_j}\right)}_{D_{\xi1}}-\underbrace{\frac{\partial\left(\overline{\rho u''_j\xi''^2}\right)}{\partial x_j}}_{T_{\xi1}}-\underbrace{2\overline{\rho u''_j\xi''}\left(\frac{\partial\tilde{\xi}}{\partial x_j}\right)}_{T_{\xi2}}+\underbrace{2\left(\overline{\dot{S}_\xi\xi}-\overline{\dot{S}_\xi}\tilde{\xi}\right)}_{T_{\xi3}}$$
$$+\underbrace{\overline{(\xi^2-\tilde{\xi}^2)\Gamma-2\tilde{\xi}(\overline{\Gamma\xi}-\overline{\Gamma}\tilde{\xi})}}_{T_{\xi4}}-\underbrace{2\bar{\rho}\tilde{\varepsilon}_\xi}_{D_{\xi2}} \tag{12.19}$$

where $\dot{S}_\xi=(\dot{S}_F-\dot{S}_O/s)/(Y_{F\infty}+Y_{O\infty}/s)$, $\dot{S}_F=(1-Y_F)\Gamma$ and $\dot{S}_O=-Y_O\Gamma$ are the evaporation contributions and $\tilde{\varepsilon}_\xi=\left[\overline{(\rho D(\partial\xi''/\partial x_j)(\partial\xi''/\partial x_j))}\right]/\bar{\rho}$ is the scalar dissipation rate of mixture fraction. The first term in the left-hand side of Eq. 12.19 is the transient term, whilst the second term is the mean advection term. The first term in the right-hand side of Eq. 12.19 is the molecular diffusion term $D_{\xi1}$, whilst the second term is the turbulent transport term $T_{\xi1}$. The term denoted as $T_{\xi2}$ represents the generation/destruction of $\widetilde{\xi''^2}$ by the mean scalar gradient $\partial\tilde{\xi}/\partial x_j$. The term denoted as $T_{\xi3}$ is the evaporation rate contribution term, the term $T_{\xi4}$ arises due to the two-way coupling because of the evaporation of the droplets, and the term $D_{\xi2}$ accounts for the molecular dissipation of $\widetilde{\xi''^2}$. In the context of second-moment closure, the terms $T_{\xi1}$, $T_{\xi3}$, $T_{\xi4}$ and $D_{\xi2}$ are unclosed and it is necessary for these terms to be modelled. The term $T_{\xi2}$ can be considered to be not unclosed in the context of second-moment closure, but its evaluation is dependent upon the accurate modelling of the turbulent flux $\overline{\rho u''_j\xi''}$. The modelling of the terms $T_{\xi1}$, $T_{\xi3}$, $T_{\xi4}$ and $D_{\xi2}$ will be addressed in Sect. 12.4.3 of this chapter.

12.2.3 Co-variance $\widetilde{Y''_F\xi''}$ Transport Equation

Similar to Eqs. 12.18 and 12.19, the transport equation of $\widetilde{Y''_F\xi''}$ for droplet-laden mixture combustion can be derived using the conservation equations of fuel mass fraction, mixture fraction and momentum components and given in the following form:

$$\frac{\partial\left(\bar{\rho}\widetilde{Y''_F\xi''}\right)}{\partial t}+\frac{\partial\left(\bar{\rho}\tilde{u}_j\widetilde{Y''_F\xi''}\right)}{\partial x_j}=\underbrace{\frac{\partial}{\partial x_j}\left(\bar{\rho}\tilde{D}\frac{\partial\widetilde{Y''_F\xi''}}{\partial x_j}\right)}_{D_{Y\xi1}}-\underbrace{\frac{\partial\left(\overline{\rho u''_jY''_F\xi''}\right)}{\partial x_j}}_{T_{Y\xi1}}-\underbrace{\overline{\rho u''_j\xi''}\left(\frac{\partial\tilde{Y}_F}{\partial x_j}\right)}_{T_{Y\xi2}}-\underbrace{\overline{\rho u''_jY''_F}\left(\frac{\partial\tilde{\xi}}{\partial x_j}\right)}_{T_{Y\xi3}}$$
$$+\underbrace{\left(\overline{\dot{\omega}_F\xi}-\overline{\dot{\omega}_F}\tilde{\xi}\right)}_{T_{Y\xi4}}+\underbrace{\overline{\dot{S}_{Y\xi}-\Gamma\xi\left(1-\tilde{Y}_F\right)-Y_F\overline{\Gamma}\tilde{\xi}-\tilde{Y}_F\dot{S}_\xi}}_{T_{Y\xi5}}-\underbrace{2\bar{\rho}\tilde{\varepsilon}_{Y\xi}}_{D_{Y\xi2}} \tag{12.20}$$

where $\overline{\dot{S}_{Y\xi}} = \overline{\Gamma\xi(1 - Y_F)} + \overline{Y_F \dot{S}_\xi} + \overline{\Gamma\xi Y_F}$ is a term which arises due to evaporation and $\widetilde{\epsilon_{Y\xi}} = \left[\overline{(\rho D(\partial Y_F''/\partial x_j)(\partial \xi''/\partial x_j))}\right]/\bar{\rho}$ is the cross scalar dissipation rate. The first term in the left-hand side of Eq. 12.20 is the transient term, whilst the second term is the mean advection term. The first term in the right-hand side of Eq. 12.20 is the molecular diffusion term $D_{Y\xi 1}$, whilst the second term is the turbulent transport term $T_{Y\xi 1}$. The term denoted as $T_{Y\xi 2}$ $(T_{Y\xi 3})$ represents the generation/destruction of $\widetilde{Y_F''\xi''}$ by the mean scalar gradient $\partial \widetilde{Y}_F/\partial x_j$ $(\partial \tilde{\xi}/\partial x_j)$. The term denoted as $T_{Y\xi 4}$ is the reaction rate contribution term, whereas the term $T_{Y\xi 5}$ arises due to the evaporation of the droplets and the term $D_{Y\xi 2}$ accounts for the molecular dissipation of $\widetilde{Y_F''\xi''}$. In the context of second-moment closure, the terms $T_{Y\xi 1}$, $T_{Y\xi 3}$, $T_{Y\xi 4}$ and $D_{Y\xi 2}$ are unclosed and thus need modelling. The terms $T_{Y\xi 2}$ and $T_{Y\xi 3}$ can be considered to be not unclosed in the context of second-moment closure, but their accurate evaluations are dependent upon the accurate modelling of the turbulent fluxes $\overline{\rho u_j''\xi''}$ and $\overline{\rho u_j'' Y_F''}$, respectively. The modelling of the terms $T_{Y\xi 1}$, $T_{Y\xi 4}$, $T_{Y\xi 5}$ and $D_{Y\xi 2}$ will be addressed in Sect. 12.4.4 of this chapter.

12.3 Attributes of DNS Data and Numerical Implementation

A widely used three-dimensional compressible DNS code SENGA (Jenkins and Cant 1999; Wandel et al. 2009; Wandel 2013, 2014; Neophytou et al. 2011, 2012; Wacks et al. 2016a, b; Wacks and Chakraborty 2016a, b) which solves the standard conservation equations of mass, momentum, energy and species in non-dimensional form has been employed by Wacks et al. (2016a) and Wacks and Chakraborty (2016a). In this framework, the spatial discretization for the internal grid points has been carried out using a tenth-order central difference scheme, but the order of differentiation drops gradually to a one-sided second-order scheme at the non-periodic boundaries (Jenkins and Cant 1999). A low-storage third-order explicit Runge–Kutta scheme has been used for time advancement (Wray 1990). A rectangular computational domain of size $63.35D_0/S_{b(\phi_g=1)} \times 42.17D_0/S_{b(\phi_g=1)} \times 42.17D_0/S_{b(\phi_g=1)}$ has been considered for the investigation by Wacks et al. (2016a) and Wacks and Chakraborty (2016a, b) where D_0 is the unburned gas diffusivity. For the thermochemistry used by Wacks et al. (2016a) and Wacks and Chakraborty (2016a, b) the Zel'dovich flame thickness $D_0/S_{b(\phi_g=1)}$ is equal to about $0.625\delta_{th}$ where $\delta_{th} = (T_{ad(\phi_g=1)} - T_0)/\max(|\nabla\hat{T}|)_L$ is the unstrained thermal laminar flame thickness of the stoichiometric laminar flame, and the subscript L refers to the values in an unstrained laminar premixed flame for the stoichiometric mixture. The simulation domain for the computational analysis considered by Wacks and Chakraborty (2016a, b) was discretized using a Cartesian grid of size $384 \times 256 \times 256$ which resolves both the flame thickness, δ_{th}, and the Kolmogorov length scale, η. The boundaries in the mean direction of flame propagation (i.e. x-direction) were taken to be partially non-reflecting, whereas the transverse (i.e. y and z) directions were assumed to be periodic. The Navier–Stokes Characteristic Boundary Condition (NSCBC) (Poinsot and Lele 1992) technique has been used for specifying the non-periodic boundary conditions. The droplets were distributed uniformly in space throughout the y- and z-directions and in the region $0.0 \leq xS_{b(\phi_g=1)}/D_0 \leq 16.53$ ahead of the flame. The initial conditions for the reacting flow field has been generated based on the steady laminar solution obtained for the desired initial values of droplet diameter, a_d, and droplet equivalence ratio, ϕ_d. This initial condition has been generated using software called COSILAB following Neophytou and Mastorakos (2009), where the one-dimensional governing equations for the gas and liquid phases are solved in a coupled manner for spray flames where fuel is supplied in the form of monodisperse droplets on the unburned gas side of the flame. The turbulent velocity fluctuations have been initialized using a standard pseudo-spectral method (Rogallo 1981), and this field is superimposed on the steady laminar spray flame solution generated using COSILAB. For the present analysis, the unburned gas temperature is $T_0 = 300$ K, which gives rise to a heat release parameter $\tau = (T_{ad(\phi_g=1)} - T_0)/T_0 = 6.54$. For all simulations, the fuel was supplied purely in the form of monodisperse droplets with non-dimensional diameters $a_d/\delta_{th} = 0.06, 0.08, 0.10$ for different values of droplet equivalence ratio: $\phi_d = 1.0, 1.25, 1.5, 1.7$ at a distance $16.53 D_0/S_{b(\phi_g=1)}$ from the point in the laminar flame at which $\hat{T} = 400\,K$, which corresponds to a non-dimensional temperature $T \approx 0.05$. The droplet number density ρ_N at $t = 0$ varies between $1.16 \leq (\rho_N)^{1/3}\delta_{th} \leq 2.27$ in the region $0.0 \leq xS_{b(\phi_g=1)}/D_0 \leq 16.53$. In all cases, the liquid volume fraction remains much smaller than 0.01. In all cases, droplets were supplied at the left-hand-side boundary to maintain a constant ϕ_d ahead of the flame. The droplets evaporate as they approach the flame front, and the droplet diameter decreases by at least 25% by the time it reaches the most reactive region of the flame, such that the volume of even the largest droplets is now less than half that of the cell volume, which validates the sub-grid point source treatment of droplets adopted for flame–droplet interactions analysed here since this study is concerned primarily with regions where reaction rate is non-negligible. The droplet to cell size for the analysis by Wacks et al. (2016a) and Wacks and Chakraborty (2016a, b) was consistent with several previous analyses (Reveillon and Vervisch 2000; Wang and Rutland 2005; Sreedhara and Huh 2007; Wandel et al. 2009; Wandel 2013, 2014; Neophytou et al. 2011, 2012).

The simulations were carried out for normalized root-mean-square (rms) turbulent velocities $u'/S_{b(\phi_g=1)} = 4.0$ and 7.5 with a non-dimensional longitudinal integral length scale $L_{11}/\delta_{th} = 2.5$. The ratio of droplet diameter to the Kolmogorov scale for this database is $a_d/\eta \approx 0.3, 0.4, 0.5$ for $a_d/\delta_{th} \approx 0.06, 0.08, 0.1$, respectively, for initial $u'/S_{b(\phi_g=1)} = 7.5$, and these ratios are larger in magnitude for the cases with $u'/S_{b(\phi_g=1)} = 4.0$. The aforementioned ratios of a_d/η remain comparable to several previous analyses (Reveillon and Vervisch 2000; Wang and Rutland 2005; Sreedhara and Huh 2007; Wandel et al. 2009; Wandel 2013, 2014; Neophytou et al. 2011, 2012). The mean-normalized inter-droplet distance s_d/η for the database considered here ranges between 0.0220 and 0.0432 (i.e. $0.0220 < s_d/\eta < 0.0432$) for initial $u'/S_{b(\phi_g=1)} = 7.5$ cases. All simulations have been carried out until $t_{final} = \max(3t_{turb}, 4t_{chem})$, where $t_{turb} = L_{11}/u'$ is the initial turbulent eddy turnover time and $t_{chem} = D_0/S^2_{b(\phi_g=1)}$ is the chemical timescale. This simulation time is either comparable to or greater than the simulation duration used in a number of recent DNS analyses (Reveillon and Vervisch 2000; Wang and Rutland 2005; Sreedhara and Huh 2007; Wandel et al. 2009; Wandel 2013, 2014; Neophytou et al. 2011, 2012), which significantly contributed to the fundamental understanding of turbulent combustion. It was shown by Wacks et al. (2016a) that the volume-integrated reaction rate, flame surface area and burning rate per unit area were not changing rapidly when the statistics have been extracted.

The Reynolds-/Favre-averaged value of a general quantity Q (i.e. $\bar{Q}$ and $\tilde{Q}$) is evaluated by ensemble averaging the relevant quantity Q over the y-z plane at a given x location. The statistical convergence of the Reynolds-/Favre-averaged values has been assessed by comparing the values obtained on full sample size with the corresponding values based on distinct half of the available sample size, and a satisfactory level of agreement has been obtained.

12.4 Result and Discussion

This section will outline the behaviour of the flames considered in the present study. Subsequently, the modelling of fuel mass fraction variance $\widetilde{Y_F''^2}$, mixture fraction variance $\widetilde{\xi''^2}$ and co-variance $\widetilde{Y_F''\xi''}$ will be addressed.

12.4.1 *Flame Behaviour*

Figure 12.1a–c shows the instantaneous distributions of normalized fuel mass fraction Y_F/Y_{Fst}, mixture fraction ξ and non-dimensional temperature $T = (\hat{T} - T_0)/(T_{(ad(\phi_g=1))} - T_0)$ where $\hat{T}$ is the instantaneous dimensional temperature fields in the central $x-z$ plane for $a_d/\delta_{th}=0.08$ and $\phi_d=1.0$ at $t=t_{chem}$, where the black dots indicate the droplets which reside immediately adjacent to the plane. The droplets shrink due to evaporation as they approach the flame, but may not completely evaporate until after passing through the flame. The evaporating droplets absorb latent heat from the background gas (this occurs on both sides of the flame, but is more noticeable on the burned gas side). In many cases, the evaporation of droplets is not complete on their arrival at the flame front, the reaction takes place predominantly under fuel-lean conditions, and, therefore, the heat release due to combustion and the resultant burned gas temperature are lower than the adiabatic flame temperature of the stoichiometric mixture (i.e. $T<1.0$) (Wacks et al. 2016a, Wacks and Chakraborty (2016a, b). The predominant fuel-lean combustion in these cases suggests a slow combustion process and low values of Damköhler number Da (i.e. $Da=(L_{11}S^2_{b(\phi_g=1)})/(u'D_0)<1$ (Wacks et al. 2016a, Wacks and Chakraborty 2016a).

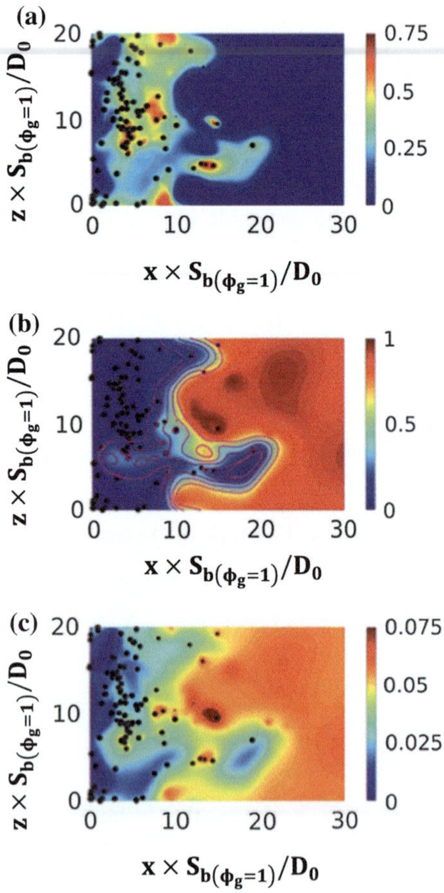

Fig. 12.1 Instantaneous fields of **a** normalized fuel mass fraction, Y_F/Y_{Fst}, **b** normalized temperature T with c isolines (left to right $c=0.1, 0.3, 0.5, 0.7, 0.9$) in magenta and **c** mixture fraction ξ fields at the central $x-z$ plane at $t=4.0t_{chem}$ for case $a_d/\delta_{th}=0.08$, $\phi_d=1.0$. Droplets are shown by black dots (not to scale). The stoichiometric mixture fraction is $\xi_{st}=0.0621$

12.4.2 *Modelling of Fuel Mass Fraction Variance* $\widetilde{Y_F''^2}$

The fuel mass fraction variance $\widetilde{Y_F''^2}$ can be modelled by either an algebraic expression or by a modelled transport equation. We will consider both modelling approaches in this section.

12.4.2.1 Algebraic Modelling of Fuel Mass Fraction Variance $\widetilde{Y_F''^2}$

Mura et al. (2007) developed an algebraic model for $\widetilde{Y_F''^2}$ for turbulent stratified gaseous mixture combustion by using an approximated Favre joint PDF between Y_F and ξ (i.e. $\tilde{P}(Y_F, \xi) = \rho P(Y_F, \xi)/\bar{\rho}$) in the following manner:

$$\begin{aligned}\tilde{P}(Y_F, \xi) = {} & \lambda_w \tilde{P}(\xi|Y_{max})\delta(Y_F - Y_{max}(\xi)) \\ & + (1 - \lambda_w)\tilde{P}(\xi|Y_{min})\delta(Y_F - Y_{min}(\xi)) + O(1/Da)\end{aligned} \tag{12.21}$$

where $\tilde{P}(\xi|Y_F)$ is the Favre PDF of ξ conditional on Y_F and the quantities $Y_{max}(\xi) = \xi$ and $Y_{min}(\xi) = A(\xi)(\xi - \xi_{st})$ are maximum and minimum values of Y_F, according to the Burke–Schumann relations (Poinsot and Veynante 2001), where $A(\xi) = H(\xi - \xi_{st})/(1 - \xi_{st})$ with $H(\xi - \xi_{st})$ being a Heaviside function. For $Da \gg 1$, the last term on the right-hand side of Eq. 12.21 disappears and λ_w is unlikely to depend on ξ, which gives $\tilde{P}(\xi|Y_{max}) = \tilde{P}(\xi|Y_{min})$ (Mura et al. 2007).

The contours of Favre joint PDFs of Y_F and ξ at $\tilde{c} = 0.1, 0.3, 0.5$ for $a_d/\delta_{th} = 0.10$, $\phi_d = 1.0$ are shown in Fig. 12.2a–c, and the same qualitative behaviour has been observed for other cases. Figure 12.2a–c indicates that $\tilde{P}(Y_F, \xi)$, in these cases, cannot be accurately approximated by discrete delta functions as suggested by Eq. 12.21 due to low Da effects.

Figure 12.3a–f shows the variations of $\tilde{Y}_F$, $\tilde{Y}_{max} = \tilde{\xi}$ and $\tilde{Y}_{min} = \widetilde{H(\xi - \xi_{st})}/(1 - \xi_{st})$ with $\tilde{c}$ across the flame brush for all cases. In all cases considered here, $\tilde{Y}_F$ remains smaller than Y_{Fst} across the flame brush indicating predominant fuel-lean combustion. Moreover, $\tilde{Y}_F$ remains small at $\tilde{c} = 0$ as the fuel is supplied purely in the form of droplets and gaseous fuel becomes available only with the evaporation of droplets as the flame is approached. In all cases, $\tilde{Y}_F$ attains a peak value somewhere in the middle of the flame brush before decreasing towards the burned gas side due to the consumption of the gaseous fuel. The peak value of $\tilde{Y}_F$ reduces and shifts towards the burned gas side of the flame brush for an increasing value of a_d as the evaporation rate is slower for the larger droplets. For $\phi_d = 1.0$ cases, the value of $\tilde{Y}_F$ vanishes at $\tilde{c} = 1$ as the fuel droplets have predominantly been evaporated within the flame brush and the fuel is subsequently consumed. However, for $\phi_d = 1.7$ cases, the droplets continue to evaporate towards the burned gas side of the flame brush and evaporation increases towards the

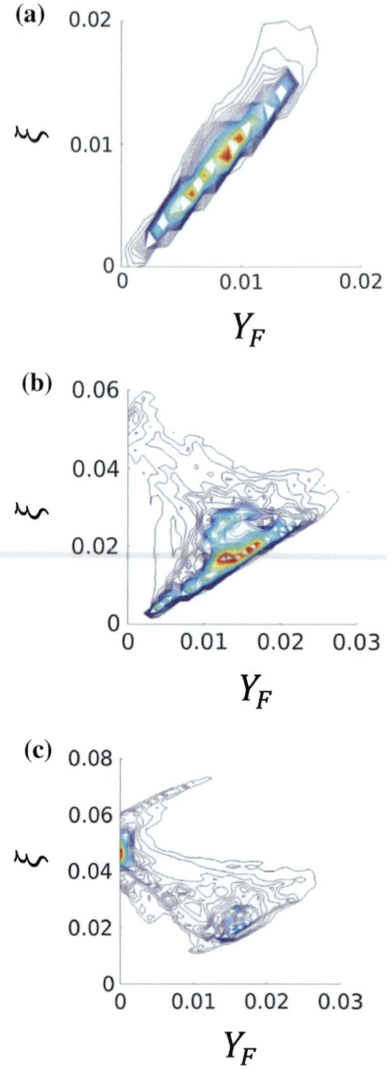

Fig. 12.2 Contours of Favre joint PDF $\tilde{P}(Y_F, \xi)$ at **a** $\tilde{c} = 0.1$, **b** $\tilde{c} = 0.3$ and **c** $\tilde{c} = 0.5$ for case $a_d/\delta_{th} = 0.10, \phi_d = 1.0$. Value rises from blue to red

reaction zone of the flame (i.e. $0.7 < \tilde{c} < 0.9$). As such, $\tilde{Y}_F$ begins to rise again towards $\tilde{c} = 1$ for these cases. The value of $\tilde{Y}_{max} = \tilde{\xi}$ has been found to rise with $\tilde{c}$ across the flame brush as the fuel droplets evaporate, but it remains smaller than ξ_{st} for all $\phi_d = 1.0$ cases. However, towards the burned gas side of the flame brush $\tilde{Y}_{max} > \xi_{st}$ for $\phi_d = 1.7$ cases, although the extent of this reduces as a_d increases. The value of $\tilde{Y}_{min} \approx 0.0$ for all $\phi_d = 1.0$ cases due to predominant fuel-lean combustion, but $\tilde{Y}_{min} > 0.0$ has been observed towards the burned gas side of the flame brush for

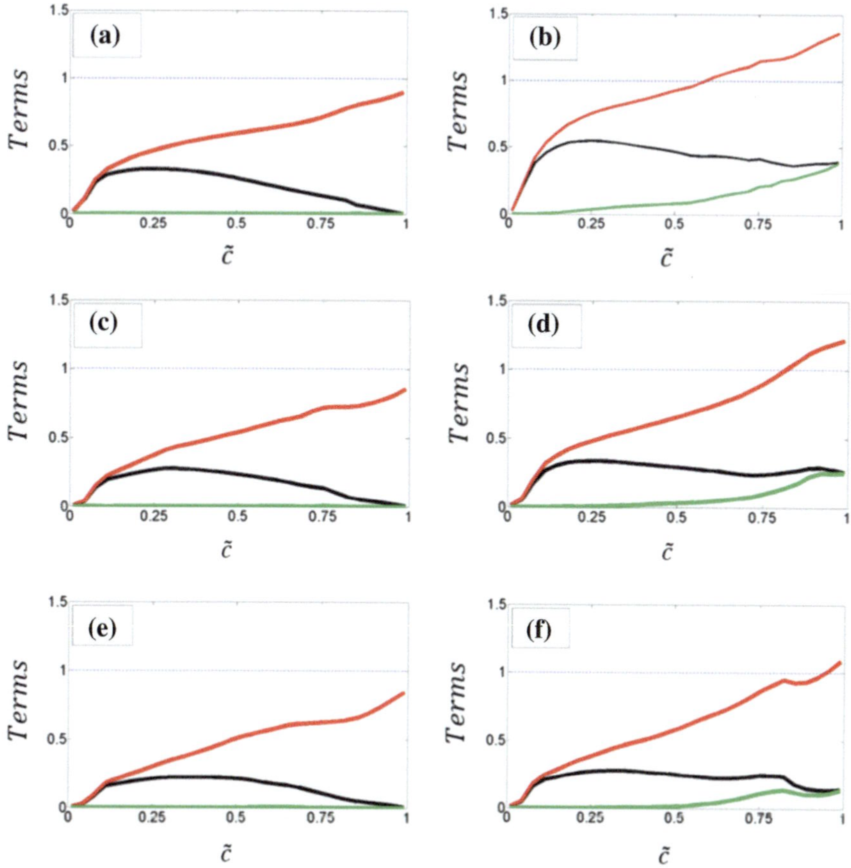

Fig. 12.3 Variations of $\tilde{Y}_F/Y_{Fst}$, $\tilde{Y}_{max}/Y_{Fst}$ and $\tilde{Y}_{min}/Y_{Fst}$ [black, red, green] with $\tilde{c}$ across the flame brush for **a** $a_d/\delta_{th}=0.06$, $\phi_d=1.0$; **b** $a_d/\delta_{th}=0.06$, $\phi_d=1.7$; **c** $a_d/\delta_{th}=0.08$, $\phi_d=1.0$; **d** $a_d/\delta_{th}=0.08$, $\phi_d=1.7$; **e** $a_d/\delta_{th}=0.10$, $\phi_d=1.0$; and **f** $a_d/\delta_{th}=0.10$, $\phi_d=1.7$. The horizontal blue line indicates the stoichiometric value

$\phi_d=1.70$ cases. The variations of $\tilde{Y}_F$, $\tilde{Y}_{max}$ and $\tilde{Y}_{min}$ play key roles in the algebraic closure proposed by Mura et al. (2007) for $\widetilde{Y_F''^2}$ under the assumption of $Da \gg 1$:

$$\widetilde{Y_F''^2} = (\tilde{Y}_{max}-\tilde{Y}_F)(\tilde{Y}_F-\tilde{Y}_{min}) + \underbrace{\left[(\tilde{Y}_F-\tilde{Y}_{min})/(\tilde{Y}_{max}-\tilde{Y}_{min}) + \left\{(\tilde{Y}_{max}-\tilde{Y}_F)/(\tilde{Y}_{max}-\tilde{Y}_{min})\right\}\tilde{A}^2\right]\widetilde{\xi''^2}}_{Q_{s1}} \quad (12.22)$$

Equation 12.22 is obtained using $\tilde{Y}_F = \iint Y_F \tilde{P}(Y_F, \xi) dY_F d\xi$ and $\widetilde{Y_F''^2} = \iint (Y_F - \tilde{Y}_F)^2 \tilde{P}(Y_F, \xi) dY_F d\xi$ which also imply that $\lambda_W = (\tilde{Y}_F - \tilde{Y}_{min})/(\tilde{Y}_{max} - \tilde{Y}_{min})$. The second term on the right-hand side of Eq. 12.22 originates due to mixture inhomogeneity. Figure 12.3a–f shows the variations of $\widetilde{Y_F''^2}$ and $\widetilde{\xi''^2}$ with $\tilde{c}$ across the flame brush, which show that $\widetilde{\xi''^2}$ remains larger than $\widetilde{Y_F''^2}$ in all cases across the flame brush (most noticeably for high values of a_d), except for the case with small a_d and ϕ_d where $\widetilde{Y_F''^2} > \widetilde{\xi''^2}$ towards the centre of the flame brush. There is a clear effect on the general behaviour of $\widetilde{Y_F''^2}$ and $\widetilde{\xi''^2}$ due to a_d, as evaporation of the fuel droplets continues to occur towards the burned gas side of the flame brush for the large droplet cases. Figure 12.4a–f shows that Eq. 12.22 captures neither the quantitative nor qualitative behaviour of $\widetilde{Y_F''^2}$. As discussed previously, Fig. 12.2a–c shows that $\tilde{P}(Y_F, \xi)$ cannot be adequately represented by Eq. 12.21 here and therefore Eq. 12.22 does not satisfactorily predict $\widetilde{Y_F''^2}$.

Mura et al. (2007) defined a segregation factor S as:

$$S = \left\{ \widetilde{Y_F''^2} - \left[(\tilde{Y}_F - \tilde{Y}_{min})/(\tilde{Y}_{max} - \tilde{Y}_{min}) \right.\right.$$
$$\left.\left. + \{ (\tilde{Y}_{max} - \tilde{Y}_F)/(\tilde{Y}_{max} - \tilde{Y}_{min}) \} \tilde{A}^2 \right] \widetilde{\xi''^2} \right\} / \left[(\tilde{Y}_{max} - \tilde{Y}_F)(\tilde{Y}_F - \tilde{Y}_{min}) \right] \Big\} \tag{12.23}$$

One obtains negative values of $Q_{s2} = \widetilde{Y_F''^2} - Q_{s1}$ and S, if $Q_{s1} = \left\{ \left[(\tilde{Y}_F - \tilde{Y}_{min})/(\tilde{Y}_{max} - \tilde{Y}_{min}) + \{ (\tilde{Y}_{max} - \tilde{Y}_F)/(\tilde{Y}_{max} - \tilde{Y}_{min}) \} \tilde{A}^2 \right] \widetilde{\xi''^2} \right\}$ becomes greater than $\widetilde{Y_F''^2}$. Figure 12.4a-f shows the variations of Q_{s1} and Q_{s2} with $\tilde{c}$ across the flame brush which indicate that Q_{s2} assumes negative values towards the unburned gas side of the flame brush in all cases and the extent of this negative value increases with increasing a_d. Note that ξ is not strictly a passive scalar in spray combustion, because evaporation leads to an extra source/sink term in its transport equation. Therefore, the Burke–Schumann relations (Poinsot and Veynante 2001) (where ξ is strictly a passive scalar), based on which the expressions of $\tilde{Y}_{max}, \tilde{Y}_{min}$ and S have been obtained, may not be valid for spray combustion.

Based upon the above analysis, it can be inferred that it may be necessary to consider a modelled transport equation for $\widetilde{Y_F''^2}$ for the combustion of turbulent droplet-laden mixtures in the absence of a satisfactory algebraic closure.

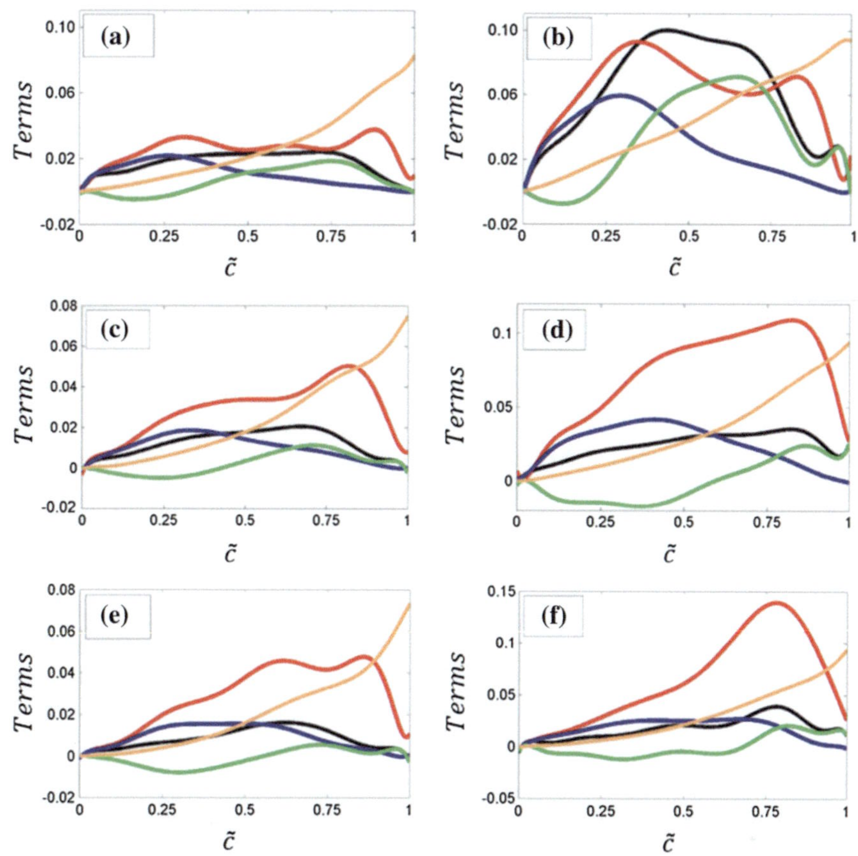

Fig. 12.4 Variations of $\widetilde{Y_F''^2}/Y_{Fst}^2$, $\widetilde{\xi''^2}/Y_{Fst}^2$, Q_{S1}/Y_{Fst}^2, Q_{S2}/Y_{Fst}^2 and $(\text{model of } \widetilde{Y_F''^2}/Y_{Fst}^2) \times 0.1$ based on Eq. 12.22 [black, red, blue, green, tan] with $\tilde{c}$ across the flame brush for **a** $a_d/\delta_{th} = 0.06$, $\phi_d = 1.0$; **b** $a_d/\delta_{th} = 0.06$, $\phi_d = 1.7$; **c** $a_d/\delta_{th} = 0.08$, $\phi_d = 1.0$; **d** $a_d/\delta_{th} = 0.08$, $\phi_d = 1.7$; **e** $a_d/\delta_{th} = 0.10$, $\phi_d = 1.0$; and **f** $a_d/\delta_{th} = 0.10$, $\phi_d = 1.7$

12.4.2.2 Modelled Transport Equation for $\widetilde{Y_F''^2}$

It has been shown in Sect. 4.1.1 that the available algebraic models do not satisfactorily capture the behaviour of $\widetilde{Y_F''^2}$ and, as such, a modelled transport equation for $\widetilde{Y_F''^2}$ may be required. This section outlines the development of a modelled transport equation for $\widetilde{Y_F''^2}$ in the context of the combustion of turbulent droplet-laden mixtures. The development of a modelled transport equation has previously been considered by Wacks et al. (2016b), but has been included here for completeness.

Statistical Behaviour of the Terms of $\widetilde{Y_F''^2}$ Transport Equation

The variations of the terms $T_{Y1}, T_{Y2}, T_{Y3}, T_{Y4}$ and $(-D_{Y2})$ (as presented in Eq. 12.18) with $\tilde{c}$ across the flame brush are shown in Fig. 12.5, which indicates that nonzero values are observed at $\tilde{c}=0$ because of mixture inhomogeneity for all terms except T_{Y3}. Figure 12.5 shows that the magnitudes of T_{Y1} and T_{Y2} remain small in comparison to the other terms. The reaction rate contribution term T_{Y3} has been found to be a significant contributor to $\widetilde{Y_F''^2}$ transport, but its magnitude remains smaller than the evaporation and molecular dissipation terms (i.e. T_{Y4} and $-D_{Y2}$). The reaction rate term T_{Y3} assumes predominantly positive values for the major part of the flame brush, although some negative values have been observed towards $\tilde{c}=0$ and $\tilde{c}=1$. The magnitude of T_{Y3} has been found to increase with increasing ϕ_d due to stronger a chemical reaction arising from the greater availability of gaseous fuel. The evaporation term T_{Y4} has been found to be a leading-order source term in all cases, whereas the dissipation rate term $(-D_{Y2})$ remains the leading-order sink across the flame brush for all cases. The magnitudes of T_{Y4} and $(-D_{Y2})$ remain large at the leading edge (i.e. $\tilde{c}\approx 0$), but they diminish with increasing $\tilde{c}$ before increasing again due to stronger evaporation within the flame until a "peak" value is obtained within the flame brush before falling again towards the trailing edge (i.e. $\tilde{c}\approx 1$) of the flame as the droplets are evaporated and the fuel is consumed within the flame brush. Figure 12.5 shows that the magnitudes of T_{Y4} and $(-D_{Y2})$ are significantly affected by both ϕ_d and a_d. The magnitudes of T_{Y4} and $(-D_{Y2})$ increase with increasing ϕ_d due to a higher amount of evaporated fuel vapour in the gaseous phase which in turn increases $|\nabla Y_F|$. The evaporation rate increases with decreasing a_d, and therefore, the magnitudes of T_{Y4} and $(-D_{Y2})$ reach their "peak" values closer to the leading edge of the flame brush for the smaller droplets which show faster evaporation rate. These observations as well as the underlying physics must be accounted for when modelling these unclosed terms. The mean scalar gradient term T_{Y2} has been found to exhibit both positive and negative values indicating a combination of gradient (i.e. $\overline{\rho u_1'' Y_F''}\left(\partial \widetilde{Y}_F/\partial x_1\right)<0$) and counter-gradient (i.e. $\overline{\rho u_1'' Y_F''}\left(\partial \widetilde{Y}_F/\partial x_1\right)>0$) transport. As the term T_{Y2} is closed in the context of second-moment closure, the modelling of T_{Y1}, T_{Y3}, T_{Y4} and $(-D_{Y2})$ will be addressed next in this chapter. However, the accuracy of the evaluation of the term T_{Y2} depends on the modelling of turbulent scalar flux $\overline{\rho u_j'' Y_F''}$ which is beyond the scope of the present chapter. Interested readers are referred to Malkeson and Chakraborty (2012) for further discussion on closures of $\overline{\rho u_j'' Y_F''}$.

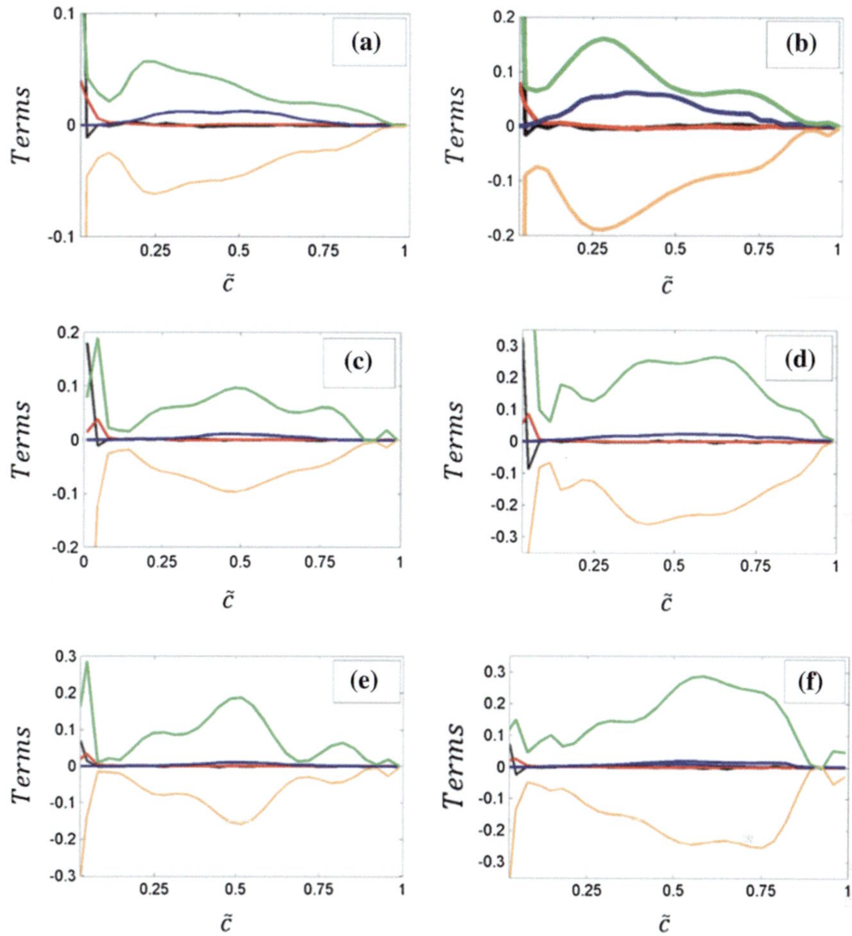

Fig. 12.5 Variations of $T_{Y1}, T_{Y2}, T_{Y3}, T_{Y4}$ and $-D_{Y2}$ [black, red, blue, green, tan] with $\tilde{c}$ across the flame brush for **a** $a_d/\delta_{th} = 0.06$, $\phi_d = 1.0$; **b** $a_d/\delta_{th} = 0.06$, $\phi_d = 1.7$; **c** $a_d/\delta_{th} = 0.08$, $\phi_d = 1.0$; **d** $a_d/\delta_{th} = 0.08$, $\phi_d = 1.7$; **e** $a_d/\delta_{th} = 0.10$, $\phi_d = 1.0$; and **f** $a_d/\delta_{th} = 0.10$, $\phi_d = 1.7$. All terms are normalized by $D_0/\rho_0 Y_{Fst}^2 S_{b(\phi_g=1)}^2$ where ρ_0 is the unburned gas density

Modelling of T_{Y1}

The modelling of T_{Y1} translates to the modelling of the unclosed turbulent fluxes (i.e. $\overline{\rho u_i'' Y_F''^2}$) of $\widetilde{Y_F''^2}$. For statistically planar flames, $\overline{\rho u_i'' Y_F''^2}$ remains the only nonzero component of $\overline{\rho u_i'' Y_F''^2}$. Often, the quantity $\overline{\rho u_i'' Y_F''^2}$ is modelled using the gradient hypothesis in the following manner (Ribert et al. 2005, 2006):

$$\overline{\rho u_i'' Y_F''^2} = -\frac{\mu_t}{\sigma_Y}\frac{\partial \widetilde{Y_F''^2}}{\partial x_i} \tag{12.24}$$

where $\mu_t = 0.09\bar{\rho}\tilde{k}^2/\tilde{\varepsilon}$ is the eddy viscosity, σ_Y is a turbulent Schmidt number, $\tilde{k}$ is the turbulent kinetic energy, and $\tilde{\varepsilon} = \overline{\mu(\partial u_i''/\partial x_j)(\partial u_i''/\partial x_j)}/\bar{\rho}$ is its dissipation rate. However, it is known that $\overline{\rho u_1'' Y_F''^2}$ can exhibit counter-gradient behaviour and therefore it will be desirable to have a model which can predict both gradient and counter-gradient transport (Malkeson and Chakraborty 2010b). Previously, an expression for $\overline{\rho u_1'' Y_F''^2}$ was proposed by using Eq. 12.24 and the identity $\overline{\rho u_1'' Y_F''^2} = \bar{\rho}\int(u_1 - \tilde{u}_1)(Y_F - \tilde{Y}_F)^2\tilde{P}(Y_F, \xi)du_1 dY_F d\xi$ in the following manner (Malkeson and Chakraborty 2010b; Wacks et al. 2016b), which is capable of predicting both gradient and counter-gradient transport:

$$\overline{\rho u_1'' Y_F''^2} = \left[2(\tilde{Y}_{Fmax} - \tilde{Y}_F) - (\tilde{Y}_{Fmax} - \tilde{Y}_{Fmin})\right]\left[\overline{\rho u_1'' Y_F''} - \lambda_W \overline{\rho u_1'' \xi''} - (1-\lambda_W)\tilde{A}\overline{\rho u_1'' \xi''}\right] + \lambda_W \overline{\rho u_1'' \xi''^2} + (1-\lambda_W)\tilde{A}\overline{\rho u_1'' \xi''^2} \tag{12.25}$$

The predictions of $\overline{\rho u_1'' Y_F''^2} = -(\mu_t/\sigma_Y)\partial\widetilde{Y_F''^2}/\partial x_1$ and Eq. 12.25 are shown in Fig. 12.6 along with the variations of $\overline{\rho u_1'' Y_F''^2}$ obtained from DNS data. Figure 12.6 shows that $-(\mu_t/\sigma_Y)\partial\widetilde{Y_F''^2}/\partial x_1$ satisfactorily captures the quantitative behaviour of $\overline{\rho u_1'' Y_F''^2}$ in all cases for $\sigma_Y = 1.0$. Although Eq. 12.25 captures the general qualitative behaviour, it overpredicts the magnitude of $\overline{\rho u_1'' Y_F''^2}$ for the major part of the flame brush. As Eq. 12.25 is derived based on Eq. 12.21 (which is strictly valid for $Da \gg 1$ combustion), it is perhaps not surprising that Eq. 12.25 does not provide accurate predictions for the $Da < 1$ cases considered here. To address this shortcoming, Eq. 12.25 was modified in the following manner (Malkeson and Chakraborty 2010b; Wacks et al. 2016b):

$$\overline{\rho u_1'' Y_F''^2} = \left[2(\tilde{Y}_{Fmax} - \tilde{Y}_F)S_{Mod}^m - (\tilde{Y}_{Fmax} - \tilde{Y}_{Fmin})\right](2S_{Mod}/S_{Mod} + 1) \times \left[\overline{\rho u_1'' Y_F''} - \lambda_W \overline{\rho u_1'' \xi''} - (1-\lambda_W)\tilde{A}\overline{\rho u_1'' \xi''}\right] + \lambda_W \overline{\rho u_1'' \xi''^2} + (1-\lambda_W)\tilde{A}\overline{\rho u_1'' \xi''^2} \tag{12.26}$$

where $m = 0.1\xi(1-\xi_{st})/[\xi_{st}(1-\xi)]$ is a model parameter, and $S_{Mod} = \max(0, S)$ is the modified segregation factor in order to eliminate unphysical negative values of the segregation factor (see Fig. 12.4a–f). Equation 12.26 becomes identical to Eq. 12.25 for $Da \gg 1$ combustion, where $S_{mod} = 1.0$. Figure 12.6 shows that Eq. 12.26 satisfactorily captures the behaviour of $\overline{\rho u_1'' Y_F''^2}$ throughout the flame brush.

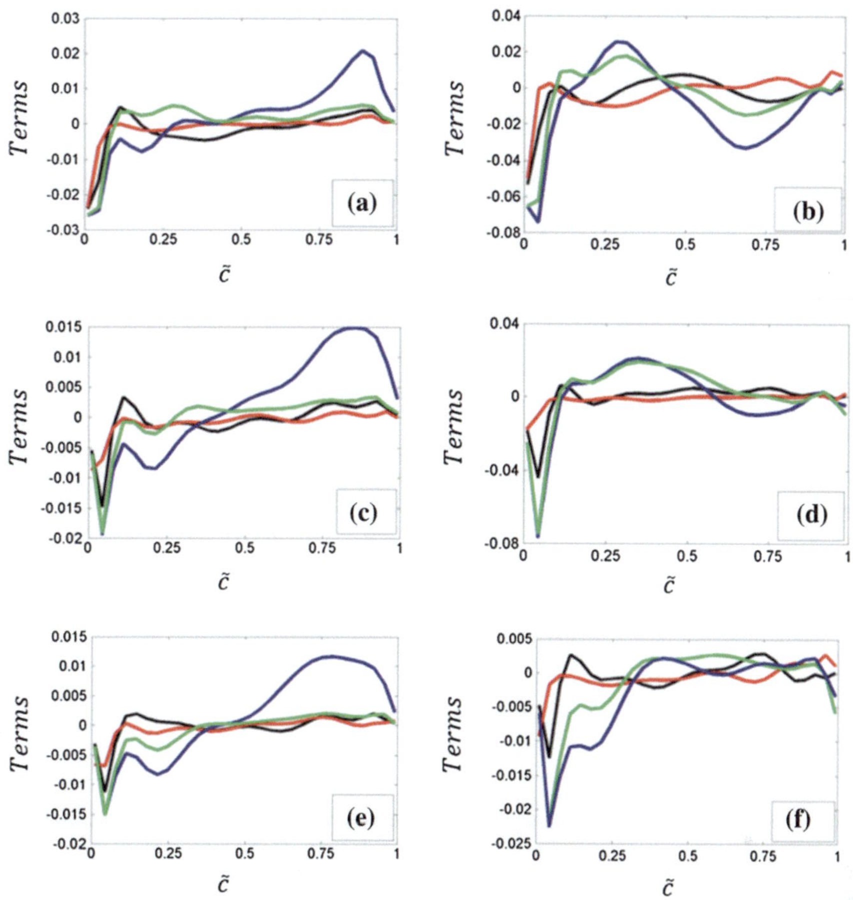

Fig. 12.6 Variations of $\overline{\rho u_1'' Y_F''^2}$ and the predictions of $[-\mu_t \partial \widetilde{Y_F''^2}/\partial x_1]$, Eqs. 12.25 and 12.26 [black, red, blue, green] with $\tilde{c}$ across the flame brush for **a** $a_d/\delta_{th} = (a)0.06$, $\phi_d = 1.0$; **b** $a_d/\delta_{th} = (a)0.06$, $\phi_d = 1.7$; **c** $a_d/\delta_{th} = 0.08$, $\phi_d = 1.0$; **d** $a_d/\delta_{th} = 0.08$, $\phi_d = 1.7$; **e** $a_d/\delta_{th} = 0.10$, $\phi_d = 1.0$; and **f** $a_d/\delta_{th} = 0.10$, $\phi_d = 1.7$. All terms are normalized by $\rho_o Y_{Fst}^2 S_{b(\phi_g = 1)}$

Modelling of T_{Y3}

The modelling of T_{Y3} is dependent upon the accurate modelling of the quantity $\Omega_Y = \left[\overline{\dot{\omega}_F Y_F} - \overline{\dot{\omega}_F}\tilde{Y}_F\right]$. Ribert et al. (2005) and Robin et al. (2006) proposed a closure for $\Omega_Y = \left[\overline{\dot{\omega}_F Y_F} - \overline{\dot{\omega}_F}\tilde{Y}_F\right]$ using the following presumed distribution of $\tilde{P}(Y_F, \xi)$:

$$\tilde{P}(Y_F, \xi) = \alpha_4 \tilde{P}_1(Y_F)\delta(\xi - \xi_{41}) + (1 - \alpha_4)\tilde{P}_2(Y_F)\delta(\xi - \xi_{42}) \tag{12.27}$$

where $\tilde{P}_1(Y_F) = \beta_4\delta(Y_F - Y_{F11}) + (1-\beta_4)\delta(Y_F - Y_{F12})$ and $\tilde{P}_2(Y_F) = \gamma_4\delta(Y_F - Y_{F21}) + (1-\gamma_4)\delta(Y_F - Y_{F22})$ and α_4, β_4 and γ_4 are defined as:

$$\alpha_4 = (\xi^{max} - \xi)/(\xi^{max} - \xi^{min}) \tag{12.28a}$$

$$\beta_4 = (Y_{F1}^{max} - \tilde{Y}_{F1})/(Y_{F1}^{max} - Y_{F1}^{min}) \tag{12.28b}$$

$$\gamma_4 = (Y_{F2}^{max} - \tilde{Y}_{F2})/(Y_{F2}^{max} - Y_{F2}^{min}) \tag{12.28c}$$

According to Robin et al. (2006), the quantities ξ_{41}, ξ_{42}, Y_{F1} and Y_{F2} are:

$$\xi_{41} = \tilde{\xi} - \left[((1-\alpha_4)/\alpha_4)\widetilde{\xi''^2}\right]^{1/2} \tag{12.29a}$$

$$\xi_{42} = \tilde{\xi} + \left[(\alpha_4/(1-\alpha_4))\widetilde{\xi''^2}\right]^{1/2} \tag{12.29b}$$

$$Y_{F1} = \tilde{Y}_F - \left[((1-\alpha_4)/\alpha_4)\widetilde{Y_F''^2}\right]^{1/2} \tag{12.29c}$$

$$Y_{F2} = \tilde{Y}_F + \left[(\alpha_4/(1-\alpha_4))\widetilde{Y_F''^2}\right]^{1/2} \tag{12.29d}$$

The quantities Y_{F11}, Y_{F12}, Y_{F21} and Y_{F22} in the expressions for $\tilde{P}_1(Y_F)$ and $\tilde{P}_2(Y_F)$ are given by (Robin et al. 2006):

$$Y_{F11} = \tilde{Y}_{F1} - \left[((1-\beta_4)/\beta_4)\widetilde{Y_{F1}''^2}\right]^{1/2} \tag{12.30a}$$

$$Y_{F12} = \tilde{Y}_{F1} + \left[(\beta_4/(1-\beta_4))\widetilde{Y_{F1}''^2}\right]^{1/2} \tag{12.30b}$$

$$Y_{F21} = \tilde{Y}_{F2} - \left[((1-\gamma_4)/\gamma_4)\widetilde{Y_{F2}''^2}\right]^{1/2} \tag{12.30c}$$

$$Y_{F22} = \tilde{Y}_{F2} + \left[(\gamma_4/(1-\gamma_4))\widetilde{Y_{F2}''^2}\right]^{1/2} \tag{12.30d}$$

The quantities ξ^{max} and ξ^{min} are the maximum and minimum values of ξ within the domain of definition, and Y_{F1}^{max} and Y_{F1}^{min} (Y_{F2}^{max} and Y_{F2}^{min}) are the maximum and minimum values of fuel mass fraction, respectively, at ξ_{41} (ξ_{42}). The variances $\widetilde{Y_{F1}''^2}$ and $\widetilde{Y_{F2}''^2}$ are evaluated using the following relations (Robin et al. 2006):

$$\widetilde{Y_F''^2} + \widetilde{Y_F}^2 = \alpha_4(\widetilde{Y_{F1}''^2} + \widetilde{Y_{F1}}^2) + (1-\alpha_4)(\widetilde{Y_{F2}''^2} + \widetilde{Y_{F2}}^2) \tag{12.31a}$$

$$\widetilde{Y''^2_{F1}}/\widetilde{Y''^2_{F2}} = (Y^{max}_{F1} - \tilde{Y}_{F1})(\tilde{Y}_{F1} - Y^{min}_{F1})/[(Y^{max}_{F2} - \tilde{Y}_{F2})(\tilde{Y}_{F2} - Y^{min}_{F2})] \quad (12.31b)$$

Therefore, according to Eq. 12.27, Ω_Y can be expressed as (Robin et al. 2006):

$$\begin{aligned}\Omega_Y = {} & \alpha_4\beta_4\dot{\omega}_A Y_{F11} + \alpha_4(1-\beta_4)\dot{\omega}_B Y_{F12} + (1-\alpha_4)\gamma_4\dot{\omega}_C Y_{F21} \\ & + (1-\alpha_4)(1-\gamma_4)\dot{\omega}_D Y_{F22} - [\alpha_4\beta_4\dot{\omega}_A + \alpha_4(1-\beta_4)\dot{\omega}_B \\ & + (1-\alpha_4)\gamma_4\dot{\omega}_C + (1-\alpha_4)(1-\gamma_4)\dot{\omega}_D]\tilde{Y}_F]\end{aligned} \quad (12.32)$$

where $\dot{\omega}_A$ and $\dot{\omega}_B$ ($\dot{\omega}_C$ and $\dot{\omega}_D$) are the fuel reaction rates when the fuel mass fraction values are given by Y_{F11} and Y_{F12} (Y_{F21} and Y_{F22}), respectively, at a mixture fraction ξ_{41} (ξ_{42}) (Robin et al. 2006).

The predictions of $\Omega_Y = \overline{\dot{\omega}_F Y_F} - \overline{\dot{\omega}_F}\tilde{Y}_F$ according to Eq. 12.32 are given in Fig. 12.7, which indicates that this model does not adequately capture both qualitative and quantitative variations of Ω_Y obtained from DNS data. However, a reasonable level of agreement is observed towards the burned gas side of the flame brush for $\phi_d = 1.0$. It should be stressed, however, that the model by Robin et al. (2006) was proposed in the context of turbulent gaseous stratified flames with $Da \gg 1$, where $\tilde{P}(Y_F, \xi)$ can be approximated by Eq. 12.21. It has already been shown in Fig. 12.2 that $\tilde{P}(Y_F, \xi)$ cannot sufficiently be approximated by Eq. 12.21. Furthermore, Eq. 12.21 implicitly assumes the Burke–Schumann relations (Poinsot and Veynante 2001) and considers ξ to be a passive scalar, which are not strictly valid in spray flames. Thus, shortcomings of this model for turbulent spray flames with predominantly $Da < 1$ combustion are not unexpected.

Modelling of T_{Y4}

Figure 12.5 shows that the evaporation term T_{Y4} is a leading-order contributor to $\widetilde{Y''^2_F}$ transport for all cases considered here. As a first attempt at modelling the term, T_{Y4} has been modelled in the following manner:

$$T_{Y4} = \bar{\rho} C_{TY4}(\tilde{\varepsilon}/\tilde{k})\widetilde{\xi''^2} \quad (12.33)$$

where C_{TY4} is a model parameter. Figure 12.8 shows that $\bar{\rho} C_{TY4}(\tilde{\varepsilon}/\tilde{k})\widetilde{\xi''^2}$ predicts T_{Y4} satisfactorily when C_{TY4} is taken to be $C_{TY4} = 9.0$.

Modelling of D_{Y2}

Equation 12.18 indicates that the modelling of $(-D_{Y2})$ is dependent upon the accurate evaluation of $\tilde{\varepsilon}_Y$. A linear relaxation model is often used in the following manner:

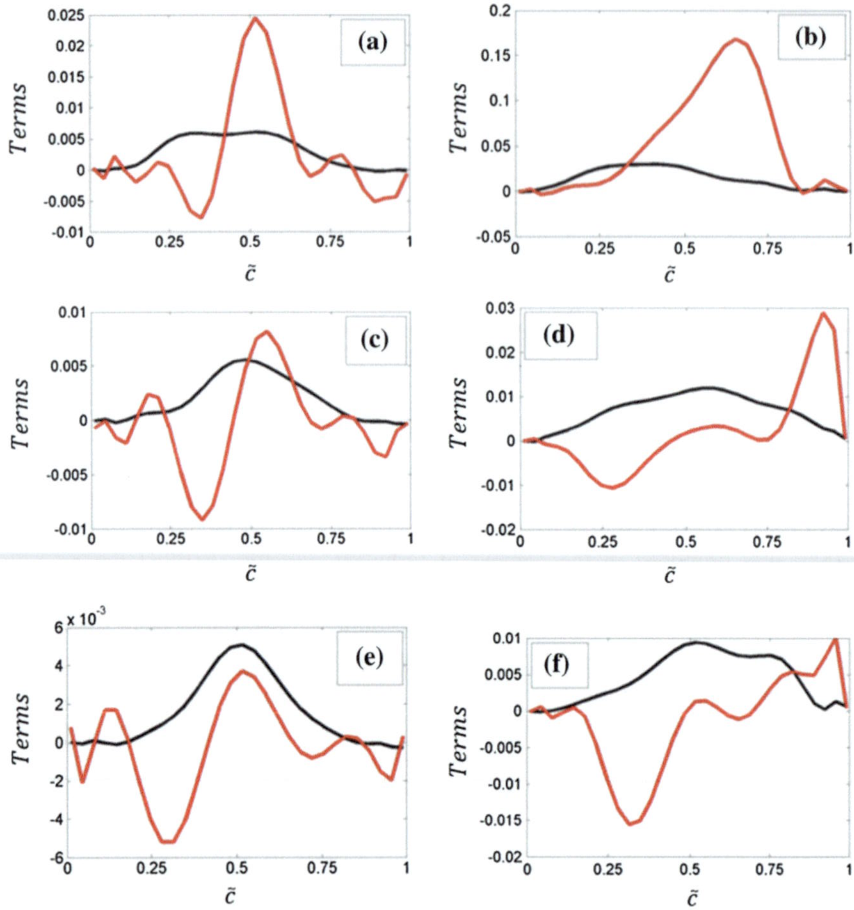

Fig. 12.7 Variations of Ω_Y [black] and the prediction of Eq. 12.32 [red] with $\tilde{c}$ across the flame brush for **a** $a_d/\delta_{th} = 0.06$, $\phi_d = 1.0$; **b** $a_d/\delta_{th} = 0.06$, $\phi_d = 1.7$; **c** $a_d/\delta_{th} = 0.08$, $\phi_d = 1.0$; **d** $a_d/\delta_{th} = 0.08$, $\phi_d = 1.7$; **e** $a_d/\delta_{th} = 0.10$, $\phi_d = 1.0$; and **f** $a_d/\delta_{th} = 0.10$, $\phi_d = 1.7$. All terms are normalized by $D_0/\rho_0 Y_{Fst}^2 S_{b(\phi_g=1)}^2$

$$\tilde{\varepsilon}_Y = C_{LRY}(\tilde{\varepsilon}/\tilde{k})\widetilde{Y_F''^2} \quad (12.34)$$

where C_{LRY} is a model parameter, which is denoted here as the LR-SDY model. Mura et al. (2007) proposed an alternative model for $\tilde{\varepsilon}_Y$ for turbulent stratified flames as:

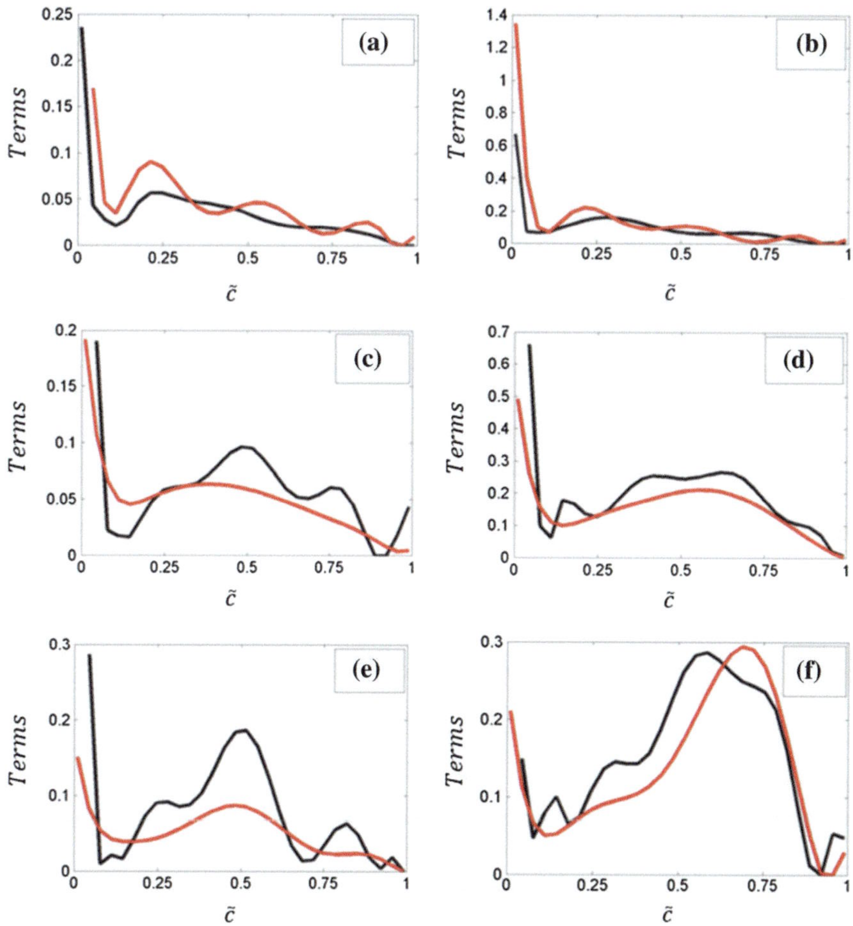

Fig. 12.8 Variations of T_{Y4} [black] and the prediction of $\bar{\rho} C_{T_{Y4}} (\tilde{\varepsilon}/\tilde{k}) \widetilde{\xi''^2}$ (for $C_{T_{Y4}} = 9.0$) [red] with $\tilde{c}$ across the flame brush for **a** $a_d/\delta_{th} = 0.06$, $\phi_d = 1.0$; **b** $a_d/\delta_{th} = 0.06$, $\phi_d = 1.7$; **c** $a_d/\delta_{th} = 0.08$, $\phi_d = 1.0$; **d** $a_d/\delta_{th} = 0.08$, $\phi_d = 1.7$; **e** $a_d/\delta_{th} = 0.10$, $\phi_d = 1.0$; and **f** $a_d/\delta_{th} = 0.10$, $\phi_d = 1.7$. All terms are normalized by $D_0/\rho_0 Y_{Fst}^2 S_{b(\phi_g=1)}^2$

$$\begin{aligned}\tilde{\varepsilon}_Y = S_{Mod}\left(-\overline{\rho D \frac{\partial \tilde{Y}_F}{\partial x_k}\frac{\partial \tilde{Y}_F}{\partial x_k}} - \frac{\bar{\dot{\omega}}_F}{2}(\tilde{Y}_{max} - \tilde{Y}_F + \tilde{Y}_{min}) + \left[\overline{\dot{\omega}_F Y_F} - \bar{\dot{\omega}}_F \tilde{Y}_F\right]\right)\frac{1}{\bar{\rho}} \\ + S_{Mod}\left(\frac{\tilde{Y}_F - \tilde{Y}_{min}}{\tilde{Y}_{max} - \tilde{Y}_{min}} + \frac{\tilde{Y}_{max} - \tilde{Y}_F}{\tilde{Y}_{max} - \tilde{Y}_{min}}\widetilde{A^2}\right) \times \tilde{\varepsilon}_\xi + (1 - S_{Mod}) C_Y \frac{\tilde{\varepsilon}}{\tilde{k}} \widetilde{Y_F''^2}\end{aligned} \quad (12.35)$$

where C_Y is a model parameter. It should also be noted that a modified segregation factor S_{Mod} (See Sect. 12.4.2.1) has been considered here instead of simply the segregation factor S as used by Mura et al. (2007). The predictions of Eq. 12.35 with $\bar{\dot{\omega}}_F$ and $\Omega_Y = \left[\overline{\dot{\omega}_F Y_F} - \bar{\dot{\omega}}_F \tilde{Y}_F\right]$ extracted from the DNS data and according to

Eq. 12.35, respectively, are shown in Fig. 12.9 where they are referred to as SDRY-M1 and SDRY-M2, respectively. Figure 12.9 shows that whilst the LR-SDY model captures the general qualitative behaviour of $\tilde{\varepsilon}_Y$, the model parameter C_{LRY} needs to be tuned in order to obtain satisfactory quantitative agreement, whereas the SDRY-M1 and SDRY-M2 models follow the general behaviour of $\tilde{\varepsilon}_Y$ across the flame brush reasonably well when C_Y is taken to be $C_Y = 6.0$. It should be noted that the predictions of the SDRY-M1 and SDRY-M2 models cannot be separated from each other for $a_d/\delta_{th} = 0.08$ and 0.1 cases (see Fig. 12.9b, c), which suggests that the contributions of $\overline{\dot{\omega}_F}$ and $\Omega_Y = \left[\overline{\dot{\omega}_F Y_F} - \overline{\dot{\omega}_F}\tilde{Y}_F\right]$ do not play major roles in the prediction of Eq. 12.35 in these cases, where the evaporation rates (and Damköhler number) are smaller than in the $a_d/\delta_{th} = 0.06$ cases. Alternatively, a modelled transport equation for $\tilde{\varepsilon}_Y$ is often considered (Robin et al. 2006; Malkeson and Chakraborty 2011), but development of a modelled transport equation in the context of the combustion of turbulent droplet-laden mixtures is beyond the scope of the current analysis. Interested readers are referred to Malkeson et al. (2016) for further discussion on the transport equation-based closure of $\tilde{\varepsilon}_Y$.

12.4.3 *Modelling of Mixture Fraction Variance* $\widetilde{\xi''^2}$

The mixture fraction variance $\widetilde{\xi''^2}$ is typically modelled using a modelled transport equation, which will be discussed in this section.

12.4.3.1 Modelled Transport Equation for $\widetilde{\xi''^2}$

This section outlines the development of a modelled transport equation for $\widetilde{\xi''^2}$ in the context of the combustion of turbulent droplet-laden mixtures.

Statistical Behaviour of the Terms $\widetilde{\xi''^2}$ Transport Equation

The variations of the terms $T_{\xi 1}, T_{\xi 2}, T_{\xi 3}, T_{\xi 4}$ and $(-D_{\xi 2})$ (as presented Eq. 12.18) with $\tilde{c}$ across the flame brush are shown in Fig. 12.10a–f for all cases considered in the current analysis. The turbulent transport term $T_{\xi 1}$ exhibits both positive and negative values across the flame brush but remains small in comparison to the other leading-order terms. The term $T_{\xi 2}$ arising due to the mean scalar gradient $\partial\tilde{\xi}/\partial x_1$ exhibits both positive and negative values across the flame brush but remains small in comparison to the leading-order terms. It can be seen from Fig. 12.10a–f that $T_{\xi 3}$ exhibits relatively large positive across the flame brush and is a leading-order term

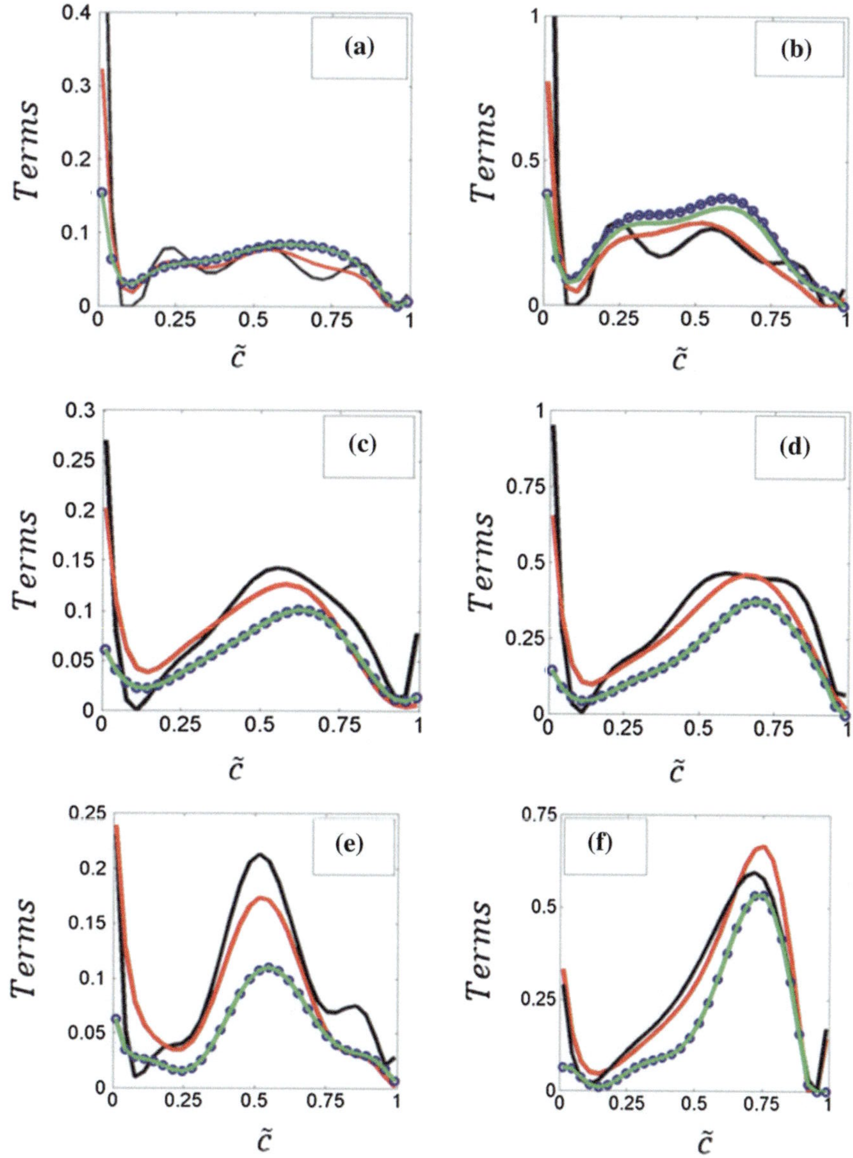

Fig. 12.9 Variations of $\tilde{\epsilon}_Y$ [black] and the predictions of SDY-LR, SDR-M1 and SDR-M2 [red, blue with circles, green] with $\tilde{c}$ across the flame brush for **a** $a_d/\delta_{th} = 0.06$, $\phi_d = 1.0$; **b** $a_d/\delta_{th} = 0.06$, $\phi_d = 1.7$; **c** $a_d/\delta_{th} = 0.08$, $\phi_d = 1.0$; **d** $a_d/\delta_{th} = 0.08$, $\phi_d = 1.7$; **e** $a_d/\delta_{th} = 0.10$, $\phi_d = 1.0$; and **f** $a_d/\delta_{th} = 0.10$, $\phi_d = 1.7$. All terms are normalized by $D_0/Y_{Fst}^2 S_{b(\phi_g=1)}^2$

in all cases considered in the current analysis. It should be noted from Fig. 12.10a–f that the magnitude of $T_{\xi 3}$ is shown to increase for increasing values of a_d and ϕ_d as well as its relative magnitude in comparison to the other terms (i.e. as a_d and ϕ_d increase, so does the relative importance of $T_{\xi 3}$. It can be seen from Fig. 12.10a–f that $T_{\xi 4}$ exhibits positive values across the flame brush but remains small in comparison to the leading-order term $T_{\xi 3}$. It should be noted from Fig. 12.10a–f that the magnitude of $T_{\xi 3}$ is shown to increase for increasing values of a_d and ϕ_d as well as its relative magnitude in comparison to the other terms (i.e. as a_d and ϕ_d increase, so does the relative importance of $T_{\xi 3}$). The observed increase in the relative importance of $T_{\xi 3}$ with increasing values of a_d and ϕ_d is not unexpected from a physical perspective as greater level fluctuations of mixture fraction can be expected from the evaporation of larger droplets with greater equivalence ratio. It should be noted from Fig. 12.10a–f that the magnitude of $T_{\xi 4}$ increases with increasing values of a_d and ϕ_d. The dissipation rate term $(-D_{\xi 2})$ is shown in Fig. 12.10a–f to be negative across the flame brush with larger magnitude values towards the unburned gas side of the flame brush (i.e. $\tilde{c} \to 0$) but falling to small values for the remainder of the flame brush. These observed behaviours of $T_{\xi 3}$ and $T_{\xi 4}$ are consistent with previous findings by Reveillon and Vervisch (2000) and Sreedhara and Huh (2007). It should be noted from Fig. 12.10a–f that $(-D_{\xi 2})$ is a leading-order term for the smaller a_d and ϕ_d case (see Fig. 12.10a) but becomes relatively less important for the larger a_d and ϕ_d cases where the term $T_{\xi 3}$ dominates. These observations with regard to the relative magnitudes and importance of the unclosed terms of the $\widetilde{\xi''^2}$ transport equations are consistent with previous analyses (Reveillon and Vervisch 2000; Sreedhara and Huh 2007). The modelling of the unclosed terms of the $\widetilde{\xi''^2}$ transport equation (i.e. $T_{\xi 1}, T_{\xi 3}, T_{\xi 4}$ and $(-D_{\xi 2})$ will be addressed in the following sections.

Modelling of $T_{\xi 1}$

The modelling of $T_{\xi 1}$ translates to the modelling of the unclosed turbulent fluxes (i.e. $\overline{\rho u_i'' \xi''^2}$) of $\widetilde{\xi''^2}$. For statistically planar flames, $\overline{\rho u_1'' \xi''^2}$ remains the only nonzero component of $\overline{\rho u_i'' \xi''^2}$. The quantity $\overline{\rho u_1'' \xi''^2}$ is often modelled using the gradient hypothesis in the following manner:

$$\overline{\rho u_i'' \xi''^2} = -(\mu_t/\sigma_\xi)\partial\widetilde{\xi''^2}/\partial x_i \tag{12.36}$$

where σ_ξ is a turbulent Schmidt number. The variations of $\overline{\rho u_1'' \xi''^2}$ with $\tilde{c}$ across the flame brush are shown in Fig. 12.11a–f. It can be seen from Fig. 12.11a–f that $\overline{\rho u_1'' \xi''^2}$ exhibits relatively large negative values towards the unburned gas side of the flame brush (i.e. as $\tilde{c} \to 0$) before becoming positive at around $\tilde{c} = 0.1$ for $a_d/\delta_{th} = 0.06$ flames, but this transition value increases for larger values of a_d/δ_{th} to

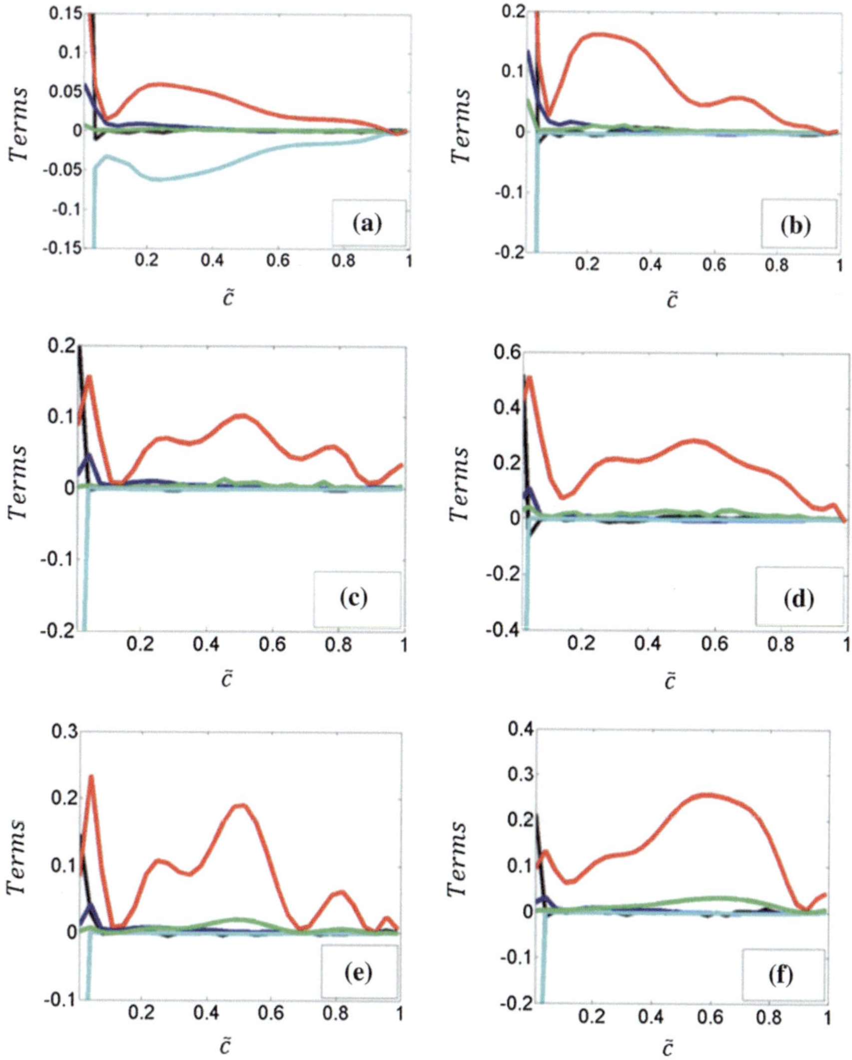

Fig. 12.10 Variations of $T_{\xi 1}, T_{\xi 2}, T_{\xi 3}, T_{\xi 4}$ and $-D_{\xi 2}$ [black, blue, red, green, cyan] with $\tilde{c}$ across the flame brush for **a** $a_d/\delta_{th} = 0.06$, $\phi_d = 1.0$; **b** $a_d/\delta_{th} = 0.06$, $\phi_d = 1.7$; **c** $a_d/\delta_{th} = 0.08$, $\phi_d = 1.0$; **d** $a_d/\delta_{th} = 0.08$, $\phi_d = 1.7$; **e** $a_d/\delta_{th} = 0.10$, $\phi_d = 1.0$; and **f** $a_d/\delta_{th} = 0.10$, $\phi_d = 1.7$. All terms are normalized by $D_0/\rho_0 \xi_{st}^2 S_{b(\phi_g = 1)}^2$

approximately $\tilde{c} = 0.3$. Moving towards the burned gas side of the flame brush, $\overline{\rho u_1'' \xi''^2}$ remains positive for the remainder of the flame brush for the $\phi_d = 1.0$ flames but can become negative towards the burned gas side of the flame brush in $\phi_d = 1.7$ flames. Any proposed model for $\overline{\rho u_1'' \xi''^2}$ must be able to capture these observed

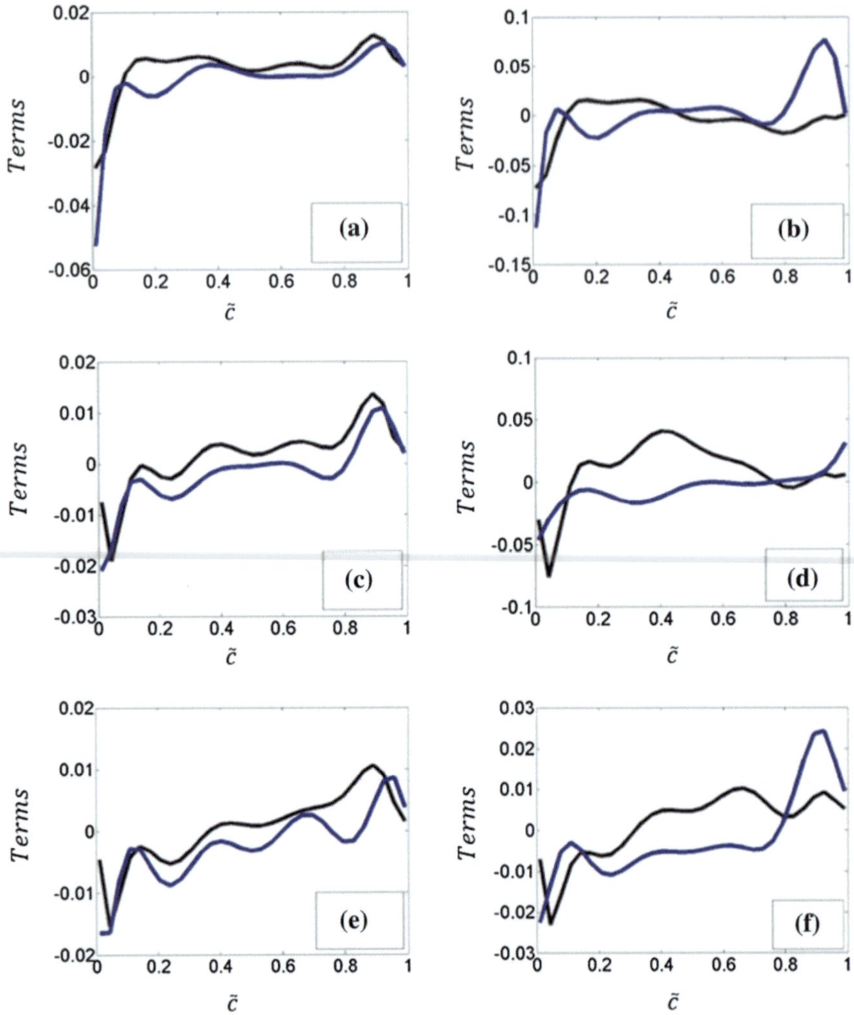

Fig. 12.11 Variations of $\overline{\rho u_1'' \xi''^2}$ and the predictions of $[-(\mu_t/\sigma_\xi)\partial\widetilde{\xi''^2}/\partial x_1]$ $(\sigma_\xi = 0.5)$ and Eq. 12.36 [black, blue] with $\tilde{c}$ across the flame brush for **a** $a_d/\delta_{th} = 0.06$, $\phi_d = 1.0$; **b** $a_d/\delta_{th} = 0.06$, $\phi_d = 1.7$; **c** $a_d/\delta_{th} = 0.08$, $\phi_d = 1.0$; **d** $a_d/\delta_{th} = 0.08$, $\phi_d = 1.7$; **e** $a_d/\delta_{th} = 0.10$, $\phi_d = 1.0$; and **f** $a_d/\delta_{th} = 0.10$, $\phi_d = 1.7$. All terms are normalized by $\rho_o \xi_{st}^2 S_{b(\phi_g = 1)}$

behaviours. The performances of the model given by Eq. 12.36 are shown in Fig. 12.11a–f where $\sigma_\xi = 0.5$. It is evident from Fig. 12.11a–f that Eq. 12.36 captures the general quantitative and qualitative behaviour of $\overline{\rho u_1'' \xi''^2}$ in all cases considered in the current analysis.

Modelling of $T_{\xi 3}$

The variations of $T_{\xi 3}$ with $\tilde{c}$ across the flame brush are shown in Fig. 12.12a–f. It can be seen from Fig. 12.12a–f that $T_{\xi 3}$ exhibits relatively large positive values towards the unburned gas side of the flame brush (i.e. as $\tilde{c} \to 0$) before falling to smaller values across the flame brush. It should be noted from Fig. 12.12a–f that the magnitude of $T_{\xi 3}$ is shown to increase for increasing values of ϕ_d. It has been found that the behaviour of $T_{\xi 3}$ can be adequately captured by the following expression:

$$T_{\xi 3} = \bar{\rho} C_{T\xi 3} (\tilde{\varepsilon}/\tilde{k}) \widetilde{\xi''^2} \tag{12.37}$$

where $C_{T\xi 3}$ is a model parameter. The performances of the model given by Eq. 12.37 are shown in Fig. 12.12a–f, where $C_{T\xi 3} = 10$. It is evident from Fig. 12.11a–f that Eq. 12.37 captures the general quantitative and qualitative behaviour of $T_{\xi 3}$ in all cases considered in the current analysis. It should be noted that the modelling of this term was inherently considered in the previous analyses (Reveillon and Vervisch 2000; Sreedhara and Huh 2007). In these studies (Reveillon and Vervisch 2000; Sreedhara and Huh 2007), the contributions of $T_{\xi 3}$ were accounted for within source terms that formed a modelled transport equation for $\widetilde{\xi''^2}$. The modelling approach considered the extraction of statistical correlations, and the proposed models were dependent upon conditional source term information as well as the use of presumed PDFs of ξ. The analysis of the models proposed by Reveillon and Vervisch (2000) and Sreedhara and Huh (2007) is beyond the scope of the current analysis and thus are not discussed here. Interested readers are referred to Wacks and Chakraborty (2016b) for discussion on the modelling of presumed PDFs of ξ for the DNS database considered here.

Modelling of $T_{\xi 4}$

The variations of $T_{\xi 4}$ with $\tilde{c}$ across the flame brush are shown in Fig. 12.13a–f. It can be seen from Fig. 12.13a–f that $T_{\xi 4}$ typically exhibits relatively large positive values towards the unburned gas side of the flame brush (i.e. as $\tilde{c} \to 0$) before falling to smaller values, which decrease to negligible magnitudes towards the burned gas side (i.e. as $\tilde{c} \to 1$). It should be noted from Fig. 12.13a–f that the magnitude of $T_{\xi 4}$ is shown to increase for increasing values of a_d/δ_{th}. It has been found that the behaviour of $T_{\xi 4}$ can be adequately captured by the following expression:

$$T_{\xi 4} = \bar{\rho} C_{T\xi 4} (\tilde{\varepsilon}/\tilde{k}) \widetilde{\xi''^2} \tag{12.38}$$

where $C_{T\xi 4}$ is a model parameter. The performances of the model given by Eq. 12.38 are shown in Fig. 12.13a–f, where $C_{T\xi 4} = 0.75$. It is evident from Fig. 12.13a–f that Eq. 12.38 captures the general quantitative and qualitative

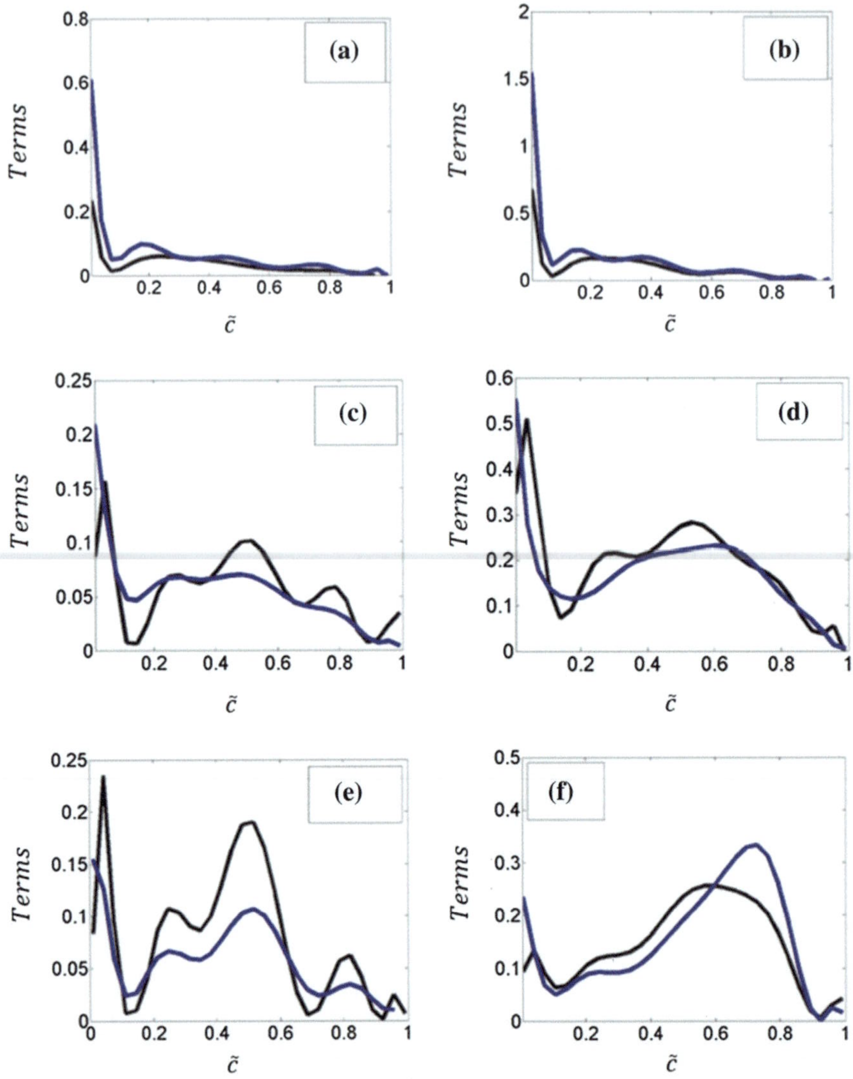

Fig. 12.12 Variations of $T_{\xi 3}$ [black] and the prediction of Eq. 12.37 [blue] (where $C_{T\xi 3} = 10$) with $\tilde{c}$ across the flame brush for **a** $a_d/\delta_{th} = 0.06$, $\phi_d = 1.0$; **b** $a_d/\delta_{th} = 0.06$, $\phi_d = 1.7$; **c** $a_d/\delta_{th} = 0.08$, $\phi_d = 1.0$; **d** $a_d/\delta_{th} = 0.08$, $\phi_d = 1.7$; **e** $a_d/\delta_{th} = 0.10$, $\phi_d = 1.0$; and **f** $a_d/\delta_{th} = 0.10$, $\phi_d = 1.7$. All terms are normalized by $D_0/\rho_0 \xi_{st}^2 S_{b(\phi_g=1)}^2$

behaviour of $T_{\xi 4}$ in all cases considered in the current analysis. Similar to the modelling of $T_{\xi 3}$ (see Sect. 12.4.3.1.3), the modelling of $T_{\xi 4}$ was inherently considered in the previous analyses (Reveillon and Vervisch 2000; Sreedhara and Huh

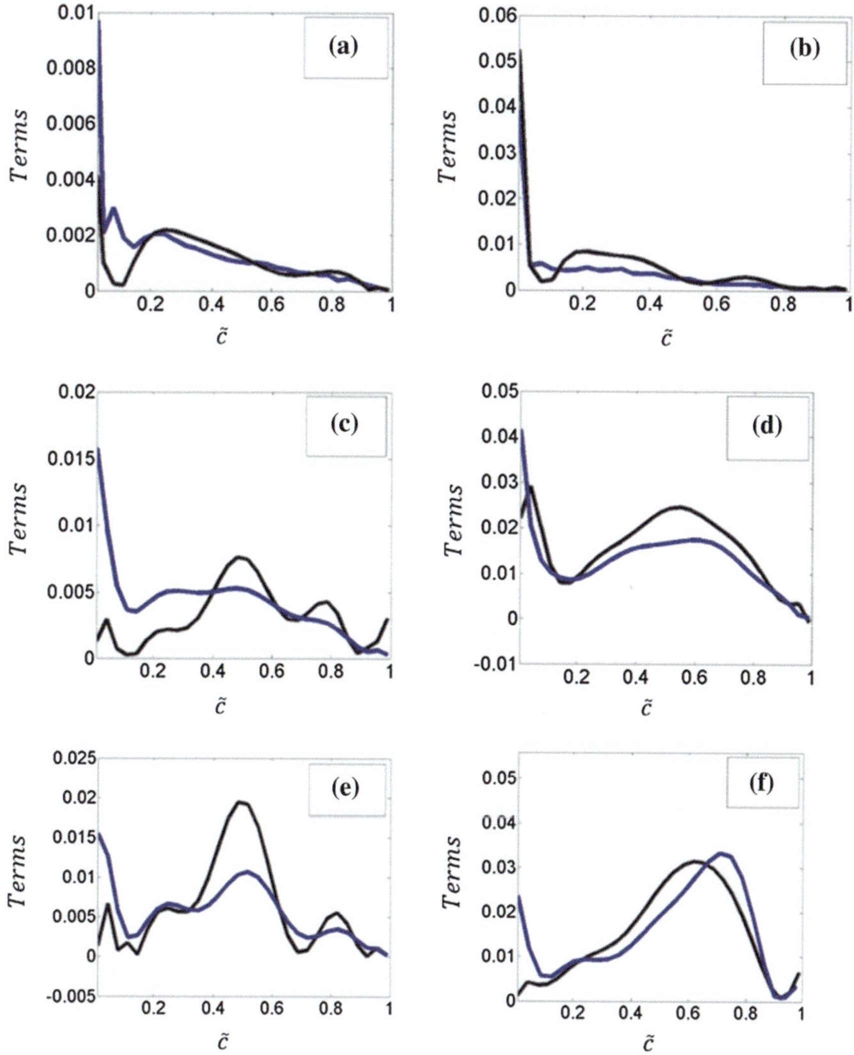

Fig. 12.13 Variations of $T_{\xi 4}$ [black] and the prediction of $\bar{\rho}C_{T_{\xi 4}}(\tilde{\varepsilon}/\tilde{k})\widetilde{\xi''^2}$ (for $C_{T_{\xi 4}} = 0.75$) [blue] with $\tilde{c}$ across the flame brush for **a** $a_d/\delta_{th} = 0.06$, $\phi_d = 1.0$; **b** $a_d/\delta_{th} = 0.06$, $\phi_d = 1.7$; **c** $a_d/\delta_{th} = 0.08$, $\phi_d = 1.0$; **d** $a_d/\delta_{th} = 0.08$, $\phi_d = 1.7$; **e** $a_d/\delta_{th} = 0.10$, $\phi_d = 1.0$; and **f** $a_d/\delta_{th} = 0.10$, $\phi_d = 1.7$. All terms are normalized by $D_0/\rho_0\xi_{st}^2 S_{b(\phi_g=1)}^2$

2007). The analysis of the models proposed by Reveillon and Vervisch (2000) and Sreedhara and Huh (2007) is beyond the scope of the current analysis and thus will not be discussed further in this chapter.

Modelling of $D_{\xi 2}$

Equation 12.19 indicates that the modelling of $(-D_{\xi 2})$ is dependent upon the accurate evaluation of $\tilde{\varepsilon}_{\xi}$. A linear relaxation model is often used in the following manner:

$$\tilde{\varepsilon}_{\xi} = C_{LR\xi}(\tilde{\varepsilon}/\tilde{k})\widetilde{\xi''^2} \tag{12.39}$$

where $C_{LR\xi}$ is a model parameter. The performances of the model given by Eq. 12.39 are shown in Fig. 12.14a–f, where $C_{LR\xi} = 5$. It is evident from Fig. 12.14a–f that Eq. 1.39 captures the general quantitative and qualitative behaviour of $\tilde{\varepsilon}_{\xi}$ in all cases considered in the current analysis. Alternatively, a modelled transport equation for $\tilde{\varepsilon}_{\xi}$ is often considered (Malkeson and Chakraborty 2011), but development of a modelled transport equation in the context of the combustion of turbulent droplet-laden mixtures is yet to be addressed in the existing literature.

12.4.4 *Modelling of Co-variance of Fuel Mass Fraction and Mixture Fraction* $\widetilde{Y_F''\xi''}$

The co-variance of fuel mass fraction and mixture fraction $\widetilde{Y_F''\xi''}$ can be modelled by either an algebraic expression or by a modelled transport equation. We will consider both modelling approaches in this section.

12.4.4.1 Algebraic Modelling of Co-variance $\widetilde{Y_F''\xi''}$

The variations of $\widetilde{Y_F''\xi''}$ with $\tilde{c}$ across the flame brush are shown in Fig. 12.15a–f for all cases considered in this chapter. It is evident from Fig. 12.15a–f that the statistical behaviour of $\widetilde{Y_F''\xi''}$ varies depending upon a_d and ϕ_d. For the $\phi_d = 1.0$ cases (see Fig. 12.15a, c and e), the same general behaviour is observed with positive values of $\widetilde{Y_F''\xi''}$ towards the unburned gas side of the flame brush before becoming negative for the remainder of the flame brush with a transition from positive to negative values at $\tilde{c} \approx 0.4$. It is also evident from Fig. 12.15, for the $\phi_d = 1.0$ cases, that the droplet diameter a_d does not appear to have any significant effect on the behaviour of $\widetilde{Y_F''\xi''}$. For the $\phi_d = 1.7$ cases (see Fig. 12.15b, d and f), it can be seen that $\widetilde{Y_F''\xi''}$ exhibits predominantly positive values across the flame brush except for a small section for the $a_d = 0.1$, $\phi_d = 1.7$ case towards the middle of the flame brush.

It is also evident from Fig. 12.15b, d and f that the magnitude of the positive values observed towards the burned gas side of the flame brush increases with

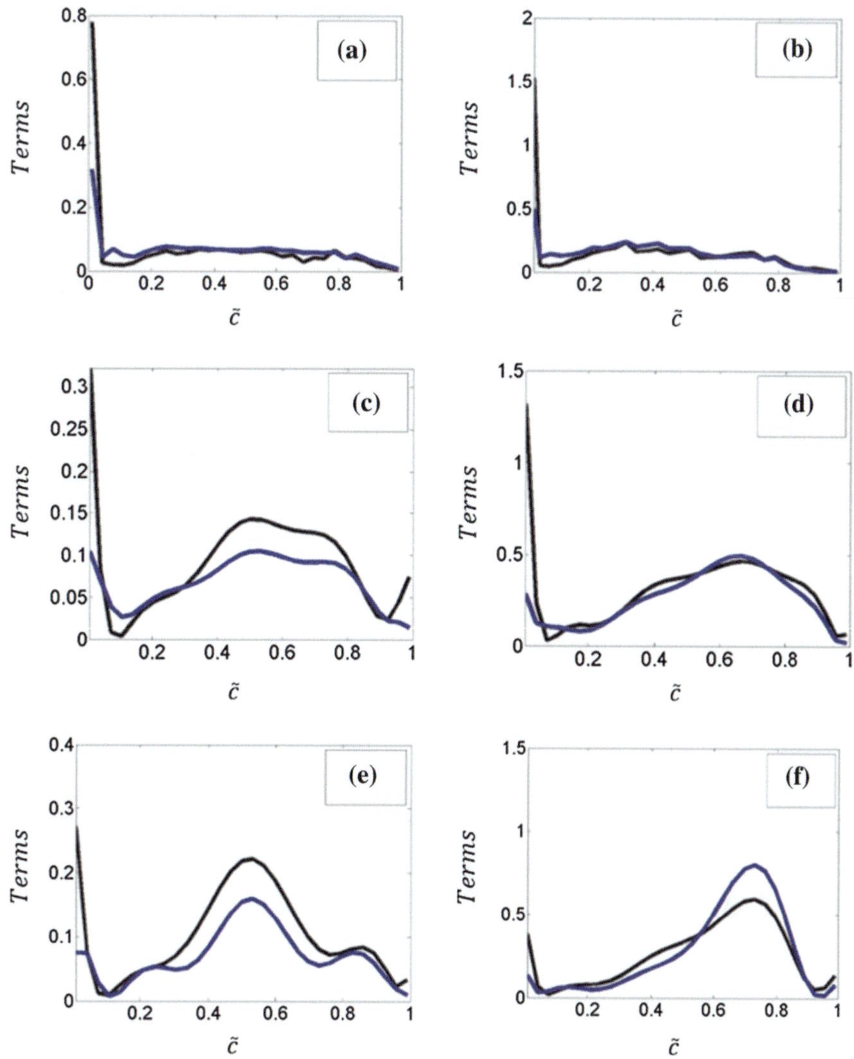

Fig. 12.14 Variations of $\tilde{\varepsilon}_\xi$ [black] and the predictions of Eq. 12.39 [blue] with $\tilde{c}$ across the flame brush for **a** $a_d/\delta_{th} = 0.06$, $\phi_d = 1.0$; **b** $a_d/\delta_{th} = 0.06$, $\phi_d = 1.7$; **c** $a_d/\delta_{th} = 0.08$, $\phi_d = 1.0$; **d** $a_d/\delta_{th} = 0.08$, $\phi_d = 1.7$; **e** $a_d/\delta_{th} = 0.10$, $\phi_d = 1.0$; and **f** $a_d/\delta_{th} = 0.10$, $\phi_d = 1.7$. All terms are normalized by $D_0/\xi_{st}^2 S_{b(\phi_g=1)}^2$

increasing droplet equivalence ratio ϕ_d. Any proposed algebraic models for $\widetilde{Y_F''\xi''}$ must capture the observed behaviours of the corresponding quantity obtained from the DNS data.

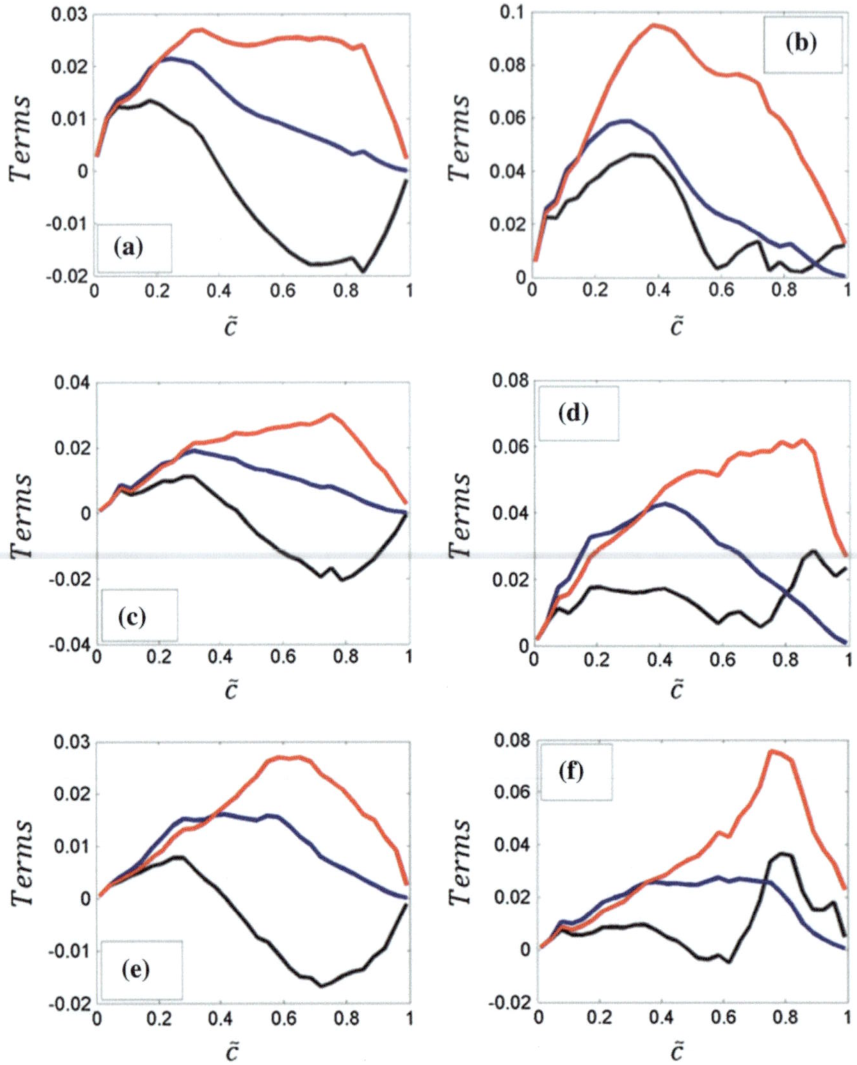

Fig. 12.15 Variations of $\widetilde{Y''_F \xi''}$, Eq. 12.40 an Eq. 12.41 [black, red, blue] with $\tilde{c}$ across the flame brush for **a** $a_d/\delta_{th} = 0.06$, $\phi_d = 1.0$; **b** $a_d/\delta_{th} = 0.06$, $\phi_d = 1.7$; **c** $a_d/\delta_{th} = 0.08$, $\phi_d = 1.0$; **d** $a_d/\delta_{th} = 0.08$, $\phi_d = 1.7$; **e** $a_d/\delta_{th} = 0.10$, $\phi_d = 1.0$; and **f** $a_d/\delta_{th} = 0.10$, $\phi_d = 1.7$. All terms are normalized by $Y_{Fst}\xi_{st}$

Using presumed approximated Favre joint PDF between Y_F and ξ (i.e. $\tilde{P}(Y_F, \xi) = \rho P(Y_F, \xi)/\bar{\rho}$) for turbulent stratified mixture combustion as outlined in Sect. 12.4.2.1, Mura et al. (2007) proposed a model for $\widetilde{Y''_F \xi''}$ in the following manner:

$$\widetilde{Y_F''\xi''} = \left[(\tilde{Y}_F - \tilde{Y}_{min})/(\tilde{Y}_{max} - \tilde{Y}_{min}) + (\tilde{Y}_{max} - \tilde{Y}_F)/(\tilde{Y}_{max} - \tilde{Y}_{min})\tilde{A}\right] \times \widetilde{\xi''^2} \tag{12.40}$$

Alternatively, Ribert et al. (2005) modelled $\widetilde{Y_F''\xi''}$ as a function of $\widetilde{Y_F''^2}$ and $\widetilde{\xi''^2}$ in the following manner:

$$\widetilde{Y_F''\xi''} = \sqrt{\widetilde{Y_F''^2} \times \widetilde{\xi''^2}} \tag{12.41}$$

Figure 12.15a–c shows the performance of the models given by Eqs. 12.40 and 12.41 in comparison to $\widetilde{Y_F''\xi''}$ extracted from the DNS data for all cases considered in the current chapter. It is evident from Fig. 12.15a-c that Eqs. 12.40 and 12.41 capture neither the qualitative nor the quantitative behaviour of $\widetilde{Y_F''\xi''}$ across the flame brush. Therefore, it may be necessary to solve a modelled transport equation for $\widetilde{Y_F''\xi''}$ for turbulent spray flames.

12.4.4.2 Modelled Transport Equation for $\widetilde{Y_F''\xi''}$

It has been shown in Sect. 12.4.4.1 that the available algebraic models do not satisfactorily capture the behaviour of $\widetilde{Y_F''\xi''}$ and, as such, a modelled transport equation for $\widetilde{Y_F''\xi''}$ may be required. This section outlined the development of a modelled transport equation for $\widetilde{Y_F''\xi''}$ in the context of the combustion of turbulent droplet-laden mixtures.

Statistical Behaviour of the Terms of the $\widetilde{Y_F''\xi''}$ Transport Equation

The variations of the terms $T_{Y\xi1}, T_{Y\xi2}, T_{Y\xi3}, T_{Y\xi4}, T_{Y\xi5}$ and $(-D_{Y\xi2})$ (as presented Eq. 12.20) with $\tilde{c}$ across the flame brush are shown in Fig. 12.16a–f for all cases considered in this chapter. The turbulent transport term $T_{Y\xi1}$ exhibits both positive and negative values across the flame brush but remains small in comparison to the leading-order terms. The term $T_{Y\xi2}$, arising due to the mean scalar gradient $\partial\widetilde{Y}_F/\partial x_1$, exhibits both positive and negative values across the flame brush but remains small in comparison to the leading-order terms. The term $T_{Y\xi3}$ arising due to the mean mixture fraction gradient $\partial\tilde{\xi}/\partial x_1$ exhibits both positive and negative values across the flame brush but remains small in comparison to the leading-order terms. The reaction rate term $T_{Y\xi4}$ remains predominantly negative across the flame brush but becomes positive towards the burned gas side of the flame brush. It should be noted that $T_{Y\xi4}$ remains small in comparison to the other leading-order terms. The droplet evaporation term $T_{Y\xi 5}$ is shown to be positive across the flame brush and is a leading-order contributor. In all cases considered in this chapter, the term $T_{Y\xi 5}$ exhibits large positive values towards the unburned gas side of the flame brush (i.e. $\tilde{c} \to 0$) but decays rapidly before increasing again within flame brush prior to assuming negligible magnitudes on the burned gas side (i.e. $\tilde{c} \to 1$). It is evident from Fig. 12.16 that the magnitude of $T_{Y\xi 5}$ increases with increasing values of a_d and ϕ_d. The dissipation term $-D_{Y\xi 2}$ is negative across the flame brush and acts as a leading-order term. In all cases considered in this chapter, the term $D_{Y\xi 2}$ exhibits large negative values towards the fresh gas side of the flame brush (i.e. $\tilde{c} \to 0$), but its magnitude decays rapidly before increasing again within flame brush prior to assuming negligible magnitudes towards the burned gas side of the flame brush (i.e. $\tilde{c} \to 1$). It is evident from Fig. 12.16 that the magnitude of $D_{Y\xi 2}$ increases with increasing values of a_d and ϕ_d. Therefore, satisfactory modelling of $T_{Y\xi 5}$ and $-D_{Y\xi 2}$ will be necessary for the accurate prediction of $\widetilde{Y_F''\xi''}$ transport.

Modelling of $T_{Y\xi 1}$

The modelling of $T_{Y\xi 1}$ translates to the modelling of the unclosed turbulent fluxes (i.e. $\overline{\rho u_i'' Y_F'' \xi''}$) of $\widetilde{Y_F''\xi''}$. For statistically planar flames, $\overline{\rho u_1'' Y_F'' \xi''}$ remains the only nonzero component of $\overline{\rho u_i'' Y_F'' \xi''}$. The quantity $\overline{\rho u_1'' Y_F'' \xi''}$ is often modelled using the gradient hypothesis as:

$$\overline{\rho u_1'' Y_F'' \xi''} = -(\mu_t/\sigma_{Y\xi})\partial \widetilde{Y_F''\xi''}/\partial x_1 \tag{12.42}$$

where $\sigma_{Y\xi}$ is a turbulent Schmidt number. The variations of $\overline{\rho u_1'' Y_F'' \xi''}$ with $\tilde{c}$ for all cases are shown in Fig. 12.17a–f which show both positive and negative values across the flame brush. Figure 12.17a–f shows that $-(\mu_t/\sigma_{Y\xi})\partial \widetilde{Y_F''\xi''}/\partial x_1$ captures the general qualitative behaviour of $\overline{\rho u_1'' Y_F'' \xi''}$ (for $\sigma_{Y\xi} = 1$), but quantitative predictions are poor.

Alternatively, an algebraic expression for $\overline{\rho u_1'' Y_F'' \xi''}$ can be derived using Eq. 12.21 and the identity $\overline{\rho u_1'' Y_F'' \xi''}/\bar{\rho} = \int (u_1 - \tilde{u}_1)(Y_F - \tilde{Y}_F)(\xi - \tilde{\xi})\tilde{P}(u_1, Y_F, \xi)\, du_1 dY_F d\xi$ in the following manner (Malkeson and Chakraborty 2013):

$$\widetilde{u_1'' Y_F'' \xi''} = \lambda_W \widetilde{u_1''\xi''^2} + (1 - \lambda_W)\tilde{A}\widetilde{u_1''\xi''^2} + \left[(\widetilde{u_1'' Y_F''} - \lambda_W \widetilde{u_1''\xi''} - (1 - \lambda_W)\tilde{A}\widetilde{u_1''\xi''})/(\tilde{Y}_{max} - \tilde{Y}_{min})\right](\widetilde{\xi''^2} - \tilde{A}\widetilde{\xi''^2}) \tag{12.43}$$

The prediction of Eq. 12.43 is shown in Fig. 12.17a–f, which indicate that the model captures the general qualitative and quantitative behaviour of $\overline{\rho u_1'' Y_F'' \xi''}$.

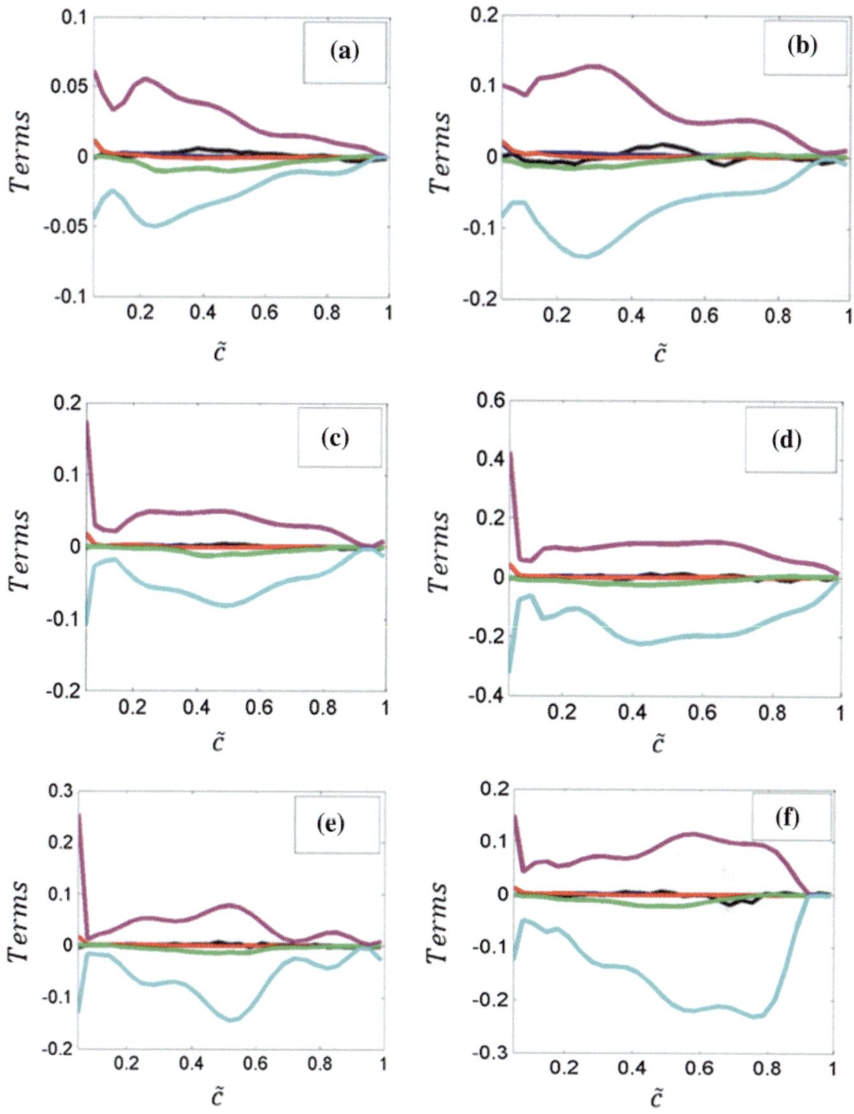

Fig. 12.16 Variations of $T_{Y\xi 1}, T_{Y\xi 2}, T_{Y\xi 3}, T_{Y\xi 4}, T_{Y\xi 5}$ and $-D_{Y\xi 2}$ [black, blue, red, green, pink, cyan] with $\tilde{c}$ across the flame brush for **a** $a_d/\delta_{th} = 0.06$, $\phi_d = 1.0$; **b** $a_d/\delta_{th} = 0.06$, $\phi_d = 1.7$; **c** $a_d/\delta_{th} = 0.08$, $\phi_d = 1.0$; **d** $a_d/\delta_{th} = 0.08$, $\phi_d = 1.7$; **e** $a_d/\delta_{th} = 0.10$, $\phi_d = 1.0$; and **f** $a_d/\delta_{th} = 0.10$, $\phi_d = 1.7$. All terms are normalized by $D_0/\rho_0 \xi_{st}^2 S_{b(\phi_g=1)}^2$

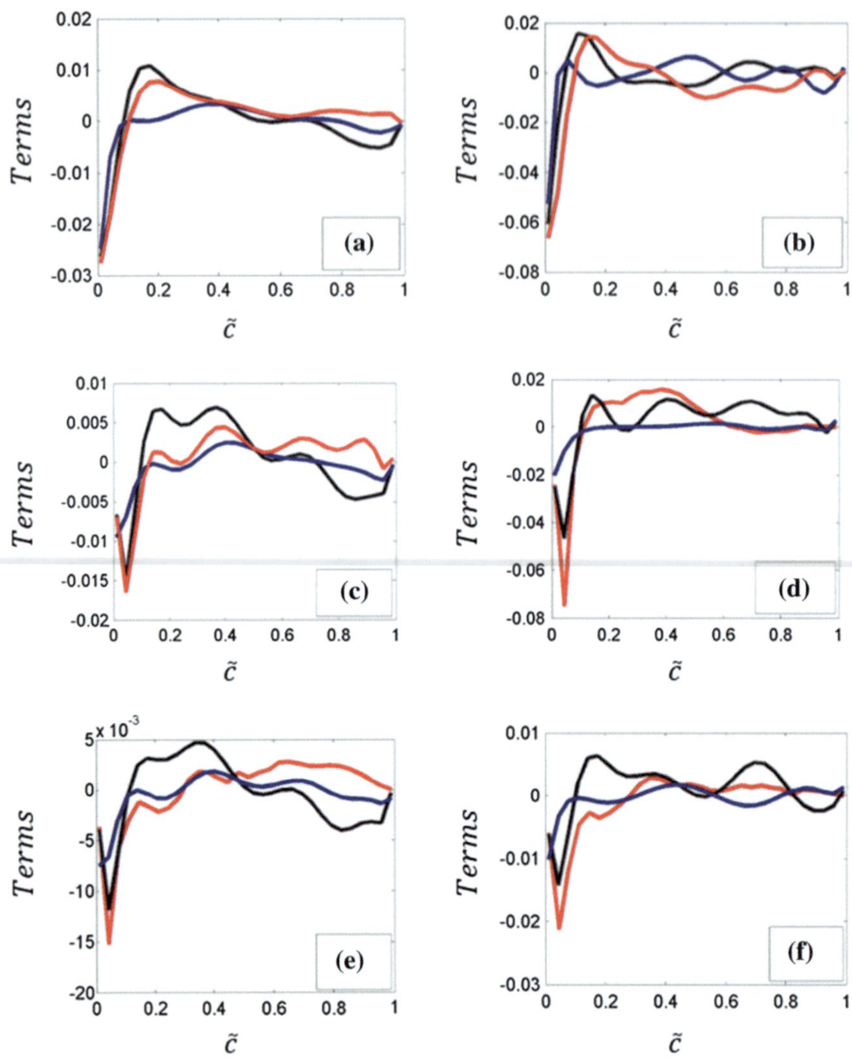

Fig. 12.17 Variations of $\overline{\rho u_1'' Y_F'' \xi''}$ and the predictions of $[-(\mu_t/\sigma_\xi)\partial \widetilde{Y_F'' \xi''}/\partial x_1]$ $(\sigma_{Y\xi} = 0.5)$ and Eq. 12.43 [black, blue, red] with $\tilde{c}$ across the flame brush for **a** $a_d/\delta_{th} = 0.06$, $\phi_d = 1.0$; **b** $a_d/\delta_{th} = 0.06$, $\phi_d = 1.7$; **c** $a_d/\delta_{th} = 0.08$, $\phi_d = 1.0$; **d** $a_d/\delta_{th} = 0.08$, $\phi_d = 1.7$; **e** $a_d/\delta_{th} = 0.10$, $\phi_d = 1.0$; and **f** $a_d/\delta_{th} = 0.10$, $\phi_d = 1.7$. All terms are normalized by $\rho_o \xi_{st}^2 S_{b(\phi_g = 1)}$

Modelling of $T_{Y\xi 4}$

Following the presumed distribution of $\tilde{P}(Y_F, \xi)$ by Robin et al. (2006), as outlined in Sect. 12.4.4.2.3, a model can be proposed for $T_{Y\xi 4}$ in the following manner:

$$T_{Y\xi 4} = \alpha_4\beta_4\dot{\omega}_A\xi_{41} + \alpha_4(1-\beta_4)\dot{\omega}_B\xi_{42} + (1-\alpha_4)\gamma_4\dot{\omega}_C\xi_{41} + (1-\alpha_4)(1-\gamma_4)\dot{\omega}_D\xi_{42} - [\alpha_4\beta_4\dot{\omega}_A + \alpha_4(1-\beta_4)\dot{\omega}_B + (1-\alpha_4)\gamma_4\dot{\omega}_C + (1-\alpha_4)(1-\gamma_4)\dot{\omega}_D]\xi \tag{12.44}$$

It is evident from Fig. 12.18 that Eq. 12.44 does not capture the behaviour of $T_{Y\xi 4}$ obtained from the DNS data and further analysis will be required to identify a satisfactory model for $T_{Y\xi 4}$.

Modelling of $T_{Y\xi 5}$

It should be noted that there are no existing models for the droplet evaporation term $T_{Y\xi 5}$. However, the behaviour of T_5 has been shown to be captured by the following expression:

$$T_{Y\xi 5} = \bar{\rho} C_{TY\xi 5}(\tilde{\varepsilon}/\tilde{k})\widetilde{\xi''^2} \tag{12.45}$$

where $C_{TY\xi 5} = 5.0$. The variations of $T_{Y\xi 5}$ with $\tilde{c}$ for all cases are shown in Fig. 12.19a–f across the flame brush. It is evident from Fig. 12.19a–f that Eq. 12.45 adequately captures the qualitative and quantitative behaviour of $T_{Y\xi 5}$ across the flame brush.

Modelling of $D_{Y\xi 2}$

The transport equation of $\widetilde{Y_F''\xi''}$ indicates that the modelling of $D_{Y\xi 2}$ is dependent upon the accurate evaluation of $\widetilde{\varepsilon_{Y\xi}}$. A linear relaxation model is often considered to model $\widetilde{\varepsilon_{Y\xi}}$ in the following manner:

$$\widetilde{\varepsilon_{Y\xi}} = C_{Y\xi LR}(\tilde{\varepsilon}/\tilde{k})\widetilde{Y_F''\xi''} \tag{12.46}$$

where $C_{Y\xi LR}$ is a model parameter. The model for $\widetilde{\varepsilon_{Y\xi}}$ given by Eq. 12.46 will henceforth be denoted as CDM.

Alternatively, Mura et al. (2007) proposed a model for $\widetilde{\varepsilon_{Y\xi}}$ as:

$$\widetilde{\varepsilon_{Y\xi}} = S_{Mod}\left(-\overline{\rho D}\frac{\partial \widetilde{Y}_F}{\partial x_k}\frac{\partial \xi}{\partial x_k} + \frac{[\overline{\dot{\omega}_F\xi} - \overline{\dot{\omega}_F}\tilde{\xi}]}{2}\right)\frac{1}{\rho} + S_{Mod}\left(\frac{\widetilde{Y}_F - \tilde{Y}_{min}}{\tilde{Y}_{max} - \tilde{Y}_{min}} + \frac{\tilde{Y}_{max} - \widetilde{Y}_F}{\tilde{Y}_{max} - \tilde{Y}_{min}}\tilde{A}\right) \times \widetilde{\varepsilon_\xi} + (1 - S_{Mod})C_{Y\xi}\left(\frac{\tilde{\varepsilon}}{\tilde{k}}\right)\widetilde{Y_F''\xi''} \tag{12.47}$$

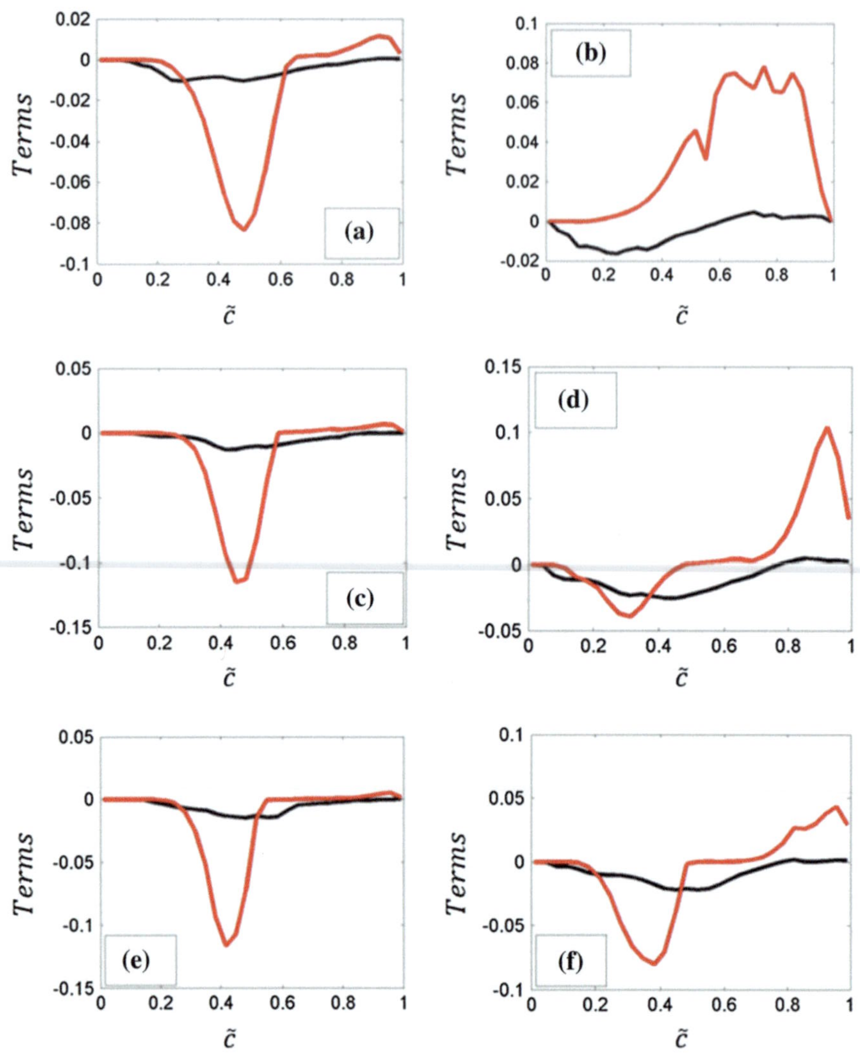

Fig. 12.18 Variations of $T_{Y\xi4}$ [black] and the prediction of Eq. 12.44 [red] with $\tilde{c}$ across the flame brush for **a** $a_d/\delta_{th} = 0.06$, $\phi_d = 1.0$; **b** $a_d/\delta_{th} = 0.06$, $\phi_d = 1.7$; **c** $a_d/\delta_{th} = 0.08$, $\phi_d = 1.0$; **d** $a_d/\delta_{th} = 0.08$, $\phi_d = 1.7$; **e** $a_d/\delta_{th} = 0.10$, $\phi_d = 1.0$; and **f** $a_d/\delta_{th} = 0.10$, $\phi_d = 1.7$. All terms are normalized by $D_0/\rho_0 \xi_{st}^2 S_{b(\phi_g=1)}^2$

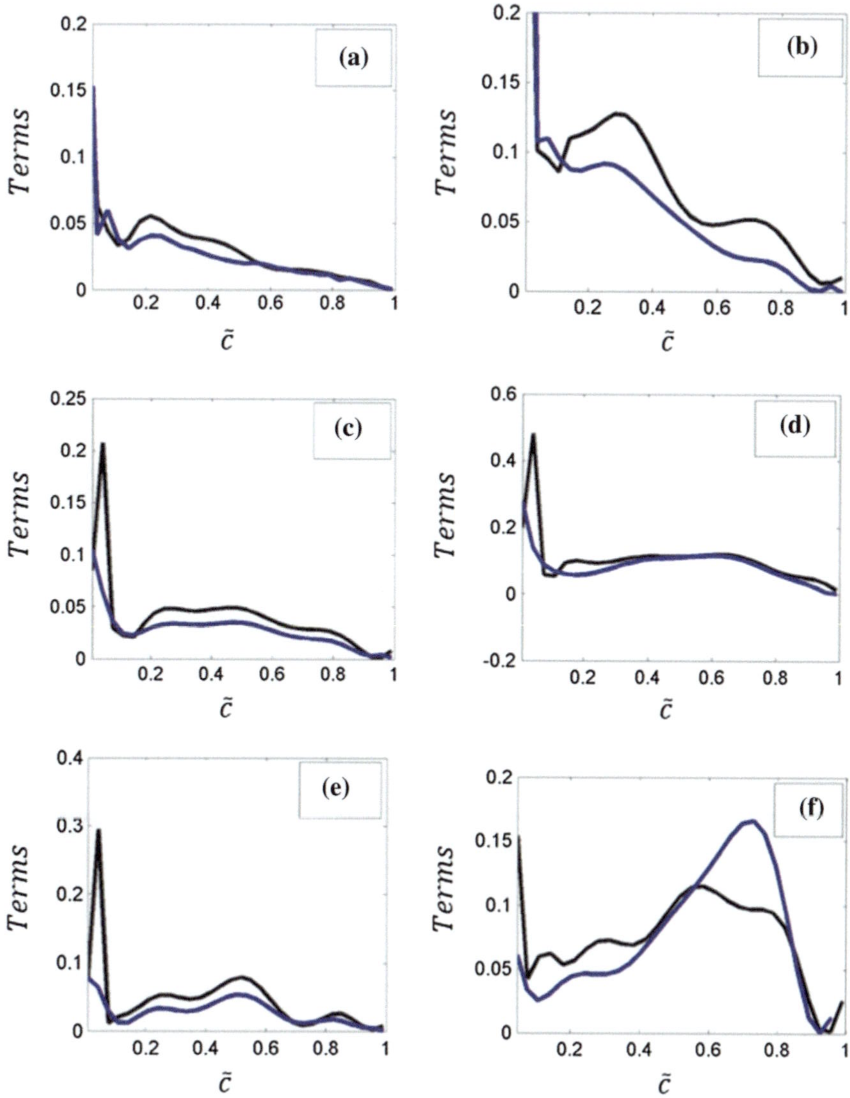

Fig. 12.19 Variations of $T_{Y\xi 5}$ [black] and the prediction of $\bar{\rho}C_{T_{Y\xi 5}}(\tilde{\varepsilon}/\tilde{k})\widetilde{\xi''^2}$ (for $C_{T_{Y\xi 5}} = 5.0$) [blue] with $\tilde{c}$ across the flame brush for **a** $a_d/\delta_{th} = 0.06$, $\phi_d = 1.0$; **b** $a_d/\delta_{th} = 0.06$, $\phi_d = 1.7$; **c** $a_d/\delta_{th} = 0.08$, $\phi_d = 1.0$; **d** $a_d/\delta_{th} = 0.08$, $\phi_d = 1.7$; **e** $a_d/\delta_{th} = 0.10$, $\phi_d = 1.0$; and **f** $a_d/\delta_{th} = 0.10$, $\phi_d = 1.7$. All terms are normalized by $D_0/\rho_0\xi_{st}^2 S_{b(\phi_g=1)}^2$

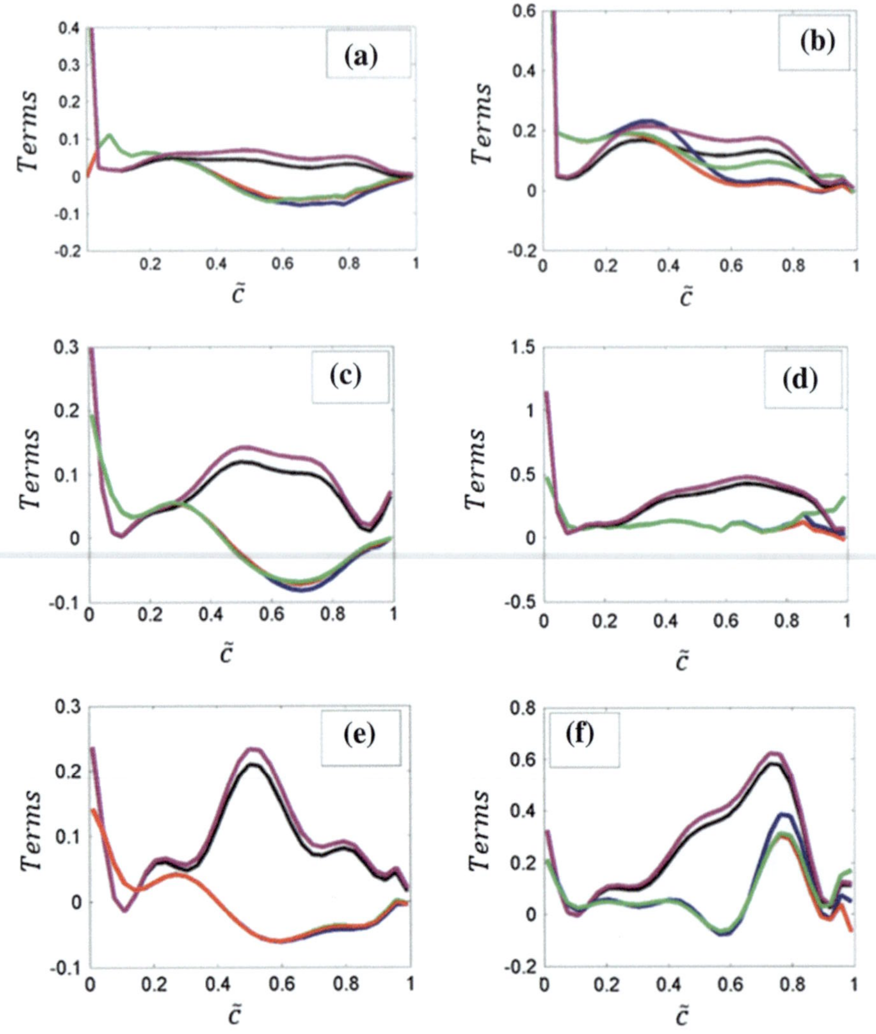

Fig. 12.20 Variations of $\tilde{\varepsilon}_{Y\xi}$ and the predictions of Eqs. 12.46 and 12.47 ($\widetilde{\varepsilon_\xi}$ and $[\overline{\dot{\omega}_F\xi} - \overline{\dot{\omega}_F}\tilde{\xi}]$ from DNS), Eq. 12.47 ($\widetilde{\varepsilon_\xi}$ and $[\overline{\dot{\omega}_F\xi} - \overline{\dot{\omega}_F}\tilde{\xi}]$ from $\widetilde{\varepsilon_\xi} = C_\xi(\tilde{\varepsilon}/\tilde{k})\widetilde{\xi''^2}$ (with $C_\xi = 1.0$) and Eq. 12.48, respectively) and $\widetilde{\varepsilon_{Y\xi}} = \sqrt{\widetilde{\varepsilon_\xi} \times \widetilde{\varepsilon_Y}}$ [black, blue, red, green, pink] with $\tilde{c}$ the flame brush for **a** $a_d/\delta_{th} = 0.06$, $\phi_d = 1.0$; **b** $a_d/\delta_{th} = 0.06$, $\phi_d = 1.7$; **c** $a_d/\delta_{th} = 0.08$, $\phi_d = 1.0$; **d** $a_d/\delta_{th} = 0.08$, $\phi_d = 1.7$; **e** $a_d/\delta_{th} = 0.10$, $\phi_d = 1.0$; and **f** $a_d/\delta_{th} = 0.10$, $\phi_d = 1.7$. All terms are normalized by $D_0/\xi_{st}^2 S_{b(\phi_g=1)}^2$

where $C_{Y\xi}$ is a model parameter. In Eq. 12.47, $\widetilde{\varepsilon_\xi}$ and $\left[\overline{\dot{\omega}_F\xi} - \overline{\dot{\omega}_F}\tilde{\xi}\right]$ are unclosed and require modelling. The quantity $\widetilde{\varepsilon_\xi}$ can be modelled using $\widetilde{\varepsilon_\xi} = C_\xi(\tilde{\varepsilon}/\tilde{k})\widetilde{\xi''^2}$ (see Sect. 12.4.3.1.5), and $\left[\overline{\dot{\omega}_F\xi} - \overline{\dot{\omega}_F}\tilde{\xi}\right]$ can be modelled in the same manner as $T_{4Y\xi}$. It should also be noted that a modified segregation factor S_{Mod} (See Sect. 12.4.2.1) has been used here instead of simply the segregation factor S as used by Mura et al. (2007).

Alternatively, Nguyen et al. (2010) proposed a model for $\widetilde{\varepsilon_{Y\xi}}$ as a function of $\widetilde{\varepsilon_{Y\xi}}$ and $\widetilde{\varepsilon_{Y\xi}}$, in the following manner (see Fig. 12.20):

$$\widetilde{\varepsilon_{Y\xi}} = \sqrt{\widetilde{\varepsilon_\xi} \times \widetilde{\varepsilon_Y}} \tag{12.48}$$

The predictions of CDM, Eq. 12.47 with $\widetilde{\varepsilon_\xi}$ and $\left[\overline{\dot{\omega}_F\xi} - \overline{\dot{\omega}_F}\tilde{\xi}\right]$ from DNS, Eq. 12.47 using $\widetilde{\varepsilon_\xi} = C_\xi(\tilde{\varepsilon}/\tilde{k})\widetilde{\xi''^2}$ (with $C_\xi = 1.0$) and $\left[\overline{\dot{\omega}_F\xi} - \overline{\dot{\omega}_F}\tilde{\xi}\right]$ according to Eq. 12.44, as well as $\widetilde{\varepsilon_{Y\xi}} = \sqrt{\widetilde{\varepsilon_\xi} \times \widetilde{\varepsilon_Y}}$ are shown in Fig. 12.20, which shows that both the CDM model and Eq. 12.47 perform poorly in all cases considered, whereas $\widetilde{\varepsilon_{Y\xi}} = \sqrt{\widetilde{\varepsilon_\xi} \times \widetilde{\varepsilon_Y}}$ (i.e. Equation 12.48) captures the qualitative and quantitative behaviour of $\widetilde{\varepsilon_{Y\xi}}$ obtained from the DNS data.

12.5 Concluding Remarks

It is evident from the above discussion that DNS data can be suitably processed to provide fundamental physical insights in turbulent combustion phenomena and can be used to assess the performance of existing models for such phenomena. Moreover, these models can, when necessary, be modified and model parameters can be identified such that the proposed models not only capture the underlying physics of the unclosed term but can also accurately predict the qualitative and quantitative behaviour. The current analysis is done in the context of RANS, but a similar analysis can be done for LES modelling as well where DNS data need to be explicitly filtered (i.e. convolution operation is performed) to assess sub-grid closures. This type of analysis is known as a priori DNS analysis, which plays a key role in the model assessment and development in the context of both RANS and LES.

In this chapter, DNS data of turbulent combustion of droplet-laden mixtures have been utilized to assess the statistical behaviours of the transport of the fuel mass fraction variance $\widetilde{Y_F''^2}$, mixture fraction variance $\widetilde{\xi''^2}$ and co-variance $\widetilde{Y_F''\xi''}$. The effects of the droplet diameter a_d and the droplet equivalence ratio ϕ_d have been assessed on the behaviours of the terms of the $\widetilde{Y_F''^2}$, $\widetilde{\xi''^2}$ and $\widetilde{Y_F''\xi''}$ transport equations. Moreover, the models for the unclosed terms of the $\widetilde{Y_F''^2}$, $\widetilde{\xi''^2}$ and $\widetilde{Y_F''\xi''}$ transport equations have been developed and assessed in comparison to the corresponding quantities obtained from DNS. As discussed in Sect. 12.4 of the chapter, satisfactory models have been identified for the unclosed terms of the $\widetilde{Y_F''^2}$, $\widetilde{\xi''^2}$ and $\widetilde{Y_F''\xi''}$ transport equations except for the reaction rate contribution terms in the $\widetilde{Y_F''^2}$ and $\widetilde{Y_F''\xi''}$ transport equations (i.e. T_{Y3} and $T_{Y\xi4}$). Therefore, further model developments will be required to identify suitable closures for these terms. It should be noted, however, that many of the models analysed here were not proposed in the context of turbulent spray combustion and, therefore, shortcomings of these models are not unexpected. As previously discussed, DNS data can be an excellent tool for analysing the fundamental physical behaviour of reacting flow phenomena by isolating the effects of simulation parameters and offer valuable insights. However, the modelling proposed solely based on a priori analysis of DNS data (as done in this chapter) needs to be considered with care. It is worth noting that some quantities (e.g. $\tilde{k}$ and $\tilde{\varepsilon}$), which are used as input parameters in the sub-models for the variance and co-variance transports, are itself unclosed and need closures. In a priori DNS analysis, as done here in this chapter, exact values of $\tilde{k}$ and $\tilde{\varepsilon}$ are extracted from DNS data. However, in actual RANS simulations, the quantities such as $\tilde{k}$ and $\tilde{\varepsilon}$ are modelled and thus the modelling inaccuracies of these quantities, in turn, affect the predictions of sub-models of the variance and co-variance transports. Thus, the sensitivity of the models for the variance and co-variance transports on the modelling inaccuracies of $\tilde{k}$ and $\tilde{\varepsilon}$ cannot be analysed solely based on a priori analysis. Moreover, DNS data correspond to flow situations with moderate values of turbulent Reynolds number. Furthermore, the methodology adopted in this chapter simplifies the chemical kinetics and makes point source assumption of the droplets without simulating the atomization process. Thus, the models proposed here need to be implemented in actual RANS simulations for a range of configurations for which experimental data is available for a posteriori assessment.

Acknowledgements The authors are grateful to EPSRC and N8/ARCHER for financial and computational assistance.

References

Aggarwal SK (1998) A review of spray ignition phenomena: present status and future research. Prog Energy Combust Sci 24:565–600

Aggarwal SK, Sirignano WA (1985) Unsteady spray flame propagation in a closed volume. Combust Flame 62:69–84

Ashayek FM, Jacobs GB (2001) Temperature-dependent reaction in droplet-laden homogeneous turbulence. Numer Heat Trans A 39:101–121

Ballal DR, Lefebvre AH (1981) Flame propagation in heterogeneous mixtures of fuel droplets, fuel vapour and air. Proc Combust Inst 18:312–328

Bilger RW (1988) The structure of turbulent non-premixed flames. Proc Combust Inst 23:475–488

Burgoyne JH, Cohen L (1954) The effect of drop size on flame propagation in liquid aerosols. Proc Roy Soc Lond A 225:375–392

Clift R, Grace JR, Weber ME (1978) Bubbles, drops, and particles. Academic Press, New York

Faeth GM (1987) Mixing, transport and combustion sprays. Prog Energy Combust Sci 13:293–345

Fujita, A, Watanabe, H, Kurose, R, Komori, S (2013) Two-dimensional direct numerical simulation of spray flames – Part 1: Effects of equivalence ratio, fuel droplet size and radiation, and validity of flamelet model. Fuel 104:515–525

Greenberg JB, Silverman I, Tambour Y (1998) On droplet enhancement of the burning velocity of laminar premixed spray flames. Combust Flame 113:271–273

Hayashi S, Kumagai S, Sakai T (1976) Propagation velocity and structure of flames in droplet-vapor-air mixtures. Combust Sci Technol 15:169–177

Heywood JB (1998) Internal combustion engine fundamentals, 1st edn. McGraw Hills

Jenkins KW, Cant RS (1999) DNS of turbulent flame kernels. In: Knight D, Sakell L (eds) Proceedings of the Second AFOSR conference on DNS and LES. Rutgers University, Kluwer Academic Publishers, pp 192–202

Lawes M, Saat A (2011) Burning rates of turbulent iso-octane aerosol mixtures in spherical flame explosions. Proc Combust Inst 33:2047–2054

Lefebvre AH (1998) Gas turbine combustion, 2nd edn. Taylor & Francis, Ann Arbor, MI

Li K, Zhou LX (2012) Studies of the effect of spray inlet conditions on the flow and flame structures of ethanol-spray combustion by Large-Eddy simulation. Numer Heat Trans A 62: 44–59

Luo K, Pitsch H, Pai MG et al (2011) Direct numerical simulations and analysis of three-dimensional n-heptane spray flames in a model swirl combustor. Proc Combust Inst 33:2143–2152

Malkeson SP, Chakraborty N (2010a) A-priori direct Numerical Simulation analysis of algebraic models of variances and scalar dissipation rates for Reynolds Averaged Navier Stokes Simulations for low Damköhler number turbulent partially-premixed combustion. Combust Sci Technol 182:960–999

Malkeson SP, Chakraborty N (2010b) The modeling of fuel mass fraction variance transport in turbulent stratified flames: a direct numerical simulation study. Numer Heat Trans A 58 (3):187–206

Malkeson SP, Chakraborty N (2011) Statistical analysis of scalar dissipation rate transport in turbulent partially premixed flames: a direct numerical simulation study. Flow Turb Combust 86(1):1–44

Malkeson SP, Chakraborty N (2012) A-priori DNS modelling of the turbulent scalar fluxes for low Damköhler number stratified flames. Combust Sci Technol 184:1680–1702

Malkeson SP, Chakraborty N (2013) A-priori direct numerical simulation modelling of co-variance transport in turbulent stratified flames. Flow Turb Combust 90(2):243–267

Malkeson SP, Wacks DH, Yi L, Chakraborty N (2016) Analysis of fuel mass fraction dissipation rate transport in turbulent flame-droplet interaction: a direct numerical simulation study, ILASS2016, Brighton, UK, 4–7th Sept 2016

Miller RS, Bellan J (1999) Direct numerical simulation of a confined three-dimensional gas mixing layer with one evaporating hydrocarbon-droplet-laden stream. J Fluid Mech 384:293–338

Mura A, Robin V, Champion M (2007) Modeling of scalar dissipation in partially premixed turbulent flames. Combust Flame 149:217–224

Neophytou A, Mastorakos E (2009) Simulations of laminar flame propagation in droplet mists. Combust Flame 156:1627–1640

Neophytou A, Mastorakos E, Cant RS (2011) Complex chemistry simulations of spark ignition in turbulent sprays. Proc Combust Inst 33:2135–2142

Neophytou A, Mastorakos E, Cant RS (2012) The internal structure of igniting turbulent sprays as revealed by complex chemistry DNS. Combust Flame 159:641–664

Nomura H, Koyama M, Miyamoto H et al (2000) Microgravity experiments of flame propagation in ethanol droplet-vapor-air mixture. Proc Combust Inst 28:999–1005

Nguyen PD, Vervisch L, Subramanian V, Domingo P (2010) Combust Flame 157:43

Poinsot TJ, Lele S (1992) Boundary conditions for direct simulations of compressible viscous flows. J Comput Phys 101:104–129

Poinsot T, Veynante D (2001) Theoretical and numerical combustion. RT Edwards, Philadelphia, USA
Reveillon J, Vervisch L (2000) Spray vaporization in non-premixed turbulent combustion modelling: a single droplet model. Combust Flame 121:75–90
Reveillon J, Demoulin FX (2007) Evaporating droplets in turbulent reacting flows. Proc Combust Inst 31:2319–2326
Ribert G, Champion M, Gicquel O et al (2005) Modeling of nonadiabatic turbulent premixed reactive flows including tabulated chemistry. Combust Flame 141:271–280
Robin V, Mura A, Champion M et al (2006) A multi-Dirac presumed PDF model for turbulent reacting flows with variable equivalence ratio. Combust Sci Technol 178:1843–1870
Rogallo RS (1981) Numerical experiments in homogeneous turbulence. NASA Technical Memorandum 81315, NASA Ames Research Center, California
Sreedhara S, Huh KY (2007) Conditional statistics of nonreacting and reacting sprays in turbulent flows by direct numerical simulation. Proc Combust Inst 31:2335–2342
Silverman I, Greenberg JB, Tambour Y (1993) Stoichiometry and polydisperse effects in premixed spray flames. Combust Flame 93:97–118
Szekely GA Jr, Faeth GM (1983) Effects of envelope flames on drop gasification rates in turbulent diffusion flames. Combust Flame 49:255–259
Tarrazo E, Sanchez A, Linan A et al (2006) A simple one-step chemistry model for partially premixed hydrocarbon combustion. Combust Flame 147:32–38
Tolpadi AK, Agarwal SK, Mongia HC (2000) An advanced spray model for application to the prediction of gas turbine combustor flow fields. Numer Heat Trans A 38:325–340
Wacks DH, Chakraborty N, Mastorakos E (2016a) Statistical analysis of turbulent flame-droplet interaction: a direct numerical simulation study. Flow Turb Combust 96:573–607
Wacks D, Malkeson S, Chakraborty N (2016b) Statistical behaviour of fuel mass fraction variance transport in turbulent flame-droplet interaction: a direct numerical simulation analysis. Numer Heat Trans Part A Appl 70(10):1087–1100
Wacks DH, Chakraborty N (2016a) Flame structure and propagation in turbulent flame-droplet interaction: a direct numerical simulation analysis. Flow Turb Combust 96:1053–1081
Wacks DH, Chakraborty N (2016b) Statistical analysis of the reaction progress variable and mixture fraction gradients in flames propagating into droplet mist: a direct numerical simulation analysis. Combust Sci Technol 188:2149–2177
Wandel A, Chakraborty N, Mastorakos E (2009) Direct numerical simulation of turbulent flame expansion in fine sprays. Proc Combust Inst 32:2283–2290
Wandel A (2013) Extinction predictors in turbulent sprays. Proc Combust Inst 34(1):1625–1632
Wandel A (2014) Influence of scalar dissipation on flame success in turbulent sprays with spark ignition. Combust Flame 161:2579–2600
Wang Y, Rutland CJ (2005) Effects of temperature and equivalence ratio on the ignition of n-heptane fuel spray in turbulent flow. Proc Combust Inst 30:893–900
Watanabe H, Kurose R, Hwang S-M et al (2007) Characteristics of flamelets in spray flames formed in a laminar counterflow. Combust Flame 148:234–248
Watanabe H, Kurose R, Komori S et al (2008) Effects of radiation on spray flame characteristics and soot formation. Combust Flame 152:2–13
Wray AA (1990) Minimal storage time advancement schemes for spectral methods, unpublished report. NASA Ames Research Center, California

Part V
Droplet and Spray Dynamics

Chapter 13
Dynamics of Droplet Break-Up

Binita Pathak, Ranganathan Kumar and Saptarshi Basu

13.1 Introduction and Background

Droplet break-up and atomization is ubiquitous in a plethora of industrial applications. Typical spray-based industrial processes such as surface coating, drying, ink-jet printing, powder and food processing involve a cluster of droplets or sprays exposed to specific environmental conditions (James et al. 2003; Basu et al. 2012). The fate of each drop is determined by various forces acting upon it which usually causes severe deformation and disintegration of the droplets. Interactions between multiple drops with the gas phase lead to drop collision, coalescence, and break-up, which determine the final drop size distribution in the spray. Droplet atomization particularly acts as the rate-determining process in internal combustion engines and gas turbines (Hsiang and Faeth 1995). Fragmentation of droplet increases surface area and enhances the rate of mixing and ensures uniform burning in the combustor (Ranger and Nicholls 1969). There are several experimental, theoretical, and numerical studies on spray dynamics and instabilities when the fluid is subjected to extreme ambient conditions (Rajamanickam and Basu 2017; Saha et al. 2012; Dukowicz 1980; Reitz 1987; Nakamura et al. 2005). However, physical understanding of a single droplet behavior is also crucial as it forms the fundamental unit of many practical applications. C K Law explored the boiling and bursting phenomenon of a single multicomponent droplet (Law 1978). Avedesian and Andres modeled micro-explosions due to heating of a single emulsion droplet (Avedisian and Andres 1978). Jackson and Avedesian identified such micro-explosions during

B. Pathak · S. Basu (✉)
Mechanical Engineering Department, Indian Institute of Science, Bengaluru, Karnataka, India
e-mail: sbasu@mecheng.iisc.ernet.in

R. Kumar
Department of Mechanical and Aerospace Engineering, University of Central Florida, Orlando, Florida, USA

S. Basu et al. (eds.), *Droplets and Sprays*, Energy, Environment, and Sustainability,
https://doi.org/10.1007/978-981-10-7449-3_13

burning of emulsified droplets (Jackson and Avedisian 1998). There are several other studies related to different types of atomization and break-up and the effect of various parameters on droplet shattering rate, time and daughter droplet sizes which are prevalent in the literature (Beck and Watkins 2002; Yoon and DesJardin 2006; Link et al. 2004).

Disintegration of droplets has been widely investigated using different experimental facilities. Atomization can be induced by external perturbations such as aerodynamics, electrical, acoustics to name a few (McDonald 1954; Kim and Nakajima 1999; Elrod et al. 1989). The shearing effect of high-speed external flow field was studied by exposing droplets to gas phase in wind tunnels and shock tubes (Cao et al. 2007; Wierzba 1990). Relative velocity between gas phase and liquid phase initiates instabilities in the form of waves at the gas–liquid interface (Hirahara and Kawahashi 1992). These shear-induced disturbances grow and lead to complete rupture of the droplet. There have been extensive studies and reviews in the literature which expound various mechanisms for break-up of free-falling drops in gas flows (Krzeczkowski 1980; Haas 1964; Lane 1951; Magarvey and Taylor 1956). Therefore, only some of the important observations and analysis related to free-falling drops are included for completeness of the chapter (Sect. 13.2.1).

Development and growth of instabilities in the form of waves were also observed on the surface of single droplets placed on externally perturbed surfaces (James et al. 2003a, b; Deepu et al. 2013; Vukasinovic et al. 2007). James et al. analysed the influence of vibrations upon sessile droplet using a mathematical model (James et al. 2003). The surface instabilities grow with increase in forcing amplitude before ejecting secondary droplets at critical condition. The entire drop ultimately atomizes into a spray of tiny daughter droplets when excited at an appropriate frequency and amplitude. Additionally, evaporation of liquid phase in droplets of colloidal suspension leads to the formation of intricate final structures (Tsapis et al. 2005; Yunker et al. 2011). Such sessile droplet evaporation has severe implications in surface patterning, coating and pharmaceutics, among others (Tsapis et al. 2005; Shaikeea et al. 2016). Therefore, the entire lifetime of a single droplet is a multiscale problem which includes evaporation, interfacial interactions such as collision and coalescence, followed by initiation and growth of instabilities causing shape distortion and the final break-up. The purpose of this chapter is to provide an extensive review of the important observations and analysis regarding the droplet atomization and break-up dynamics in sessile mode (Sect. 13.2.2). Evaporation is not considered in any detail in the current chapter.

Acoustic levitation is another appropriate way which provides contact-free environment to study deformation and disintegration of droplets. However, only a few studies related to levitated droplets are available in the literature (Lee et al. 1991, 1994; Anilkumar et al. 1993; Basu et al. 2013). Lee et al. studied the effects of acoustic pressure by manipulating the amplitude of the sound wave (Lee et al. 1991, 1994). Severe shape deformation was observed due to acoustic radiation stress which leads to complete rupture of the droplet. Saha et al. studied the influence of external heating upon acoustically levitated fuel droplets (Basu et al. 2013). Combined effect of acoustics and heating induces various higher-order modes of deformation and break-up in droplets which depends upon the heating rate and the properties of the fluid system (Pathak and Basu 2016a, b; Tijerino et al. 2013; Basu et al. 2013). Section 13.2.3 of the Chapter incorporates the different modes of deformation and break-up in acoustically levitated droplets in detail. The importance of droplet atomization and break-up is highlighted in the Conclusion Sect. 13.3.

13.2 Different Aspects of Droplet Break-Up

13.2.1 Free-Falling Droplets

One of the most commonly observed phenomena of free-falling droplets is rain drops impinging upon the surface of aircrafts and missiles (Ranger and Nicholls 1969). Understanding the mechanism of deformation and disintegration in free-falling drops is a crucial step toward the process of damage control in aerodynamics and propulsion. The effect of fluid properties and various forces upon droplet disintegration and rupture has been investigated by many researchers (Bartz et al. 2011; Hinze 1955; Hanson et al. 1963). In addition, the impact of droplets on a deep pool has applications in cleaning up oil spills, spray cooling, painting, etc. Castillo-Orozco et al. (2015) conducted experimental and numerical studies to demonstrate the effect of liquid properties on instabilities leading to Rayleigh jet breakup and the formation of daughter drops subsequent to pinch-off.

13.2.1.1 Different Break-Up Mechanisms

Fragmentation of liquid droplets subjected to high-speed stream can be studied by exposing a stream of droplets into a test chamber. Hirahara et al. studied the dynamics by subjecting water and silicone droplets into hot air flow in a shock tube. The stream of droplets was generated at the exit of a capillary tube by perturbing a fluid jet and the behavior of droplets was recorded using shadowgraph (back-lighting) and a high-speed camera (Fig. 13.1). The initial size of droplets was controlled by applying oscillations of suitable frequencies to the capillary tube. The effect of gravity is usually negligible in such experimental configurations.

Ranger and Nicholls studied the influence of various parameters upon droplet shattering in a convective flow field (Ranger and Nicholls 1969). The conditions (temperature, pressure, velocity) inside the test section were controlled to understand the action of aerodynamic forces by blowing continuous stream of air/gas perpendicular to the droplets. The rate and time of disintegration was also examined by sudden exposure of the droplets to shock waves in air moving over the droplets. Different types of break-up mechanism usually observed in such experiments were (Zhao et al. 2011, 2013; Dai and Faeth 2001): (a) bag break-up, (b) stamen mode, (c) bag-stamen, and (d) shear-stripping mode (Fig. 13.2). Combination of these modes also occurred depending upon the governing conditions.

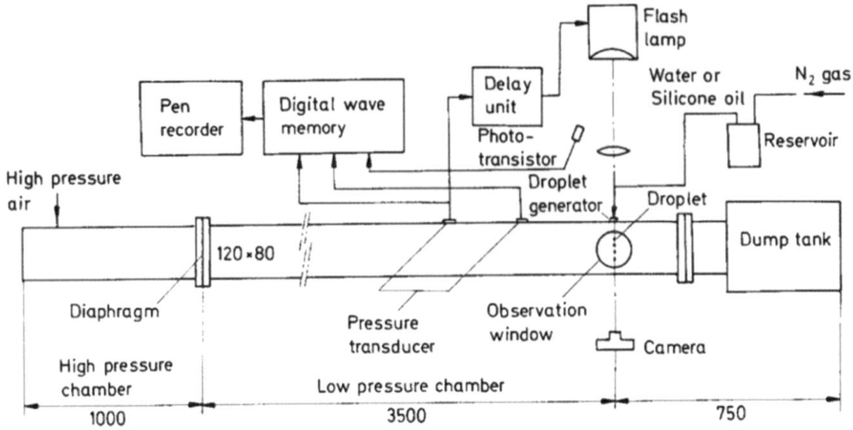

Fig. 13.1 Schematic of the experimental setup for the study of dynamics of droplet subjected to high-speed stream (Hirahara and Kawahashi 1992)

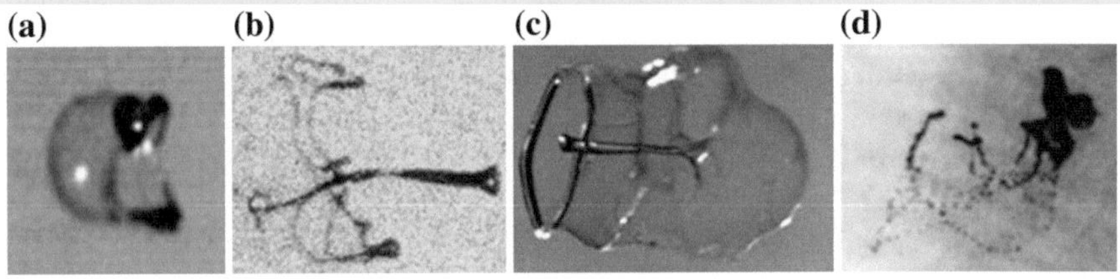

Fig. 13.2 Different mechanism of droplet break-up **a** bag mode (Zhao et al. 2011), **b** stamen mode (Hirahara and Kawahashi 1992), **c** bag-stamen (Zhao et al. 2013), and **d** shear-stripping break-up (Dai and Faeth 2001)

The aerodynamic forces in a gas stream cause deformation of the droplet which is followed by the subsequent break-up phenomenon. The deformation stage depends upon the ambient conditions (pressure) and surface tension of the droplet (Hirahara and Kawahashi 1992). Some of the non-dimensional parameters governing the behavior of droplets are: (a) Weber number $\left(We = \frac{\rho_l V^2 d}{\sigma}\right)$, (b) Bond number $\left(Bo = \frac{a_d \rho_l d^2}{4\sigma}\right)$, and (c) Ohnesorge number $\left(Oh = \frac{\mu_l}{(\rho_l \sigma d)^{\frac{1}{2}}}\right)$ [ρ_l, μ_l and σ are the density, viscosity, and surface tension of the droplet of diameter "d." a_d is acceleration of the droplet. "V" is the velocity of the gas phase]. Both We and Oh play crucial role in the break-up phenomenon. Weber number determines the strength of disrupting hydrodynamic forces and the stabilizing force of surface tension of the droplet (Cao et al. 2007). Break-up usually occurs when the Weber number exceeds a critical value (Krzeczkowski 1980). Cao et al. also investigated break-up in droplets (water and ethanol) subjected to high-speed uniform jet flow

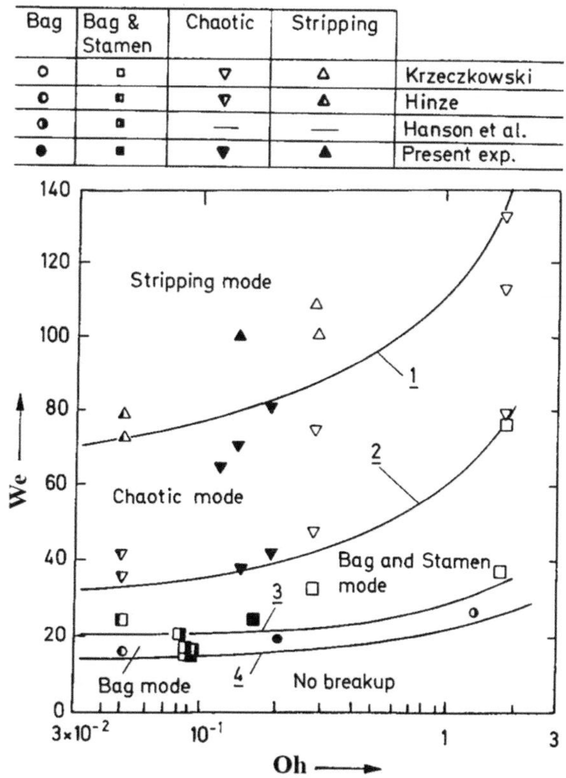

Fig. 13.3 Regime map showing different modes of droplet break-up (Hirahara and Kawahashi 1992)

(Cao et al. 2007). They identified that Weber number is the only dominant parameter determining the break-up regime in low viscous fluids. Viscosity has significant effects upon the break-up phenomenon, thereby increasing the break-up timescale as observed for the silicone droplets (Hirahara and Kawahashi 1992). The effect of viscosity is incorporated in the Ohnesorge number. The different break-up modes can be determined in a regime map based upon the values of Weber number and Ohnesorge number, respectively (Fig. 13.3) (Cao et al. 2007; Hirahara and Kawahashi 1992; Krzeczkowski 1980). The modes are determined with much clarity at low Weber numbers. The break-up mechanisms are much more complex and chaotic with the increasing Weber numbers.

13.2.1.2 Influence of External Flow Field

As stated in the earlier section, deformation of droplet acts as a precursor to different break-up mechanism. The boundary conditions at the interface between the liquid phase (droplet) and the gas phase were derived by Krzeczkowski (1980) as:

$$\sigma\left(\frac{1}{r_1}+\frac{1}{r_2}\right) = P_l - P_g \tag{13.1}$$

$$\tau_l = \tau_g \tag{13.2}$$

$$V_l = V_g \tag{13.3}$$

where P, τ, and V are the pressure, shear stress, and velocity of the phases [*l* and g represents liquid phase and gas phase, respectively]. r_1 and r_2 are the radii of curvature of droplet. The ratio of viscosities $\left(\frac{\mu_l}{\mu_g}\right)$ and velocities $\left(\frac{V_l}{V_g}\right)$ of the two phases was found to be equally important in determining the mechanism of deformation and break-up of the droplets in gas stream.

Non-uniform pressure distribution $\left(\Delta P \sim \frac{1}{2}\rho_g V_r^2\right)$(maximum pressure on upstream surface) causes flattening of the droplet into a disk or ellipsoidal shape. Dynamic pressure of the flow field (gas phase) has a prominent effect upon break-up phenomena (Kim and Nakajima 1999). High ambient dynamic pressure causes rapid break-up of the droplets. The phenomenon is initiated by the growth of small disturbances on the surface of droplet which destroys the stability of the droplet. The droplet is gradually flattened into ellipsoidal shape due to external pressure. Distortion occurs when the dynamic pressure $\left(P_d = C_D \frac{1}{2}\rho_g V_r^2\right)$ overcomes the restoring force of surface tension of the droplet $\left(P_d = \frac{4\sigma}{d}\right)$ (Lane 1951). Thereafter, the droplet undergoes ballooning (hollow sphere or bag) and subsequent rupture at the rim in bag-type break-up mechanism. Bag break-up occurs when the relative velocity (V_r) exceeds a critical value. The droplet blows downstream forming a hollow bag which bursts and causes fragmentation of the rim into tiny droplets (Zhao et al. 2011). Bag-stamen mechanism is characterized by the formation of a jet along with the ballooning of the disc-shaped droplet. Shear-stripping break-up mechanism is characterized by stripping of tiny droplets from the periphery (Fig. 13.2d). The droplet initially follows a parabolic trajectory due to constant acceleration which is followed by an increased acceleration due to faster decrease in droplet mass (due to stripping) and reduction in the drag forces. The governing momentum equation for a droplet of diameter "d" is given as (Ranger and Nicholls 1969):

$$C_D \frac{1}{2}\rho_g V_r^2 A = m_d a \tag{13.4}$$

where "V_r" is the relative velocity of the droplet of frontal area, $A = \frac{\pi d^2}{4}$ and mass "m_d." C_D is the drag coefficient. ρ_g is the ambient density. Acceleration of the droplet is estimated as: $a = \frac{3}{4}\frac{C_D}{d}\frac{\rho_g}{\rho_l}V_r^2$. The displacement of droplet was estimated by

Ranger and Nicholls by assuming a constant acceleration such that (Ranger and Nicholls 1969):

$$x = \frac{1}{2}at^2 = \frac{3}{8}\frac{C_D}{d}\frac{\rho_g}{\rho_l}V_r^2t^2 \qquad (13.5)$$

Equation (13.5) was validated by using experimental data in which droplets of various sizes were exposed to sudden shock waves of different strengths (Ranger and Nicholls 1969). It was found that the droplet break-up time depends upon the droplet diameter (d), velocity (V_r) and the liquid to gas density ratio $\left(\frac{\rho_g}{\rho_l}\right)$, and therefore, the timescale was derived as: $t \propto \frac{d}{V_r}\frac{\rho_g}{\rho_l}$. It should be noted that the shock do not have any direct impact upon the break-up phenomena. The shock exerts a high-speed convective flow upon the droplet. The shear stress exerted by the high-speed flow causes stripping of droplets from the boundary layer formed on the droplet surface. Therefore, the phenomenon of break-up can be considered to be similar to that of a droplet in a high-speed flow field. The rate of disintegration can be obtained by determining the total mass flux in the boundary layer which can be computed using experimentally obtained variation in droplet shape and velocity.

13.2.2 *Sessile Droplets*

13.2.2.1 Influence of Vibrating Surface

Vibration-induced atomization of droplet into numerous tiny drops is a convenient way to study sprays. The effect of external forcing parameters upon the dynamics of droplet can be studied by placing the droplet on a vibrating diaphragm and gradually increasing the forcing amplitude and frequency (James et al. 2003). The experimental configuration used by James et al. for such atomization study is shown in Fig. 13.4. The diaphragm was actuated by a signal generator connected to an amplifier. The amplifier applied sinusoidal voltage to the piezo attached to the diaphragm thereby inducing oscillations to the diaphragm. The accelerometer attached to piezo monitors the acceleration of the diaphragm. The droplets of appropriate sizes can be placed on the surface of the diaphragm using a micro-pipette.

External perturbation induces initial disturbances (in form of waves) on the droplet surface. The onset of instability on the surface depends upon the critical acceleration of the waves. These disturbances grow and form ligaments/spikes which ultimately undergo break-up at the tip. At small forcing frequency, low-order axisymmetric modes were excited and a single droplet ejection occurred due to break-up at the tip of the spikes (James et al. 2003). Moreover, no break-up occurred if the amplitude of forcing wave was also small. Alternatively,

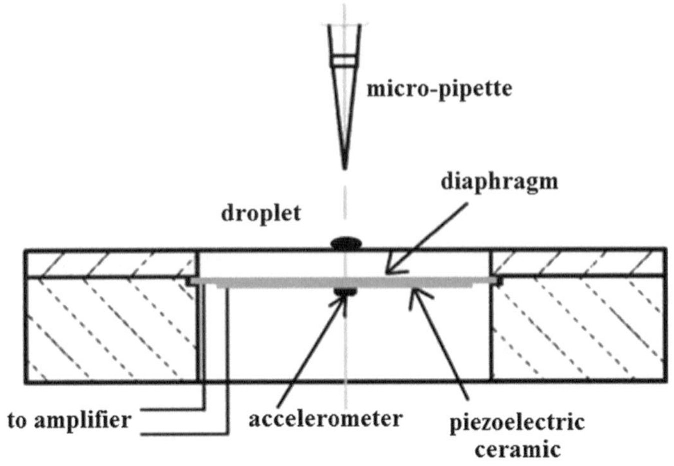

Fig. 13.4 Experimental setup for the vibration-induced atomization of droplet (James et al. 2003)

higher-order axisymmetric modes of excitation were detected at high forcing frequencies which lead to multiple break-up of the spikes.

Formation of free surface waves was first reported by Faraday on the surface of a vertically excited liquid column (Faraday 1831). The frequency of such waves equals half of the excitation frequency. The surface waves on liquid column receive energy and grow to form ligaments which undergo subsequent break-up into droplets. A single spike was formed at low forcing frequency due to dominant gravity waves which broke immediately at the tip (Fig. 13.5a) (Goodridge et al. 1999, 1996). On the other hand, many liquid spikes were formed due to multiple unstable modes at high forcing frequency due to dominant capillary waves which ejected numerous droplets (Fig. 13.5b). The crossover wave number $\left(k_0 = \sqrt{\frac{g\rho}{\sigma}}\right)$ and frequency $\left(\omega_0 = \left(\frac{4g^3\rho}{\sigma}\right)^{\frac{1}{4}}\right)$ determines the gravity wave regimes at $\omega < \omega_0$ and the capillary wave regime at $\omega > \omega_0$, respectively (Goodridge et al. 1996). The transition of surface instabilities to droplet ejecting state occurs beyond certain threshold conditions. The threshold acceleration (a_{th}) for droplet ejection in liquid column surface depends upon interfacial surface tension of the liquid (σ) and amplitude (ω) of the forced oscillations (for high-frequency disturbances) which can be expressed as: $a_{th} \sim \omega^{\frac{4}{3}}\left(\frac{\sigma}{\rho}\right)^{\frac{1}{3}}$. The mode of droplet ejection varied rapidly from periodic to turbulent with an increase in the forcing frequency in the experiments (Fig. 13.5b, c).

Spikes of highly viscous fluid (viscosity, μ) did not break immediately and formed ligaments of longer lengthscale (Fig. 13.5d). These ligaments subsequently broke up beyond the critical condition (explained later). Goodridge determined the

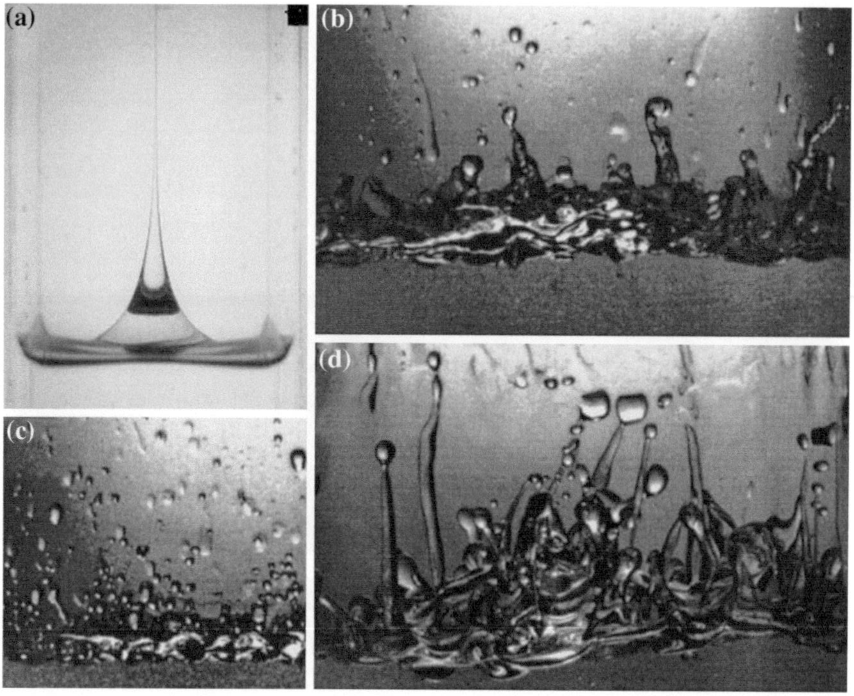

Fig. 13.5 **a** Droplet ejection from a single spike in gravity regime of glycerin–water solution column excited at 7.1 Hz. Multiple ejections on the surface of **b** water column at 20 Hz and **c** 30 Hz, and **d** glycerin-water solution droplet at 20 Hz (Goodridge et al. 1996, 1997)

relationship for the threshold acceleration as: $a_{th} \sim \omega^{\frac{3}{2}}\left(\frac{\mu}{\rho}\right)^{\frac{1}{2}}$ (Goodridge et al. 1997). Similar scaling orders for the threshold governing parameters were also derived for ligaments formed on the surface of sessile droplets placed on externally perturbed substrates (in subsequent section). Higher-order modes of oscillation and atomization can be induced in a droplet which was observed in experimental as well as numerical studies using such sessile droplet configuration (James et al. 2003, b; Deepu et al. 2013).

James et al. studied the various modes of vibration which can be induced by gradually increasing the forcing amplitude which gets altered at higher frequency of the forcing wave (James et al. 2003). Low excitation amplitudes generated axisymmetric standing waves of the same frequency (equivalent to the forcing frequency) on the surface of the droplet (Fig. 13.6b). An azimuthal mode of instability was also triggered along the contact line when the driving amplitude exceeded a threshold value. Further increase in the excitation amplitude complicated the surface waves (Fig. 13.6d). Subsequently, numerous craters and spikes were formed which was followed by bursting into many tiny droplets (Fig. 13.6e and f). Ejection of droplets depends upon the excitation amplitudes.

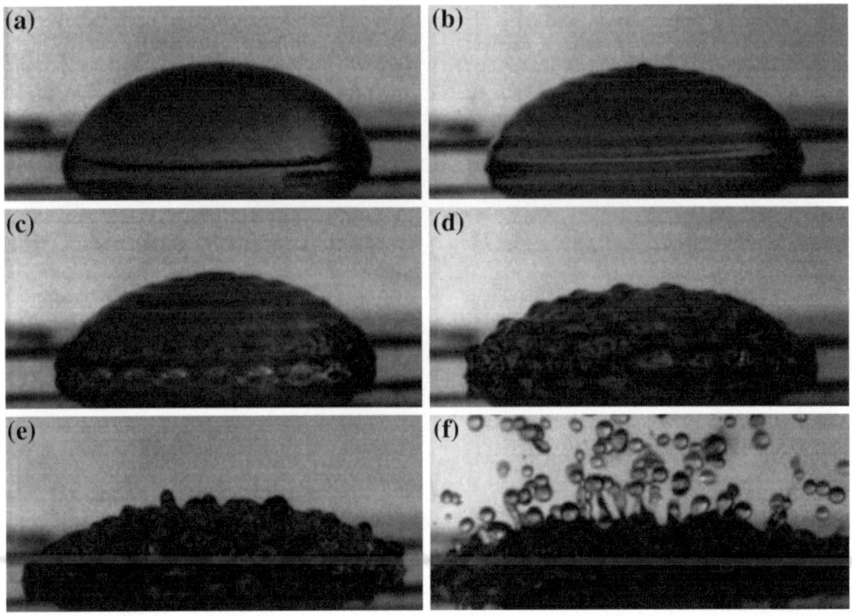

Fig. 13.6 Sequential growth of the instabilities on the surface of a 100 μL water droplet by gradually increasing the amplitude of the forcing wave oscillating at a frequency of 987 Hz: **a** initial droplet, **b** axisymmetric standing waves of the same frequency on the droplet surface, **c** azimuthal waves on the surface, **d** time-dependent waves spread over the entire surface, **e** appearance of craters and spikes, and **f** ejection of numerous droplets (James et al. 2003)

High amplitude of the forcing oscillations deforms the droplet leading to the formation of a crater at the droplet centre (at $t = 1.2$ in Fig. 13.7). Droplet tends to resist the deformation on account of its surface tension which causes the crater to enclose toward the centre. Pressurized liquid moves upward with high velocity forming a liquid spike which undergoes subsequent break-up into daughter droplets ($t = 1.6$ in Fig. 13.7).

13.2.2.2 Scaling Analysis

Vukasinovic et al. determined the threshold parameters for droplet ejection using glycerin–water droplets for various capillary (Ca) and Ohnesorge numbers (Oh) (Vukasinovic et al. 2007). For low capillary numbers (Ca << 1), the threshold acceleration (a_c) was assumed to be independent of the fluid viscosity and was scaled as: $a_c \sim f^{\frac{4}{3}}\left(\frac{\sigma}{\rho}\right)^{\frac{1}{3}}$ [f is forcing frequency, σ and ρ are the surface tension and density of the droplet, respectively,] and the droplet diameter was scaled as: $\widehat{d} = \frac{d f^{\frac{2}{3}}}{\left(\frac{\sigma}{\rho}\right)^{\frac{1}{3}}}$ (Vukasinovic et al. 2007). As explained in the previous section, collapse of the

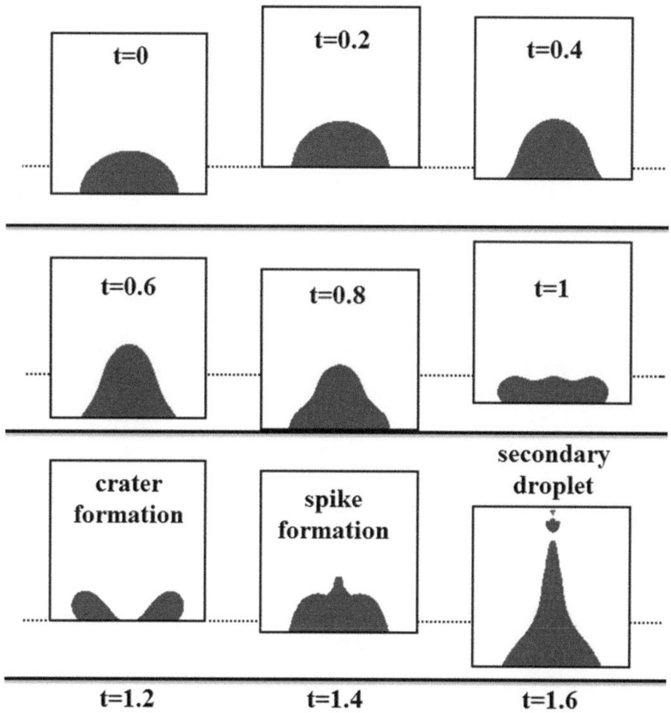

Fig. 13.7 Numerical results showing the various stages of droplet ejection from the surface of a sessile droplet (time is in forcing units). The solid and dotted lines represents ground state and maximum displacement states of the base, respectively (James et al. 2003)

crater provides initial momentum for growth of the ligament on droplet surface which eventually breaks due to capillary pinch-off mechanism (Ca $\leq$ 1). Multiple droplets were ejected from the ligament of low capillary number due to continuous droplet pinch-off at the tip [(a) Ca = 0.0069 in Fig. 13.8]. Increase in capillary number (viscous liquids) causes an increase in break-up length and time of the ligaments. Viscosity causes significant damping of the capillary instability and increase in the spike length [(b) Ca = 0.604 in Fig. 13.8]. These ligaments elongate and undergo subsequent break-up at the base. The lower end of ligament retracts and undergoes tip-break-up. Regimes with intermediate capillary number showed either tip-break-up or base-break-up. Dimensional analysis of the forcing parameters established the relationship as: $f^* = \frac{f\nu^3}{\left(\frac{\sigma}{\rho}\right)^2}$ (f is the forcing frequency) such that the tip-break-up was dominant for non-dimensionalized frequency $f^* < 0.002$ and base break-up was usually obtained for $f^* > 0.02$.

The growth and break-up of the ligaments is governed by inertial, viscous and surface tension forces which are characterized by the Ohnesorge number $\left(\text{Oh} = \frac{\mu}{\rho l \sigma}\right)$ and the Weber number $\left(\text{We} = \frac{\rho l v_i^2}{\sigma}\right)$, respectively. The initial velocity

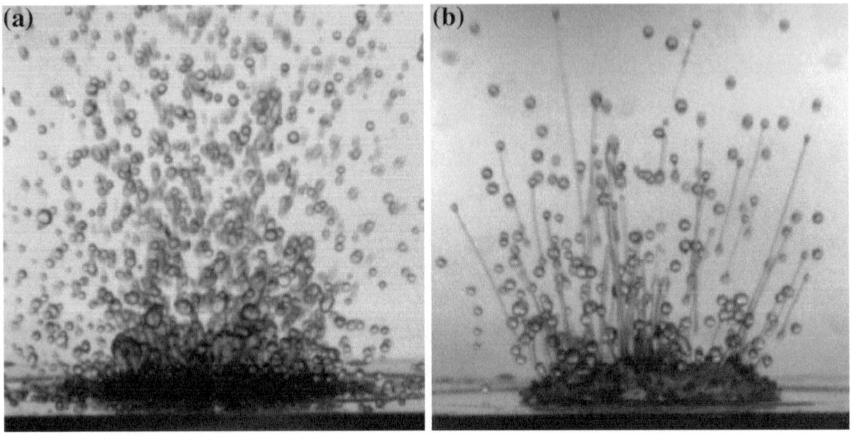

Fig. 13.8 Ejection of numerous droplets from the surface of a 0.1 ml droplet of **a** water [Ca = 0.0069] and **b** glycerin–water solution [Ca = 0.604] (Vukasinovic et al. 2007)

of ligament has no significant influence upon the time required for growth and break-up of the ligaments (not shown here). Therefore, break-up phenomenon is mostly governed by fluid properties and the ligament diameter. Moreover, the radius of ligaments is determined by the dimension of the craters formed on droplet surface. On the other hand, the size of craters depends upon the wavelength on the free surface of the droplet which is a function of properties of the fluid and the forcing frequency (Vukasinovic et al. 2007). Therefore, the ligament break-up time (t_b) is determined by the driving frequency and the fluid properties. The timescale was expressed by Vukasinovic et al. as (Vukasinovic et al. 2007):

$$t_b^* = \frac{t_b}{\nu^3}\left(\frac{\sigma}{\rho}\right)^2 \tag{13.6}$$

The timescale was validated using experimental data (Fig. 13.9a). The data (t_b^*) showed universal behavior against non-dimensionalized frequency $\left(f^*\right)$ which established the relation (Vukasinovic et al. 2007):

$$t_b \sim \frac{\nu^{\frac{3}{4}}}{f^{\frac{3}{4}}\left(\frac{\sigma}{\rho}\right)^{\frac{1}{2}}} \tag{13.7}$$

Both surface tension and driving frequency reduce t_b. Increase of frequency reduces the wavelength on the droplet surface and produces ligaments of smaller diameter which are more prone to break-up. Surface tension increases the local pressure differences which intensify the capillary break-up leading to a reduction in t_b. The flow rate in the ligament dampens due to viscosity, thereby increasing the break-up timescale (t_b). The experimentally estimated maximum length of the

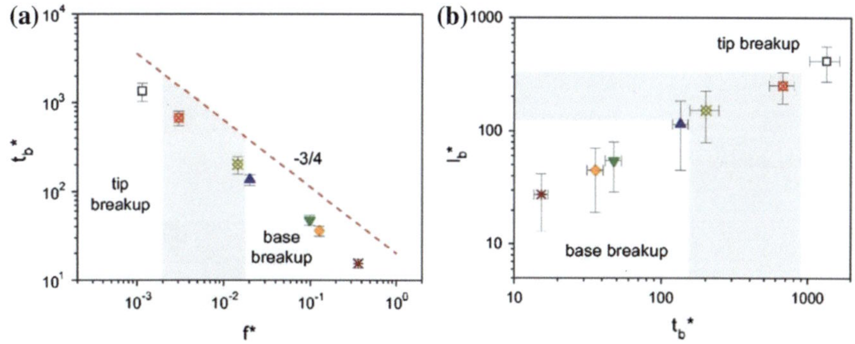

Fig. 13.9 **a** Variation of ligament break-up timescale (t_b) non-dimensionalized using $\left[\frac{\nu^3}{\left(\frac{\sigma}{\rho}\right)^2}\right]$ and **b** lengthscale (l_b) non-dimensionalized with $\left[\frac{\nu^2}{\left(\frac{\sigma}{\rho}\right)}\right]$ (Vukasinovic et al. 2007)

ligaments (l_b) was non-dimensionalized by Vukasinovic et al. using $\left[\frac{\nu^2}{\left(\frac{\sigma}{\rho}\right)}\right]$ and plotted against time t_b^* (Fig. 13.9b) which established the relation:

$$l_b \sim \frac{\nu^{0.65}}{f^{0.45}\left(\frac{\sigma}{\rho}\right)^{0.1}} \tag{13.8}$$

Increase in viscosity has a prominent effect upon the length of the ligament. The length increases due to increase in the break-up time period. Deepu et al. included the effects of acoustic pressure in droplets (water and glycerol mixture) excited by external perturbation (Deepu et al. 2013). The droplet was placed on the emitter surface which generated the acoustic wave. The droplet initially spread and the ultrasonic energy from the oscillating surface lead to the development of multi-modal capillary waves on the droplet surface (Fig. 13.10). These waves lead to the formation of ligaments which ultimately broke into the secondary droplets. Spreading of droplet was actuated by the radial component of acoustic force. The radial component also assisted in ligament growth although the growth was mainly supported by the z-component. Non-uniform velocity due to deformation of the droplet (two-lobe shape) alongside the acoustic streaming resulted in shear-induced instability at the surface. The onset of instability was defined by the Weber number criterion such that: $We = \frac{\rho h v^2}{\sigma} > 1$(h, ρ and σ are height, density, and surface tension of the droplet. "ν" is the rate of flow/spreading. High viscosity dampened spreading of the droplet due to low flow rate and suppressed the growth of capillary instability modes. Therefore, Weber number never exceeded the critical value $\left(We = \frac{\rho h v^2}{\sigma} < 1\right)$. Therefore, droplets of highly viscous fluids did not undergo

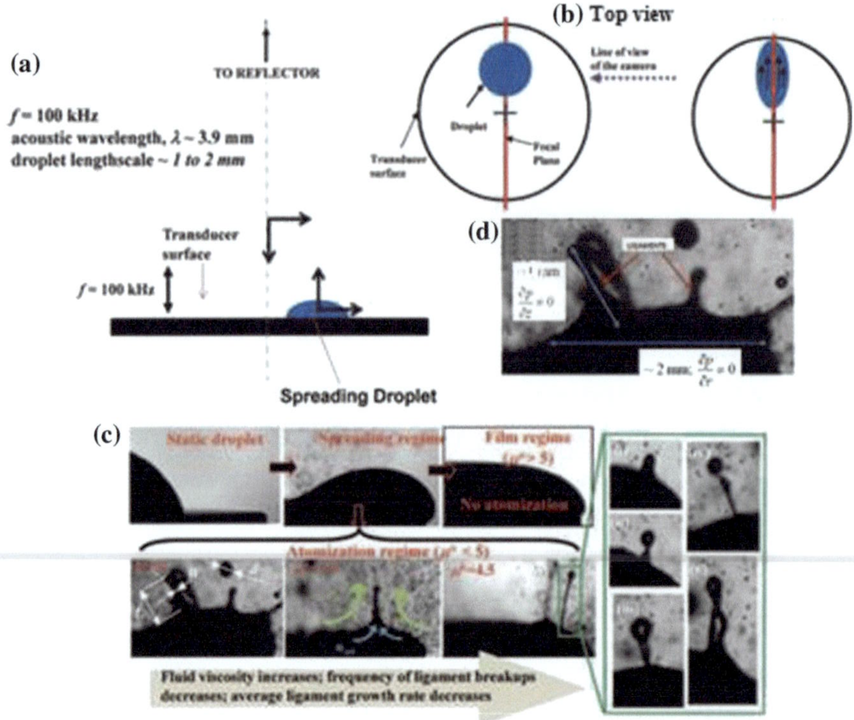

Fig. 13.10 **a** Schematic showing (**a**) the side view and **b** the top view of droplet resting on the surface of droplet placed on the surface of transducer, **c** different regimes of droplet spreading, ligament inception, grwoth and break-up [break-up mechanisms observed are (ii) tip-break-up, (iii) base break-up, (iv) tip-base break-up, and (v) Rayleigh instability break-up], and **d** close-up view of ligament and the pressure variation across the ligament (Deepu et al. 2013)

any atomization. The inertia for ligament growth is balanced by the viscous force (F_v) acting upon the longitudinal section of ligaments (of area ~ Wl_v):

$$F_v \sim \frac{\mu l_v^2}{t_v} \tag{13.9}$$

where l_v and t_v are the respective lengthscale and timescale for viscosity-driven ligaments. The ligament inertia scales as: $F_i \sim \frac{\rho l_v^4}{t_v^2}$. Therefore, viscous dissipation is dominant if $F_v > F_i$ which implies: $l_v \leq \sqrt{\frac{\mu}{\rho} t_v}$.

Ligaments of low viscous fluids, on the other hand, extended much longer prior to break-up due to higher initial velocity as compared to that of highly viscous fluids. Majority of the ligaments suffered Plateau–Rayleigh type break-up which was verified by Deepu et al. by estimating the critical aspect ratio at the point of break-up (ratio of the length to width at the half length): $\frac{l_{max}}{W_{half}} \geq 3.14$ (Deepu et al.

2013). Other mechanisms such as tip-break-up and base break-up were also observed in the ligaments (Fig. 13.10c).

13.2.3 *Acoustically Levitated Droplets*

As stated earlier, levitation is an excellent way to study instabilities in droplets without any surface-contact effects. The instabilities induced due to external perturbations lead to shape oscillations and atomization of the droplets. Acoustic levitation utilizes a transducer vibrating at a fixed frequency which drives a sound wave and gets reflected at the reflector (at the top) thereby producing a standing wave.

13.2.3.1 Influence of Acoustic Pressure

The droplets stabilized at the nodes of the standing wave can be perturbed by regulating the intensity of sound field used for levitation (Lee et al. 1991; Anilkumar et al. 1993; Lee et al. 1994). The droplet usually stabilizes slightly below at a distance z_c from pressure node of the wave due to gravity (Fig. 13.11). The inhomogeneous pressure field ($P_a = P_p - P_e$) surrounding the droplet tends to deform it into an ellipsoidal shape. The force due to surface tension (F_{ST}) acts as a restoring force of the droplet to retain the spherical shape. Therefore, the stability is maintained provided that:

$$P_p - P_e = \frac{\sigma}{R_c} \tag{13.10}$$

where σ is the surface tension of the droplet of radius of curvature (R_c) against Bernoulli pressure (P_p and P_e are the pressure at the pole and the equator, respectively). ($P_p - P_e$) can be used an appropriate scale to account for the non-uniform pressure distribution (Basu et al. 2013).

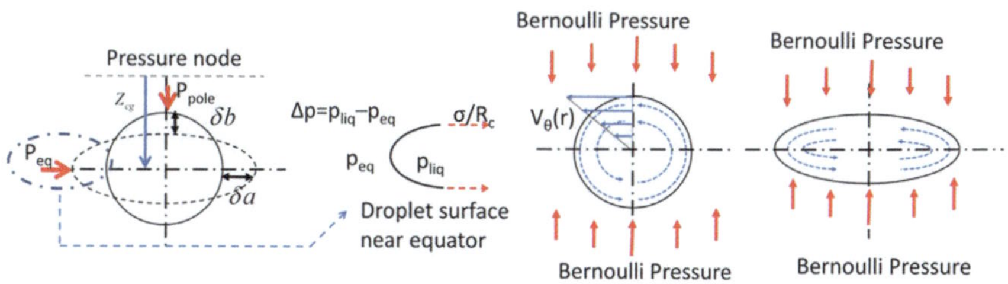

Fig. 13.11 Pressures acting on the droplet surface stabilized at the node of an acoustic levitator (Basu et al. 2013)

Perturbation of the external pressure field initiates oscillations in the aspect ratio of droplet ($a = x/y$, x and y are the polar and equatorial radii of the droplet). The increase in aspect ratio disturbs the pressure field surrounding the droplet which causes further change in the aspect ratio. The pressure at the pole (P_p) and equator (P_e) is given as:

$$P_p = P_m \sin(k_z(z_c - y)) \tag{13.11}$$

$$P_e = P_m \sin(k_z z_c) J_0(k_r x) \tag{13.12}$$

where J_0 is Bessel function of zeroth order and k_r, k_z are the wave numbers in radial and axial directions, respectively. Differentiating Eqs. (13.11) and (13.12) with respect to the droplet aspect ratio (a) and using $a = x/y$ and the droplet volume $\left(V = \frac{4}{3}\pi x^2 y\right)$ such that $\frac{dx}{da} = \frac{x}{3a}$ and $\frac{dy}{da} = -\frac{2y}{3a}$ gives:

$$\delta P_p = [P_m k_z \cos(k_z(z_c - y))] \frac{2y}{3a} \delta a \tag{13.13}$$

$$\delta P_e = -[P_m k_r \sin(k_z z_c) J_1(k_r x)] \frac{x}{3a} \delta a \tag{13.14}$$

Equations (13.13) and (13.14) signify that the increase in aspect ratio causes an increase in the pressure at the pole and a decrease in pressure at the equator, thereby increasing the net pressure ($P_p - P_e$) acting upon the droplet. The radius of curvature of the droplet $\left(R_c = \frac{y^2}{x}\right)$ decreases due to increase in the aspect ratio. The net pressure, radius of curvature, and surface tension ultimately determines stability of the droplet.

The rate of deformation of the droplet can be monitored by changing the frequency of the wave (Fig. 13.12). The curvature of the flattened droplet (ellipsoidal shape) is altered forming a thin concave membrane with a gradual rise in the wave frequency. The liquid flows and is collected in the periphery of the membrane forming a donut-like structure (Fig. 13.12). Subsequently, the membrane vibrates and undulations of short wavelengths (capillary ripples) develop at the center of the structure which spreads radially outward. The periphery of the membrane then moves inward and encloses forming a shell. The momentum gained at the closing end leads to the formation of two jets which grow in vertically opposite directions (Fig. 13.12a). The jet ultimately causes bursting of the shell-type structure. This type of break-up is known as secondary atomization and is prominent in larger droplets (water droplet diameter ≥ 4 mm). Intense membrane vibration can even lead to atomization of the structure into daughter droplets without showing any prominent jet-like behavior. Droplets of smaller sizes form thin liquid membrane which undergoes subsequent rupture into tiny droplets.

Loss of stability of droplet is initiated by the development of small-scale Kelvin–Helmholtz instabilities which acts as precursor to the growth of capillary waves on the surface and leads to complete rupture of the droplet. At equilibrium, the

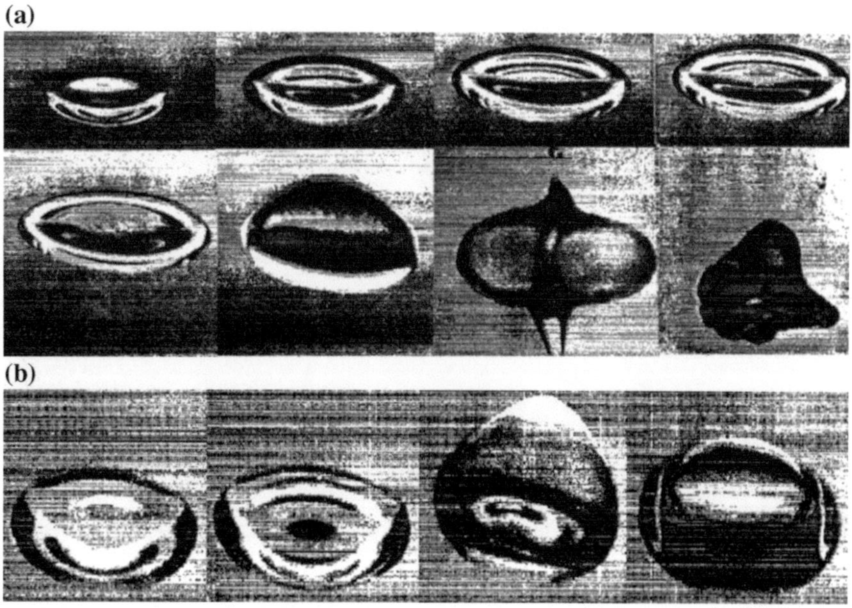

Fig. 13.12 Evolution of a water droplet in acoustic levitator with a gradual increase in the intensity: **a** flattening, adopting a concave shape, increase in concavity, formation of ripples at the center of the membrane type structure, ripples spreading throughout the structure, closing of the membrane forming a hollow structure, jet formation and the final break-up. **b** The stages of deformation in highly viscous fluid (glycerin) droplets (Lee et al. 1991)

radiation pressure at the top and bottom surface of the droplet, suction stress due to Bernoulli effect of acoustic velocity and the surface tension of the droplet are all balanced. The compression of droplet surface and the suction at the rim cause flattening of the droplet. The rate of flattening increases with an increase in the sound intensity (Fig. 13.12).

Viscosity acts as a damping parameter in the process of atomization. Droplets of highly viscous fluid (glycerin) undergo flattening into thin sheet which balloons out into a stable shell-type structure (Fig. 13.12b). No ripples and jets are formed on the surface of such droplets. Viscous droplets of smaller size show severe flattening and subsequent disintegration into tiny droplets.

13.2.3.2 Combined Effect of Acoustic Pressure and External Heating

Pure Fluid Droplets

Dynamics of an acoustically levitated droplet is altered under the influence of external heating. The instabilities imposed due to combined effect of heat and acoustics were studied by Basu et al. (2012, 2013). The droplets were stabilized at

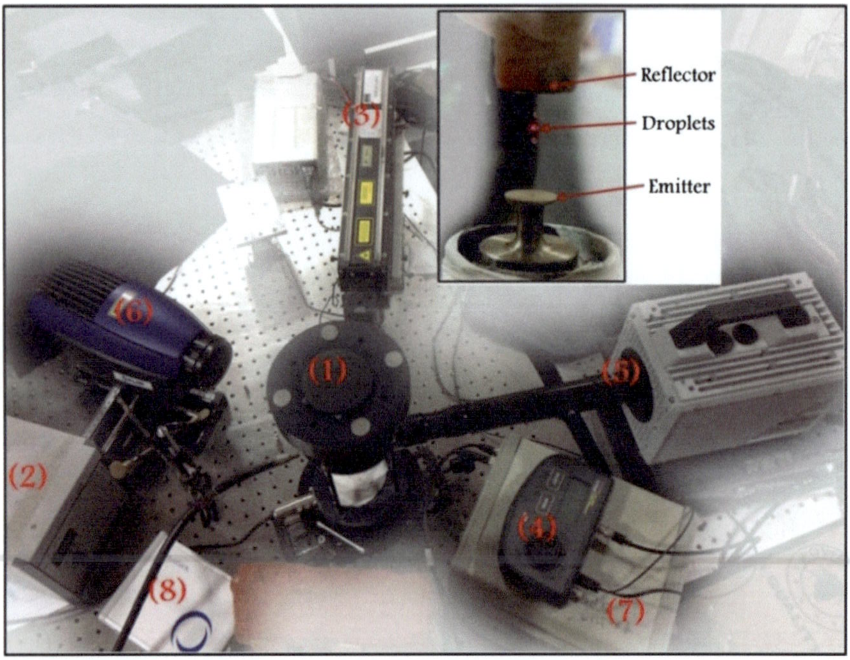

Fig. 13.13 Experimental setup for acoustically levitated droplets [(1) acoustic levitator (Tec5 levitator, 100 kHz frequency, 154 dB), (2) levitator controller, (3) CO2 laser, (4) laser controller, (5) high-speed camera (Photron Fastcam SA5) with zoom lens assembly, (6) IR camera (FLIR SC5200), (7) delay generator, and (8) Scott LLS white light] [inset: droplets suspended at the nodes of the levitator]

the nodes of an acoustic levitator and heated with a CO_2 laser. The dynamics of droplet and the temperature were monitored using a high-speed camera (back lightened) and IR camera, respectively. Both the cameras synchronized using a delay generator (Fig. 13.13).

The inhomogeneous distribution of acoustic pressure tends to deform the droplet which is resisted on account of its surface tension. Transient heating increases the temperature of the droplet which causes a reduction in the surface tension of the droplet. The dominant acoustic pressure $\left(P_{pole} - P_{eq} = \frac{\sigma}{R_c}\right)$ morphs the droplet into an ellipsoidal shape [where, P_{pole} and P_{eq} are pressure at the poles and the equator, respectively. σ and R_c are surface tension and radius of curvature of the droplet]. Two modes of atomization (primary and secondary) can be induced by heating the droplets (of low vapor pressure). The primary mode of atomization is triggered due to Kelvin–Helmholtz (KH) instability. Tiny droplets are ejected mostly in the equatorial region of the droplet (Fig. 13.14a). Sharp velocity gradient between the droplet and the surrounding air leads to maximum shear at the equatorial region. The liquid layer is subsequently atomized when the Weber number

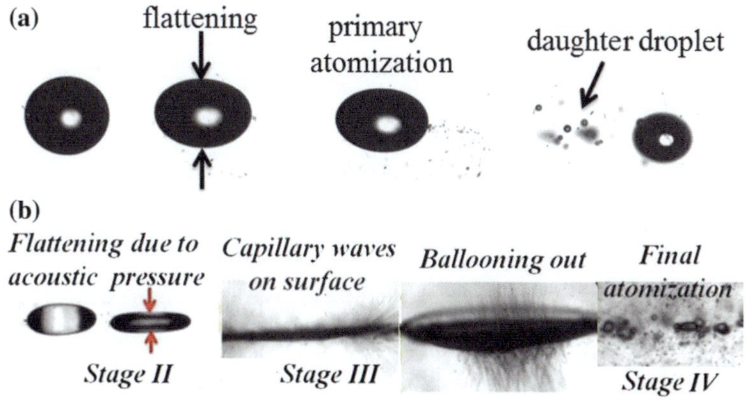

Fig. 13.14 Different stages of **a** primary atomization and **b** secondary atomization in droplets of low vapor pressure fluids (dodecane) (Pathak and Basu 2016a)

$\left[\mathrm{We_{eff}} = 2P_0\gamma_0\frac{R_c}{\sigma}\left(\frac{u_{eff}}{c_0}\right)^2\right]$ of the droplet exceeds the critical value $\left[\mathrm{We_{crit}} = 2P_0\gamma_0\frac{R_c}{\sigma}\left(\frac{u_{crit}}{c_0}\right)^2\right]$ (where P_0 is the atmospheric pressure, γ_0 is the ratio of specific heats, c_0 is sonic velocity, u_{eff} is the streaming velocity. The critical velocity is $u_{crit}^2 = \left[\frac{2(\rho_l+\rho_0)}{\rho_l\rho_0}\right]\sqrt{\sigma g(\rho_l - \rho_0)}$. ρ_l and ρ_0 represents density of liquid phase and gas phase, respectively). The droplet do not show any shape oscillations.

The secondary mode of atomization is triggered due to non-uniform pressure surrounding the droplet. The KH instability acts as precursor to the secondary instability which occurs when a critical criterion is attained by the droplets (Basu et al. 2013). External heating alters the properties of the droplet (surface tension and viscosity) which plays a significant role in the onset of such instability. If the viscous damping is not significant, then the irreversible flattening leads to catastrophic break-up of the droplet (Fig. 13.14b).

Bicomponent Fuel Droplets

The deformation and break-up modes can also be altered by changing the composition of the constituents in bicomponent droplets (Pathak and Basu 2016a). Droplets of low volatile fluid (dodecane) usually undergo severe shape deformation and subsequently secondary atomization as explained in the previous section. A droplet of highly volatile fluid (benzene), on the other hand, evaporates completely due to latent heating and shows no shape oscillation in its lifetime. Distinctly different modes (ligaments and bubbles) were identified in bicomponent droplets (dodecane–benzene) due to wide variation in the properties of the two components. It should be noted that dodecane and benzene were the two components used as a

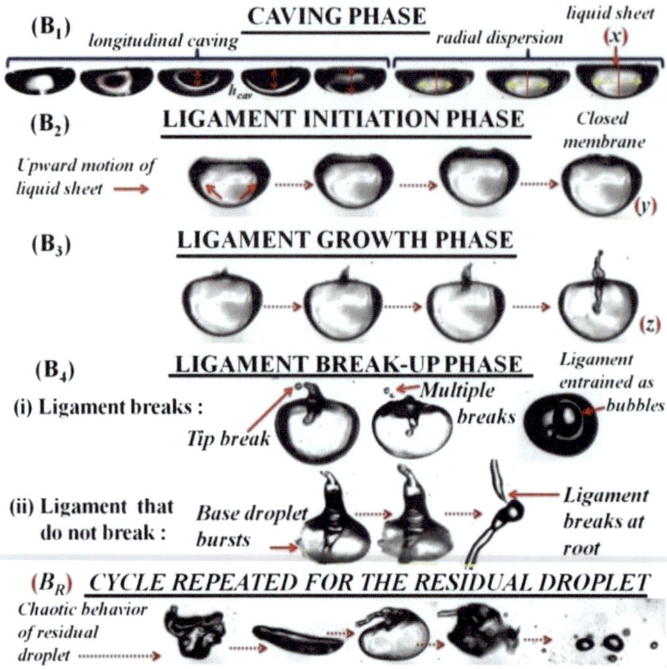

Fig. 13.15 Evolution of bicomponent droplets with low concentration of the highly volatile component (benzene <70% by volume) (Pathak and Basu 2016a)

sample case. The observations and the physical mechanism are explained in the subsequent sections.

Droplets with Lower Concentration of the Highly Volatile Component

Flattening of droplets due to imbalance of acoustic force ($F_{acc} = P_{acc}\pi r_d^2$) and the restoring force of surface tension ($F_{ST} = \sigma\pi r_d$) can be observed in bicomponent droplets although the rate of flattening was sufficiently reduced. The net force ($F_{net} = F_{acc} - F_{ST}$) acting upon the droplet results in caving of the top surface into a bowl-shaped structure with trapped liquid inside (Fig. 13.15 (caving phase)).

The governing equation for the caving phase is given as:

$$\frac{d(m_c v_c)}{dt_c} = P_a \pi r_d^2 - 2\pi r_d \sigma \qquad (13.15)$$

where m_c and v_c denote the mass and rate $\left(v_c = \frac{dh_c}{dt_c}\right)$ of the liquid caving inward the droplet of radius r_d. m_c $\left(= \frac{2}{3}\pi\rho_L h_c r_c^2\right)$ can be estimated assuming an ellipsoidal shape of the cavity (h_c and r_c represent the instantaneous depth and radius of the cavity formed). Scaling analysis of Eq. (13.15) estimates the rate of caving as:

$$v_c \sim \left[\frac{P_a r_d^2 - 2\sigma r_d}{\frac{2}{3}\rho r_c^2}\right] \tag{13.16}$$

The rate of cavity formation v_c estimated using Eq. (13.16) showed reasonable match with the experimental data $\left(v_c = \frac{dh_c}{dt_c}\right)$ which validated the governing Eq. (13.15) (Pathak and Basu 2016a). The volumetric growth of cavity $\left(V_c = \frac{2}{3}\pi h_c r_c^2\right)$ can be estimated assuming an ellipsoidal shape of the cavity and non-dimensionalized with the droplet volume $\left(V_d = \frac{2}{3}\pi d_{d,0} r_d^2\right)$ at the point of initiation of caving (V_c/V_d) as:

$$\frac{V_c}{V_d} = \left[\frac{1}{d_{d,0}}\right]\left(\frac{r_c}{r_d}\right)^2 v_c t_c \tag{13.17}$$

Substituting v_c using Eqs. (13.16) and (13.17) estimates the timescale (t_c) as:

$$t_c = \left(\frac{r_d}{r_c}\right)^2 \left[\frac{\frac{2}{3}\rho r_c^2}{P_a r_d^2 - 2\sigma r_d}\right]^{0.5} d_{d,0} \tag{13.18}$$

The volume of cavity calculated from experiments (V_c) (non-dimensionalized with V_d) was plotted against time (Fig. 13.16) which was non-dimensionalized with t_c estimated using Eq. (13.18) (Pathak and Basu 2016a). The data showed a universal behavior across various concentrations which validated the timescale estimated in Eq. (13.18).

The inward caving due to net pressure causes pressure upsurge $\left(P_c = \frac{1}{2}\rho v_c^2\right)$ inside the droplet (bowl-shaped) and forces the liquid to move upward along the curvature $\left(F_e = P_c 4\pi r_e^2\right)$. A first-order scaling approximation of the force balance equation is given as:

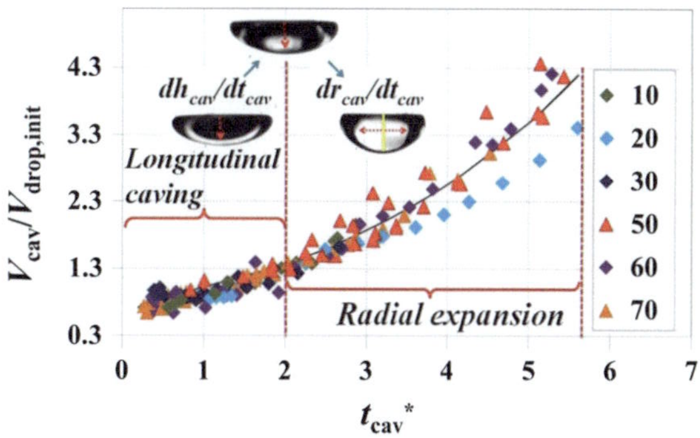

Fig. 13.16 Temporal variation of the cavity volume normalized with the droplet volume at the point of initiation of caving (time is normalized using t_c estimated using Eq. (13.17)) (Pathak and Basu 2016a)

$$\frac{d(m_e v_e)}{dt_e} = P_c 4\pi r_e^2 - 2\pi r_e \sigma \quad (13.19)$$

The rate at which the bowl-shape droplet (of radius r_e) encloses (v_e) can be estimated as: $v_e \sim \frac{0.5\rho V_c^2 4\pi r_e^2 - 2\sigma r_e}{m_e} t_e$. The closing of membrane (bowl-shaped) leads to the initiation of two ligaments growing with velocity (v_l) in vertically opposite directions. The ligament grows owing to the thrust that builds up due to closing of the membrane (bowl-shaped) ($F_e \sim 0.5\rho v_e^2 A_e$). It should be noted that the various forces acting upon the ligaments are about an order less than F_e which indicates that F_e is the only force that assist in the ligament growth. Applying momentum balance to the ligament growth phase $\left[F_e = \frac{d(m_l v_l)}{dt_l}\right]$ estimates the ligament length (l_l) to be $l_l \sim \left(\frac{F_e t_l^2}{\rho \pi r_l^2}\right)^{\frac{1}{2}}$. The maximum length of the ligament measured in experiments (l_{max}) was non-dimensionalized with l_l and was found out to be of order 1 which validated the theoretical estimation (Fig. 13.17) (Pathak and Basu 2016a). The thrust developed at the closing end may not be equally distributed between the two ligaments which resulted in slight divergence from the value of 1 in the graph.

Subsequently, the ligaments undergo break-up at a single point or multiple points if the energy at the closing end (of the bowl-shape) was sufficiently high ($6\times10^{-9}J - 2\times10^{-8}J$). These ligaments usually undergo Plateau–Rayleigh break-up which was validated by calculating the aspect ratio (ratio of the maximum ligament length to the width at half of the maximum length) which was greater than pi (~3.14) $\left[\frac{l_{max}}{W} > 3.14\right]$.

Droplets with Higher Concentration of the Highly Volatile Component

Increase in concentration of the highly volatile component (benzene volume fraction ≥ 0.7) suppresses flattening and caving of the droplets. The precursor stage of pure evaporation was followed by boiling and bursting of the droplet. Large difference in the rate of evaporation of the two components (dodecane and benzene) and preferential evaporation of benzene from the surface lead to a concentration

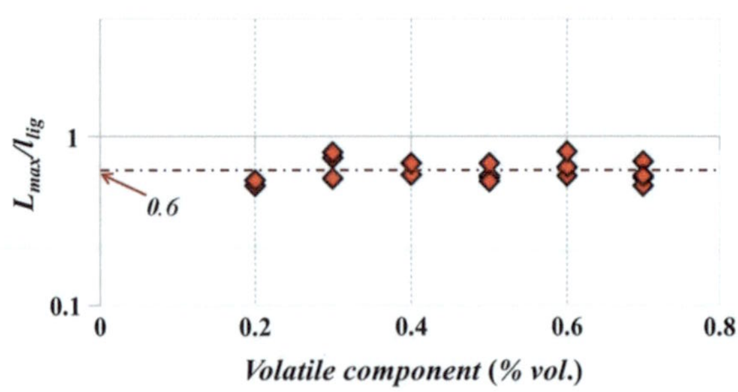

Fig. 13.17 Maximum length of the ligament plotted for different benzene concentration. The length was normalized using $\left(\frac{F_e t_l^2}{\rho \pi r_l^2}\right)^{\frac{1}{2}}$ (Pathak and Basu 2016a)

gradient throughout the droplet volume (Pathak and Basu 2016a). Benzene remains trapped within the droplet volume due to low rate of mass diffusion. Additionally, high rate of heat transport [high liquid-phase Lewis number, $Le_L \sim O(10^2)$] triggers boiling of the trapped benzene as the temperature exceeds the boiling point ($\geq 80°C$). Boiling was perceived due to inception of bubble inside the droplet. The total amount of heat absorbed by the droplet is utilized as latent heat for boiling:

$$\dot{m}h_{fg} = IA_d \tag{13.20}$$

where $\dot{m} = \rho_b \frac{dV_b}{dt_b}$ is the rate of vaporization of benzene (V_b: volume of bubble in time t_b. ρ_b and h_{fg} represents density and enthalpy of vaporization of benzene). Volume of the bubble (V_b) can be estimated by integrating Eq. (13.20) over bubble lifetime (t_b) as:

$$V_b = \frac{IA_d t_b}{\rho_b h_{fg}} \tag{13.21}$$

Equation (13.21) can be non-dimensionalized using parent droplet volume (V_d) and the timescale can be estimated as $t^* = \frac{tIA_d}{\rho_b h_{fg} V_d}$. The timescale t^* was validated using experimental data $\left(\frac{V_b}{V_d}\right)$ across various concentration which collapsed to a universal behavior (Fig. 13.18). The expanding bubble increases pressure inside the

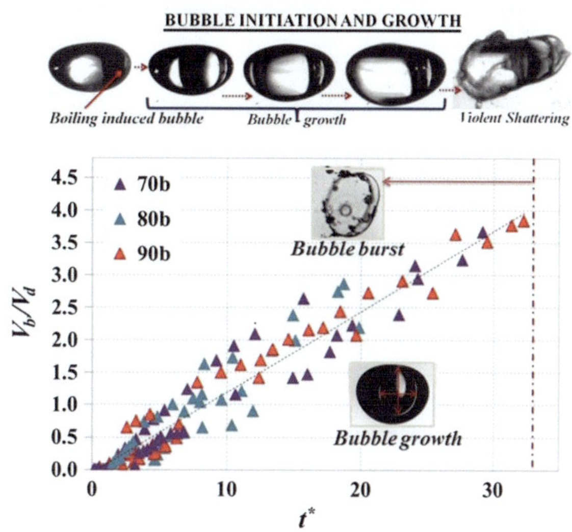

Fig. 13.18 Temporal variation of the bubble growth (volume is non-dimensionalized using parent droplet volume). Time is non-dimensionalized using $\frac{h_{fg}\rho_b V_d}{IA_d}$ [inset: bubble-induced boiling and bursting in bicomponent droplets for higher concentration of the highly volatile component] (Pathak and Basu 2016a)

droplet. Bursting occurs when interfacial tension cannot resist the pressure upsurge within the droplet.

Functionalized Droplets

Vigorous atomization of droplet is essential in many spray-based applications. Addition of nano-sized particles alters the combustion characteristics of traditional fuels (Gan and Qiao 2011; Gan et al. 2012). The efficiency and stability of the process is improved due to efficient mixing. Additionally, nanoadditives reduce the emission of pollutants in the system (Mehta et al. 2014). Another promising application of nanoparticles is in spray-drying technologies which results in powders with improved properties. The regimes of two different modes of atomization in pure fuel droplets explained in Sect. 13.2.3.2 can be altered by adding nanoparticles in the mixture. A new mode of break-up due to boiling and bursting was introduced in fuels with suspended nano-sized particles (Fig. 13.19).

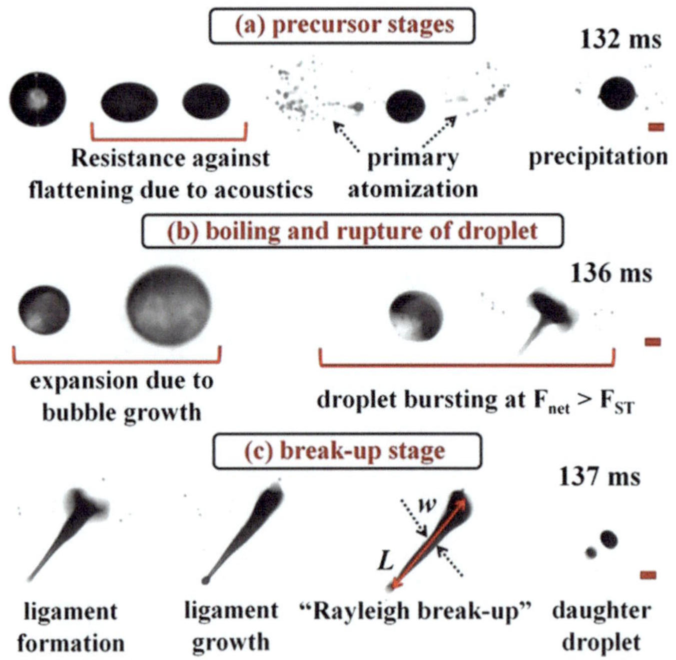

Fig. 13.19 Different stages of boiling mode of atomization observed in particle suspended droplets (mostly for initial particle concentration, $\phi \geq 1$wt%) which depicts **a** precursor stages, **b** initiation of boiling and bursting of droplet and **c** formation of ligament which grows longitudinally and undergoes Rayleigh break-up into daughter droplets. [*L* and *w* are the length and width (at half length) of the ligaments, respectively] [scale represents 100 μm] (Pathak and Basu 2016b)

Stability of nanosuspension is crucial in the spray process which necessitates careful preparation of the solution. Nano-sized particles (alumina, Al_2O_3, 21 nm diameter) can be added to the pure fuel (dodecane) and stirred vigorously to avoid formation of any agglomerates. The suspension should be sonicated as it affects the interparticle forces and reduces agglomeration. Addition of surfactant (TWEEN 85, HLB ~ 11) promotes stability of the solution. Surfactant to particle ratio can be as 2:1 to attain maximum solution stability.

Nanoparticles improve the heat absorbing capacity of the droplets. Increase of temperature due to transient heating $\left(\frac{dT}{dt} \sim 800\,°Cs^{-1} - 2000\,°Cs^{-1}\right.$ for $\psi = 0 - 5\text{wt.}\,\%$ at $I = 9W)$ causes faster reduction in surface tension of the droplets. As stated earlier, force of surface tension is the only restoring capacity of droplets against the external deforming forces. These droplets are more susceptible to deformation and secondary atomization (SA). Therefore, addition of particles induces SA in smaller droplets which usually undergo only primary atomization in absence of particles (critical size for SA, d_{min} is reduced). Increase in the external heating rate also reduces the critical size required for catastrophic SA.

The boiling mode of atomization is triggered for droplets of sizes "d" such that $d_{min} < d < d_{crit}$ (d_{min} and d_{crit} vary depending upon the heating rates and concentration of the added particles in solution) (Pathak and Basu 2016b). Evaporation of liquid enhances the rate of collision among particles leading to the formation of aggregates which acts as nucleation points inside the droplet. Continuous external heating enhances temperature and leads to boiling of liquid in contact with the aggregates. It should be noted that although surface tension (σ) reduces due to transient heating, high rate of evaporation causes significant decrease in radius of curvature (R_c). The net effect maintains stability of the droplet against deformation $\left(We = \frac{P_a R_c}{\sigma} < 1\right)$.

The heat absorbed by the droplet $\left(\dot{Q}_{abs} = f\frac{IA_d}{A_{beam}}\right)$ is employed for transient heating of the droplet $\left(\dot{Q}_t = m_d C_p \frac{dT_d}{dt}\right)$ and subsequent liquid-phase evaporation $\left(\dot{Q}_v = h_{fg}\frac{dm_d}{dt}\right)$. Heat absorbed by particles enhances temperature of the particles and leads to boiling of liquid inside the droplet $\left(\dot{Q}_b = \rho_l h_{fg}\frac{dV_{agg}}{dt}\right)$. The initial bubble size can be considered to be equivalent to the size of aggregates formed at the point of initiation of boiling $\left(d_{agg} = r_p f_r^{\frac{1}{D-3}}\right)$ [r_p is radius of particles and f_r is fractal dimension]. The temperature attained is estimated as:

$$T_p = T_s + \frac{\sigma T_s}{h_{fg}\rho_v d_{agg}} \tag{13.22}$$

The initiation of boiling is followed by gradual growth of bubble in time t_b which can be estimated by assuming that the total heat absorbed by the droplet is utilized for boiling:

$$\dot{Q}_{abs} = \frac{d\left(\frac{4}{3}\pi r_b^3 h_{fg}\right)}{dt_b} \tag{13.23}$$

The timescale of bubble growth (radius R_b at time "t_b") can be calculated by integrating Eq. (13.23) such as:

$$t_b = \frac{\frac{4}{3}\pi\left(R_b^3 - r_{agg}^3\right)\rho_l h_{fg}}{f\frac{IA_d}{A_{beam}}} \tag{13.24}$$

Increase of pressure due to expanding bubble is estimated using Rayleigh–Plasset equation as:

$$P_{int} - P_\infty = \rho_\infty\left[R\frac{d^2R}{dt^2} + \frac{3}{2}\left(\frac{dR}{dt}\right)^2 + 4\frac{\mu_\infty}{R\rho_\infty}\frac{dR}{dt} + \frac{2\sigma}{\rho_\infty R}\right] \tag{13.25}$$

where P, μ, and ρ are the pressure, viscosity, and density ("∞" indicates surrounding liquid).

The liquid compressed by the growing bubble inside forms a thin layer between the surface of the bubble and the droplet. Presence of this thin liquid layer complicates the calculation of the interfacial tension (σ). The bubble bursting occurs when the interfacial tension can no longer resist the internal pressure. Bursting of the entire droplet produces ligament (of length l_{lig} and radius r_{lig}) which undergoes elongation and subsequent break-up into daughter droplets (Fig. 13.19). The net residual energy at the point of bursting provides the impetus for the ligament growth:

$$\frac{1}{2}\pi r_{lig}^2 l_{lig}\rho v_{lig}^2 \sim \left[P_{int} - P_\infty - \left(\frac{F_{ST}}{A_b}\right)\right]\Delta V_b \tag{13.26}$$

The maximum ligament length at the point of break-up can be estimated using Eq. (13.26) [$l_{lig} \sim O(10^{-4})$ m]. The aspect ratio of the ligaments $\left(\frac{l_{lig}}{W_{lig}}\right)$ is greater than pi (~3.14) which satisfies the criterion for Plateau–Rayleigh type of break-up.

13.2.4 Droplet Impact on Liquid Pools

In the study of free-falling droplet impact on a large pool, spreading, splashing, bouncing, crater depth, crown formation, cavity evolution, and bubble entrainment are important and have been visualized (Worthington 1908; Huang 1983; Meckenstock et al. 2014; Aziz and Chandra 2000). Here, the Rayleigh jet pinch-off and formation of secondary drops are important considerations; therefore, Ohnesorge number (ratio of viscosity to surface tension force, $Oh = \mu/\sqrt{\sigma\rho R}$), is an appropriate parameter to describe the interfacial liquid/liquid interactions (Aryafar and Kavehpour 2006; Paulsen 2013). Weber number (inertia to surface tension force $We = \rho U^2 R/\sigma$) and Reynolds number (ratio of inertia to viscous force, $Re = \rho UR/\mu$) are used to describe the crown splash, where μ, σ, and ρ are the dynamic viscosity, surface tension, and density of the liquids, R is the initial droplet radius, and U is the impact velocity. Weber number incorporates the effect of droplet velocity (based on releasing height). It is suggested that the role of gravity is the most important before the rupture (Walls et al. 2015; Bordoloi and Longmire 2012). Bond number (effect of gravity versus surface tension force, $Bo = \Delta\rho g R^2/\sigma$) could be another important parameter; however, if gravity is incorporated indirectly in these parameters through droplet velocity, Bo may not play a direct or significant role. Thus, Ohnesorge number, Weber number, and Reynolds number are the main parameters that play a strong role in determining the dynamics of instability on the liquid surface due to droplet impact.

Using Volume of Fluid (VOF) method, Castillo-Orozco et al. (Castillo-Orozco et al. 2015) implemented continuum surface force model in the OpenFOAM platform to compare the secondary droplet pinch-off with experiments using silicone oil (Fig. 13.20).

As the droplet impinges on the pool of the same liquid at a high velocity, the impact causes large disturbances to the pool which forms a deep crater followed by

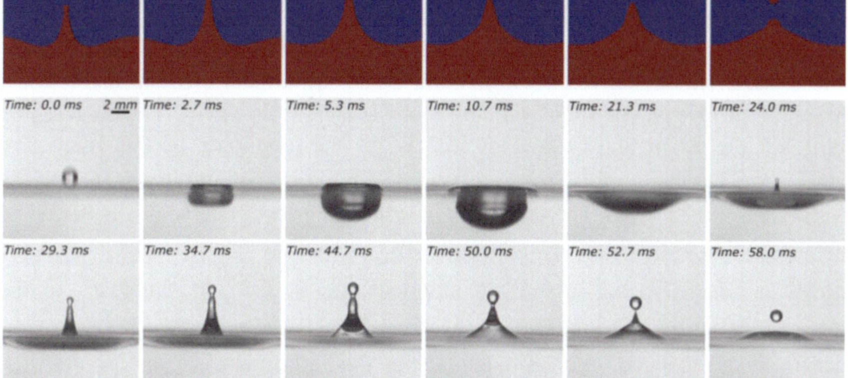

Fig. 13.20 Qualitative comparison of the numerical and experimental results. Time evolution of the Rayleigh jet for the impact of a silicone 5 cSt droplet with its own pool (Re = 324, We = 135, Oh = 0.036) (Castillo-Orozco et al. 2015)

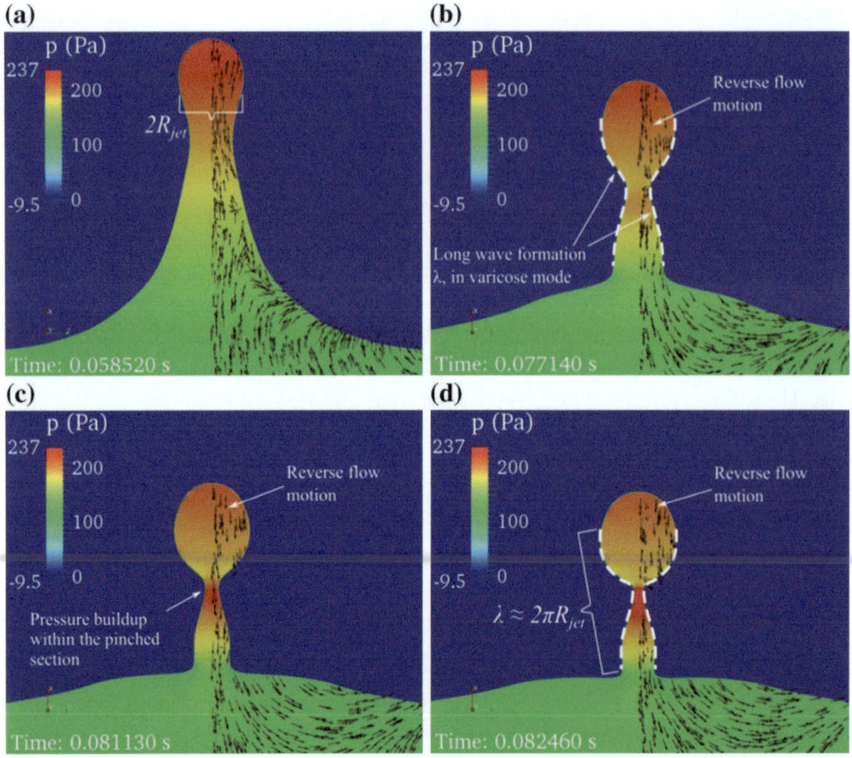

Fig. 13.21 Pressure buildup in the pinched region after the impact of a silicone oil 1.8 mm droplet (10 cSt) with impact velocity of 2.5 m/s ($We = 261$, $Oh = 0.0716$). Pressure field is shown on each image and velocity vectors of the central Rayleigh jet are shown on right side of each image. Some vectors have been removed for clarity (Castillo-Orozco et al. 2015)

a Rayleigh jet. The Rayleigh jet has the potential to breakup due to Rayleigh–Plateau instability, as shown in Fig. 13.21 below. The Rayleigh–Plateau instability occurs when surface waves begin to form under the influence of surface tension. When the surface waves are of varicose mode and are long waves, a pinched region forms on the jet as shown in Fig. 13.21b. As the amplitude of the long wave $\left(\lambda = 2\pi R_{jet}\right)$ begins to grow, pressure begins to build up within the pinched region as shown in Fig. 13.21c. The jet breaks up once the growth rate peaks. It should be noted that the wavelength is approximately the circumference of the jet and the growth rate for a jet has a capillary timescale, $t_{cap} \sim \sqrt{\rho R_{jet}^3/\sigma}$, where R_{jet} is the radius of the jet (Eggers and Villermaux 2008). The current experiments suggested that R_{jet} is of the same order of magnitude as the initial radius of the droplet (R_o), therefore $t_{cap} \sim \sqrt{\rho R_o^3/\sigma}$. The viscosity of the jet plays no significant role on the range of unstable wavelengths. However, at higher Oh, viscosity can slow down the growth rate of the unstable waves with a viscous timescale of $t_{visc} \sim \mu R_{jet}/\sigma$ (Eggers

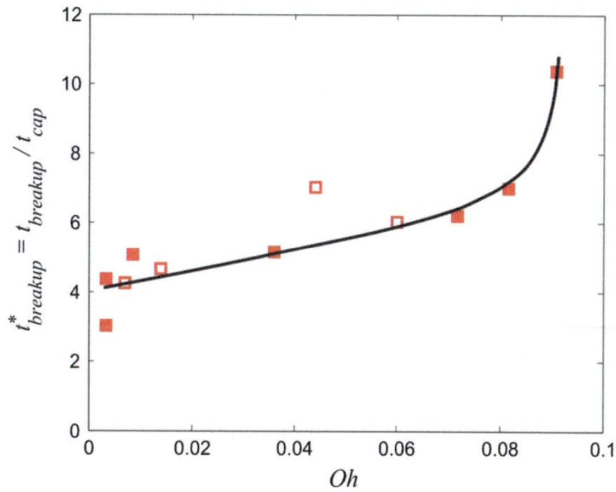

Fig. 13.22 Normalized breakup time of the Rayleigh jet versus Ohnesorge number. The filled and unfilled symbols represent the experimental and numerical data, respectively (Castillo-Orozco et al. 2015)

and Villermaux 2008). Additionally, the ratio of t_{visc} to t_{cap} represents the Ohnesorge number. Therefore, *Oh* is an indicator of how much the breakup time has lagged. The capillary timescale and breakup time are nearly of the same order in the reported experiments, even though the former is approximately 3 to 10 times smaller than the latter. Thus, the capillary timescale is the appropriate scale to normalize breakup time. The breakup time is calculated from the time at which the jet emerges from the interface until it reaches the maximum height where it pinches off (Fig. 13.22). The non-dimensional time, $t^*_{breakup}$, is seen to increase linearly with *Oh* up to $Oh = 0.06$, beyond which the curve becomes nonlinear.

Castillo-Orozco et al. (2015) experiments were performed using distilled water, potassium hydroxide (KOH), ethanol, ethylene glycol, silicone oils 5, 10, 13, 14, 16, 18, and 20 cSt. The silicone oils have similar surface tension (~20 mN/m) and densities (~940 kg/m^3), but different viscosities. If the impact velocity is high enough, a central Rayleigh jet will form. Under certain conditions, the tip of the jet pinches off due to Rayleigh–Plateau instability and the secondary droplet forms. Weber number identifies the boundaries of transitions between no breakup and Rayleigh jet and later into crown splash (Fig. 13.23). The Weber number at which this transition forms is called the critical Weber number. When viscous effects are under consideration to study instability, *Oh* is a more appropriate parameter as it isolates the property effects more. *Re* has also been used to classify the morphologies of crown droplets on a *We-Re* map (Zhang et al. 2010). In Fig. 13.23a and b, regime maps for Rayleigh jet breakup, crown splash and subsequent formation of secondary droplets are plotted both as *We vs Re* and We *vs Oh,* respectively. Each Ohnesorge number represents a distinct fluid, whereas the variation in Weber number is due to changes in release height of the droplet.

For a combination of $Oh \leq 0.091$ and Weber numbers beyond the critical value, the jet breakup leads to one or multiple secondary droplets that are on the order of $0.5D_o$ to $2D_o$. At low impact *We* number, the kinetic energy cannot

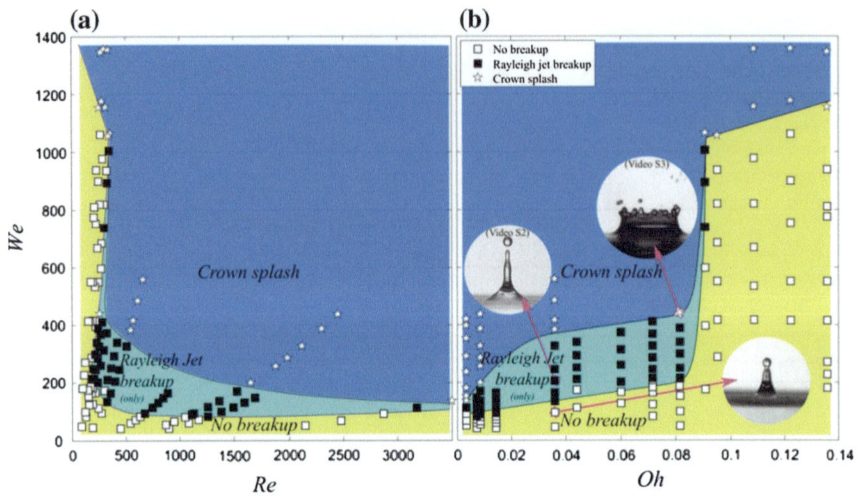

Fig. 13.23 **a** and **b** Regime maps for Rayleigh jet breakup and subsequent secondary droplets formation based on Re and Oh, respectively. Filled markers represent the cases where breakup took place and single or multiple secondary droplets were observed. Blank and star symbols represent no breakup and crown splash, respectively. Ohnesorge number of 0.007, 0.014, 0.044, and 0.060 was obtained from numerical simulations. The rest of the cases were obtained from experimental results (Castillo-Orozco et al. 2015)

overcome the surface tension forces. Depending on the impact velocity, in some cases, both Rayleigh jet breakup and crown splash can occur. It was found that high-impact velocity initiates the crown formation, which may disintegrate into numerous secondary droplets. Since high viscosity fluids tend to dampen the impact, they mitigate the occurrence of crown splash except at higher impact velocities.

13.3 Conclusions

Fundamental understanding of droplet atomization and break-up is important due to its widespread applications. Different modes of break-up can be induced by external factors which initiates small-scale perturbations on the surface of the droplet. The perturbation in form of waves develops in space and time and leads to the formation of spikes or ligaments which grows and causes complete rupture of the droplet. The onset of surface instabilities depends upon the external forcing parameters and physiochemical properties of the droplet (surface tension, viscosity, and density) which are characterized by Ohnesorge number and Weber number. Different types of break-up include single point tip-break-up or multiple points Plateau-Rayleigh break-up in droplets of different fluids. Acoustically levitated droplet demonstrates two major modes of atomization (primary and secondary) which are triggered based upon the critical Weber number criterion. Inclusion of bicomponents with large difference in volatilities of the two components in such levitated droplets leads to bubble-induced boiling and bursting. Boiling is induced due to diffusional entrapment of the highly volatile component in the droplets with highest concentration of the low volatile component at the surface. Alternatively, addition of nano-sized particles in low volatile fluid droplets acts as nucleation points and induces boiling and bursting due to heterogeneous boiling inside the contact-free droplets. Droplet atomization and break-up are an outcome of several complicated dynamics. Although a significant number of studies related to a single droplet break-up exist in literature, complete understanding of the different mechanisms requires rigorous experimental evidence, supported by numerical analysis.

References

Anilkumar AV, Lee CP, Wang TG (1993) Stability of an acoustically levitated and flattened drop: an experimental study. Phys Fluids A 5(11):2763–2774

Aryafar H, Kavehpour HP (2006) Drop coalescence through planar surfaces. Phys Fluids 18 (7):072105

Avedisian CT, Andres RP (1978) Bubble nucleation in superheated liquid—liquid emulsions. J Colloid Interface Sci 64(3):438–453

Aziz SD, Chandra S (2000) Impact, recoil and splashing of molten metal droplets. Int J Heat Mass Transf 43(16):2841–2857

Bartz FO, Guildenbecher DR, Schmehl R, Koch R, Bauer HJ, Sojka PE (2011) Model comparison for single droplet fragmentation under varying accelerations. In: 24th European conference on liquid atomization and spray systems, Sept 2011

Basu S, Saha A, Kumar R (2012) Thermally induced secondary atomization of droplet in an acoustic field. Appl Phys Lett 100(5):054101, 1–4

Basu S, Saha A, Kumar R (2013a) Criteria for thermally induced atomization and catastrophic breakup of acoustically levitated droplet. Int J Heat Mass Transf 59:316–327

Basu S, Tijerino E, Kumar R (2013b) Insight into morphology changes of nanoparticle laden droplets in acoustic field. Appl Phys Lett 102:141602

Beck JC, Watkins AP (2002) On the development of spray submodels based on droplet size moments. J Comput Phys 182(2):586–621

Bordoloi AD, Longmire EK (2012) Effect of neighboring perturbations on drop coalescence at an interface. Phys Fluids 24(6):062106

Cao XK, Sun ZG, Li WF, Liu HF, Yu ZH (2007) A new breakup regime of liquid drops identified in a continuous and uniform air jet flow. Phys Fluids 19(5):057103, 1–7

Castillo-Orozco E, Davanlou A, Choudhury PK, Kumar R (2015) Droplet impact on deep liquid pools: rayleigh jet to formation of secondary droplets. Phys Rev E 92:053022

Dai Z, Faeth GM (2001) Temporal properties of secondary drop breakup in the multimode breakup regime. Int J Multiph Flow 27(2):217–236

Deepu P, Basu S, Kumar R (2013) Dynamics and fracture of ligaments from a droplet on a vibrating surface. Phys Fluids 25(8):082106, 1–19

Dukowicz JK (1980) A particle-fluid numerical model for liquid sprays. J Comput Phys 35 (2):229–253

Eggers J, Villermaux E (2008) Physics of liquid jets. Rep Prog Phys 71(3):036601

Elrod SA, Hadimioglu B, Khuri-Yakub BT, Rawson EG, Richley E, Quate C, Lundgren TS (1989) Nozzleless droplet formation with focused acoustic beams. J Appl Phys 65(9):3441–3447

Faraday M (1831) On the forms and states of fluids on vibrating elastic surfaces. Philos Trans R Soc London 52:319

Gan Y, Qiao L (2011) Combustion characteristics of fuel droplets with addition of nano and micron-sized aluminum particles. Combust Flame 158(2):354–368

Gan Y, Lim YS, Qiao L (2012) Combustion of nanofluid fuels with the addition of boron and iron particles at dilute and dense concentrations. Combust Flame 159(4):1732–1740

Goodridge CL, Shi WT, Lathrop DP (1996) Threshold dynamics of singular gravity-capillary waves. Phys Rev Lett 76(11):1824, 1–4

Goodridge CL, Shi WT, Hentschel HGE, Lathrop DP (1997) Viscous effects in droplet-ejecting capillary waves. Phys Rev E 56(1):472, 1–4

Goodridge CL, Hentschel HGE, Lathrop DP (1999) Breaking faraday waves: critical slowing of droplet ejection rates. Phys Rev Lett 82(15):3062

Haas FC (1964) Stability of droplets suddently exposed to a high velocity gas stream. AIChE J 10 (6):920–924

Hanson AR, Domich EG, Adams HS (1963) Shock tube investigation of the breakup of drops by air blasts. Phys Fluids 6(8):1070–1080

Hinze JO (1955) Fundamentals of the hydrodynamic mechanism of splitting in dispersion processes. AIChE J 1(3):289–295

Hirahara H, Kawahashi M (1992) Experimental investigation of viscous effects upon a breakup of droplets in high-speed air flow. Exp Fluids 13(6):423–428

Hsiang LP, Faeth GM (1995) Drop deformation and breakup due to shock wave and steady disturbances. Int J Multiph Flow 21(4):545–560

Huang JC (1983) A review of the state-of-the-art of oil spill fate/behavior models. In: International oil spill conference proceedings, vol 1983. American Petroleum Institute, p 313

Jackson GS, Avedisian CT (1998) Combustion of unsupported water-in-n-heptane emulsion droplets in a convection-free environment. Int J Heat Mass Transf 41(16):2503–2515

James AJ, Smith MK, Glezer A (2003a) Vibration-induced drop atomization and the numerical simulation of low-frequency single-droplet ejection. J Fluid Mech 476:29–62

James AJ, Vukasinovic B, Smith MK, Glezer A (2003b) Vibration-induced drop atomization and bursting. J Fluid Mech 476:1–28

Kim JH, Nakajima T (1999) Aerodynamic influences on droplet atomization in an electrostatic spray. JSME Int J Ser B 42(2):224–229

Krzeczkowski SA (1980) Measurement of liquid droplet disintegration mechanisms. Int J Multiph Flow 6(3):227–239

Lane WR (1951) Shatter of drops in streams of air. Ind Eng Chem 43(6):1312–1317

Law CK (1978) Internal boiling and superheating in vaporizing multicomponent droplets. AIChE J 24(4):626–632

Lee CP, Anilkumar AV, Wang TG (1991) Static shape and instability of an acoustically levitated liquid drop. Phys Fluids A 3(11):2497–2515

Lee CP, Anilkumar AV, Wang TG (1994) Static shape of an acoustically levitated drop with wave–drop interaction. Phys Fluids 6(11):3554–3566

Link DR, Anna SL, Weitz DA, Stone HA (2004) Geometrically mediated breakup of drops in microfluidic devices. Phys Rev Lett 92(5):054503, 1–4

Magarvey RH, Taylor BW (1956) Free fall breakup of large drops. J Appl Phys 27(10):1129–1135

McDonald JE (1954) The shape and aerodynamics of large raindrops. J Meteorol 11(6):478–494

Meckenstock RU, von Netzer F, Stumpp C, Lueders T, Himmelberg AM, Hertkorn N, Schmitt-Kopplin P, Harir M, Hosein R, Haque S, Schulze-Makuch D (2014) Water droplets in oil are microhabitats for microbial life. Science 345(6197):673–676

Mehta RN, Chakraborty M, Parikh PA (2014) Nanofuelscombustion, engine performance and emissions. Fuel 120:91–97

Nakamura M, Akamatsu F, Kurose R, Katsuki M (2005) Combustion mechanism of liquid fuel spray in a gaseous flame. Phys Fluids 17(12):123301, 1–14

Pathak B, Basu S (2016a) Deformation pathways and breakup modes in acoustically levitated bicomponent droplets under external heating. Physical Review E 93(3):033103, 1–12

Pathak B, Basu S (2016b) Phenomenology of break-up modes in contact free externally heated nanoparticle laden fuel droplets. Phys Fluids 28(12):123302, 1–14

Paulsen JD (2013) Approach and coalescence of liquid drops in air. Phys Rev E 88(6):063010

Rajamanickam K, Basu S (2017) Insights into the dynamics of spray–swirl interactions. J Fluid Mech 810:82–126

Ranger AA, Nicholls JA (1969) Aerodynamic shattering of liquid drops. AIAA J 7(2):285–290

Reitz R (1987) Modeling atomization processes in high-pressure vaporizing sprays. At Spray Technol 3(4):309–337

Saha A, Lee JD, Basu S, Kumar R (2012) Breakup and coalescence characteristics of a hollow cone swirling spray. Phys Fluids 24(12):124103, 1–21

Shaikeea A, Basu S, Hatte S, Bansal L (2016) Insights into vapor-mediated interactions in a nanocolloidal droplet system: evaporation dynamics and affects on self-assembly topologies on macro-to microscales. Langmuir 32(40):10334–10343

Tijerino E, Basu S, Kumar R (2013) Nanoparticle agglomeration in an evaporating levitated droplet for different acoustic amplitudes. J Appl Phys 113:034307

Tsapis N, Dufresne ER, Sinha SS, Riera CS, Hutchinson JW, Mahadevan L, Weitz DA (2005) Onset of buckling in drying droplets of colloidal suspensions. Phys Rev Lett 94(1):018302

Vukasinovic B, Smith MK, Glezer A (2007) Mechanisms of free-surface breakup in vibration-induced liquid atomization. Phys Fluids 19(1): 012104, 1–15

Walls PL, Henaux L, Bird JC (2015) Jet drops from bursting bubbles: How gravity and viscosity couple to inhibit droplet production. Phys Rev E 92(2):021002

Wierzba A (1990) Deformation and breakup of liquid drops in a gas stream at nearly critical Weber numbers. Exp Fluids 9(1):59–64

Worthington AM (1908) A study of splashes. Longmans, Green, London, p 129

Yoon SS, DesJardin PE (2006) Modelling spray impingement using linear stability theories for droplet shattering. Int J Numer Meth Fluids 50(4):469–489

Yunker PJ, Still T, Lohr MA, Yodh AG (2011) Suppression of the coffee-ring effect by shape-dependent capillary interactions. Nature 476(7360):308–311

Zhang LV, Brunet P, Eggers J, Deegan RD (2010) Wavelength selection in the crown splash. Phys Fluids 22(12):122105

Zhao H, Liu HF, Xu JL, Li WF (2011) Experimental study of drop size distribution in the bag breakup regime. Ind Eng Chem Res 50(16):9767–9773

Zhao H, Liu HF, Xu JL, Li WF, Lin KF (2013) Temporal properties of secondary drop breakup in the bag-stamen breakup regime. Phys Fluids 25(5):054102

Chapter 14
Intermittency: A State that Precedes Thermoacoustic Instability

Samadhan A. Pawar and R. I. Sujith

Abstract Thermoacoustic instability is a plaguing problem in confined combustion systems, where self-sustained periodic oscillations of ruinous amplitudes that cause serious damage and performance loss to propulsive and power generating systems occur. In this chapter, we review the recent developments in understanding the transition route to thermoacoustic instability in gaseous combustion systems and describe a detailed methodology to detect this route in a two-phase flow combustion system. Until now, in these combustion systems, the transition to such instabilities has been reported as Hopf bifurcation, wherein the system dynamics change from a state of fixed point to limit cycle oscillations. However, a recent observation in turbulent gaseous combustion system has shown the presence of intermittency that precedes the onset of thermoacoustic instability. Intermittency is a dynamical state of combustion dynamics consisting of a sequence of high amplitude bursts of periodic oscillations amidst regions of relatively low amplitude aperiodic oscillations. Here, we discuss the process of transition to thermoacoustic instability in the two-phase flow system due to change in the control parameter, a location of flame inside the duct. As the flame location is varied, the system dynamics is observed to change from a region of low amplitude aperiodic oscillations to high amplitude self-sustained limit cycle oscillations through intermittency. The maximum amplitude of such intermittent oscillations witnessed during the onset of intermittency is much higher than that of limit cycle oscillations. We further describe the use of various tools from dynamical systems theory in identifying the type of intermittency in combustion systems.

Keywords Spray combustion · Thermoacoustic instability · Type-II intermittency

S. A. Pawar · R. I. Sujith (✉)
Department of Aerospace Engineering, IIT Madras, Chennai 600036, India
e-mail: samadhanpawar@ymail.com

R. I. Sujith
e-mail: sujith@iitm.ac.in

S. Basu et al. (eds.), *Droplets and Sprays*, Energy, Environment, and Sustainability,
https://doi.org/10.1007/978-981-10-7449-3_14

14.1 Introduction

Thermoacoustic instabilities have been a topic of immense research for combustion community for several decades. Initially, such instabilities were observed in liquid and solid rocket motors in the mid of the twentieth century, and henceforth were reported in other propulsion engines (such as ramjet and afterburner), land-based gas turbine engines, boilers, industrial furnaces, etc. (Young 1995; Lieuwen and Yang 2005). These are self-sustained, high amplitude periodic oscillations in acoustic pressure generated mainly due to complex interaction between unsteady heat release rate from the flame and natural acoustic modes of the combustor (Lieuwen 2012).

The fluctuations in acoustic pressure lead to fluctuations in heat release rate either by modulating of the flame surface area, changing the fuel–air mixing ratio, varying the chemical kinetics or by affecting the underlying hydrodynamic flow-field of the combustor (Ducruix et al. 2003) (see Fig. 14.1). This oscillatory heat release, in turn, adds energy to acoustic field of the system, thus completing a feedback loop. According to Lord Rayleigh (1878), when the fluctuations in the heat release rate happen in phase with the acoustic pressure oscillations, a positive feedback is developed between the coupled processes of the combustor. Furthermore, when the net driving of the acoustic oscillations overcomes the overall damping in the system, the fluctuations in the system dynamics get amplified exponentially and saturate to high amplitude limit cycle oscillations when the driving matches with the nonlinear damping (Lieuwen and Yang 2005).

The intense growth in pressure amplitude may have serious damage to mechanical components of the combustor such as liners, fuel injectors, turbine blades, due to increased vibrations, or may degrade its performance (Culick and Kuentzmann 2006). Various control strategies have been developed over the years to either reduce the strength of such instabilities (McManus et al. 1993; Candel 2002; Culick and Kuentzmann 2006) or to forewarn their occurrences (Lieuwen 2005; Gotoda et al. 2014; Nair et al. 2013; Nair and Sujith 2014).

Although much progress has been made in understanding the physical mechanisms that cause thermoacoustic instability, their reliable prediction is still elusive.

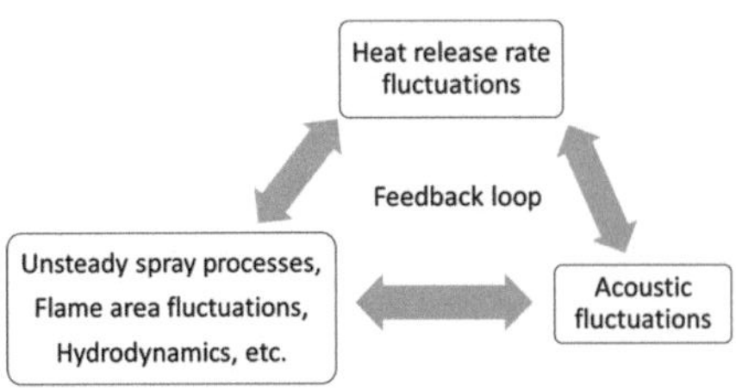

$$\underbrace{\int_V \int_T p'(x,t)\dot{q}'(x,t)\,dt\,dV}_{\text{Acoustic driving}} \geq \underbrace{\int_V \int_T \sum_i L_i(x,t)\,dt\,dV}_{\text{Acoustic damping}}$$

Fig. 14.1 Elementary processes leading to self-sustained thermoacoustic instability. The equation on the right side represents the criterion for the onset of thermoacoustic instability. When driving in the system is greater than damping, acoustic oscillations in the system grow to a state of limit cycle oscillations

In this context, the study on the transition of a thermoacoustic system from stable to unstable operation has received a growing interest in recent times. Here, stable operation corresponds to the state of combustion noise, and unstable operation represents the state of thermoacoustic instability (Smith and Zukoski 1985; Sterling and Zukoski 1987; Lieuwen 2002). The dynamics of a turbulent combustor observed during combustion noise comprises of low amplitude aperiodic oscillations, whereas that during thermoacoustic instability observed mostly in the form of high amplitude periodic (limit cycle) oscillations. In the case of laminar combustion systems, combustion noise corresponds to a steady state, while thermoacoustic instability may exhibit various forms such as quasiperiodic, period-2, period-k or chaotic oscillations in addition to limit cycle (Kabiraj et al. 2012; Kashinath et al. 2014).

The transition of combustor operation from combustion noise to thermoacoustic instability has been reported to occur via a Hopf bifurcation (Ananthkrishnan et al. 2005; Subramanian 2011; Sujith et al. 2016), wherein at a particular value of the control parameter, a sudden generation of periodic oscillations happens in the system dynamics from its steady-state value (Lieuwen 2002). However, a recent study on a turbulent combustor by Nair et al. (2014) showed that such transition of the combustor operation to thermoacoustic instability is not sudden, but happens gradually through the intermittent occurrence of bursts of periodic oscillations. They referred to such a transitional state as intermittency, wherein the system dynamics transitions alternately between low amplitude aperiodic oscillations and high amplitude periodic oscillations (see Fig. 14.2). It is to be noted that the amplitude of bursts observed during intermittency in their system was much lower than that of the limit cycle oscillations. Henceforth, various studies on gaseous combustion systems report the observation of intermittency in their respective systems (Gotoda et al. 2014; Unni and Sujith 2015; Kabiraj et al. 2016; Domen et al. 2015; Murugesan and Sujith 2015).

In thermoacoustics, the term intermittency was first introduced by Kabiraj and Sujith (2012). They reported the presence of high amplitude bursts of chaotic oscillations amidst regions of low amplitude periodic oscillations prior to a flame blowout. Furthermore, Unni and Sujith (2017) observed the presence of intermittency in

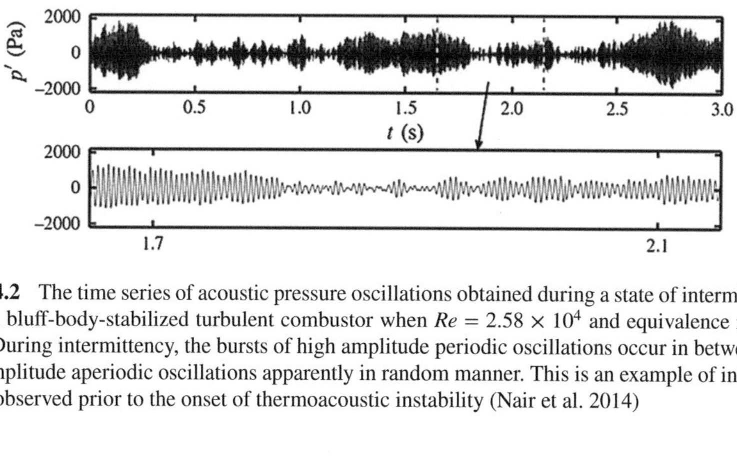

Fig. 14.2 The time series of acoustic pressure oscillations obtained during a state of intermittency from a bluff-body-stabilized turbulent combustor when $Re = 2.58 \times 10^4$ and equivalence ratio = 0.85. During intermittency, the bursts of high amplitude periodic oscillations occur in between the low amplitude aperiodic oscillations apparently in random manner. This is an example of intermittency observed prior to the onset of thermoacoustic instability (Nair et al. 2014)

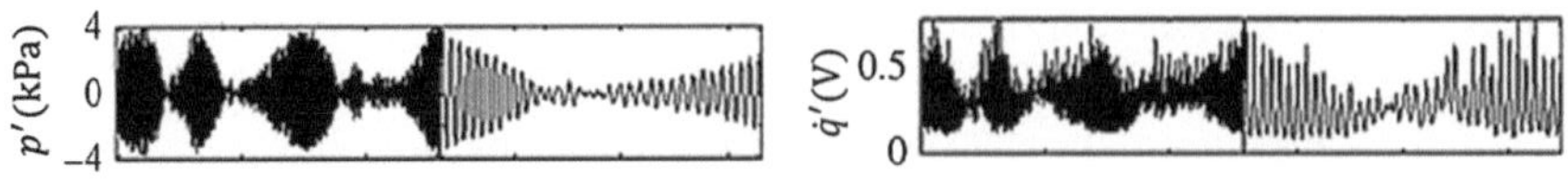

Fig. 14.3 The variation of acoustic pressure and heat release rate fluctuations obtained during a state of intermittency observed prior to a flame blowout in a bluff-body-stabilized turbulent combustor. The flow conditions were $Re = 2.28 \times 10^4$ and equivalence ratio = 0.47. During intermittency, both acoustic pressure and heat release rate fluctuations show apparently random switching between bursts of high amplitude periodic oscillations amongst the regions of low amplitude aperiodic oscillations (Unni and Sujith 2015)

turbulent combustor prior to the flame blowout, in which the bursts comprised of high amplitude periodic oscillations in between the regions of low amplitude aperiodic oscillations (refer Fig. 14.3).

The presence of intermittency is common in a variety of physical systems such as turbulent flows, Rayleigh–Bénard convection, plasma, lasers, and electronic circuits. Intermittency in dynamical systems theory is described as an irregular alternation of the behaviour of the system from periodic to chaotic oscillations (Hilborn 2000). Traditionally, there are three basic types of intermittency (i.e. type-I, type-II and type-III) which can cause the transition of system behaviour from periodic to chaotic state (Pomeau and Manneville 1980). In addition to these, various other types of intermittencies such as on-off, type-V, type-X, eyelet, and crisis-induced exist in the literature of dynamical systems (Elaskar and Del Río 2017). The type of intermittency is based on the specific kind of bifurcation a system has to undergo while transitioning from one dynamics to another. In this chapter, we will restrict ourselves only to three types: type-I, type-II and type-III intermittencies. Herein, type-I intermittency is associated with the saddle-node bifurcation, type-II intermittency with the subcritical-Hopf bifurcation and type-III intermittency with the period-doubling bifurcation.

Characterization of the type of intermittency is based on a few statistical methods such as (i) probability distribution of the duration between two consecutive bursts, and statistical variation of mean of this duration with control parameter, (ii) first return map and (iii) recurrence plot analysis (Pomeau and Manneville 1980; Schuster and Just 2006; Klimaszewska and Żebrowski 2009). Establishing the type of intermittency from experiments is necessary for improving the understanding of the phenomena and also in validating the models that predict the onset of thermoacoustic instability. In thermoacoustics, Kabiraj and Sujith (2012) observed type-II intermittency in the gaseous laminar premixed combustor (see Fig. 14.4), whereas Nair and Sujith (2013) mentioned the existence of type-II or type-III intermittency in the gaseous flame turbulent combustor. Furthermore, Unni and Sujith (2017) compared the intermittent oscillations observed in the turbulent combustor prior to thermoacoustic instability and flame blowout and confirmed the presence of type-II intermittency in both the cases. In addition, the presence of intermittency prior to thermoacoustic instability as well as flame blowout can aid in devising different

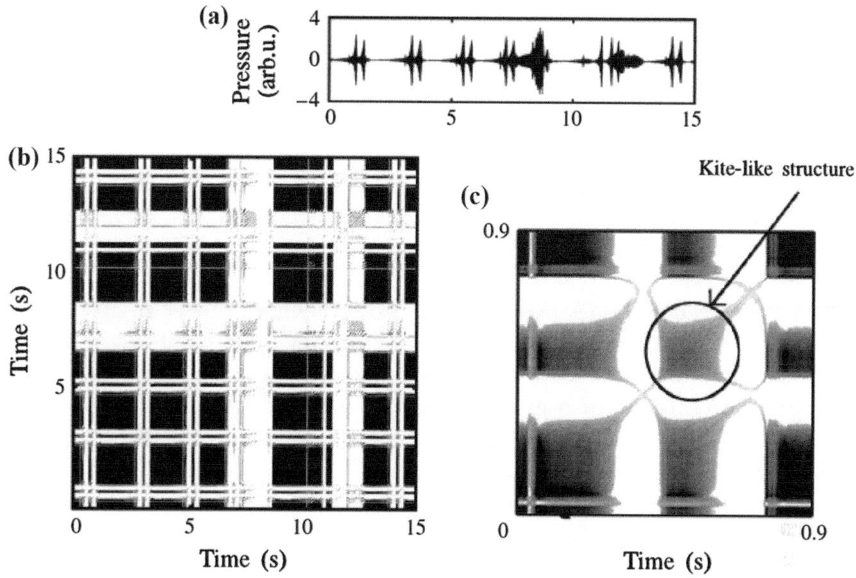

Fig. 14.4 **a** Time series of intermittency observed prior to flame blowout in a laminar combustor. Intermittency consists of bursts of chaotic oscillations amongst regions of low amplitude apparently periodic oscillations. **b** The recurrence plot of the intermittency signal depicting the black patches (corresponding to non-bursts regions) amongst the white patches (corresponding to burst regions). **c** An enlarged view of recurrence plot shows a kite-like structure with a long elongation at right upper corner depicting the property of type-II intermittency (Kabiraj and Sujith 2012)

control strategies that will forewarn the onset of such phenomena (Nair et al. 2013; Nair and Sujith 2014; Nair et al. 2014; Unni and Sujith 2015).

In spite of these recent advances in gaseous combustion systems in understanding the route to thermoacoustic instability, presence of such studies in spray combustion systems is negligible. Most combustion systems in practice utilize liquid fuels as a source of energy. Therefore, it is important to understand the phenomena of onset of thermoacoustic instabilities in such systems.

Spray combustors are susceptible to thermoacoustic instability because of the inherent unsteady processes involved in the combustion of such two-phase mixtures. Understanding the combustion of spray systems is complex as it has characteristics of premixed and diffusion flames at any time in different spatial regions of the combustor. The possible sources contributing to unsteadiness in the spray combustion are the methods of fuel injection, atomization and evaporation of droplets, mixing of fuel vapours with air and finally the burning of this mixture in the hot environment of the combustor (Young 1995; Culick 1988). Thermoacoustic instabilities in spray combustors are generally classified as injection-coupled instabilities, where the pressure oscillations in the combustor affect the injection pressure difference across the spray nozzle, thereby modulating the fuel flow rate entering into the combustor. The modulation in fuel flow rate fluctuates the equivalence ratio of combustion, which in turn modulates the heat release rate in the system (Young 1995). Factors affecting the onset of thermoacoustic instability can be understood by simplifying the studies of spray–acoustic interaction. For instance, most of the studies in spray–acoustic interaction have been performed on single droplets. The presence of acoustic oscillations is simulated by imposing the external perturbations on the liquid droplets. The presence of acoustic field modifies the spray characteristics such as droplet diameter, velocity, evaporation and burning rates. Further, the response of droplets to the acoustic field depends highly on their spatial location inside the spray (Chishty 2005).

The experimental studies (Kumagai and Isoda 1955; Tanabe et al. 2000; Saito et al. 1996; Okai et al. 2000; Dattarajan et al. 2006) on the effect of acoustic oscillations on a single pendant drop showed that the increase in amplitude of acoustic forcing increases the rate of evaporation and burning of the droplet. Furthermore, for a moving droplet, the application of axial acoustic perturbations reduces its terminal velocity (Sujith et al. 1997). The modelling study by Duvvur et al. (1996) and Lei and Turan (2009) revealed that the vapourization of droplet acts as a main driving force in the generation of thermoacoustic instabilities, whereas the study by Anderson et al. (1998) indicated that the mean size of the droplets, special distribution of droplets of variable sizes in the combustor and periodicity in atomization of the spray also have an important effect on the onset of thermoacoustic instability. Besides, the position of spray in the combustor affects the intensity of acoustic oscillations generated in the system (Carvalho et al. 1989; Dubey et al. 1998). The presence of high amplitude acoustic oscillations reduces the length of the spray either by decreasing the droplets size (through increased evaporation rate) or by increasing the mean drag on the droplets Sujith 2005.

In summary, the main aim of these previous studies (numerical as well as experimental) was focused on characterizing the effect of acoustic oscillations (self-excited or externally forced) on the physical processes of the spray combustion observed during the state of thermoacoustic instability. On the contrary, the characterization of the route through which such instabilities are developed in spray combustor systems has not yet received much attention.

Recently, Pawar et al. (2016) performed a detailed study on the characterization of intermittency route to thermoacoustic instability in a laboratory-scale spray combustion system. The system comprised of a unified design of needle spray injector that generates a uniform droplet spray having velocity field only in the axial direction. Various tools from dynamical systems theory were used to understand dynamics of the combustor acquired at various locations of the flame inside the duct. In this chapter, we discuss the key results from their study on characterization of intermittency route to thermoacoustic instability and identification of type of intermittency in the spray combustion dynamics. Finally, the main takeaways from this study are summarized at the end of the chapter.

14.2 Tools from Nonlinear Dynamics

Dynamical systems are systems which evolve with time. The dynamics of such systems at every instant of time can be correctly determined from the knowledge of a fixed number of the dynamical variables, also known as state variables (Hilborn 2000). The time evolution of such variables can be expressed in the form $\mathrm{X} = \dot{\phi}(X)$, where the dot represents the time derivative. For a given set of initial conditions and the functional form (ϕ), the future dynamical states of the variable can be correctly determined from the above equation.

14.2.1 Phase Space Reconstruction

In order to describe the state of a practical dynamical system completely, infinite number of state variables are required. In contrast, the number of state variables available from such systems are limited and in the limiting case it is one. According to Takens′ delay embedding theorem (Takens 1981), it is possible to obtain all variables that are required to describe the state of a system using a time series of single observable. The dynamics of such systems can then be embedded into a higher-dimensional phase space using an appropriate value of embedding parameters such as the optimum time delay (τ) and the minimum embedding dimension (d) (Abarbanel et al. 1993). Suppose we have a single variable time series as $x(t) = x_1, x_2, x_3, \ldots\ldots.., x_N$. Then the delay vectors are constructed as $Y_i(d) = [x(i), x(i+\tau), x(i+2\tau), \ldots\ldots.., x(i+(d-1)\tau)]$, where $i = 1, 2, 3, \ldots\ldots., N-(d-1)\tau$. The plot of these delayed vectors represents the time evolution of every state of a given dynamical system in the d-dimensional phase space. The construction of phase space is highly dependent on the choice of τ and d.

14.2.1.1 Selection of Optimum Time Delay (τ) and Minimum Embedding Dimension (d)

The conventional way to select the optimum value of time delay (τ) is to choose the first local minimum of the average mutual information (AMI) (Fraser and Swinney 1986). The choice of time delay should be such that the resulting vectors obtained from lagging the samples of a given signal are independent of each other (Kantz and Schreiber 2004). If the time delay is very small, it will produce highly correlated vectors; on the contrary, if the time delay is very large, the resulting vectors would be completely uncorrelated with each other (Abarbanel et al. 1993). Now, consider a time series data of sampled variable $x(t)$ as $x(t_1), x(t_2), \ldots\ldots., x(t_i)$, and the corresponding delayed signal is $x(t_1+\tau), x(t_2+\tau), \ldots\ldots\ldots., x(t_i+\tau)$. The AMI of such signal is then calculated as

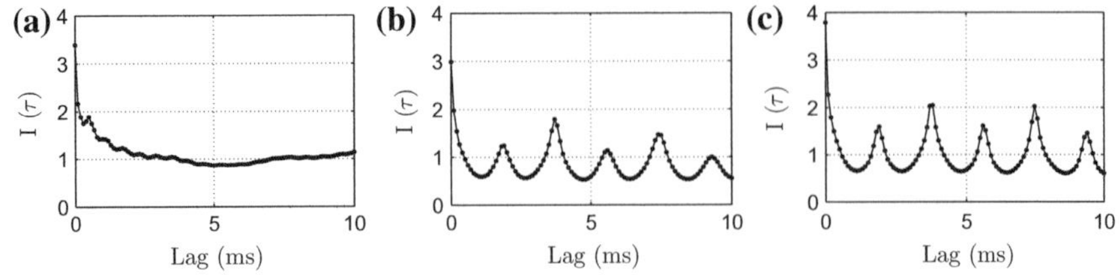

Fig. 14.5 The plots of average mutual information (AMI) corresponding to three signals acquired during the states of **a** combustion noise (x_f = 520 mm), **b** intermittency (x_f = 590 mm) and **c** thermoacoustic instability (x_f = 680 mm). The first local minima of the plot showing variation of AMI versus lag (shown by arrow) is chosen as the optimum value of time delay required for the reconstruction of phase space

$$I(\tau) = \sum_{x(t_i),\, x(t_i+\tau)} P(x(t_i), x(t_i + \tau)) \log_2 \left[\frac{P(x(t_i), x(t_i + \tau))}{P(x(t_i))P(x(t_i + \tau))} \right]$$

where $P(x(t_i))$ and $P(x(t_i + \tau))$ are the marginal probabilities, and $P(x(t_i), x(t_i + \tau))$ is the joint probability of occurrence of $x(t_i)$ and $x(t_i + \tau)$. Figure 14.5 represents the plots of average mutual information corresponding to three states of the combustion dynamics as combustion noise, intermittency and limit cycle oscillations.

The embedding dimension d is usually found from a method of false nearest neighbours (FNN) (Kennel et al. 1992). In this method, the percentage of closeness (measured in terms of distances) of the neighbouring points of the phase space trajectory is calculated in a given dimensional phase space, and then the value of this closeness is compared for the next dimensional space. If the value of the percentage is unchanged with an increase in the dimension, we assume that the attractor of the given signal is unfolded and the minimum embedding dimension required for the reconstruction of the phase space is achieved. The unfolding of all trajectories of the attractor is necessary to remove the possibility of false neighbours in the reconstructed phase space. If the neighbours are false, the changing of dimension from d to d+1 results in change in the number of neighbours of a given state point on the phase space trajectory. We, however, use another method suggested by Cao (1997), an optimization of FNN method, to find the embedding dimension.

If $Y_i(d)$ is the ith reconstructed vector of d-dimensional space obtained using Takens′ delay embedding theorem (Takens 1981), we can define a new quantity $a(i, d)$ such that

$$a(i, d) = \frac{||Y_i(d + 1) - Y_{n(i,d)}(d + 1)||}{||Y_i(d) - Y_{n(i,d)}(d)||}$$

where $i = 1, 2, 3, ..., (N - d * \tau)$, and $||.||$ is the Euclidian norm. Further, $Y_i(d+1)$ is the ith reconstructed vector in the $(d+1)$-dimensional space, and $n(i, d)$ is the nearest neighbour of $Y_i(d)$ in the d-dimensional space. Then, the mean of all $a(i, d)$ can be calculated as

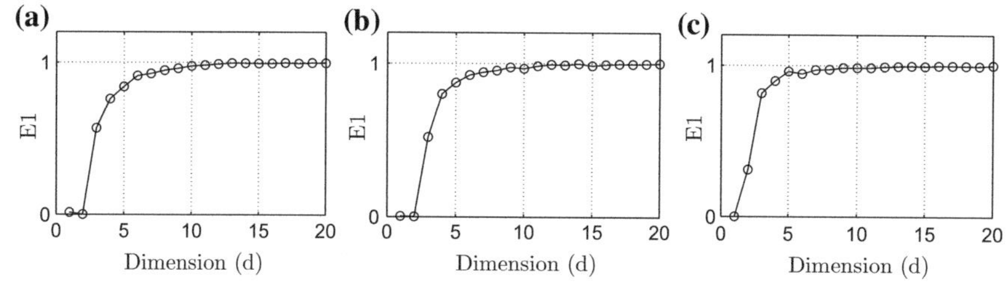

Fig. 14.6 The minimum embedding dimension is selected from the Caos method for three signals corresponding to the states of **a** combustion noise ($x_f = 520$ mm), **b** intermittency ($x_f = 590$ mm) and **c** thermoacoustic instability ($x_f = 680$ mm). The minimum value of embedding dimension d is chosen as the next dimension at which E1 saturates for the first time in the plot

$$E(d) = \frac{1}{(N-d)} \sum_{i=1}^{N-d\tau} a(i,d)$$

The variation of $E(d)$ due to change in the dimension from d to $d+1$ is expressed in terms of another quantity $E1(d)$ as

$$E1(d) = E(d+1)/E(d)$$

When $E1(d)$ is observed to stop changing after a dimension d_0, then the dimension $d_0 + 1$ can be chosen as the minimum embedding dimension required for the reconstruction of the phase space. Figure 14.6 shows the variation of $E1$ with the embedding dimension for the different states of combustion dynamics as combustion noise, intermittency and limit cycle oscillations.

14.2.2 Recurrence Plots and Quantification Analysis

Recurrence is a fundamental property of a deterministic dynamical system (Eckmann et al. 1987). The time evolution of the system dynamics can be captured by plotting the recurrence plot (RP) of a given time series. RP not only aids in visualizing the dynamics of a given system in a lower-dimensional space but also helps in detecting the hidden dynamical patterns present in the signal (Eckmann et al. 1987; Marwan et al. 2007). The concept of construction of RP is based on the process of recurrence of a phase space trajectory in the embedded space (see Fig. 14.7). Thus, in order to obtain RP for any given signal, we first need to reconstruct the phase space using the Takens′ delay embedding theorem (Takens 1981), and then from the choice of an appropriate value for the distance threshold, recurrence of the trajectory is quantified. The choice of size of the threshold is further dependent on the problem in hand. This threshold could be some fraction of size of the attractor (maximum or mean distance of the attractor) or could be a fixed number of nearest neighbours. Larger the size of

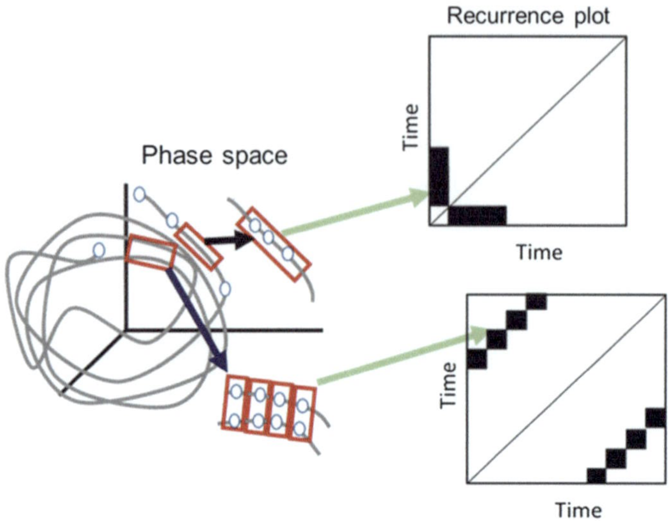

Fig. 14.7 The schematic of the construction of RP plot from the reconstructed phase space. The diagonal lines in the RP depict that two trajectories are running parallel to each other, whereas the vertical lines in RP demonstrate that more than one points of the same trajectory are recurring at the same time

threshold, more will be the recurrence of points in the phase space, hence we always need to choose an appropriate value of the threshold. The recurrence of the phase space trajectory can be quantified from following equation

$$R_{i,j} = \Theta(\varepsilon_i - ||x_i - x_j||); \quad i,j = 1, 2, ..., N_1$$

where $x_{i,j}$ are the delayed vectors, Θ is the Heaviside step function, $N_1 = N-(d-1)\tau$ is the total number of delayed vector, ε_i is a predefined threshold. Whenever the neighbouring trajectory of a given state point in the phase space falls within the threshold, it is marked as 1; otherwise, it is marked as 0 in the recurrence matrix. Thus, the recurrence plot is a graphical representation of black and white points, where black points correspond to $R_{i,j} = 0$ and white points correspond to $R_{i,j} = 1$.

Since RP is a visual inspection tool and its results are mostly qualitative in nature, Zbilut and Webber (1992) and Webber and Zbilut (1994) developed a recurrence quantification analysis (RQA) to establish robust conclusions from this method. RQA technique is mainly based on the quantification of the structural patterns such as diagonal lines, and vertical or horizontal lines present in the RP (Marwan et al. 2007). This quantification of structural patterns helps either in identifying the bifurcation points occurring in the complex dynamical systems or in deciding the deterministic nature of the system dynamics (Marwan et al. 2007). Based on this quantification, various statistical measures are developed, namely recurrence rate (*REC*), determinism (*DET*), laminarity (*LAM*), entropy (*ENTR*), longest diagonal line (L_{max}), longest vertical line (V_{max}) and trapping time (*TT*). Out of these measures, we will be discussing the use two quantifiers, namely percentage of determinism (*%DET*) and maximum value of diagonal line (L_{max}) in this chapter. We found that these measures are better than others in identifying the random or deterministic characteristics of the system dynamics (Marwan et al. 2007). Their values exhibit peak for periodic dynamics and have lower values for aperiodic dynamics. Further, these measures have been successfully used in the past for the detection of chaos-periodic transition in the same time series (Marwan et al. 2007).

Determinism (*DET*): It represents the quantification of lines that are parallel to main diagonal line in RP. It is the ratio of number of black points in RP that form a diagonal line to the total number of black points present in RP.

$$DET = \frac{\sum_{i=l_{min}}^{N} lP(l)}{\sum_{i=1}^{N} lP(l)}$$

where $P(l)$ is the histogram of the lengths l that form diagonal lines in RP and $l_{min} =$ 2 points. Determinism is a measure of predictability of the dynamical system and shows a higher value close to 1 for the regular behaviour and lower value close to zero for the irregular behaviour.

Longest diagonal line (L_{max}): It is the length of the longest diagonal line, other than the main diagonal line, present in RP. The inverse of L_{max} is called divergence. It is also referred as an estimator for the largest positive Lyapunov exponent of the dynamical system (Marwan et al. 2007). Since the largest positive Lyapunov exponent is a measure of divergence of the phase space trajectory, its value is more for chaotic oscillations and less for periodic oscillations. The features of RQA in the detection of the determinism of the system dynamics are discussed in the results section.

14.3 Experimental Set-up

An experimental rig developed for studying the route to thermoacoustic instability in two-phase flow combustor is shown in Fig. 14.8. The detailed explanation of the experimental set-up description is available in Pawar et al. (2016). The experimental set-up consists of mainly four parts, i.e. fuel supply head, air input unit, needle injector and combustor. The fuel supply head is fitted at the top of the combustion system, whose main purpose is to store the fuel prior to entering into the combustor and also aid in equally distributing it through all needles of the injector. It is then followed by the air input unit, which is designed to supply air uniformly from all sides of the combustor. The combustor is a 1000 mm long quartz glass tube with inner diameter of 50 mm and thickness of 2.5 mm. Ethanol is used as a combustion fuel and is supplied from a gravity flow tank placed above the set-up. The pressurized air (1 bar gauge pressure) is provided by an air compressor.

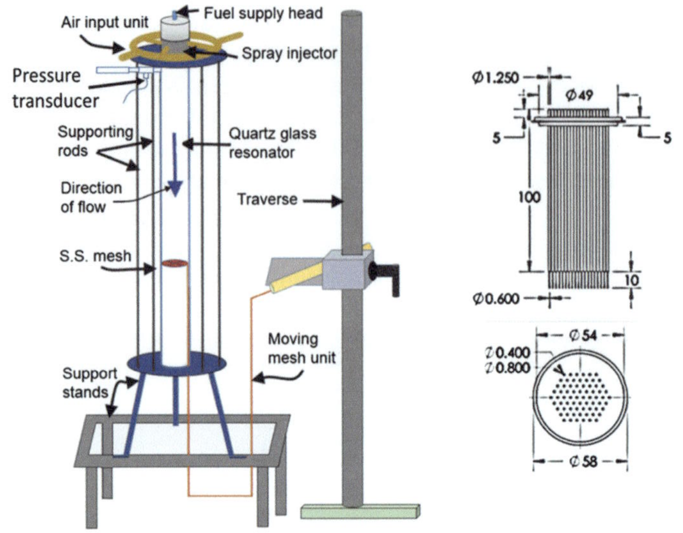

Fig. 14.8 Schematic of the experimental set-up of the spray combustion system is shown on the left side. The front and top views of the needle injector used in the generation of droplet spray are shown on the right side (Pawar et al. 2015). All dimensions are in mm

The injector consists of 70 stainless steel needles of 0.8 mm inner diameter, 1.25 mm outer diameter and 100 mm length. The size of the droplets is further reduced by fastening 10 mm long capillary tubes of size 0.4 mm inner diameter and 0.6 mm outer diameter at the end of primary needles of the injector. The needles are placed on the metal plate in such a way that each needle is fixed at the apex of the equilateral triangle of side 4 mm. This placing of needles helps in generating a uniform fuel flow across the combustor, and also in avoiding the interactions amongst droplets such as collision and coalescence, once they are injected.

Secondary atomization is achieved through two different size stainless steel wire meshes packed together as a single mesh unit. The mesh unit also serves as a function of anchoring the flame at a specific location inside the combustor. The relative position of the mesh unit from the bottom end of the combustor, defined as x_f, is varied as a control parameter.

Experiments were conducted at a constant equivalence ratio of 0.49 (uncertainty of 0.03). This is achieved by maintaining the fuel flow rate at 8 ccm and the air flow rate at 100 slpm. Fuel–air mixture is ignited from the bottom of the combustor using a pilot flame. Initially, during ignition, the mesh is lowered down to x_f = 110 mm and once the flame is anchored at the mesh, it is moved upward to x_f = 480 mm. At this position of the flame, the combustor is preheated at least for 7 min in every experiment before the start of data acquisition. Preheating minimizes the variation of temperature of the combustor wall from the time of ignition and ensures a constant wall temperature throughout the experiment.

Unsteady acoustic pressure fluctuations in the combustor are measured using a data acquisition system that consists of a pressure transducer, a signal conditioner and a data acquisition card. The pressure transducer (PCB piezotronics 103B02, the sensitivity of 223.4 mv/kPa, a resolution of 0.142 Pa) is fixed on a T-mount at a location of 140 mm from the injector plate, as shown in Fig. 14.8. The transducer is connected to the data acquisition card (NI PCI-6221) through a signal conditioner (PCB 480E09). A manual traverse arrangement is employed to move the mesh unit which, in turn, determines the location of the flame inside the combustor. The location of the flame holder (mesh unit) is varied in a step sizes of 10 mm throughout the experiment (least count on the traverse is 1 mm). Data is acquired for 30 s at the rate of 10 kHz at each location of the flame holder. A settling time of 30 s is ensured before the acquisition of data at every location of the flame. This settling time is necessary to avoid the transients associated with the change in the flame location.

14.4 Results and Discussion

14.4.1 Intermittency Route to Thermoacoustic Instability

In this section, we characterize the variation in the system dynamics due to change in the location of flame inside the combustor as a system parameter. The flame location is systematically varied from bottom to top half of the duct using a traverse mechanism. The dynamics of the combustor is analysed from the recordings of acoustic pressure acquired at each location of flame inside the system. Figure 14.9 displays a sequence of time series plots of the acoustic pressure fluctuations acquired at different locations of the flame inside the duct. With variation in the flame location, we notice that dynamics of the system transitions from a state of low amplitude aperiodic oscillations (stable operation) (Fig. 14.9a) to a state of high amplitude limit cycle oscillations (unstable operation) (Fig. 14.9p) via a regime of intermittency (Figs. 14.9b–o). Intermittency is the occurrence of a series of bursts of periodic oscillations in between the relatively silent regions of aperiodic oscillations (see Fig. 14.9q). During the onset of intermittency ($x_f = 540$ mm), we notice that the bursts are short-lived, large amplitude periodic oscillations. As the flame location is further increased, we note an increase in the size (duration) of the bursts which ultimately culminate in the limit cycle oscillations exhibiting sustained periodic oscillations in the pressure signal.

The amplitude spectrum of the acoustic pressure signals obtained during three major states of the combustion dynamics, i.e. combustion noise, intermittency and thermoacoustic instability is shown in Figs. 14.10a–c, respectively. We observe two low amplitude bands of broadband frequencies during the state of combustion noise. In these bands, the lower frequency band ($f_{dominant} = 97.9$ Hz) possibly corresponds to the heat release rate fluctuations present in the flame, whereas the larger frequency band ($f_{dominant} = 272.7$ Hz) shows the frequencies associated with the natural acoustic

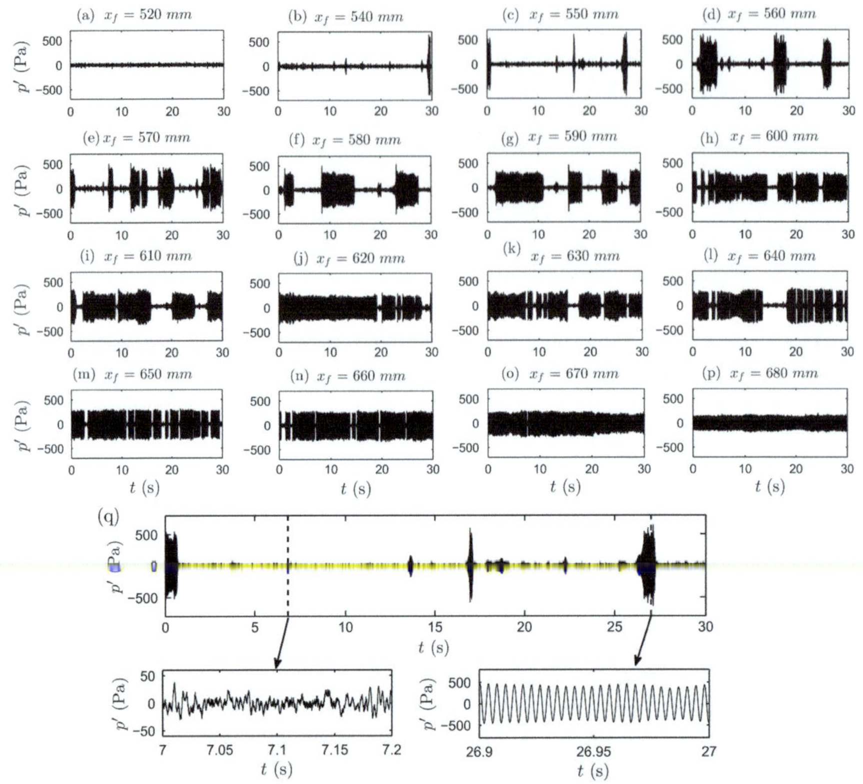

Fig. 14.9 Time series data of the acoustic pressure fluctuations acquired at different locations of flame in the combustor. **a** Combustion noise state ($x_f = 520$ mm), **b–o** intermittency state ($x_f = 540$ mm to $x_f = 670$ mm) and **p** limit cycle state ($x_f = 680$ mm). These plots show the intermittency route to limit cycle oscillations in a spray combustion system. **q** A magnified portion of intermittency signal obtained at $x_f = 550$ mm shows the low amplitude aperiodic oscillations and the burst of high amplitude periodic oscillations present in the signal (Pawar et al. 2015)

mode of the duct. As the system dynamics enter into a state of intermittency and limit cycle, a sharp frequency peak corresponding to the periodic oscillations is observed in the frequency range of 260 to 280 Hz. This frequency range matches nearly with the value of second longitudinal mode of the acoustic oscillations developed in a close–open configuration of the set-up ($f_n = 3c/4L$, where $c = 360$ m/s, $L = 1$ m). The transition of the system dynamics due to variation in the system parameter is further qualitatively analysed using a bifurcation diagram.

The bifurcation analysis of the acoustic pressure fluctuations is presented in Fig. 14.11. It shows the variation of the global maxima (p'_{max}) (Fig. 14.11a) and the root-mean-square (Fig. 14.11b) values of the acoustic pressure signal with change in the location of flame holder inside the combustor. The plot is divided into three regions: region I—combustion noise, region II—intermittency and region III—

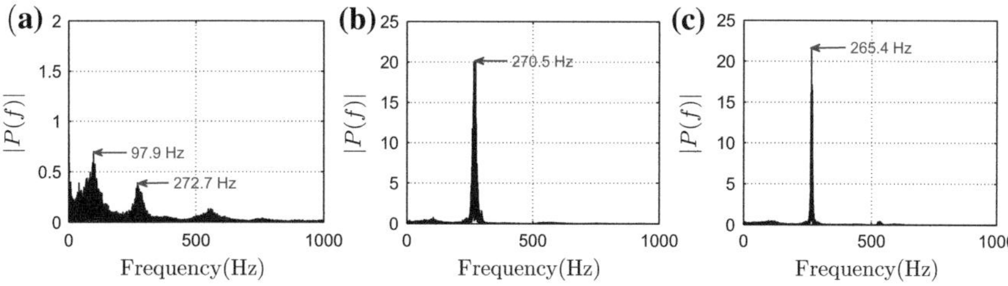

Fig. 14.10 The amplitude spectrum of the acoustic pressure signals acquired during the state of **a** combustion noise ($x_f = 520$ mm), **b** intermittency ($x_f = 590$ mm) and **c** limit cycle ($x_f = 680$ mm) oscillations in the combustor

thermoacoustic instability. During stable operation of the combustor, up to $x_f = 530$ mm, low amplitude aperiodic oscillations are observed (see region I of Figs. 14.11a, b). At $x_f = 540$ mm, as the system dynamics enters into a zone of intermittency, a jump in the maximum amplitude of the acoustic pressure signal is noticed. This sudden hike in the pressure amplitude is due to the occurrence of high amplitude bursts of periodic oscillations from the background of low amplitude aperiodic oscillations. This point is further referred to as the onset of intermittency in the system dynamics. The intermittent state of oscillations is observed from $x_f = 540$ mm to $x_f = 670$ mm, as shown in region II of Fig. 14.11a. Finally, for $x_f = 680$ mm and $x_f = 690$ mm (see region III of Figs. 14.11a, b), the system enters into a region of limit cycle oscillations, where sustained periodic oscillations are observed in the system dynamics. We found that during the onset of intermittency, the maximum amplitude of bursts (in the start of region II of Fig. 14.11a) is much higher in magnitude than that of the limit cycle oscillations (region III of Fig. 14.11a). We notice a decrease in maximum amplitude of acoustic pressure oscillations from the point of onset of intermittency to that of limit cycle oscillations as seen in Fig. 14.11a and also in the time series plots presented in Fig. 14.9. However, this behaviour of decrease in the pressure amplitude is not effectively seen in the plot of variation of p'_{rms} with flame location (see Fig. 14.11b). This is because p'_{rms} averages out the values of high amplitude pressure oscillations that occur for a short duration amongst the lengthy regions of low amplitude aperiodic oscillations during the onset of intermittency.

Nair et al. (2014) reported the observation of intermittency route to thermoacoustic instability in their turbulent gaseous flame combustor. They noticed that the root-mean-square values of the acoustic pressure fluctuations (p'_{rms}) increase continuously from the point of onset of intermittency to limit cycle oscillations due to change in Reynolds number of the flow in the system. In stark contrast with Nair et al. (2014), we witness that the amplitude (local maxima as well as root-mean-square values) of intermittent oscillations of acoustic pressure fluctuations is much higher than that observed for limit cycle oscillations. Historically, it was considered that limit cycle oscillations are dangerous to the combustion systems. The problems associated with high amplitude limit cycle oscillations have been described by various researchers in the past (McManus et al. 1993; Culick and Kuentzmann 2006).

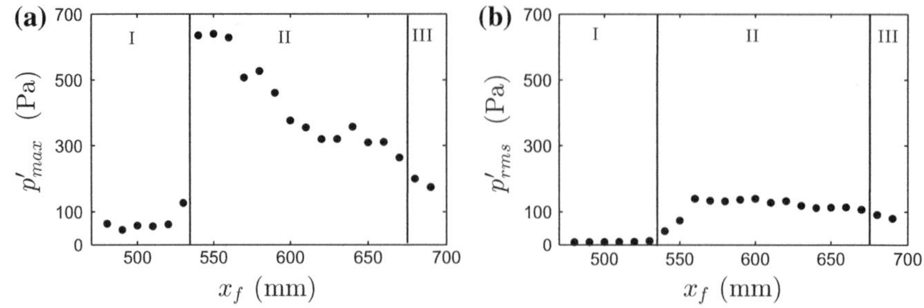

Fig. 14.11 The variation of **a** a global maximum (p'_{max}) and **b** a root-mean-square (p'_{rms}) value of the acoustic pressure signal versus the location of flame (x_f) inside the combustor. The plots are divided into three distinct regions of system dynamics as (I) low amplitude aperiodic oscillations, (II) intermittency and (III) limit cycle oscillations (Pawar et al. 2015)

The continuous loading of the system during the sustained instability state may lead to severe vibrations of mechanical parts of the combustor that either induce thermal and hoop stresses in the combustor walls, cause compressor surge or damage turbine blades. The accumulation of these stresses over a period of time may lead to catastrophic failure of the engine during its operation. However, the effects of high amplitude intermittent loads on the structural properties of the engine need to be investigated in future studies.

14.4.2 *Qualitative and Quantitative Analysis of Intermittency Route*

In this section, the dynamics of the acoustic pressure fluctuations acquired at different states of the combustor operation is explored through tools from dynamical systems theory. Thermoacoustic systems are nonlinear in nature, wherein the nonlinearity stems mainly from the coupled interaction between the acoustic field of the duct and the unsteady heat release present in the flame (Lieuwen 2002). Using the approach of nonlinear dynamics, we can visualize the single variable time series of the experimental signal into a higher-dimensional reconstructed phase space. Such phase space is reconstructed using delay embedding theorem proposed by Takens (1981). Figure 14.12 shows three-dimensional phase portraits of the acoustic pressure fluctuations (p') measured at three states of the combustor operation which are combustion noise, intermittency and thermoacoustic instability. In Fig. 14.12a, the phase portrait of p' acquired during the combustion noise state is shown. As the acoustic pressure oscillations are aperiodic in nature, the phase portrait corresponding to this state displays a clutter of trajectories. During intermittency, it is observed that the phase space trajectory continuously switches between high amplitude periodic (outer regular circle) and low amplitude aperiodic (centre disorder region)

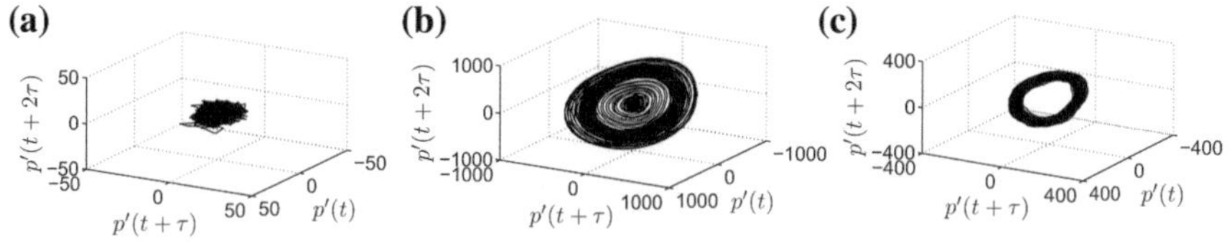

Fig. 14.12 Reconstructed phase portrait of a portion of the pressure time series (0.5 s) obtained during three different dynamical states system dynamics such as **a** combustion noise (x_f = 520 mm; $\tau = 3, d = 11$), **b** intermittency (x_f = 590 mm; $\tau = 10, d = 9$) and **c** thermoacoustic instability (x_f = 680 mm; $\tau = 10, d = 9$) (Pawar et al. 2015)

oscillations as shown in Fig. 14.12b. During limit cycle oscillations, the phase space of the signal displays a thick circular structure (see Fig. 14.12c), in which the thickness of the circle is caused by the amplitude variation of the acoustic pressure oscillations observed during limit cycle state.

We further characterize the nature (random or chaotic) of acoustic pressure fluctuations obtained during the combustion noise state. Such classification of the signal observed during this state is necessary to develop an appropriate model required for the analysis of the combustor dynamics. Various methods are available in the literature to distinguish chaotic signal from a noisy one. These include Lyapunov exponent, 0–1 test, determinism test (Kantz and Schreiber 2004). However, in the present analysis, we discuss the use of an approach based on RPs. The advantage of this method is that it requires a shorter time series, unlike other cases, to analyse the data. Further it helps in better visualization of divergence of the phase space trajectory compared to the other methods.

14.4.2.1 Detection of Chaos

In this section, the difference between the random process and the chaotic process is examined by using RPs and the measures based on its quantification. Towards this purpose, we have performed a comparison test of the experimental data obtained at $x_f = 520$ mm and the data generated by random shuffling of the experimental signal using a method of surrogate analysis (Theiler et al. 1992). This test aids in distinguishing the nonlinear behaviour of the real-world data from the linear process by using an appropriate null hypothesis. In order to reject the hypothesis, a significance of the test against the null hypothesis is examined. Many realizations of the null hypothesis are generated, and the significance of the test is estimated empirically. This check is important in showing that the underlying dynamics present during the aperiodic state of oscillations is a deterministic chaos and not a random noise.

The structural patterns present in RPs further represent the typical behaviour of the phase space trajectory. The long lines parallel to the main diagonal exemplify a periodic (deterministic) process, whereas homogeneously distributed black points indicate white noise (random) process. On the other hand, a chaotic process is shown

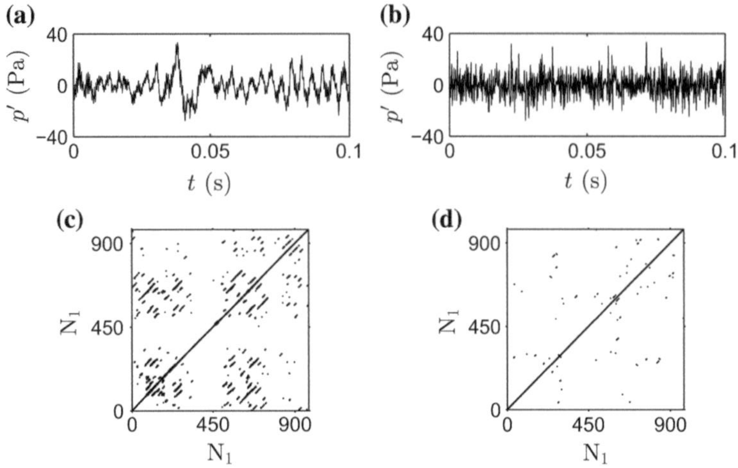

Fig. 14.13 **a** The recurrence plots of the pressure signal measured at $x_f = 520$ mm and **b** the signal obtained by random shuffling of the pressure data in a. **c** The short diagonal lines parallel to main diagonal confirm the underlying dynamics is chaotic, and **d** grainy structure with random distribution of single points in RP shows the dynamics is random. $N = 1000$, $\tau = 3$, $d = 10$, $\varepsilon = 30\%$ of the mean diameter of the attractor, and $N_1 = 973$

by the apparently random distribution of short diagonal lines along with a few isolated black points in RP (Marwan et al. 2007). The characteristic of chaotic oscillations is that the neighbouring trajectories in the phase space of such signals remain nearby for a short duration and then diverge exponentially at far distances. Hence, when the trajectories are adjacent, they recur inside the recurrence threshold and do not recur after the exponential divergence. Conversely, in the case of random signals, every event of the signal is independent of the previous one. Hence, for such signals, the possibility of recurrence of the phase trajectory is very low.

The RPs of the aperiodic oscillations observed during combustion noise state (Fig. 14.13a) and the signal generated by random shuffling of data points of this signal by surrogate analysis (Fig. 14.13b) are shown in Fig. 14.13. During the random shuffling of data points, the statistical properties of the signal such as sample mean, variance and autocorrelation function are maintained same as that of the original signal used for the construction. In Fig. 14.13c, we notice the presence of short (broken) diagonal lines along with few single points in RP suggesting the possibility of chaotic oscillations during the combustion noise state. In the contrast, Fig. 14.13d shows the grainy structure with the apparently random distribution of isolated black points in RP confirming the noisy behaviour of the oscillations in the signal.

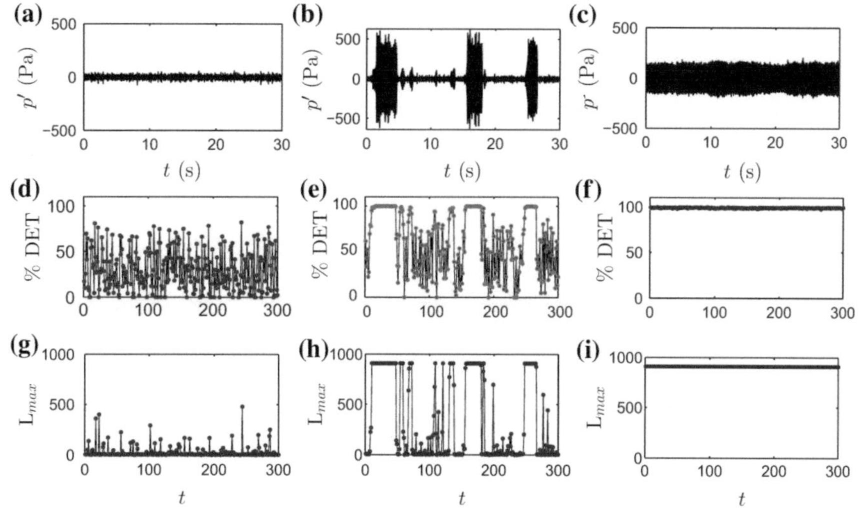

Fig. 14.14 **a–c** The pressure time signal obtained at $x_f = 520$ mm (chaotic), $x_f = 560$ mm (intermittency) and $x_f = 680$ mm (limit cycle) respectively. **d–i** The variations of $\%DET$ and L_{max} along the time axis for a time window of 1000 data points. The values of $\%DET$ and L_{max} show apparently random fluctuations for chaotic oscillations and a constant line at relatively maximum value for periodic oscillations. The values of $\%DET$ are always much higher than zero, confirming that the underlying dynamics during low amplitude state is deterministic. $N = 300000$, $\tau = 10$, $d = 10$, $\varepsilon = 25\%$ of the mean diameter of the attractor and a data window of 1000 points

14.4.2.2 Detection of Dynamical Transitions from Chaos to Period Using RQA

We further use measures of recurrence quantification analysis, mainly based on the characterization of diagonal patterns from RP (i.e. $\%DET$ and L_{max}), to investigate the transition points of the system dynamics from aperiodic to periodic oscillations and vice versa. In order to do this, we divide the entire time series into non-overlapping time windows of 1000 data points. The values of $\%DET$ and L_{max} are then calculated for each time window and plotted against the number of windows for the given signal. Figure 14.14 shows the variation of these measures obtained for combustion noise (Fig. 14.14a), intermittency (Fig. 14.14b) and thermoacoustic instability (Fig. 14.14c) states. The length of the diagonal line has a direct connection with determinism property of the system dynamics. The points along the diagonal line show evolution of the similar situations in the future time. Moreover, the short diagonal lines indicate the short-term predictability and long diagonal lines demonstrate the long-term predictability of the system dynamics.

The values of $\%DET$ are sufficiently higher than zero for every time window indicating the possibility of chaotic oscillations Marwan et al. 2007 in the underlying dynamics of aperiodic oscillations observed during the combustion noise state (see Figs. 14.14d, g). The values of $\%DET$ and L_{max} reach a maximum during the periodic phase of oscillations. During the state of intermittency, the values of recurrence quantifiers display lower value for aperiodic oscillations and reach a maximum value during periodic oscillations (see Figs. 14.14e, h). For a proper value of the recurrence threshold, the temporal variation of these recurrence measures will clearly manifest the transition points of the system dynamics from aperiodic to periodic oscillations, and vice versa, in the time series displaying intermittency. This is observed from the strong change in $\%DET$ and L_{max} values across the transition point shown in Figs. 14.14e, h. The ability of these RQA measures in discerning even low amplitude periodic oscillations from the low amplitude aperiodic oscillations present during intermittency is also illustrated in Figs. 14.14e, h. However, such transitions from aperiodic to periodic oscillations or vice versa, for the bursts of low amplitude periodic oscillations, are difficult to identify from the visual inspection of the phase portraits of such signals. For the periodic oscillations (or thermoacoustic instability) of any amplitude (low or high) in the pressure signal, the values of $\%DET$ and L_{max} stay at maxima throughout the signal, as seen in Figs. 14.14f, i. Thus, the periodic oscillations observed during thermoacoustic instability establish the highest predictability by showing only continuous diagonal lines in the RP.

The transition of system dynamics from combustion noise to thermoacoustic instability can be quantitatively explained by plotting the time average values of $\%DET$ and L_{max} with control parameter as shown in Fig. 14.15. The values of $\%DET$ and L_{max} are calculated for every time window of 1000 data points and averaged over the entire signal. The average value of $\%DET$ is close to 1 for oscillations during thermoacoustic instability, as the observed dynamics during this state is completely periodic; on the contrary, it is around 0.2 for the combustion noise state, as the observed dynamics is aperiodic. During intermittency, the value of this measure lies in between these extreme values. Similar behaviour is observed in the variation of average value of L_{max}. The increase in the average values of $\%DET$ and L_{max} with control parameter indicates the increase in predictability of the system dynamics. Other types of RQA measures, namely recurrence rate, entropy and trapping time

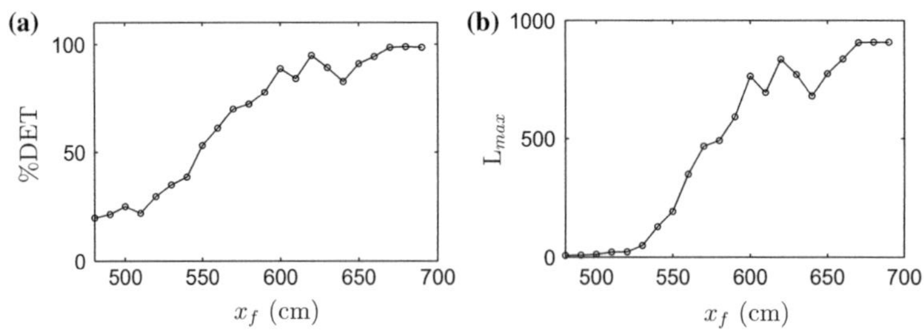

Fig. 14.15 **a, b** The variation of average values of the percentage of determinism ($\%DET$) and the maximum diagonal line (L_{max}) with location of control parameter shows a smooth transition of system behaviour from low amplitude chaotic oscillations to sustained limit cycle oscillations. The maximum value of $\%DET$ and L_{max} at x_f = 680 mm and x_f = 690 mm indicate that the system is having highest determinism value ($\tau = 10$, $d = 10$, $\varepsilon = 25\%$ of the mean diameter of the attractor)

have been previously used by Nair et al. (2014) to forewarn the impending instability of the practical gas turbine combustor. Controlling the operation of combustor at the onset of first burst in the pressure signal, by detecting the changes in $\%DET$ and L_{max} plots, will help in protecting the system components from any further damage.

14.4.3 Detection of Type of Intermittency

In the following analysis of intermittency, a low amplitude aperiodic regions present between the consecutive bursts are defined as the turbulent phases and the regions of high amplitude bursts of periodic oscillations are called as the laminar phases. The terms turbulent phase and laminar phase are completely different from the terminology of turbulent flow and laminar flow used in the fluid mechanics. The length of the turbulent phases (T) present in the unsteady pressure signal is determined to quantify the type of intermittency. Furthermore, the conventions of laminar and turbulent phase used in the present analysis may contradict with that used in the literature of dynamical systems theory, wherein the bursts of chaotic oscillations are described as turbulent phases and the regions of periodic oscillations observed in between the bursts are described as laminar phases.

Figure 14.16 shows a portion of the acoustic pressure signal acquired during a state of intermittent dynamics. For the calculation of length of turbulent phase (T), we adopt a method suggested by Hammer et al. (1994). Using this approach, an amplitude threshold is expressed in terms of fraction of maximum pressure such that threshold = $max(P)/(2^n)$, where P is the maximum value of the pressure amplitude of the signal and $1 < n < 6$. Here, we present the results corresponding to $n = 4$, which is equal to a pressure amplitude threshold of 54 Pa. The duration in the acoustic pressure signal having an amplitude below the selected threshold (i.e. the signal length between consecutive bursts) is defined as the length of the turbulent phase, as described in Fig. 14.16. The turbulent phase starts when the last waveform

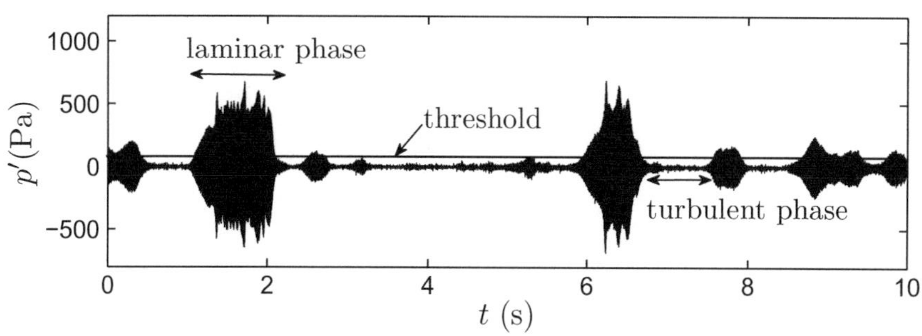

Fig. 14.16 The intermittency signal consists of low amplitude turbulent phases in between consecutive high amplitude laminar phases. A pressure threshold demarks the high amplitude periodic oscillations from low amplitude aperiodic oscillations (Pawar et al. 2015)

of the periodic burst falls below the threshold and ends when the first waveform of the next burst exceeds the threshold. These turbulent lengths are binned to calculate their probability distribution (using histogram) near the critical point. The critical point (x_{f0}) is the value of control parameter corresponding to the onset of intermittency. In our case, the onset of intermittency was observed at $x_{f0} = 540$ mm. In order to characterize the type of intermittency, a separate experiment was performed to acquire a sufficient number of bursts near the critical point. The acquisition of such lengthy data is necessary to confirm the statistical convergence of the obtained results. Towards this purpose, the data was acquired for 300 s at a sampling rate of 3 kHz.

14.4.3.1 Distribution of the Length of the Turbulent Phase

The histogram of the length of the turbulent phase present in the acoustic pressure signal acquired at $x_f = 540$ mm is shown in Fig. 14.17. An optimum pressure amplitude threshold of 54 Pa was chosen to ensure that no aperiodic oscillations are captured inside the threshold. Figure 14.17 shows an exponential decrease in the probability distribution of the length of the turbulent phase. Such distribution is consistent with the distribution shown for type-II and type-III intermittencies (Xiao et al. 2009; Frank and Schmidt 1997; Sacher et al. 1989; Schuster and Just 2006). Therefore, this distribution of turbulent lengths eliminates the possibility of type-I intermittency, which is characterized by a bimodal distribution (Schuster and Just 2006; Feng et al. 1996).

Figure 14.17b shows the variation of the average length of the turbulent phase with the normalized control parameter. Since the appearance of bursts is apparently random during intermittency, the average length of the turbulent phases $< T >$ is used as a representative of the total lengths of turbulent phases existing in the corresponding pressure signal. The normalized flame location is given by $r = (\tilde{x}_f - \tilde{x}_{f0})/\tilde{x}_{f0}$, where $\tilde{x}_f$ $(= x_f/L)$ and $\tilde{x}_{f0}$ $(= x_{f0}/L)$ are the locations of the

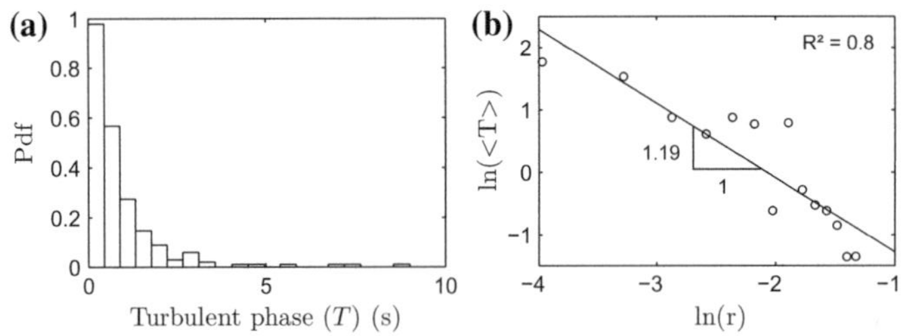

Fig. 14.17 **a** The distribution of the length of the turbulent phases obtained at $x_f = 540$ mm shows an exponential tail. **b** A log–log plot of the average value of the length of the turbulent phase $< T >$ versus the normalized flame location (r) shows a scaling law behaviour of type-II intermittency Pawar et al. (2015)

flame holder, measured from open end of the combustor (x_f) and that corresponding to the onset of intermittency (x_{f0}) respectively, normalized by the length of the duct (L). The mean duration of the turbulent phase plotted against each location of the flame holder on a log–log plot shows a power law behaviour. The variation of $< T >$ with r shows a linear relationship of slope -1.19, which is close to the theoretical value predicted for type-II and type-III intermittencies, i.e. $< T > \sim r^{-1}$ (Pomeau and Manneville 1980; Sacher et al. 1989; Chatterjee and Mallik 1996; Ringuet et al. 1993); on the contrary, for type-I intermittency, it is $< T > \sim r^{-1/2}$ (Pomeau and Manneville 1980; Feng et al. 1996). This statistical distribution again rules out the possibility of type-I intermittency in our experimental data. The difference in the values of slope obtained from theory (-1) and from experiments (-1.19) could be the result of finite length of the experimental signal, unavoidable noise in experiments and also in measurements, and challenge in detecting the length of turbulent phases accurately.

14.4.3.2 Analysis Using First Return Map

Finally, a method of first return map is used to differentiate type-II intermittency from type-III intermittency (Sacher et al. 1989). It is a two-dimensional plot between first local maxima of the signal versus next local maxima. Here, we plot the local maxima, say X_n, of the oscillations in laminar phases present in between the consecutive turbulent phases, is plotted against the next local maxima, i.e. X_{n+1}. In the first return map, the scatter of points along the main diagonal line is observed (Fig. 14.18a), whereas the evolution of these points shows a spiral-like structure in the plot (Figs. 14.18b, c). Such spiralling structure of the phase space trajectory is a typical characteristic property of type-II intermittency (Frank and Schmidt 1997; Sacher et al. 1989; Ringuet et al. 1993). In Fig. 14.18b, the evolution of some of the initial points from Fig. 14.18a is plotted to show the spiralling behaviour of the phase space trajectory. For type-I intermittency, the first return map shows an open channel between the diagonal line and the phase space trajectory (Schuster and Just 2006; Feng et al. 1996), whereas for type-III intermittency, it is represented by a two-fold trajectory crossing the diagonal line with increasing trend for sub-harmonic amplitude and decreasing trend for fundamental amplitude (Schuster and Just 2006; Griffith et al. 1997), which are different from that shown in Fig. 14.18. This confirms the presence of type-II intermittency in the present study.

14.4.3.3 Investigation Based on Recurrence Plot

Another way to identify the type of intermittency is based on RP analysis (Klimaszewska and Żebrowski 2009). The occurrence of different types of intermittency will result in various characteristic patterns in the RP. The recognition of the structure in RP will help in determining the type of intermittency. We have used a recurrence quantification analysis software package developed by Hasson et al. (2008)

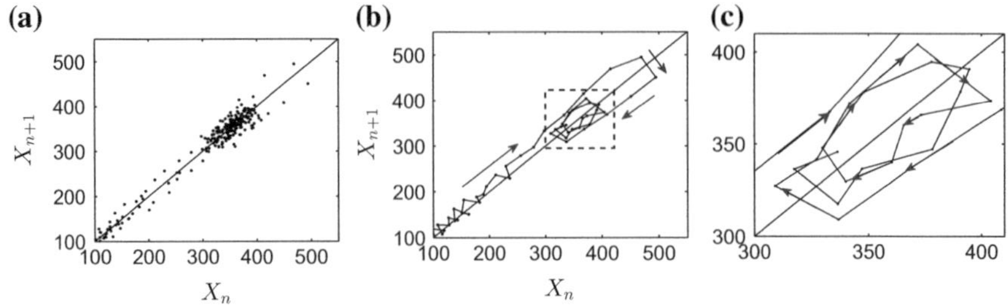

Fig. 14.18 **a** First return a map of the intermittency signal shows a scatter of points along the diagonal line. **b**, **c** The time evolution of these points indicates a spiralling behaviour of the trajectory, which confirms the possibility of type-II intermittency in the system dynamics (Pawar et al. 2015)

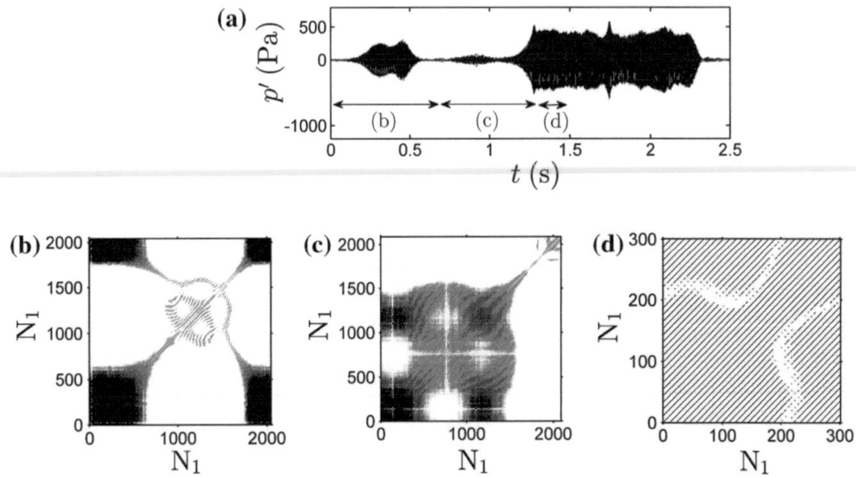

Fig. 14.19 **a** A portion of the signal displaying intermittency. **b**–**d** Recurrence plots (RP) obtained from different regions of the intermittency signal. **c** RP shows a kite-like structure of type-II intermittency. Data points (N) = 7500, $\tau = 3$, $d = 10$, $\varepsilon = 20\%$ of the mean diameter of the attractor (Pawar et al. 2015)

to obtain RPs of the experimental signal. Figures 14.19b–d show the RPs of small portions of the time series displaying intermittency (see Fig. 14.19a) obtained at x_f = 540 mm. Figure 14.19c shows a kite-like structure, representative of type-II intermittency, observed during the transition of system dynamics from low amplitude aperiodic to high amplitude periodic oscillations. A sinusoidal shape at the right upper corner of the kite with white perforations inside the kite (Fig. 14.19c) is the characteristic of type-II intermittency. Figure 14.19d shows the diagonal lines parallel to main diagonal line representing the periodic behaviour present in the burst. Kabiraj and Sujith (2012) and Unni and Sujith (2017) reported the presence of such kite-like structures, observed prior to blowout of the flame, for confirming type-II intermittency in gaseous laminar premixed and turbulent partially premixed combustors, respectively.

14.5 Conclusion

In this chapter, we discuss the observation of intermittency route to thermoacoustic instability in a laboratory-scale spray combustion system due to variation in the flame location as the control parameter. The intermittency is characterized by the presence of high amplitude bursts of periodic oscillations amongst regions of low amplitude aperiodic fluctuations. The analysis of recurrence plot and recurrence quantification of the experimental data proves that the low amplitude aperiodic fluctuations possibly exhibit chaotic behaviour. Furthermore, the transitions of the system dynamics, from aperiodic to periodic oscillations and vice versa, can be clearly detected by the recurrence quantification measures such as $\%DET$ and L_{max} during intermittency. The characterization of type of intermittency demonstrates the presence of type-II intermittency in the dynamics of the spray combustion system.

The bifurcation study of the acoustic pressure oscillations demonstrates the presence of higher amplitude bursts during intermittency compared to that during limit cycle oscillations. The maximum amplitude of such bursts can grow as high as three times that of the sustained periodic oscillations. The presence of these high amplitude intermittent loads during intermittency may be more dangerous to the system components than the low amplitude continuous cyclic loads observed during limit cycle oscillations. A sudden jump in the pressure amplitude can cause fracture of the brittle materials or can lead to deformation of the ductile materials. Sometimes, the damage caused by a single burst cannot be recognized easily but the continuous exposure of different amplitude intermittent loads over a long-term operation can lead to fatigue failure of the system parts. As a result, the intermittent oscillations wherein the high amplitude bursts occur pose a major challenge to the engine operator to protect the system before reaching a state of sustained limit cycle oscillations. Such behaviour of intermittent oscillations might be present in the practical combustion systems and should be considered during the construction of models of the thermoacoustic systems.

Acknowledgements We gratefully acknowledge Prof. M. V. Panchagnula for the valuable guidance provided during the course of development of the experimental set-up. We are grateful to Mr. R. Vishnu and Dr. M. Vadivukkarasan for the help provided in performing the experiments and for the fruitful discussion on the obtained results. This work has been supported by Defence Research & Development Organization (DRDO), India. We would like to thank Dr. V. Ramanujachari and Mr. S. Manikandan from RIC, DRDO.

References

Abarbanel HD, Brown R, Sidorowich JJ, Tsimring LS (1993) The analysis of observed chaotic data in physical systems. Rev Mod Phys 65(4):1331

Ananthkrishnan N, Deo S, Culick FE (2005) Reduced-order modeling and dynamics of nonlinear acoustic waves in a combustion chamber. Combust Sci Technol 177(2):221–248

Anderson W, Miller K, Ryan H, Pal S, Santoro R, Dressler J (1998) Effects of periodic atomization on combustion instability in liquid-fueled propulsion systems. J Propul Power 14(5)

Candel S (2002) Combustion dynamics and control: progress and challenges. Proc Combust Inst 29(1):1–28

Cao L (1997) Practical method for determining the minimum embedding dimension of a scalar time series. Phys D Nonlin Phenomena 110(1–2):43–50

Carvalho J, Ferreira M, Bressan C, Ferreira J (1989) Definition of heater location to drive maximum amplitude acoustic oscillations in a rijke tube. Combust Flame 76(1):17–27

Chatterjee S, Mallik A (1996) Three kinds of intermittency in a nonlinear mechanical system. Phys Rev E 53(5):4362

Chishty WA (2005) Effects of thermoacoustic oscillations on spray combustion dynamics with implications for lean direct injection systems. Ph.D. Thesis, Virginia Tech

Culick F (1988) Combustion instabilities in liquid-fuelled propulsion systems

Culick F, Kuentzmann P (2006) Unsteady motions in combustion chambers for propulsion systems. Technical Report, Nato Research and Technology Organization Neuilly-SUR-SEINE (France)

Dattarajan S, Lutomirski A, Lobbia R, Smith O, Karagozian A (2006) Acoustic excitation of droplet combustion in microgravity and normal gravity. Combust Flame 144(1):299–317

Domen S, Gotoda H, Kuriyama T, Okuno Y, Tachibana S (2015) Detection and prevention of blowout in a lean premixed gas-turbine model combustor using the concept of dynamical system theory. Proc Combust Inst 35(3):3245–3253

Dubey R, McQuay M, Carvalho J (1998) An experimental and theoretical investigation on the effects of acoustics on spray combustion. In: Symposium (International) on Combustion, vol. 27, pp 2017–2023. Elsevier

Ducruix S, Schuller T, Durox D, Candel S (2003) Combustion dynamics and instabilities: elementary coupling and driving mechanisms. J Prop Power 19(5):722–734

Duvvur A, Chiang C, Sirignano W (1996) Oscillatory fuel droplet vaporization-driving mechanism for combustion instability. J Prop Power 12(2):358–365

Eckmann JP, Kamphorst SO, Ruelle D (1987) Recurrence plots of dynamical systems. EPL (Europhysics Letters) 4(9):973

Elaskar S, Del Río E (2017) New advances on chaotic intermittency and its applications. Springer

Feng D, Zheng J, Huang W, Yu C, Ding W (1996) Type-i-like intermittent chaos in multicomponent plasmas with negative ions. Phys Rev E 54(3):2839

Frank M, Schmidt M (1997) Time series investigations on an experimental system driven by phase transitions. Phys Rev E 56(3):2423

Fraser AM, Swinney HL (1986) Independent coordinates for strange attractors from mutual information. Phys Rev A 33(2):1134

Gotoda H, Shinoda Y, Kobayashi M, Okuno Y, Tachibana S (2014) Detection and control of combustion instability based on the concept of dynamical system theory. Phys Rev E 89(2), 022,910

Griffith T, Parthimos D, Crombie J, Edwards DH (1997) Critical scaling and type-iii intermittent chaos in isolated rabbit resistance arteries. Phys Rev E 56(6):R6287

Hammer PW, Platt N, Hammel SM, Heagy JF, Lee BD (1994) Experimental observation of on-off intermittency. Phys Rev Lett 73(8):1095

Hasson CJ, Van Emmerik RE, Caldwell GE, Haddad JM, Gagnon JL, Hamill J (2008) Influence of embedding parameters and noise in center of pressure recurrence quantification analysis. Gait Post 27(3):416–422

Hilborn RC (2000) Chaos and nonlinear dynamics: an introduction for scientists and engineers. Oxford University Press on Demand

Kabiraj L, Paschereit CO, Gutmark E (2016) Analysis of combustion oscillations in a staged mldi burner using decomposition methods and recurrence analysis

Kabiraj L, Saurabh A, Wahi P, Sujith R (2012) Route to chaos for combustion instability in ducted laminar premixed flames. Chaos Interdisc J Nonlin Sci 22(2), 023,129

Kabiraj L, Sujith R (2012) Nonlinear self-excited thermoacoustic oscillations: intermittency and flame blowout. J Fluid Mech 713:376–397

Kantz H, Schreiber T (2004)Nonlinear time series analysis, vol. 7. Cambridge University Press

Kashinath K, Waugh IC, Juniper MP (2014) Nonlinear self-excited thermoacoustic oscillations of a ducted premixed flame: bifurcations and routes to chaos. J Fluid Mech 761:399–430

Kennel MB, Brown R, Abarbanel HD (1992) Determining embedding dimension for phase-space reconstruction using a geometrical construction. Phys Rev A 45(6):3403

Klimaszewska K, Żebrowski JJ (2009) Detection of the type of intermittency using characteristic patterns in recurrence plots. Phys Rev E 80(2), 026,214

Kumagai S, Isoda H (1955) Combustion of fuel droplets in a vibrating air field. In: Symposium (International) on Combustion, vol. 5, pp 129–132. Elsevier

Lei S, Turan A (2009) Nonlinear/chaotic analysis, modelling and control of combustion instabilities due to vaporizing sprays. Chaos, Solitons & Fract 42(3):1766–1779

Lieuwen T (2005) Online combustor stability margin assessment using dynamic pressure data. Trans ASME-A-Eng Gas Turbines Power 127(3):478–482

Lieuwen TC (2002) Experimental investigation of limit-cycle oscillations in an unstable gas turbine combustor. J Prop Power 18(1):61–67

Lieuwen TC (2012) Unsteady combustor physics. Cambridge University Press

Lieuwen TC, Yang V (2005) Combustion instabilities in gas turbine engines (operational experience, fundamental mechanisms and modeling). Progress in astronautics and aeronautics

Marwan N, Romano MC, Thiel M, Kurths J (2007) Recurrence plots for the analysis of complex systems. Phys Rep 438(5):237–329

McManus K, Poinsot T, Candel S (1993) A review of active control of combustion instabilities. Prog Energy Combust Sci 19(1):1–29

Muugesan M, Sujith R (2015) Intermittency in combustion dynamics. In: 51st AIAA/SAE/ASEE Joint Propulsion Conference, p. 3967

Nair V, Sujith R (2013) Identifying homoclinic orbits in the dynamics of intermittent signals through recurrence quantification. Chaos Interdisc J Nonlin Sci 23(3), 033–136

Nair V, Sujith R (2014) Multifractality in combustion noise: predicting an impending combustion instability. J Fluid Mech 747:635–655

Nair V, Thampi G, Karuppusamy S, Gopalan S, Sujith R (2013) Loss of chaos in combustion noise as a precursor of impending combustion instability. Int J Spray Combust Dynam 5(4):273–290

Nair V, Thampi G, Sujith R (2014) Intermittency route to thermoacoustic instability in turbulent combustors. J Fluid Mech 756:470–487

Okai K, Moriue O, Araki M, Tsue M, Kono M, Sato J, Dietrich D, Williams F (2000) Combustion of single droplets and droplet pairs in a vibrating field under microgravity. Proc Combust Inst 28(1):977–983

Pawar SA, Vishnu R, Vadivukkarasan M, Panchagnula MV, Sujith R (2015) Intermittency route to combustion instability in a laboratory spray combustor. ASME Turbo Expo Power for Land, Sea, and Air (GT2015-42919)

Pawar SA, Vishnu R, Vadivukkarasan M, Panchagnula MV, Sujith RI (2016) Intermittency route to combustion instability in a laboratory spray combustor. J Eng Gas Turbines Power 138(4), 041,505

Pomeau Y, Manneville P (1980) Intermittent transition to turbulence in dissipative dynamical systems. Commun Math Phys 74(2):189–197

Rayleigh JWS (1878) The explanation of certain acoustical phenomena. Nature 18(455):319–321

Ringuet E, Rozé C, Gouesbet G (1993) Experimental observation of type-ii intermittency in a hydrodynamic system. Phys Rev E 47(2):1405

Sacher J, Elsässer W, Göbel EO (1989) Intermittency in the coherence collapse of a semiconductor laser with external feedback. Phys Rev Lett 63(20):2224
Saito M, Hoshikawa M, Sato M (1996) Enhancement of evaporation/combustion rate coefficient of a single fuel droplet by acoustic oscillation. Fuel 75(6):669–674
Schuster HG, Just W (2006) Deterministic chaos: an introduction. Wiley
Smith DA, Zukoski EE (1985) Combustion instability sustained by unsteady vortex combustion
Sterling J, Zukoski E (1987) Longitudinal mode combustion instabilities in a dump combustor
Subramanian P (2011) Dynamical systems approach to the investigation of thermoacoustic instabilities
Sujith R (2005) An experimental investigation of interaction of sprays with acoustic fields. Exp Fluids 38(5):576–587
Sujith R, Juniper M, Schmid P (2016) Non-normality and nonlinearity in thermoacoustic instabilities. Int J Spray Combust Dynamics 8(2):119–146
Sujith R, Waldherr G, Jagoda J, Zinn B (1997) An experimental investigation of the behavior of droplets in axial acoustic fields. J Vib Acoust 119:285–292
Takens F et al (1981) Detecting strange attractors in turbulence. Lecture Notes Math 898(1):366–381
Tanabe M, Morita T, Aoki K, Satoh K, Fujimori T, Sato J (2000) Influence of standing sound waves on droplet combustion. Proc Combust Inst 28(1):1007–1013
Theiler J, Eubank S, Longtin A, Galdrikian B, Farmer JD (1992) Testing for nonlinearity in time series: the method of surrogate data. Phys D Nonlin Phenom 58(1–4):77–94
Unni VR, Sujith R (2015) Multifractal characteristics of combustor dynamics close to lean blowout. J Fluid Mech 784:30–50
Unni VR, Sujith R (2017) Flame dynamics during intermittency in a turbulent combustor. Proc Combust Inst 36(3):3791–3798
Webber CL, Zbilut JP (1994) Dynamical assessment of physiological systems and states using recurrence plot strategies. J Appl Physiol 76(2):965–973
Xiao Y, Wang Y, Lai YC (2009) Dependence of intermittency scaling on threshold in chaotic systems. Phys Rev E 80(5), 057,202
Young V (1995) Liquid rocket engine combustion instability, vol. 169. Aiaa
Zbilut JP, Webber CL (1992) Embeddings and delays as derived from quantification of recurrence plots. Phys Lett A 171(3–4):199–203

MIX
Papier aus verantwortungsvollen Quellen
Paper from responsible sources
FSC® C105338

If you have any concerns about our products,
you can contact us on
ProductSafety@springernature.com

In case Publisher is established outside the EU,
the EU authorized representative is:
Springer Nature Customer Service Center GmbH
Europaplatz 3, 69115 Heidelberg, Germany

Printed by Libri Plureos GmbH
in Hamburg, Germany